Book of Abstracts of the 57th Annual Meeting of the European Association for Animal Production

The EAAP Book of Abstracts is published under the direction of Ynze van der Honing

EAAP - European Federation of Animal Science

The European Association for Animal Production wishes to express its appreciation to the
Ministero delle Politiche Agricole e Forestali (Italy) and the
Associazione Italiana Allevatori (Italy)
for their valuable support of its activities.

Book of Abstracts of the 57th Annual Meeting of the European Association for Animal Production

Antalya, Turkey, September 17-20, 2006

Ynze van der Honing, Editor-in-chief

E. Strandberg, O. Moreira, C. Fourichon, M. Vestergaard, C. Lazzaroni, M. Gauly, C. Wenk, W. Martin-Rosset, J. Hermansen and C. Thomas

ISBN-10: 90-8686-003-6
ISBN-13: 978-90-8686-003-6
ISSN 1382-6077

First published, 2006

Wageningen Academic Publishers
The Netherlands, 2006

Preface

The 57th annual meeting of the European Association for Animal Production (EAAP) is held in Antalya, Turkey, 17 to 20 September 2006. The annual EAAP meeting gives the opportunity to present new scientific results and discuss their potential applicability in animal production practices. This year's meeting is of particular interest to participants from a wide range of animal production organisations and institutions. Discussions stimulate developments in animal production and encourage research on relevant topics. In Turkey the main theme of the meeting is "**Sustaining production systems to improve the livelihoods (health, well being, wealth) of livestock farmers**". In total 35 sessions are planned.

The book of abstracts is the main publication of the scientific contributions to this meeting; it covers a wide range of disciplines and livestock species. It contains the full programme and abstracts of the invited as well as the contributing speakers, including posters, of all 35 sessions. The number of abstracts submitted for presentation at this meeting (710) is a true challenge for the different study commissions and chairpersons to put together a scientific programme. This year, special attention has been given to the quality of the abstracts. Therefore, quite a few have been rejected and others had to be revised before publishing.

Several persons have been involved in the development of the book of abstracts. Wageningen Academic Publishers has been responsible for organising the administrative and editing work and the production of the book. The contact persons of the study commissions have been responsible for organising the scientific programme and communicating with the chairpersons and invited speakers. Their help and that of Cled Thomas in the programme of all free communications is highly appreciated.

The programme is very interesting and I trust we will have a good meeting in Antalya. I hope that you will find this book a useful reference source as well as a reminder of a good meeting during which a large number of people actively involved in livestock science and production will meet and exchange ideas.

Ynze van der Honing
Editor-in Chief

EAAP Program Foundation

Aims

EAAP aims to bring to our annual meetings, speakers who can present the latest findings and views on developments in the various fields of science relevant to animal production and its allied industries. In order to sustain the quality of the scientific program that will continue to entice the broad interest in EAAP meetings we have created the "EAAP Program Foundation". This Foundation aims to support:

- Invited speakers with a high international profile by funding part or all of registration and travel costs.
- Delegates from less favoured areas by offering scholarships to attend EAAP meetings
- Young scientists by providing prizes for best presentations

The **"EAAP Program Foundation"** is an initiative of the Scientific Committee (SC) of EAAP. The Foundation aims to stimulating the quality of the scientific program of the EAAP meetings and to ensure that the science meets societal needs. The Foundation Board of Trustees oversees theses aims and seeks to recruit sponsors to support its activities.

Sponsorships

We distinguish three categories of sponsorship: **"Gold Sponsor", "Sponsor" and "Student Award Sponsor".** The sponsors will be acknowledged in an appropriate way during the scientific sessions. The names of Student Award sponsors will be linked to the awards given to young scientists with the best presentation. "**Gold Sponsors" and "Sponsors"** will have opportunity to advertise their activities during the meeting and their support for EAAP.

Board of Trustees

Chair	*Vice-Chair*	*Treasurer/Secretary*
Prof Cled Thomas	Prof Johan van Arendonk	Dr Andrea Rosati
Chair S.C. EAAP	Wageningen University	EAAP Secretary General
	The Netherlands	Italy

The association

EAAP, (The European Federation of Animal Science), organises every year an international meeting which attracts between 800 and 1000 people. The main aims of EAAP are to promote, by means of active co-operation between its members and other relevant international and national organisations, the advancement of scientific research, sustainable development and systems of production; experimentation, application and extension; to improve the technical and economic conditions of the livestock sector; to promote the welfare of farm animals and the conservation of the rural environment; to control and optimise the use of natural resources in general and animal genetic resources in particular; to encourage the involvement of young scientists and technicians. More information on the organisation and its activities can be found at www.eaap.org

Contact and further information

If you are interested to become a sponsor of the "EAAP Program Foundation" or want to have further information, please contact the EAAP secretariat:
eaap@eaap.org
Fax +39 06 86 32 92 63

Acknowledgements

First of all, the members of the Organising and Scientific Committees of the 57th Annual Meeting of EAAP from the Ankara University, Faculty of Agriculture, would like to extend their gratitude to their Rectorate and Deanship for their sensitivity in sustaining the membership of the Department of Animal Science since 1952.

The Organising and Scientific Committees of the EAAP 57th Annual Meeting express their deep appreciation to the Scientific and Technological Research Council of Turkey (TÜBITAK) for making an important contribution to the publication expenses of the Book of Abstracts of the 57th Annual Meeting of the EAAP.

We also sincerely thank Saltur Tourism Travel Agency for providing the major part of the participation expenses of almost half of the Turkish Organizing and Scientific Committee members to the Bled (Slovenia) and Uppsala (Sweden) meetings.

European Association for Animal Production (EAAP)

President	Jim Flanagan
Secretary General	Andrea Rosati
Address	Via G. Tomassetti 3 A/1
	I-00161 Rome, Italy
Phone:	+39 06 44202639
Telefax	+39 06 86329263
E-Mail:	eaap@eaap.org

In cooperation with:

Ankara University, Faculty of Agriculture, Department of Animal Science
Akdeniz University, Faculty of Agriculture, Department of Animal Science

Organising Committee

President	S. Metin Yener	Ankara University, Faculty of Agriculture
Secretary	Fatin Cedden	Ankara University Faculty of Agriculture
Members	Ercan Efe	Kahramanmaras Sütçü Imam University, Faculty of Agriculture
	Mustafa Ertürk	Akdeniz University, Faculty of Agriculture
	Selahattin Kumlu	Akdeniz University, Faculty of Agriculture
	Hasan Rüştü Kutlu	Çukurova University, Faculty of Agriculture
	Salim Mutaf	Akdeniz University, Faculty of Agriculture
	Sezen Özkan	Ege University, Faculty of Agriculture
	M. Ihsan Soysal	Trakya University, Faculty of Agriculture
	Mehmet Ali Yildiz	Ankara University, Faculty of Agriculture

Scientific Committee

President	Mehmet Ertuğrul	Ankara University, Faculty of Agriculture
Secretary	Gürsel Dellal	Ankara University, Faculty of Agriculture
Members	Mehmet Ziya Firat	Akdeniz University, Faculty of Agriculture
	Oktay Gürsoy	Çukurova University, Faculty of Agriculture
	Mehmet Kuran	Ondokuz Mayis University, Faculty of Agriculture
	Ceyhan Özbeyaz	Ankara University, Faculty of Veterinary Medicine
	Nihat Özen	Akdeniz University, Faculty of Agriculture
	Özel Şekerden	Mustafa Kemal University, Faculty of Agriculture
	Cengiz Yalçin	Ankara University, Faculty of Veterinary Medicine

Official Organising Secretariat

SALTUR Tourism Travel agency
Atatürk Bulvan No: 175/8,
06680 Ankara-TURKEY

Tel: +90 312 419 84 80
Fax: +90 312 419 84 79
E-mail: saltur@eaap2006.gen.tr
Web site: www.eaap2006.gen.tr

58th Annual Meeting of the European Association for Animal Production

Dublin, Ireland, 26-29 August, 2007

Website: www.eaap2007.ie

Organising Committee
President: Dr. Tom Teehan
Department of Agriculture and Food
tom.teehan@agriculture.gov.ie

Organising Secretariat
Secretary: Mr. John Byrne
Department of Agriculture and Food
John.Byrne@agriculture.gov.ie

Scientific Committee
President: Prof. Maurice Boland
University College Dublin
maurice.boland@ucd.ie

Professional Conference Organisers
Ovation Group
1 Clarinda Park North
Dun Laoghaire, Co Dublin, Ireland.
Phone: +353 1 2802641
Fax: +353 1 2802665
Email: eaap2007@ovation.ie

Local Organising Committee representation
• Department of Agriculture and Food
• Teagasc
• University College Dublin
• Irish Grassland Association
• Irish Cattle Breeders Federation
• Irish Business and Employers Federation

Abstract submission
Please find the abstract form on the internet:
www.WageningenAcademic.com/eaap
Submission of the completed abstract form before March 1st , 2007

36th ICAR Session and INTERBULL Meeting

Niagara Falls, New York, USA, 16-24 June, 2008

Contact: ICAR Secretariat, Rome, Italy
Email: icar@eaap.org

Scientific Programme EAAP-2006

Sunday 17 Sept 09.30 – 13.00.	Sunday 17 Sept 14.00 – 18.00	Monday 18 Sept 08.30 – 12.30
Session 1 (L*, Ethics WG, M) Ethics of sustainability Chair: Marie (F)	**Session 5 (M*, L, P) Interaction between management, animal health and farmer health Chair: Fourichon (F)**	**Session 10 (G*, P) Genetics of reproduction and maternal traits Chair: Grandinson (SWE)**
Session 2 (N*, Ph, C, P, S, H) Nutrition and reproduction Chair: Bertoni (I)	Session 6 (H) Impact of reproduction technology on horse breeding programs Chair: Saastamoinen (Fin)	Session 11 (H*, M) Effect of management and housing on horse welfare Chair: Sondergaard (DK)
Session 3 (G) Statistical analysis of genetic data Chair: Szyda (PL)	Session 7 (Ph*, G) Advances in functional genomics Chair: Leroux (F)	Session 12 (Ph*, N) The early life of pigs - physiology and nutrition Chair: Taylor- Pickard
Session 4 (C*, ICAR) Cattle breeding and genetic resources for product quality Chair: Thaller (D)	Session 8 (P*, N) Micronutrient impact on high producing animals Chair: Torrallardona (ES)	Session 13 (M*, C) Indoor vs outdoor cattle production systems Chair: Svennersten–Sjaunja (SWE)
	Session 9 (C*, S) New developments in evaluation of carcass and meat quality in cattle and sheep Chair: Lazzaroni (I)	Session 14 (S) Awassi sheep Chair: Gursoy (TR)
		Session 15 (L) Advances in decision support concepts and tools for managing towards sustainability Chair: Herrero (Kenya)
		Session 16 (N) Impact of nutrition on the environment Chair: Crovetto (I)

PLENARY PAPER, Sunday 17th September, 08.30 – 9.15: – 'Improving meat quality and safety to meet consumer needs' *

Key - G Genetics; N, Nutrition; Ph, Physiology; P, Pig Production; C, Cattle Production; Ethics WG, EAAP Ethics Working Group; S, Sheep and Goat Production; M, Management and Health; H, Horse Production; L, Livestock Farming Systems; ICAR, International Committee on Animal Recording. (*) Denotes organising commission. **Bold** - Sessions contributing to the theme of the meeting **'Sustaining production systems to improve the livelihoods (health, well being, wealth) of livestock farmers'.**

Monday 18 Sept 14.00 – 18.00	Tuesday 19 Sept 08.30 – 12.30	Wednesday 20 Sept 08.30 – 12.30
Sessions 17 (G) Free Communications Animal Genetics Chair: Baumung	**Sessions 26 (S*, L, N) Strategies to cope with feed scarcity in harsh environments Chair: Ben Salem (Tun)**	**Session 31 (L*, C, S, M) Scale dependent opportunities and efficiency in livestock farm development Chair: Dedieu (F)**
Sessions 18 (N) Free Communications Animal Nutrition Chair: Crovetto	Session 27 (P*, Ph, M) Scientific evaluation of behavioural manipulations in pigs and poultry Chair: Chadd (UK)	Session 32 (N*, P) Product quality: nutrition and management Chair: Wenk (CH)
Sessions 19 (M) Free Communications Animal Management & Health Chair: Geers/ von Borell	Session 28 (C) Improving cattle performance and economics of dairy farming Chair: Kuipers (NL)	Session 33 (C) Free Communications Cattle Production Chair: Keane (IRL)
Sessions 20 (Ph) Free Communications Animal Physiology Chair: Sejrsen/ Vestergaard	Session 29 (G) Free Communications in Genetics Chair: Fernandez (ES)	Session 34 (G) Breeding programs and economics Chair: Simianer (D)
Sessions 21 (L) Free Communications Livestock Farming Systems Chair: Hermansen/Zervas	Session 30 (H) Dietetics, feeds and horse feeding Chair: Coenen (D)	Session 35 (Ph) Metabolic programming in growth and development Chair: Nielsen (DK)
Sessions 23 (S) Free Communications Sheep & Goat Production Chair: Schneeberger		Equine Industry in Turkey Session 36 (H) Chair: Yener (TR) followed by Free Communications Horse Production Chair: Ricard (F)
Sessions 24 (P) Free Communications Pig Production Chair: Wenk		
Sessions 25 (H) Free Communications Horse Production Chair: Ricard/ Martin-Rosset		

ROUND TABLE DISCUSSION, Tuesday 19th September, 14.00 – 15.30: 'Whole Food Chain Approach to Meat Quality and Safety' *

Key - G Genetics; N, Nutrition; Ph, Physiology; P, Pig Production; C, Cattle Production; Ethics WG, EAAP Ethics Working Group; S, Sheep and Goat Production; M, Management and Health; H, Horse Production; L, Livestock Farming Systems; ICAR, International Committee on Animal Recording. (*) Denotes organising commission. **Bold** - Sessions contributing to the theme of the meeting **'Sustaining production systems to improve the livelihoods (health, well being, wealth) of livestock farmers'.**

Commission on Animal Genetics

Dr Ducrocq	President	INRA
	France	Vincent.ducrocq@dga.jouy.inra.fr
Prof. Dr Simianer	Vice-President	University of Goettingen
	Germany	simianer@genetics-network.de
Dr Gandini	Vice-President	University of Milan
	Italy	gustavo.gandini@unimi.it
Dr Strandberg	Secretary	SLU
	Sweden	Erling.Strandberg@hgen.slu.se
Dr Szyda	Secretary	Agricultural University of Wroclaw
	Poland	szyda@karnet.ar.wroc.pl

Commission on Animal Nutrition

Prof. Crovetto	President	University of Milano
	Italy	matteo.crovetto@unimi.it
Dr Lindberg	Vice-President	SLU
	Sweden	Jan-Eric.Lindberg@huv.slu.se
Dr Ortigues-Marty	Vice-President	INRA
	France	ortigues@sancy.clermont.inra.fr
Dr Moreira	Secretary	University of the Azores
	Portugal	ocmoreira@netcabo.pt
Dr Cenkvàri	Secretary	Szent Istvan University
	Hungary	Czenkvari.Eva@aotk.szie.hu

Commission on Animal Management & Health

Prof. von Borell	President	University Halle-Wittenberg
	Germany	borell@landw.uni-halle.de
Dr Sorensen	Vice-President	DIAS
	Denmark	jantind.sorensen@agrsci.dk
Prof. Metz	Vice-President	Wageningen University
	Netherlands	jos.metz@user.aenf.wag-ur.nl
Dr Edwards	Secretary	University of Newcastle Upon Tyne
	United Kingdom	Sandra.edwards@ncl.ac.uk
Prof. Fourichon	Secretary	INRA
	France	fourichon@vet-nantes.fr

Commission on Animal Physiology

Dr Sejrsen	President	DIAS
	Denmark	Kr.Sejrsen@agrsci.dk
Dr Knight	Vice-President	Hannah Institute
	UK	Knightc@hri.sari.ac.uk
Dr Royal	Vice-President	University Liverpool
	UK	mdroyal@liverpool.ac.uk
Dr Ratky	Vice-President	Research Institute
	Hungary	jozsef.ratky@atk.hu
Dr Chilliard	Secretary	INRA
	France	chilliar@clermont.inra.fr
Dr Vestergaard	Secretary	DIAS
	Denmark	Mogens.Vestergaard@agrsci.dk

Commission on Livestock Farming Systems

Dr Gibon	President	UMR INRA
	France	gibon@toulouse.inra.fr
Prof. Peters	Vice-President	University of Berlin
	Germany	k.peters@rz.hu-berlin.de
Prof. Zervas	Vice-President	Agricultural University Athens
	Greece	gzervas@aua.gr
Dr Hermansen	Secretary	DIAS
	Denmark	john.hermansen@agrsci.dk
Dr Bernues Jal	Secretary	C.I.T.A.
	Spain	abernues@aragon.es

Commission on Cattle Production

Dr Gigli	President	ISZ
	Italy	sergio.gigli@isz.it
Dr Keane	Vice-President	TEAGASC
	Ireland	gkeane@grange.teagasc.ie
Dr Hocquette	Vice-President	INRA
	France	hocquet@clermont.inra.fr
Dr Lazzaroni	Secretary	University of Torino
	Italy	carla.lazzaroni@unito.it
Dr Kuipers	Secretary	Wageningen University
	Netherlands	Abele.Kuipers@wur.nl

Commission on Sheep and Goat Production

Dr Schneeberger	President	ETH Zentrum
	Switzerland	markus.schneeberger@inw.agrl.ethz.ch
Dr. Rihani	Vice-President	DPA
	Morocco	n.rihani@iav.ac.ma
Dr Dýrmundsson	Vice-President	Farmers Assoc.
	Iceland	ord@bondi.is
Prof. Niznikowski	Vice-President	Warsaw Agricultural University
	Poland	niznikowski@alpha.sggw.waw.pl
Prof. Gauly	Secretary	University Göttingen
	Germany	mgauly@gwdg.de
Dr Nieuwhof	Secretary	Meat and Livestock Commission
	United Kingdom	gert_nieuwhof@mlc.org.uk

Commission on Pig Production

Prof. Dr Wenk	President	ETH Zentrum
	Switzerland	caspar.wenk@inw.agrl.ethz.ch
Dr Chadd	Vice-President	Royal Agric. College
	UK	steve.chadd@royagcol.ac.uk
Dr Knap	Vice-President	PIC Deutschland
	Germany	KnaP@de.pic.co.uk
Dr Pescovicova	Secretary	Research Institute of Animal Production
	Slovak Republic	peskovic@vuzv.sk
Dr Torrallardona	Secretary	IRTA
	Spain	David.Torrallardona@irta.es

Commission on Horse Production

Dr Martin-Rosset	President	INRA
	France	wrosset@clermont.inra.fr
Dr Kennedy	Vice-President	Writtle College
	UK	mjk@writtle.ac.uk
Dr Verini	Vice-President	University Perugia
	Italy	vete2@unipg.it
Dr Koenen	Vice-President	NRS BV
	Netherlands	Koenen.E@cr-delta.nl
Dr Saastamoinen	Secretary	MTT Equines
	Finland	markku.saastamoinen@mtt.fi
Dr Sondergaard	Secretary	DIAS
	Denmark	eva.sondergaard@agrsci.dk

Session 1. Ethics of sustainability

Date: 17 September '06; 9:30 - 13:00 hours
Chairperson: Marie (F)

Session 2. Nutrition and reproduction

Date: 17 September '06; 9:30 - 13:00 hours
Chairperson: Bertoni (I)

Session 3. Statistical analysis of genetic data

Date: 17 September '06; 9:30 - 13:00 hours
Chairperson: Szyda (PL)

Poster   **Session G3 no. Page**

Session 4. Cattle breeding, genetic resources, product quality

Date: 17 September '06; 9:30 - 13:00 hours
Chairperson: Thaller (D)

Session 5. Interaction between management, animal and farmer health

Date: 17 September '06; 14:00 - 18:00 hours
Chairperson: Fourichon (F)

Session 6. Impact of reproduction technology on horse breeding programs

Date: 17 September '06; 14:00 - 18:00 hours
Chairperson: Saastamoinen (Fin)

Session 7. Advances in functional genomics

Date: 17 September '06; 14:00 - 18:00 hours
Chairperson: Leroux (F)

Session 8. Micronutrient impact on high producing animals

Date: 17 September '06; 14:00 - 18:00 hours
Chairperson: Torrallardona (E)

Session 9. New developments in evaluation of carcass and meat quality in cattle and sheep

Date: 17 September '06; 14:00 - 18:00 hours
Chairperson: Lazzaroni (I)

Session 10. Genetics of reproduction and maternal traits

Date: 18 September '06; 8:30 - 12:30 hours
Chairperson: Grandinson (S)

Theatre   **Session G10 no. Page**

Poster   **Session G10 no. Page**

Session 11. Effect of management and housing on horse welfare

Date: 18 September '06; 8:30 - 12:30 hours
Chairperson: Sondergaard (DK)

Theatre   **Session H11 no. Page**

Poster

Session 12. The early life of pigs - physiology and nutrition

Date: 18 September '06; 8:30 - 12:30 hours
Chairperson: Taylor-Pickard (UK)

Theatre

Session 13. Indoor vs outdoor cattle production systems

Date: 18 September '06; 8:30 - 12:30 hours
Chairperson: Svennersten-Sjaunja (SE)

Session 14. Awassi sheep

Date: 18 September '06; 8:30 - 12:30 hours
Chairperson: Gursoy (T)

Session 15. Advances in decision support concepts and tools for managing towards sustainability

Date: 18 September '06; 8:30 - 12:30 hours
Chairperson: Herrero (Kenya)

Theatre   **Session L15 no. Page**

Session 16. Impact of nutrition on the environment

Date: 18 September '06; 8:30 - 12:30 hours
Chairperson: Crovetto (I)

Development of technique for manufacturing activated carbon from livestock manure and its characteristics   7   102
Choi, H.C., J.I. Song, D.J. Kwon, J.H. Kwag, Y.H. Yoo, C.B. Yang, Y.T. Park, D.K. Park, K.S. Park and Y.K. Kim

Session 17. Free communications animal Genetics

Date: 18 September '06; 14:00 - 18:00 hours
Chairperson: Baumung (G)

Session 18. Free communications animal Nutrition

Date: 18 September '06; 14:00 - 18:00 hours
Chairperson: Crovetto (I)

Poster   **Session N18 no. Page**

Session 19. Free communications animal Management & Health

Date: 18 September '06; 14:00 - 18:00 hours
Chairperson: Von Borell (D)

Theatre   **Session M19 no. Page**

Poster   **Session M19 no. Page**

Session 20. Free communications animal Physiology

Date: 18 September '06; 14:00 - 18:00 hours
Chairperson: Vestergaard (DK)

Theatre   **Session Ph20 no. Page**

Poster   **Session Ph20 no. Page**

Session 21. Free communications Livestock Farming Systems

Date: 18 September '06; 14:00 - 18:00 hours
Chairperson: Hermansen (D)

Session 23. Free communications Sheep and Goat production

Date: 18 September '06; 14:00 - 18:00 hours
Chairperson: Schneeberger (CH)

Cholesterol and fat acids profile in the meat of lambs fed with diets containing
fish residue silage   66   227
Yamamoto, S.M. and C. Buzzulini

Session 24. Free communications Pig production

Date: 18 September '06; 14:00 - 18:00 hours
Chairperson: Wenk (CH)

Session 25. Free communications Horse production

Date: 18 September '06; 14:00 - 18:00 hours
Chairperson: Ricard (F)

Theatre   **Session H25 no. Page**

Poster   **Session H25 no. Page**

Session 26. Strategies to cope with feed scarcity in harsh environments

Date: 19 September '06; 8:30 - 12:30 hours
Chairperson: Ben Salem (Tun)

Theatre   **Session S26 no. Page**

Poster   **Session S26 no. Page**

Session 27. Scientific evaluation of behavioural manipulations in pigs and poultry

Date: 19 September '06; 8:30 - 12:30 hours
Chairperson: Chadd (UK)

Theatre   **Session P27 no. Page**

Poster

Session 28. Improving cattle performance and economics of dairy farming

Date: 19 September '06; 8:30 - 12:30 hours
Chairperson: Kuipers (NL)

Poster   **Session C28 no. Page**

Session 29. Free communications animal Genetics

Date: 19 September '06; 8:30 - 12:30 hours
Chairperson: Fernandez (E)

Theatre   **Session G29 no. Page**

Session 30. Dietetics, feeds and horse feeding

Date: 19 September '06; 8:30 - 12:30 hours
Chairperson: Coenen (D)

Theatre   **Session H30 no. Page**

Poster   **Session H30 no. Page**

Session 31. Scale dependent opportunities and efficiency in livestock farming development

Date: 20 September '06; 8:30 - 12:30 hours
Chairperson: Dedieu (F)

Theatre   **Session L31 no. Page**

Session 32. Product quality: nutrition and management

Date: 20 September '06; 8:30 - 12:30 hours
Chairperson: Wenk (CH)

Session 33. Free communications Cattle production

Date: 19 September '06; 8:30 - 12:30 hours
Chairperson: Keane (Ire)

Theatre   **Session C33 no. Page**

Session 34. Breeding programs and economics

Date: 20 September '06; 8:30 - 12:30 hours
Chairperson: Simianer (D)

Theatre	Session G34 no.	Page

Session 35. Metabolic programming in growth and development

Date: 20 September '06; 8:30 - 12:30 hours
Chairperson: Nielsen (DK)

Theatre   **Session Ph35 no. Page**

Poster   **Session Ph35 no. Page**

Session 36. Equine industry in Turkey

Date: 20 September '06; 8:30 - 12:30 hours
Chairperson: Yener (T)

Theatre Session H36 no. Page

Horse breeding in Turkey | 1 | 347
Ozbeyaz, C., O. Gücüyener and S.M. Yener

Turkish equestrian sports and clubs | 2 | 348
Yener, S.M., D. Alic and K. Ural

Studies on horses in Turkey | 3 | 348
Koçak, S., M. Tekerli and C. Ozbeyaz

Possibility of improving racing traits of Arabian and Thoroughbred horses in Turkey | 4 | 349
Ekiz, B., O. Kocak and A. Yilmaz

Connectedness between seven European countries for horse jumping
competition, the "Interstallion Pilot Project II" | 5 | 349
Ruhlmann, C., E. Bruns, E. Fræhr, E.P.C. Koenen, J. Philipsson, M. Pierson, K. Quinn and A. Ricard

Survival analysis of the length of competition life of Standardbred trotters in Sweden | 6 | 350
Arnason, Th.

Selection of racehorses on jumping ability based on their steeplechase race results | 7 | 350
Bokor, A., C. Blouin and B. Langlois

Genetic parameters for endurance ride in the Spanish Arab Horse | 8 | 351
Cervantes, I., M. Valera, M.D. Gómez, C. Medina and A. Molina

Genetic evaluation of show jumping performance in young Spanish sport horses | 9 | 351
Gómez, M.D., I. Cervantes, E. Bartolomé, A. Molina and M. Valera

Genetic evaluation of dressage performance in Spanish Purebred horses | 10 | 352
Valera, M. and M.D. Gómez

Development of a BAC-based physical map of the horse genome | 11 | 352
Blöcker, H., M. Scharfe, M. Jarek, G. Nordsiek, F. Schrader, C. Vogl, B. Zhu, P.J. de Jong, B.P. Chowdhary, T. Leeb and O. Distl

Genetic analysis of allergic eczema in Icelandic horses | 12 | 353
Grandinson, K., S. Eriksson, L. Lindberg, S. Mikko, H. Broström, R. Frey, M. Sundquist and G. Lindgren

Poster Session H36 no. Page

Biometrical and breeding analysis of Hafling horses in the Czech Republic | 13 | 353
Navratil, J., V. Kutilova, P. Kutilova and F. Louda

Tori Stud Farm 150, founder of Tori Horse Hetman 120 | 14 | 354
Peterson, H.

Session L1

Theatre 1

Ethical bases of sustainability

Paul B. Thompson, Michigan State University, Agricultural Economics, East Lansing, MI 48824, USA

Sustainability has proven to be a fruitful concept for interpreting and promoting a wide range of environmentally oriented goals and values. It is an implicitly normative concept: sustainability is always understood as something that societies, organizations and individuals should try to achieve, rather than a purely empirical or descriptive accounting of current or future trends. However, the literature on sustainability offers many different ways to integrate ethical norms (understood here as encompassing the full range of normative values) the conceptualization of sustainability. This paper reviews several leading theoretical approaches to the achievement of sustainability within livestock production systems, and examines how ethics either are or might be analyzed as playing a role in each approach. In particular, I will emphasize approaches to sustainability that utilize "indicators" of sustainability as a way to resolve value conflicts over environmental goals. One point of view sees each indicator representing alternative ethical norms, then utilizes empirical research to measure the relative effectiveness of alternative production systems in meeting achieving progress with respect to indicators. An alternative point of view sees indicators as themselves reflecting a consensus among parties holding different values, a consensus that should be reached before empirical work begins. In the end, it is important for livestock researchers and environmental policy analysts to continue debating the relationship between ethics and sustainability, if there is to be any hope of attaining any measure of coherence among alternative approaches.

Session L1

Theatre 2

Sustainability and precaution

K.K. Jensen, Danish Centre for Bioethics and Risk Assessment, DK-1958 Frederiksberg C., Denmark

The widely used concept of *sustainability* is seldom precisely defined, and its clarification involves making up one's mind about a range of difficult questions. A starting point is to define sustainable agricultural practice as a practice that can be continued in perpetuity. However, as economists would point out, the possibility of substitution of goods (for instance, due to technological development) makes it irrelevant to consider continuing the *same* practice in all future. But if we are to compare the sustainability of evolving practices, sustainability has to be measured on a more fundamental level.

Our Common Future suggested the measure that the present generation's need-satisfaction does not compromise the need-satisfaction of future generations. This measure needs to be refined in several dimensions. However, even given such refinements, there is still a problem. At present we do not know what substitutions will be possible in the future. This uncertainty clearly affects the prescriptions that follow from the measure of sustainability. Some authors distinguish between optimistic and pessimistic views on possibilities of substitution, the latter leading to stricter requirements than the former.

Consequently, decisions about how to make future agriculture sustainable are decisions under uncertainty. The main ethical problem is how to deal with the uncertainty. A large part of disagreements about the exact prescriptions of sustainability are disagreements about the level of risk aversion we should adopt in face of the uncertainty.

A more balanced approach to agriculture, animal farming and ethics of sustainability

G. Bengisu[1], A. Rıza Öztürkmen[2], [1]Crop Science Department, [2]Soil Science Department, Harran University, Turkey

In spite of increasing world's population, the area for agriculture is limited. In many countries agricultural land is decreasing due to misusing of lands. Therefore yield and crop quality need to be improved to feed the world's population. New intensive agricultural and animal farming systems can cause dramatic changes in the soil quality in short time. It is, therefore, important to monitor the soil quality for sustainable use and management. It is the aim to obtain economic, ecological and social sustainability of agriculture, soil and animal farming. There are several ways in which the sustainability of the ecosystem, the agro ecological system and the farm animal are interrelated. There is need of a new agenda in which research on soil ecology and plant ecology would become the foundation of agricultural research. Interdisciplinary research will be needed to develop agricultural systems that are productive, economically viable and environmentally sound. Different teams will be needed to answer different questions. New crops and livestock options will be developed for use in new integrated farming systems.

Role of the sustainability concept on orientations in livestock farming system research

A. Gibon[1] and J.E. Hermansen[2], [1]UMR 1201 Dynafor INRA – INPT/ENSAT, BP 52627, F-31326 Castanet-Tolosan cedex, France, [2]Danish Institute of Agricultural Sciences, Research Centre Foulum, Dept. of Agroecology, P.O. BOX 50, DK-8830 Tjele, Denmark

The emergence of Livestock Farming Systems research area in Europe is closely associated with the spread of sustainability as a framework concept for development in the late 1980s. The ethical attitude included into the sustainable development concept led animal scientists to search for integrated frameworks and methodologies allowing addressing livestock systems both in their human and bio-technical dimensions. In many countries research groups developed research orientations that support the view of livestock systems as complex social-bitotechnical systems. It led in particular to give important consideration to the variety of objectives of livestock production stakeholders and the decision-making systems of managers in support to sustainable development. In this paper, we give an account of the approaches developed at the level of both livestock-production units and wider complexes, such as the local region or the food chain. Current advances and main topical issues attached to research into sustainable development of livestock production are discussed in light of ethical principles that ground the research area.

Assessment of dairy production sustainability: Case of dairy farms in the district of Nabeul in north of Tunisia

N. M'hamdi[1], R. Aloulou[1], M. Ben Hamouda[2] and K. Kraiem[1]*. [1]Ecole Supérieure d'Horticulture et d'Elevage de Chott Mariem, 4042 Sousse Tunisia, [2]Ecole Supérieure d'Agriculture de Mateur, 7030 Tunisia

Using IDEA method, sustainability of 30 farms from the district of Nabeul was evaluated.Based on statistical methods (PCA, ACH), each scale of sustainability (agroecology, socioterritorial, and economic) has been characterized and has allowed to built groups. "Production system" typology is relevant for describing farms by agroecological and economical scales. The socioterritorial scale gives the limiting sustainability value for most of the farms. Inside this scale, the main way of progress relies on employment and services improvement (services, contribution to employment and collective work).Economically, sustainability is determined by level of efficiency and depends on financial independence. Socio-territorial scale is the only one which is not linked to production system and is based on farmer's way of life. On the other end, global sustainability evaluation of farm as well as creating collective references means to be able to analyze links between the three sustainability scales.

Socio-cultural issues of dairy production systems in the Netherlands assessed through farm visits with citizen panels

B.K. Boogaard*, S.J. Oosting and B.B. Bock, Wageningen University, P.O.Box 338, 6700 AH Wageningen, The Netherlands

Agricultural development in the Netherlands has reached a historical crossroad; food production as the sole objective of land-use is no longer obvious. Demand for sustainable agriculture is a general expression of collective concern. Socio-cultural sustainability is about values, subjects and processes that really matter to people. The present study had the objective to assess relevant socio-cultural issues in dairy production systems essential for design of sustainable future dairy production.

In three areas of the Netherlands two citizen panels (eight respondents per panel) visited two dairy farms. On farm, each respondent filled in a questionnaire about his/her individual perception (hearing, smelling, seeing and feeling) and by means of a digital camera, each respondent recorded ten aspects per farm which he/she found valuable to preserve for the future. Data of the six panels were analysed qualitatively, which resulted in socio-cultural issues for dairy production systems. One of the values dairy production systems have for Dutch society is "The calm and quietness of farm life", as illustrated by the quotation "*You can calm down at the farm life and learn to respect nature. The quietness is an important part of this*". This valuable aspect was mentioned by 35 respondents (74%). Other issues defined were: historical values, Dutch identity, respect for animals, respect for farmers and the Dutch landscape.

The livestock feeding management in the dairy sheep family and private sector in the North of Tunisia
S. Snoussi*, A. BenYounès and S. Zemzmi, U.R.Développement de la filière lait de brebis, Ecole Supérieure d'Agriculture, 7030 Mateur, Tunisie

This study aims to assess the food strategy and practices of the dairy sheep herds and its impact on the technical and economic performances.
A field survey was run at the 25 private householders of Bizerte area that, in spite of the difficulties known in the dairy sheep sector, continue to practice this traditional activity threatened with abandonment.
The most of the breeders use the grazing on the natural rangeland and stubble with a variable complementation in hay and in concentrates. The typology has been defined in relation to the combination of: i / the integration of green fodder in the ration, ii / the distribution of concentrate and its origin: farmer or industrial.
Five feeding systems are described. Three ones are based essentially on the grazing with a stocking rate per hectare relatively low (around 1 UGB). The 2 others that integrate the green fodders in the ration have a more elevated stocking rate (about 4 UGB) and use more important quantities in concentrate.
Even if, according to the statistical survey, there is no difference between the technical and economic results, the performances are varying between the groups (63 to 92 litres of milk per ewe). Others factors of the livestock management have more acts: weaning date, lactation length, ram breed.

Attitudes of Dutch pig farmers towards animal welfare
M.M. van Huik* and B.B. Bock, Rural Sociology Group, Wageningen University, P.O. Box 8130, 6700 EW Wageningen, The Netherlands

This paper presents partial results of the Welfare Quality research project. Using semi-structured, in-depth interviews with a purposive sample of 62 Dutch pig farmers, farmer's attitudes towards animal welfare, legislation and views on consumers and market possibilities for animal welfare products were studied. The sample of farmers was stratified according to farmer's participation in quality assurance schemes to detect what motivates farmers to choose for animal friendly production.
Two dominant approaches of animal welfare were detected. These approaches are closely related to farmers understanding of good farming. The first approach appreciates animal welfare for its importance for animal health and the zoo-technical performance of the animals. The second approach emphasizes the possibility of animals to express natural behavior. These approaches farmers use, also influence farmer's appreciation of animal welfare measures and legislation.
The type of quality assurance scheme that a farmer participates in follows the farmer's approach of animal welfare and provides the farmer with the conditions that enforces the used approach. When wishing to stimulate animal friendly production, it is not enough to merely entice farmers with the possibility to earn a premium price. A quality assurance scheme has to fit farmer's understanding of animal welfare. In order to induce farmers to switch to animal friendly production, efforts must be made to expand farmer's understanding of animal welfare and good farming.

Session N2
Theatre 1

Impact of nutrient supply on reproductive outcome in pigs
C.J. Ashworth, SAC, Roslin Biocentre, Roslin, Midlothian, EH25 9PS, UK

For many years, it has been known that alteration of the composition or quantity of the diet consumed by pigs prior to mating and/or during pregnancy can effect prenatal survival and eventual litter size. Recent research has highlighted that appropriate nutritional management affects not only the number of surviving embryos and piglets, but also their quality and viability, and has demonstrated that changes to the composition of the diet often has greater impact on reproduction than merely altering the amount of feed consumed. Significant progress has been made towards understanding the mechanisms by which the reproductive axis responds to altered feed intake and in highlighting nutritionally sensitive stages of development when altered nutrient supply has both immediate effects on reproductive systems and longer-term effects on the subsequent development of the conceptus and piglet. For example, alterations in the composition of the diet fed to pigs before mating affects not only oocyte maturation, but also oocyte quality (as assessed by the ability of an oocyte to form a viable blastocyst), embryo survival, fetal growth and litter size at birth. These changes are believed to arise from nutritionally induced changes in peripheral concentrations of intermediary metabolites and reproductive hormones that in turn affect ovarian development and ultimately the follicular environment in which the oocyte develops. This increased understanding of the impact of altered nutrient supply provides exciting new opportunities to gain further insight into the reproductive biology of the pig and to improve reproductive outcome.

Session N2
Theatre 2

A new approach to reproduction, nutrition and immune response relationships
G. Bertoni, E. Trevisi and R. Lombardelli, Istituto di Zootecnica, UCSC, 29100 Piacenza, Italy

Speaking in terms of general relationship between nutrition and reproduction, many different aspects are more or less involved according to geographical areas, species, production systems, technological levels etc. They are deficiency conditions: energy, proteins, vitamins and minerals, but also some excesses (few minerals) or toxic substances like micotoxins or plant compounds (i.e. oestrogen-like). Their relevance is different in the intensive systems for an easier making of appropriate diets. Nevertheless intensification does not reduce the nutritional risks for livestock reproduction for several reasons, namely a suspected higher susceptibility to the usual stresses and a new one: the metabolic stress. The latter is particularly relevant when early lactation and new pregnancy are close (dairy cows, does and, in some extent, sows). The negative energy balance is the main cause of metabolic stress, but the oxidative stress as well as the disease stress (pro-inflammatory cytokines) seem to be of great relevance. In dairy cows, inflammatory phenomena around calving – then an immune response in spite of clinical symptoms can be missing – are significantly related to a lower pregnancy rate, while milk yield and BCS are also reduced. The apparent paradox could be justified by pro-inflammatory cytokines which modify liver synthesis and seem to impair energy balance (increasing expenditure and reducing feed intake).

Effects of calcium salts and protected methionine supplementation on the productive and reproductive performances of high producing dairy cows in early lactation
M. Ben Salem, INRA Tunisia, Laboratory of Animal and Forage Production, Rue Hédi Karray, 2049 Ariana, Tunisia

Fifty-four high producing Holstein cows in early lactation were used in a randomized design to determine the effects of calcium salts of palm fatty acids (CSFA) and protected methionine (M) supplementation on the performances of dairy cows during the first 120 days of lactation. Cows were randomly assigned to 3 treatments. Treatments were: concentrate only (Control, C), control + added CSFA (700 g/head/day, CSFA) and control + added CSFA containing 6% (40g) of M (700 g/head/day, CSFAM). All cows were fed a basal diet consisting of tritical silage, rye grass and oat hay. Results showed that fat supplementation decreased (P < 0.05) dry matter intake and milk protein content, but increased (P < 0.05) milk yield and milk fat content. Fat corrected milk was 37.8 kg/d for the control and increased an average of 3 to 4 kg/d with CSFA and CSFAM treatments. The addition of M to CSFA did prevent the decrease in milk protein. Supplemental fat significantly increased (P < 0.05) first service and conception rates and reduced (P < 0.05) the number of services per conception. Results suggest that CSFA supplementation improves major reproductive indices and milk yield of high producing dairy cows and that adding M to CSFA helps alleviate the milk protein depression commonly observed with added fat and further improves the performances of cows during early lactation.

Sow body condition evaluation and reproduction performance in organic sow herds
A.G. Kongsted and J.E. Hermansen, Danish Institute of Agricultural Sciences, Dept. of Agroecology, PO Box 50, 8830 Tjele, Denmark*

According to the Danish legislation, organically produced pigs must be at least seven weeks of age before weaning. The long lactation length has been mentioned as a potential threat to sow welfare and reproduction due to large weight losses. The aim of this study was to evaluate the sow body condition and reproduction performance in organic sow herds.
A field study was carried out from June 2005 to June 2006 in nine organic sow herds. The herd sizes varied from 50 to 400 sows. All sows were outdoors during lactation and pregnancy. The piglets were weaned at 7 to 9 weeks of age. Back fat was measured at weaning in 5 to 10 sows randomly chosen in each of approximately 10 batches in each herd. Days from weaning to first mating, pregnancy rates and litter sizes were registered for all sows weaned during the study period.
There were large within and between herd variations in back fat at weaning. The back fat varied e.g. from 11 to 17 mm between herds (mean 13 mm). 24% of all sows had back fat below 10 mm. This varied from 6 to 38% (preliminary results). The relation between the back fat at weaning and the subsequent reproduction performance will be evaluated.

Effect of feeding flushing on fertility of oestrus synchronized Awassi ewes
Ü. Yavuzer and A.Can, Department of Animal Science, Faculty of Agriculture, Harran University, Şanlıurfa, Turkey

The aim of this research was to determine the effect of flushing on fertility of oestrus synchronized Awassi ewes. The research was carried out with 40 Awassi ewes at the different ages. The ewes were randomly allocated into two treatment groups with equal numbers. Ewes were fed according to guideline of NCR, (1985) for flushing and control groups. 1 ml of prostaglandin F_2 alpha ($PGF_{2\alpha}$) was two times injected with 10 day intervals for oestrus synchronisation during their breeding season for both groups in order to compare their pregnancy rates.

Oestrus rates of 94% and 87% were obtained from flushing and control group, respectively. Real-time transrectal ultrasonography was performed on all ewes on day 35 of pregnancy. Pregnancy rates were obtained 94% and 68% for flushing and control groups, respectively. Number of pregnant ewes per ewe exposed to the ram at the lambing was significantly higher in ewes of the flushing treatment group ($P< 0.01$) than in those of the control group.

Effect of energy restriction and management on reproduction in Belgian Blue cows
L.O. Fiems, W. Van Caelenbergh, S. De Campeneere* and D.L. De Brabander, ILVO, Animal Science Unit, Scheldeweg 68, B-9090 Melle, Belgium

Belgian Blue double-muscled cows (n=123) were used from December 1999 to April 2004 to study the effect of energy level during the indoor period (E: 100, 90, 80 or 70% of the requirements; 140 days) on reproduction. Within each E, half of the calves were suckled by their dams for 16 weeks; the other calves were reared. Service bulls were used. No oestrus synchronisation was applied. Cows were eliminated if not pregnant within 9 months postpartum. Body condition score (BCS; 6-point scale) was determined at calving.

Eight abortions and 213 parturitions were registered. Six calving were not assisted by caesarean. The occurrence of abortion (P=0.191), length of calving interval (P=0.865), calf birth weight (P=0.289) and calf sex ratio (P=0.670) were not affected by E. Fifty of the 123 cows were culled because they were not pregnant within nine months postpartum, but the number of open cows was not affected by E (P=0.922). Mean BCS at calving decreased with decreasing E (P=0.006). More open cows were culled if their BCS was more than 2 (P=0.019). There was a tendency for an increased calving interval for suckling cows (P=0.087). Calf mortality seemed to be increased by a reduced E (P=0.075), but not by rearing or suckling (P=0.522). The effect of dam BCS at calving on calf mortality was less clear (P=0.245).

Association between milk urea nitrogen and fertility of early lactation dairy cows
M. Nourozi*, M. Raisianzadeh and M. Abazari, Agricultural and Natural Resources Research Center of Khorasan Province, Mashad P.O.Box 91735-1148, Iran

The objective of this study was to evaluate the association between milk urea nitrogen (MUN) and fertility of early lactation dairy cows. The data were obtained from 10 dairy herds in the countryside of Mashhad. Reproductive and productive data and MUN measurements from cows that calved between 2001 and 2003 were included in the study. Survival analysis, using the Cox proportional hazards model, was performed and days from calving to conception or to the end of the study were used as the outcome. The mean MUN values in the first three months of lactation were used to reflect the MUN status of a cow. Parity, mean milk yield, mean percentage of milk fat and herd were included in the models as fixed effects. Cows with MUN value 12-14 and 14-16 mg/dl had higher fertility (15 and 8%, respectively) and cows with MUN value above 18 mg/dl had lower fertility (10%) than cows with MUN value below 12 mg/dl ($p<0.0001$). Our results indicate that increasing MUN levels above 18 mg/dl appears to be negatively related to dairy cow fertility, only in first parity ($p<0.01$). It is also suggested that the levels of MUN that are adversely associated with fertility might be lower than 12 and greater than 18 mg/dl.

Major nutritional influences on reproductive function in ruminants
S. Yildiz[1], M. Kaya[1], M. Cenesiz[1], O. Ucar[2], D. Blache[3], M. Uzun[1], F. Onder[1] and G.B. Martin[3], Departments of [1]Physiology and [2]Reproduction & Artificial Insemination, Faculty of Veterinary Sciences, University of Kafkas, Kars, Turkey, [3]Faculty of Natural & Agricultural Sciences, University of Western Australia, Crawley 6009, Australia

In ruminants, both energy intake and body fatness are associated with reproductive output. However, effects of different sources of feed energy have not been well documented. We have carried out studies on the effects of various sources of energy on reproductive function in prepubertal and mature sheep, and observed LH pulse frequency, leptin and growth hormone to reveal the types of reproductive and metabolic response. The effects of secondary plant compounds (eg, tannins, saponins) have generally been ignored. During the prepubertal period, we observed no effects of tanniniferous oak leaves on LH secretion, leptin and IGF-I levels in lambs if the diets were given in isonitrogenous and isoenergetical fashion. Complexity and sensitive structure of these compounds, taken together with relatively high consumption both in humans and animals, warrants further studies in this area. Finally, it is clear that body fatness is an important determinant of reproductive function but there are few data for fat-tailed sheep, so these breeds need special attention with respect to the role of body condition score in the control of the secretion of leptin, IGF-I, tonic and surge modes of LH release, as well as oestrous behaviour.

Effect of nutrition on the occurrence of puberty in horses

D. Guillaume*, G. Fleurance, J. Schneider M. Donabedian, C. Robert, G. Arnaud, M. Leveau, D. Chesneau and W. Martin-Rosset, INRA-HN-ENV, France

In horses, puberty occurs around 15 months with a great variability. Our aim was to estimate the influence of nutrition on this age.

23 foals were fed from birth to puberty 2 feeding levels (150 vs 100% INRA recommendations) to achieve moderate growth (MG: 6 males M + 5 Females F) or high growth (HG: 5 M + 6 F). Females were considered pubescent at their first increase of progesterone and males when testosterone concentrations exceeded 0.5 ng/ml for 4 subsequent blood samples.

The 6 HG-F had their first ovulation at 393 ± 11 and 4 MG-F at 415 ± 1 but the 5th MG-F at 700 days of age (not significant). The testosterone and total estrogens rates are higher for the HG-M than for the MG-M. The age at puberty was significantly earlier for HG-M:529 ± 57 than for MG-M: 754 ± 30 days. The eGH rates were significantly higher for MG than for HG group, and for F than for M.

Differences in feeding level induced different growth rates and different GH concentrations between the 2 groups. Only one MG-F has its puberty delayed of one year. For the MG-M, the age of puberty is systematically delayed for one year by the interaction of feeding and season.

Effects of dietary tallow and choline supplementation on blood metabolites of early lactation cows

Kamyar Heidarnezhad, Animal science Department, Tabriz Islamic Azad University, P.O. Box 1655, Tabriz, Iran

Twenty eight early lactation Holstein cows were assigned to different levels of tallow and tallow plus choline during wk 2 to 8 postpartum. Cows fed with seven diets (18 % CP and 1.75 Mcal/kg NEL, DM basis) for ad libitum intake, with different levels of tallow (0,2,4,6% DM) and tallow plus rumen protected choline (2+0.15, 4+0.3, 6+0.45% DM).Cows were Housed as individually. Blood samples were collected from 2wk postpartum until 4wk after artificial inoculation as third weekly. Plasma samples were analyzed for plasma urea nitrogen (PUN), non esterifies fatty acids (NEFA), very low density lipoprotein (VLDL), progesterone (p4), estradiol (E2), Glucose and cholesterol. Data analyzed by producers of SPSS. Concentration of plasma glucose,cholesterol,VLDL,NEFA,E_2 and P_4 increased by feeding with tallow plus RPC than to tallow diets.The highest values of glucose ,cholesterol,VLDL,E_2 and P_4 were in the diets containing (4 % tallow+0.3% choline).Supplemental tallow and tallow plus choline did not consistently effects on PUN. Regression analysis of data showed a positive relationship and strong correlation ($R^2=0.9$) between the supplemental tallow plus choline in diets and increase of VLDL concentration. This study indicates that supplemental tallow plus choline can increase concentrations of blood P_4,E_2 Cholesterol and VLDL. Increase of this metabolites associated with development of ovarian follicles that should be beneficial to reproductive performance in high production dairy cows.

Plasma leptin and testosterone changes during pubertal development in Holstein bulls
H. Gholami, A. Towhidi and A Zareshahne, University of Tehran, Faculty of Agriculture, Animal Science Department*

The hypothesis that leptin plays an important role in regulating GnRH secretion and ultimately in reproduction stems from several studies. Collectively, these observations suggest that leptin may play an important role as a signal linking nutritional status to the central reproductive axis in cattle. The objective of this experiment was to study the changes of plasma leptin during puberty and its relationship with testosterone and testis dimensions in Holstein bull calves. The effects of age (P<0.01), testosterone (P<0.01), BW (P<0.01), BCS (P<0.01) and testicular width and length (P<0.05) on plasma leptin concentration were significant. Mean plasma leptin concentration during the period of peripuberty was significantly higher than those during puberty (P<0.01) and postpuberty (P<0.01). Mean plasma testosterone concentration during puberty was significantly higher than that of peripuberty (P<0.05). Testosterone pulses amplitudes during puberty and postpuberty were also higher than those of puberty (P<0.01), however testosterone pulses frequencies during peripuberty were higher than those during puberty and postpuberty (P<0.01). Results indicate that in growing bull calves unlike heifers, plasma concentrations of leptin decreased during puberty, while circulating testosterone increased. Hence, it can be concluded that role of leptin in the neuroendocrinology of sexual maturation is different in bulls and heifers.

Nutritional flushing enhances oestrous behaviour and increases ovulation rate of female goats exposed to the male effect under extensive conditions in subtropical Mexico
M.A. De Santiago-Miramontes[1], M. Muñoz-Gutiérrez[3], R.J. Scaramuzzi[2] and J.A. Delgadillo[1], [1]Centro de Investigación en Reproducción Caprina, Universidad Antonio Narro, México, [2]Department of Veterinary Basic Sciences, Royal Veterinary College, UK, [3]Departamento de Biología de la Reproducción, Universidad Autónoma Metropolitana, México*

We investigated if nutritional supplementation could improve sexual activity and increase ovulation rate, in response to the male effect, of does kept in extensive conditions. Fifty does were used; half received no nutritional supplementation (NS), while the other half were supplemented (S) with 290 g of corn, 140 g of soya and 950 g of alfalfa hay per animal, daily for the seven days before 4 sexually active bucks were introduced. Over the 5 days after introduction, oestrus was stimulated in the S group (S: 23/25, NS; 15/25; $P < 0.01$), However, between days 6 and 15, oestrus (S: 23/25, NS: 21/25) was not affected by nutritional supplementation ($P > 0.05$). Over the first 5 days, the ovulation rate was higher in S (1.6 ± 0.2 corpora lutea) than in NS (1.0 ± 0.2 corpora lutea; $P < 0.05$) groups. However between days 6 and 15, the ovulation rate did not differ between S (1.44 ± 0.1) and NS (1.36 ± 0.1; P>0.05). We conclude that seven days of supplementation improves oestrous behavior and increases ovulation rate of female goats exposed to the male effect.

Effects of addition of different fats to flushing diet on ovarian artro-venous differences of fatty acids and cholesterol in fat-tailed ewes: I - In the follicular phase

H. Sadeghi-Panah and A. Zare Shahneh, Tehran University Faculty of Agriculture, Karaj, Iran*

Sixteen 6 years old Zandi ewes were assigned to four experimental diets. Diets were: 1) without supplemental fat (C), 2) containing 4.5% calcium salts of tallow (T), 3) containing 4.5% calcium salts of soybean oil (S) and 4) containing 2.25% calcium salts of tallow + 2.25% calcium salts of soybean oil (M). Diets were isoenergetic and isonitrogenous. Estrous synchronization was performed by CIDR. Two days after CIDR removal, carotid artery and ovarian vein catheterized surgically and blood samples were collected. Plasma concentrations of fatty acids and cholesterol were measured and artro-venous differences (AVD) were calculated. There was a net uptake (positive AVD) of total fatty acids in all groups. AVD of total fatty acids for group S was higher than group C. AVDs of palmitic, oleic and linolenic acids for group S were higher than other groups. AVD of stearic acid for groups S and M was higher than other groups and for group T was higher than group C. AVDs of linoleic and arashidonic acids for group S was the highest and for group C was the lowest. There were no significant differences in AVDs of capric, myristic and palmitoleic acids among experimental groups. In groups that were received supplemental fat, especially group S there was a net production of lauric acid (negative AVD). There were no significant differences in cholesterol AVD among experimental groups.

Effects of addition of different fats to flushing diet on ovarian artro-venous differences of fatty acids and cholesterol in fat-tailed ewes: II - In the luteal phase

A. Zare Shahneh[1], H. Sadeghi-Panah[1] and A. Niasari Nasalji[2], [1]Tehran University Faculty of Agriculture, Karaj, Iran, [2]Tehran University Faculty of Veterinary Medicine, Tehran, Iran*

Sixteen 6 years old Zandi ewes were assigned to four experimental diets. Diets were: 1) without supplemental fat (C), 2) containing 4.5% calcium salts of tallow (T), 3) containing 4.5% calcium salts of soybean oil (S) and 4) containing 2.25% calcium salts of tallow + 2.25% calcium salts of soybean oil (M). Diets were isoenergetic and isonitrogenous. Estrous synchronization was performed by CIDR. 11 days after CIDR removal, carotid artery and ovarian vein catheterized surgically and blood samples were collected. Plasma concentrations of fatty acids and cholesterol were measured and artro-venous differences (AVD) were calculated. There was a net uptake (positive AVD) of total fatty acids in all groups. AVD of total fatty acids for groups S and M was higher than other groups. AVD of linoleic acid for groups that were received supplemental fat was higher than group C. In groups that were received supplemental fat, especially group T there was a net production of capric acid (negative AVD). There were no significant differences in AVDs of lauric, myristic, palmitic, palmitoleic, stearic, oleic, linolenic and arashidonic acids and cholesterol among experimental groups, but in groups that were received supplemental fat, AVDs of stearic, oleic and linolenic acids nonsignificantly were higher than group C.

Ovarian functions in fat-tailed ewes fed different fats: I - In the follicular phase
H. Sadeghi-Panah*, A. Zare Shahneh, A. Nik-Khah, M. Moradi Shahrbabak, Tehran University Faculty of Agriculture, Karaj, Iran

Sixteen 6 years old Zandi ewes were assigned to four experimental diets. Diets were: 1) without supplemental fat (C), 2) containing 4.5% calcium salts of tallow (T), 3) containing 4.5% calcium salts of soybean oil (S) and 4) containing 2.25% calcium salts of tallow + 2.25% calcium salts of soybean oil (M). Diets were isoenergetic and isonitrogenous. Estrous synchronization was performed by CIDR. Two days after CIDR removal, carotid artery and ovarian vein catheterized surgically and blood samples were collected. Follicles ≥ 5 and 3-5 mm on both ovaries were counted. Ovaries were removed and weighted. Plasma concentrations of progesterone and estradiol were measured and ovarian artro-venous differences (AVD) were calculated. There was a net production (negative AVD) of progesterone and estradiol in all groups, but there were no significant differences in their AVDs among experimental groups. Number of follicles ≥ 5 mm for group S was higher than other groups, but there was no significant difference in the number of follicles 3-5 mm among experimental groups. Weight of active ovary and total of two ovary for group S were higher than groups T and M, but there was no significant difference in the weight of inactive ovary among experimental groups.

Ovarian functions in fat-tailed ewes fed different fats: I - In the luteal phase
A. Zare Shahneh[1], H. Sadeghi-Panah[1]*, A. Nik-Khah[1], M. Moradi Shahrbabak[1] and A. Niasari Nasalji[2], [1]Tehran University Faculty of Agriculture, Karaj, Iran, [2]Tehran University Faculty of Veterinary Medicine, Tehran, Iran

Sixteen 6 years old Zandi ewes were assigned to four experimental diets. Diets were: 1) without supplemental fat (C), 2) containing 4.5% calcium salts of tallow (T), 3) containing 4.5% calcium salts of soybean oil (S) and 4) containing 2.25% calcium salts of tallow + 2.25% calcium salts of soybean oil (M). Diets were isoenergetic and isonitrogenous. Estrous synchronization was performed by CIDR. 11 days after CIDR removal, carotid artery and ovarian vein catheterized surgically and blood samples were collected. Follicles ≥ 3 mm and corpus luteums (CL) on both ovaries were counted (number of CL was concerned as ovulation rate index). Corpus luteums diameter was measured. Ovaries were removed and weighted. Plasma concentrations of progesterone and estradiol were measured and ovarian artro-venous differences (AVD) were calculated. There was a net production (negative AVD) of progesterone and estradiol in all groups. In groups that were received supplemental fat, net production of progesterone was higher (AVD was lower) and net production of estradiol was lower (AVD was higher) than group C. There was no significant differences in ovulation rate, corpus luteums diameter, number of follicles and weight of ovaries among experimental groups.

Effects of zearalenone and Toxy-Nil Plus Dry on physiological responses of boar semen
A. Siukscius, J. Kutra, A. Urbsys, V. Pileckas, R. Nainiene and A. Mankeviciene, Institute of Animal Science of LVA, R. Zebenkos 12, LT-82317, Baisogala, Radviliskis distr., Lithuania

The aim of the present study was to determine the effects of the mycotoxin zearalenone as well as the detoxicating product Toxy-Nil Plus Dry (TNPD) on the quality of boar semen. Boars in the experimental groups were given feed containing zearalenone (0.57 mg/kg) for 32 days and detoxicated (1 kg TNPD per 1000 kg of feed) feeds. In a week following the administration of zearalenone-containing feeds of the boars, the volume of ejaculation had decreased by 40.8% (P<0.005) compared with the control group. The lowest initial spermatozoa motility in the semen collected during intoxication was determined in the group of boars receiving zearalenone-containing feed (3.9±1.79 points, P=0.01). On replacement of the contaminated feed, the motility of spermatozoa recovered within a week and amounted to 7.0±0.55 points. The negative effect of TNPD of spermatozoa was noted during the recovery period. The total quantity of pathologic spermatozoa in a corresponding group increased to 33.2±8.75% compared with 21.7±8.27% (P<0.05) in the pre-experimental period. The study indicated that the investigated level of zearalenone in feed negatively affects the reproductive performance of boars. The unhealthy effect of zearalenone may be significantly reduced by treating contaminated feeds with TNPD.

Interaction between nutrition and breeding season in mares
J. Salazar-ortiz and D. Guillaume*, PRC-INRA-CNRS-HN-Université, 37380 Nouzilly, France

Our aims were to estimate the influence of nutrition on the annual rhythm of reproduction in mares and on the resumption of ovulations.

During 3 years, 3 groups of 10 pony mares received different diets. The "well fed" (W) group was fed to maintain the mares in good body condition; the "restricted" (R) group, to maintain the mares very thin but in good health and the last group "variable" (V) was fed alternatively as R or W group, with the aim to mimic seasonal variations of feeds allowances. Blood samples were collected during two 24 h periods.

The mean date (± SEM in days) for the beginning, the end of ovulatory inactivity and the ratio of mares with inactivity, between second and third winter, were: 4 Jan. ± 6, 8 Apr. ± 12, (3.5/9), 7 Oct. ± 6, 6 May ± 6, (10/10), and 6 Nov. ± 6, 6 Mai ± 6, (8/10), respectively for the W, R and V groups. No postprandial peak of glucose or insulin was observed in R group. GH levels were higher in R and V than in W group. No important difference was found in melatonin levels.

The increase in feeds allowances of the V group during autumn or winter did not advance the first annual ovulation. Mares' body condition is the main cause of the initialisation and the length of the winter ovulatory inactivity.

Leptin and metabolic ways in controlling the reproduction function in animals
M.A.A. EL-Barody, Animal Production of Department, Fac. Of Agriculture, Minia Univ. Minia-Egypt

Leptin has the potential to play a role in energetic control of reproduction. Reproduction maturation fails to occur in the complete absence of functional leptin. It has not been demonstrated in a wide variety of species that above-fasting levels of leptin are necessary for normal reproductive function. High levels of leptin are not sufficient for normal reproduction when the oxidation of metabolic fuels is blocked. There are at least two possible ways that leptin might interact with the metabolic sensory system that controls reproduction. First, leptin synthesis and secretion in various tissues might be sensitive to metabolic fuel availability. If so, and these changes are shown to be necessary for estrous cycles. Then it might be concluded that leptin acts as a mediator of the metabolic signal. Second, leptin might affect reproduction indirectly by way of modulating the metabolic signal that is known to influence reproduction. Leptin treatment causes tissue-specific increases fuel uptake and oxidation and these in turn would be expected to influence the animal reproductive performance

---

Impact of melatonin and artificial photoperiod on anestrous buffalo-cows and delayed pubertal buffalo heifers
K.A. El. Battawy[1], S.G. Hassan[1], A.A. El- Menoufy[2], M. Younis[2] and R.M. Khattab[3], [1]National Research centre, Dokki, Egypt; [2]Fact- vet. Med. Cairo Univ. Giza, [3]Agric. Res. Cantre, Ministry of Agriculture

This investigation showed that melatonin administration (18mg/animal) daily to postpartum anestrous buffaloes starting from mid-July till the end of September (or animals exhibited estrus) resulted in induction of estrus, ovulation and improving the conception rate, Also administration of melatonin to the non pubertal heifers resulted in appearance of estrous symptoms and conception.
Moreover, modulation of the light/dark cycle during summer season (long photoperiod) to an induced artificial short photoperiod resulted in induction of estrus, ovulation and pregnancy in postpartum anestrous buffaloes.

Session G3   Theatre I

Towards genetical genomics in livestock

D.J. de Koning*, C. Cabrera and C.S. Haley, The Roslin Institute, Roslin, EH25 9PS, United Kingdom

Microarrays have been widely implemented across the life sciences but there is still debate on the most effective uses of such transcriptomic approaches. One of the most promising applications for unravelling complex traits is the integration of transcriptomics with genomic mapping. In genetical genomics, gene expression measurements are treated as quantitative traits and genome regions affecting expression levels are denoted as expression quantitative trait loci or eQTL. One powerful outcome of genetical genomics is the reconstruction of genetic pathways underlying complex trait variation by combining gene location with genome information and expression (co)variation and pleiotropy of eQTL. The potential of pathway reconstruction has now been demonstrated in some model species despite the small sample size in eQTL experiments to date. The successful implementation of genetical genomics in livestock genetics will require active development of the following areas: 1) genome sequencing and, crucially, annotation of livestock genomes. 2) Adaptation of pathway databases from model species and humans to livestock species. 3) Methodological development for data base mining and pathway reconstruction incorporating a genetic component.

In the build-up towards full-blown eQTL studies, we can study the effects of known candidate genes or marked QTL at the gene expression level in more focussed studies. We will illustrate this with an example in chicken, where we identified genes that are differentially expressed between genotypes of a marked body weight QTL.

Session G3   Theatre 2

Genetic variance in a quantitative-genetic single locus model incorporating genomic imprinting

Norbert Reinsch, Forschungsbereich Genetik und Biometrie, Forschungsinstitut für die Biologie landwirtschaftlicher Nutztiere, FBN, Wilhelm-Stahl-Allee2, 18196 Dummerstorf, Germany

A quantitative-genetic model with a single genomically imprinted locus with two loci is introduced. As in the standard mendelian case the differences between homozygous genotypes and homozygous and heterozygous genotypes are discribed by two parameters a and d. The difference between A_1A_2 and A_2A_1 heterozygotes (the first alle is paternally derived) is set equal to the imprinting parameter i. It is shown, that the total genetic variance is the sum of a Mendelian, an imprinting- and a dominance contribution. The imprinting component of the genetic variance has a paternal and a maternal contribution, which is different unless no imprinting exists. Without imprinting all formulas reduce to the standard Mendelian case. Results are contrasted with previous proposals from the literature and generalisations to multiple loci are discussed.

The use of asymmetric distributions on genetic evaluation: An example with pig litter size
L. Varona, N. Ibañez, R.N. Pena, J.L. Noguera and R. Quintanilla, Area de Producció Animal, Centre UdL-IRTA, 25198 Lleida

Statistical models for genetic evaluation make use of Gaussian or *t* residual distributions. However, some new statistical development allows using asymmetric distribution for the residual not controlled by the model. We have analysed a data set of litter size on pigs consisting of 2072 data for number of piglets born alive from 657 sows, with a total pedigree of 765 individuals. The model includes order of parity and herd-year-season systematic effects, Gaussian additive genetic effects and Gaussian permanent environmental effects. In addition, we use three different distributions for the residuals. 1) Symmetric Gaussian distribution, 2) Asymmetric Gaussian distribution and 3) Asymmetric *t* distribution. The three statistical models were compared using a Bayes Factor. The most suitable model corresponds to the asymmetric Gaussian distribution (Model 2). The posterior mean of heritability was 0.063 with a posterior standard deviation of 0.028. However, the resulting distributions of the residuals are strongly asymmetric, indicating that the sources of variation not controlled by the model have mostly a negative influence on the prolificacy. The asymmetry parameter can be understood as a measure of sensibility to negative environmental influences on the phenotype. The consequences on prediction of breeding values are discussed. In future research, the model can generalized with the use of hierarchical Bayesian methods allowing each systematic effect of even each individual to have a different residual distribution.

A study of canalization for slaughter weight in pigs
N. Ibáñez[1]*, L. Varona[1], D. Sorensen[2] and J.L. Noguera[1], [1]Àrea de Producció Animal - Centre UdL-IRTA, 25198 Lleida, Spain, [2]Danish Institute of Agricultural Sciences, PB50, 8830 Tjele, Denmark

A study of heterogeneous residual variance of weight at slaughter (175 days of age) in pigs is presented. Here, the heterogeneity of residual variance was associated with systematic and additive genetic effects. In addition, we evaluated the influence of the heterogeneous residual variance in the future economic performance, given a commercial price function. The results of the study showed that the residual variance is controlled by both systematic and additive genetic effects. Moreover, we also found a small negative correlation between the genetic additive effects affecting the mean and the ones controlling the residual variance. Supplementary, we showed that both, the additive genetic effect on the mean and the residual variance have an important repercussion in the future economic performance of selected individuals. The approach proposed in this study allows obtaining the information of breeding values for the mean and the residual variance at the same time. This fact opens the possibility to select by means of an index composed by both breeding values, with the objectives of increasing the mean and reducing the variance of the weight at slaughter.

Use of principal component analyses to select traits: Application to conformation traits
Y. de Haas and H.N. Kadarmideen, ETH Zurich, Institute of Animal Sciences (UNS D8), CH-8092 Zurich, Switzerland

Principal component analyses (PCA) was shown to be a useful method in deriving the most informative traits from a set of multiple correlated traits. Data on lactating heifers from Holstein (67,839), Brown Swiss (173,372) and Red & White breeds (53,784) were available, and were used to quantify genetic variation. Analysed traits were stature (ST) and heart girth (HG), body depth (BD), rump width (RW), and dairy character (DC) or muscularity (MU). Body condition score (BCS) was only scored for Holsteins. A sire model was used for all breeds, and variance components were estimated using ASREML. Heritabilities for ST were high (0.6 – 0.8), and heritabilities for the linear type traits ranged from 0.3 to 0.5, for all breeds. Based on the Eigenvalues and Eigenvectors of the genetic correlation matrices of the conformation traits, it can be concluded that for Holstein BD, RW and BCS are the most informative traits, whereas for Brown Swiss BD, ST and HG are most informative. For Red & White cattle BD, RW and HG seemed to be most informative. Principal component analyses can be used to derive a few traits from a large number of multiple correlated traits, normally found in dairy cattle data. The results from PCA on conformation traits in this study would be useful in constructing total merit indexes for Swiss dairy cows.

Transmission disequilibrium test for fine mapping based on reconstructed haplotypes
X.D. Ding[1,2]*, Q. Zhang[2] and H. Simianer[1], [1]Institute of Animal Breeding and Genetics, Albrecht-Thaer-Weg 3, 37075 Goettingen, Germany, [2]China Agricultural University, Beijing, China

A growing number of studies demonstrate that haplotype-based approaches may provide more power and accuracy in locating quantitative trait loci (QTL) and causative disease variants than single-locus methods. Usually, haplotype-based studies are likely to follow a two-step procedure: first, haplo-genotypes of individuals with phase unknown genotypes are reconstructed, and second, these reconstructed haplo-genotypes are analyzed with suitable mapping approaches. There are two principal approaches to inferring haplotypes, one is directly using population data assuming presence of linkage disequilibrium, and the other one is using family data. Recent literature has suggested that haplotype inference through close relatives can be an alternative strategy for haplotype reconstruction, which both reduces haplotype ambiguity and accounts for linkage disequilibrium and thus improves the efficiency of haplotype inference. For fine mapping, the transmission disequilibrium test (TDT) was shown to be more powerful than linkage analysis when markers are tightly linked with QTL. The family information is also used in the case of TDT. Moreover, TDT with haplotype will be more powerful than only with single locus for fine mapping. We present an extension of the TDT based on haplotype reconstruction using family information. Simulations comparing our approach with alternative algorithms demonstrate the improved efficiency of gene mapping from our new approach.

Effectiveness of a new strategy for the computation of transmission probabilities using partial pedigrees

M. Mayer*, A. Tuchscherer and N. Reinsch, Research Institute for the Biology of Farm Animals (FBN), Wilhelm-Stahl-Allee 2, 18299 Dummerstorf, Germany

A new strategy for the computation of transmission probabilities and thus conditional genotypic and gametic relationship matrices is presented for situations where marker information is missing. This strategy can deal with pedigrees of arbitrary size and is very efficient under computational aspects. A simulation study was performed, based on a data structure which is currently used in marker assisted BLUP evaluation for the trait somatic cell score in the German Holstein population, that showed the effectiveness of the proposed strategy of using partial pedigrees in the computation of transmission probabilities. In dependence on the family type and the marker genotypes there were clear shifts of the estimated transmission probabilities from 0.5 towards to the true values. The transmission probabilities computed in this way way can be used to construct a complete conditional gametic or genotypic relationship matrix.

Investigation on major genes for functional traits from dairy cattle experiment

B. Karacaören,* L.L.G. Janss, and H. N. Kadarmideen Statistical Animal Genetics Group, Institute of Animal Science, Swiss Federal Institute of Technology, ETH Centrum, Zurich CH 8092, Switzerland

The main aim of this study was to investigate existence of major gene for milk yield, milking speed, dry matter intake, and body weight recorded at different stages of lactation on first lactation dairy cows (n=320) stationed at the research farm of the Swiss Federal Institute of Technology over the period of April 1994-2004. Data were modeled as a longitudinal trait and analysed using Bayesian segregation analyses. Gibbs sampling was used to make statistical inferences on posterior distributions; inferences was based on single run of the Markov chain for each trait with 500000 samples, with each 10th sample collected due to the high correlation among the samples. Small sampling size and prediction of permanent environmental effect were constraints of this study. Posterior mean (and SD) of major gene variance for milk yield was 4.46 (2.59), for milking speed was 0.83 (1.26), for dry matter intake was 4.37 (2.34) and for body weight was 2056.43 (665.67). Highest posterior density regions for all three traits did not include zero except milking speed. With additional analyses on transmission probabilities, we could confirm the existence of major gene for milk yield, but not for milking speed, dry matter intake and body weight. Expected Mendelian transmission probabilities and their model fits were also compared.

An autonomous agent algorithm to compute the Weitzman genetic diversity for large sets of taxonomic units

H. Simianer* and M. Tietze, Institute of Animal Breeding and Genetics, Albrecht-Thaer-Weg 3, 37075 Goettingen, Germany

The computing time of the exact algorithm to compute Weitzman's genetic diversity D for a set of N operational taxonomic units (OTUs) is proportional to 2^N and never could be applied to data sets of more than 35 OTUs. Different approximations have been suggested which increase the size of the set that can be analysed, but result in a systematic underestimation of D. We show that the problem can be re-formulated in the form of a discrete mathematical optimisation problem similar to the well-known travelling salesman problem. The problem then is to find the unique sequence of so-called link OTUs which maximize D. An autonomous agent algorithm was used which mimics the strategy of ants to find paths from the colony to food. Different parameters (number of ants, pheromone concentration and degradation rates, number of independent restarts etc.) were optimised. For data sets for which the exact solution was known, the suggested algorithm converged to the correct solution in a fraction of the computing time needed for the exact algorithm. With large sets of OTUs (>100), good results are obtained, however it is impossible to say whether they match the exact value of D. We discuss criteria to judge the 'goodness' of estimates of D where the exact solution is unknown based on arguments from order statistics.

Genetic characterization by single nucleotide polymorphisms (SNPs) of Hungarian Grey and Maremmana cattle breeds

L. Pariset[1], C. Marchitelli[1], A. Crisà[1], S. Scardigli[1], S. Gigli[2], J. Williams[3], S. Dunner[4], H. Leveziel[5], A. Maroti-Agots[6] I. Bodò[7], A. Valentini[1], A. Nardone[1], Department of Animal Production, Università della Tuscia, Viterbo, Italy, [2]C.R.A., Monterotondo, Rome, Italy, [3]Roslin Institute, Roslin, Scotland, [4]Departimento de Produccion Animal, Universidad Complutense, Madrid, Spain, [5]INRA-Centre de Limoges, Unité de Génétique Moléculaire Animale, Limoges cedex, France, [6]SZIE University, Budapest, Hungary, [7]University of Agriculture, Debrecen, Hungary

Approximately 100 single nucleotide polymorphisms (SNPs) in genes candidate for meat traits were genotyped on individuals belonging to 4 different beef breeds. The Italian Maremmana and the Hungarian Grey breeds belong to the same podolic group of cattle, show a similar external conformation and recently underwent a similar demographic reduction. Marchigiana and Piemontese are two Italian meat breeds that went through different selection histories. Population genetic parameters such as allelic frequencies and heterozygosity values were assessed. Genetic distances, calculated to evaluate the possibility of recent admixture between the populations, show that the similarity between Hungarian Grey and Maremmana cattle is mainly morphological, while they show genetically distinct features. The assignment of individuals to breeds was inferred using Bayesian inference, confirming that the set of chosen SNPs is able to distinguish among populations. A method for the identification of loci that were likely subject to selection was applied, suggesting that abnormal allelic frequencies at single loci could reflect similar selection or demographic history.

Generalized linear models vs. linear models for the detection of QTL effects on within litter variance

D. Wittenburg, V. Guiard and N. Reinsch, Forschungsinstitut für die Biologie landwirtschaftlicher Nutztiere FBN, Wilhelm-Stahl-Allee 2, 18196 Dummerstorf, Germany*

Quantitative trait loci (QTL) may not only affect the mean of a trait but also its variability. A special aspect is the variability between multiple measurements of genotyped animals, for example the within litter variance of piglets birth weights. The sample variance of repeated measurements is assigned as observation for every genotyped individual. The conditional distribution of the non-normally distributed trait was approximated by a gamma distribution. To detect QTL effects in the daughter-design a generalized linear model with a log-link function was applied. Adapted test statistics were constructed to evaluate the test problem in terms of statistical power and desired error probability under the null hypothesis H0: No QTL with effect on the within litter variance is segregating. Furthermore we take a look on the estimates of the QTL effect and the QTL position. To compare the advantages of this method with more common tools of statistics a weighted regression approach was developed, taking a transformed sample variance as observation.

Application of microsatellites for parentage control and individual identification in Spanish pig breeds

R. Delgado, N. Trilla, N. Torrentó, J. Soler and J. Tibau, IRTA, 17121 Monells, Spain

A set of 593 Spanish pig boars of three breeds (Piétrain, Duroc and Landrace) from the Spanish pig breeders association (ANPS) was studied in order to estimate the level of genetic variability that can be detected by using microsatellites. The efficiency of these genetic markers as a tool for verifying the parentage and identification of individual animals in these porcine populations was established. DNA was obtained from biological samples of these selected purebred boars to investigate the genetic variability in 12 microsatellites located in 11 different chromosomes. The average expected heterozygosity was 0.553, 0.638 and 0.639 in Duroc, Landrace and Piétrain breeds, respectively. The mean Polymorphism Information Content (PIC) values of the 12 polymorphic loci ranged from 0.497 (Duroc), 0.584 (Landrace) to 0.594 (Piétrain). We have found in four microsatellites seven alleles that could be breed specific between these three breeds, with frequencies up to 0.20 and 0.40. The combined probabilities to exclude non-parents with one parental genotype unavailable were 91.99% (Duroc), 96.52% (Landrace) and 97.47% (Piétrain). Assuming a known parent-pair the set of 12 markers provided exclusion probabilities of 99.23% (Duroc), 99.81% (Landrace) and 99.88% (Piétrain). The estimated probability of exclusion of wrongly named parents was high and this panel of 12 microsatellites represents an efficient method for determining parentage and for the individual identification in porcine populations of these three Spanish breeds.

Analysis of microarray data – influence of pooling biological samples

G. Nürnberg[1]*, J. Vanselow[2], D. Koczanz[3], [1]Research Institute for the Biology of Farm Animals, Genetics&Biometry, [2]Molecular Biology D-18196 Dummerstorf, W.-Stahl-Allee 2, [3]University of Rostock, Institute of Immunology, Germany

In microarray experiments biological samples are often pooled to reduce experimental costs or to reduce biological variation. Data of a gene expression experiment comparing ovary tissue of two mouse lines, a long term selected line due to fertility and a control line, are investigated for two designs using the MAS5.0 analysing method. A design with 10 single sample arrays per line were compared to a design with six pooled arrays per line (pooling five samples per array). Because gene expression measurements are affected by biological and technical variation, we also used technical replicates of the six pooled arrays per line. The practical effect of pooling depends on the proportion of biological and technical variance, because the larger the biological variability is relative to technical variability the larger the benefit of pooled designs, if the assumption holds that the variance of a mean is reduced by the number of samples per mean. We questioned if pooled and non-pooled designs lead to same number of significant expressed genes and which significant expressed genes are in both gene lists. Using pooled arrays we found a reduced number of significant expressed genes and only 50–80% of significant expressed genes were on both lists. Reasons for these findings are discussed.

Microarray analysis of gene expression profiling in pigs with extreme glycolytic potential values of skeletal muscle

V. Russo*[1], R. Davoli[1], M. Colombo[1], S. Schiavina[1], A. Stella[2], L Buttazzoni[3] and L. Fontanesi[1], [1]DIPROVAL, Sezione di Allevamenti Zootecnici, University of Bologna, Reggio Emilia, Italy, [2]Parco Tecnologico Padano Lodi, Italy,[3] Associazione Nazionale Allevatori, Roma, Italy

Glycolytic potential (GP) is an important parameter of skeletal muscle affecting porcine meat quality traits. A higher level of GP is associated with a lower ultimate pH influencing meat colour, water holding capacity, drip loss, tenderness and yield processing. As a first step to investigate the genetic basis of GP in pigs and to identify the genes related to or affecting this parameter, microarray analysis was performed studying the expression profile of skeletal muscle tissue. GP was determined in samples of muscle *longissimus dorsi* of 277 sib-tested Italian Large White pigs taken at slaughtering. Eight pigs with extreme and divergent values for this parameter were selected in order to obtain two RNA pools each of four samples: pool high (with value of GP >145 μmol lactate equivalent g $^{-1}$ muscle wet weight) and pool low (with GP <65 μmol). The porcine 11K oligonucleotides Operon microarray was used to compare the gene expression profiles of these two extreme groups. Data analysis by linear model ANOVA allowed to identify a set of differentially expressed genes mainly coding for energy metabolism enzymes and structural protein of muscle.

Phylogenetic relationship among different horse breeds in Pakistan based on random amplified polymorphic DNA (RAPD) analysis

*F. Hassan[*1], S. Ali[1], M.S. Khan[1], M.S. Rehman[1] and Y. Zafar[2], [1]Department of Animal Breeding and Genetics, University of Agriculture, Faisalabad, Pakistan, [2]National Institute for Biotechnology and Genetic Engineering, Faisalabad, Pakistan*

Phylogenetic relationships among four indigenous horse breeds viz., Morna, Siean, Anmol and Bralanwala Kakka and eight exotic breeds viz., Thoroughbred, Cleveland Bay, Percheron, Suffolk Punch, Noriker, Hanoverian, Arab and Anglo Arab were assessed using RAPD assay. The genomic DNA bulks were surveyed using 40 arbitrary sequence primers. A total of 297 DNA fragments were generated out of 40 primers through polymerase chain reaction (PCR) with an average of 7.4 bands per primer. Fifty seven bands were polymorphic indicating 19.19 % DNA polymorphism among horse breeds. Three primers OPE-4, OPF-6 and OPL-7 could characterize indigenous breeds from exotic breeds. The dendrogram obtained using unweighted pair group method of arithmetic means (UPGMA) revealed two main clusters. Thoroughbred, Cleveland Bay, Percheron, Suffolk Punch, Noriker, Hanoverian, Arab, Anmol and Anglo Arab breeds were grouped together as cluster A. In cluster B native breeds Morna and Siean grouped together with Bralanwala Kakka showing 93.57 % similarity. This indicates that indigenous horse breeds of Pakistan have recently diverged from each other and indiscriminate crossbreeding of Thoroughbred and Arab horses have contributed significantly over years in the development of indigenous breeds.

Estimation of genetic parameters for categorical, continuous and molecular genetic data in multivariate animal threshold models using Gibbs sampling

K.F. Stock[1], O. Distl[1] and I. Hoeschele[2], [1]Institute for Animal Breeding and Genetics, University of Veterinary Medicine Hannover, 30559 Hannover, Germany, [2]Virginia Bioinformatics Institute and Department of Statistics, Virginia Polytechnic Institute and State University, Blacksburg 24061, Virginia, USA*

Simulated populations included 7 generations and 40000 animals per generation. Fixed effects, residual and additive genetic variances for one continuous trait (T1) and liabilities for four categorical traits (T2 to T5) were simulated. QTL effects were simulated for T2 with recombination rates (r) of 0.00 or 0.01 and polymorphism information content (PIC) of markers of 0.9 or 0.7. Simulated heritabilities (h^2) were 0.50 (T1), 0.25 (T3, T5) and 0.10 (T2, T4). Simulated additive genetic correlations (r_g) were ±0.20. After dichotomization trait prevalences were 0.25 (T2, T5) and 0.10 (T3, T4). Phenotypes of 10000 animals from one generation (P1) or phenotypes and genotypes of 5000 animals and their parents (G2) were used for multivariate estimations. Most biased parameters were r_{g12} and r_{g14} in P1 (-33% to -55%) and h^2_2 and r_{g45} in G2 (+36% to +52%). Correlations between true and predicted breeding values (BV) for the categorical traits did mostly not differ significantly between P1 and G2. Selection on the basis of BV from G1 was significantly less effective than selection on the basis of genotype and BV from G2. Selection response was significantly lower with PIC=0.7 and r=0.00 than with PIC=0.9 and r=0.00 or r=0.01.

Regression of the phenotypic similarity on the coancestry as heritability estimator

S.T. Rodríguez-Ramilo[1], M.A. Toro[1], A. Caballero[2] and J. Fernández[1]*, [1]Dpto. de Mejora Genética Animal. INIA. Crta. A Coruña Km. 7,5. 28040 Madrid, Spain, [2]Dpto. Bioquímica, Genética e Inmunología. Universidad de Vigo. 36310 Vigo, Spain

Although estimates of heritability require accurate information on the relationship among individuals based on pedigree data, this information is generally unavailable. However, markers can be used to estimate coancestry from molecular information. Ritland, in 1996, developed an approach to infer heritability estimates based on the regression of the phenotypic similarity on the estimated coancestry. We carried out simulations to test the accuracy of this method considering 10-120 additive, biallelic loci controlling an evaluated quantitative trait, and 10-100 codominant neutral markers. Simulations involved two population sizes (50-500) and two scenarios: (1) populations with no family structure, where contributions of parents to progeny are free and mating is random, and (2) populations with a familiar structure, where 5-50 families are obtained through equal contributions of parents to the next generation, and parents of each family are chosen at random or selected according to their phenotypic value. Our results suggest that two main factors affect the accuracy of the heritability estimator: (1) the absence of a familiar structure in the population, and (2) long-term selection on the evaluated trait. Nevertheless, for large population sizes and a considerable number of markers the proposed estimator shows a suitable behaviour.

Introducing the suitable model for analysis of the lamb body weight using Log likelihood ratio test

M.A. Abbasi[1], R. Vaez Torshizi[1], A. Nejati Javaremi[2], R. Osfoori[3], [1]Dept of Animal Science, Islamic Azad University, Khodabandeh Branch, [2]Dept of Animal Science, Tehran University, [3]Zanjan Jahad-e-Keshavarzi

The objectives of this study was to introduce the most suitable animal model for analysis of body weights at birth (BW), one month (SW1), two month (SW2), three month (SW3), four month (SW4), five month (SW5) and six month (SW6) of ages. Data from 3952 lambs born from 142 sires and 1182 dams, collected from 1992 to 2003 were used. Nine animal models were considered that differed in the (co)variance components. All traits were analyzed with DFREML software package by AI-REML algorithm. A Log likelihood ratio test was used to choose the most suitable animal models for each trait. The addition of maternal permanent environmental effect (c) with maternal genetic effect (m) already fitted did not increase -2 log L value significantly for SW5 and SW6. Maternal additive genetic effect increased significantly -2 log L of model (p<0.01), but this effect together with maternal permanent and temporary environmental effect (t) did not improved -2 log L. Therefore, when maternal permanent and temporary environmental effects are in models, maternal genetic effect can be ignore for all traits. The model including direct additive (a), maternal permanent and maternal temporary environmental effects was determined to be the most suitable model for BW, SW1, SW2 and SW5. The most suitable model for SW3 and SW4 traits was model 2 (including a and c). For SW6, model 3 (including a and m) was the best model.

Relationships between herd life and growth parameters in Holsteins

A. Nishiura*, T. Yamazaki and H. Takeda, National Agricultural Research Center for Hokkaido Region, 062-8555, Sapporo, Japan

Genetic and phenotypic correlations between herd life and growth parameters were estimated. Data was consisted of 6,955 Holstein dairy cows. Two traits were defined for herd life, true herd life (THL) and functional herd life (FHL), which were adjusted for milk production. Five models of growth curve were examined, and Bertalanffy model was selected by goodness of fit and computational difficulty. Growth parameters of body weight and other 10 body measurements from $y_n = A(1-Be^{-kn})^3$ were estimated by nonlinear regressions and provided estimates of mature value (A) and rate of maturing (k) for each cow. Estimates of phenotypic and genetic correlations between THL and mature value (A) were positive in most cases. Estimates of phenotypic and genetic correlations between FHL and rate of maturing (k) were negative in all cases. Estimates of phenotypic and genetic correlations between herd life and mature value (A) were larger than estimates between herd life and body measurements. It was suggested that late-maturing cow tend to have longer herd life than early maturing cow, and we can remove measurements errors and environmental effects on measurements to some extent by estimating mature value from many data in growth days.

Genetic and environmental trends for production traits in Holstein cattle of Iran

M. Razmkabir[1], A. Nejati-Javaremi[1], M. Moradi-Shahrbabak[1], A. Rashidi[2] and M. Sayadnejad[3], [1]Dept. of Animal Science, University of Tehran, Iran, [2]Dept. of Animal Science, University of Kurdistan, Iran, [3]National Animal Breeding Center of Iran, Karaj, Iran

Production traits are the most important traits which affect the economic efficiency of dairy cows. Estimation of genetic trends is necessary to monitor and evaluate the success of breeding programs. In this study production traits data of the Iranian dairy herds were obtained from National Animal Breeding Center of Iran. The dataset included First lactation records of milk, fat and protein yields. Records were adjusted for milking times (2X) and to 305 days in milk. Data were analyzed by Matvec program and also a multiple trait mixed linear model was fitted to obtain BLUP of the breeding values of the animals. Genetic trend was estimated as the linear regression of average breeding values on the birth year. The estimated genetic trend values for milk, fat and protein yields were 33.84 ± 2.81, 0.64 ± 0.05 and 1.00 ± 0.08 kg/year, respectively. Genetic trend for all traits was positive and significant ($p<0.01$). Average environmental trend for milk, fat and protein yields were 88.44, 3.78 and 2.06 kg/year, respectively. The absolute values of the environmental trends were greater for all traits. The result of this study show that genetic improvement for production traits was at a acceptable level in the Holstein herds of Iran.

Analysis of (co)variance components on economic traits of Baluchi sheep with REML method

M. Hosseinpour Mashhadi, Islamic Azad University Mashhad Campus, College of Agriculture, Department of Animal Science, P.O Box 91735-413, Mashhad, Iran

5913 records of Baluchi sheep breed which is native of eastern provinces of Iran, were used for estimation (co)variance components. Genetic and phenotypic (co)variances of traits were estimated with REML method under a single trait animal model with DFREML software. Fixed factors effected on traits were age of dam, sex, type of birth, year and month of birth. A covariate was used for adjustment of traits, it was number of days between birth date to date of recording of each traits Direct heritability of weights on birth, three, six and nine months, yearling weight, average daily gains from birth to three month weight and from three month to yearling weight and first recording of spring wool were estimated to be 0.14 (.032), 0.28 (0.045), 0.25 (0.05), 0.28 (0.053), 0.33 (0.055), 0.08 (0.029), 0.09 (0.029) and 0.11 (0.028) respectively. Genetic and phenotypic correlations, respectively, for birth and nine month weights 0.56 and 0.31, birth and yearling weights 0.59 and 0.31, three month and nine month weights 0.92 and 0.5, three month and yearling weights 0.94 and 0.61, nine month and yearling weights 0.95 and 0.82 and average daily gains from birth to three month weight and three month to yearling weight -0.07 and -0.17 were estimated.

Production of healthy and high quality beef

K. Nuernberg[1], D. Dannenberger[1], N. Scollan[2], I. Richardson[3], A. Moloney[4], St. Desmet[5] and K. Ender[1], [1]Research Institute for the Biology of Farm Animals, Wilhelm-Stahl-Allee 2, D-18196 Dummerstorf, Germany, [2] Institute of Grassland and Environmental Research, Aberystwyth, SY23 3EB, UK, [3]Division of Farm Animal Science, University of Bristol, Langford, Bristol, BS40 5DU, UK, [4]Teagasc, Grange Research Centre, Dunsany, Co. Meath, Ireland, [5]Department of Animal Production, University of Ghent, Proefhoevestratt 10, B-9090 Melle, Belgium

There is increasing interest in animal production practices to enhance beneficial fatty acids in meat and meat products to make them more attractive for health reasons. The aim of the studies was to characterise the effects of genotype and different feeding systems on meat quality, sensory quality, and fatty acid composition of *longissimus* muscle, in particular n-3 fatty acids and CLA*cis*-9, *trans*-11. This paper summarises the main results of various beef experiments representing the major beef production systems in different countries. Feeding grass, concentrates supplemented with linseed or ruminally protected lipids increases not only the C18:3n-3 but also the long chain n-3 fatty acids in the intramuscular fat. The most abundant CLA*cis*-9,*trans*-11 in muscle is mainly deposited in the triacylglycerol fraction and the content is changed by different feeding systems. The *trans* vaccenic acid (C18:1*trans*-11) is the major C18:1*trans* isomer in beef. The alteration of the fatty acid profile affected the meat quality to different extents. Grass feeding influenced colour and lipid shelf life positively.

Application of a panel of microsatellite markers for the genetic traceability of bovine origin products

C. Dalvit*, C. Targhetta, M. Gervaso, M. De Marchi, R. Mantovani and M. Cassandro, Department of Animal Science, University of Padova, 35020 Legnaro (PD), Italy

Aim of this study was to test the efficacy of a panel of twelve microsatellite markers for the genetic traceability of bovine products. A total of 187 animals belonging to six different cattle breeds were genotyped. The breeds involved were four beef cattle: Chianina, Marchigiana, Romagnola and Piemontese and two dairy cattle: Brown Swiss and Holstein Friesian. The genotyping was carried out on blood and meat individual samples for beef breeds and blood and milk individual samples for dairy breeds; in most cases the animals showed an identical genetic profile obtained from the different DNA sources. The probability to find two individuals sharing the same genotype was estimated considering different number of loci evidencing estimates on the order of 10^{-6} even with less than twelve microsatellites. In addition, on the bases of allelic frequencies in each breed, the attribution of samples to the breed of origin was performed with two probabilistic approach (bayesian and maximum likelihood) revealing the efficacy of this microsatellite panel. Results will be discussed considering a potential use of this method for individual and breed traceability of bovine products in response to the confidence crisis of consumers and as instrument for the valorisation of niche bovine products based on one breed origin.

Comparative analysis of selection schemes in small cattle populations

A. Montironi[1,2]*, A. Stella[1] and G. Gandini[2], Parco Tecnologico Padano, via Einstein – Polo Universitario, 26900 Lodi, Italy, [2]VSA, University of Milan, 20133 Milan, Italy

Progeny testing has traditionally been considered not applicable to small cattle populations, both for inability to obtain significant genetic gains and low profitability. Genetic improvement in small populations can contribute to sustainable utilisation of cattle genetic resources. The objective of this study was to compare, for different numbers of recorded females, genetic gain under three selection schemes: no progeny test, with selection of sires on pedigree index (NPT); progeny test of sires of sires (PT3); classical progeny test of sires of sires and sires of dams (PT5). Selection response per year in populations from 1,000 to 15,000 recorded females was deterministically analyzed with a constraint of 1% inbreeding per generation. PT3 gives higher responses from 1,000 (.12 SD) to approximately 12,000 females (.17 SD). Superiority of PT3 vs. PT5 was of 36% and 13 % respectively with 2,000 and 5,000 females. In populations larger than 12,000 females, superiority of PT5 increases with population size. With low numbers of females (<2,000) NPT yields genetic responses intermediate between those of PT5 and PT3. Results suggest the application of progeny testing in populations as small as few thousand cows, but costs analysis is worth investigating.

Identification of BLAD and DUMPS as genetic disorders using PCR-RFLP in Holstein bulls reared in Turkey

H. Meydan, F. Ozdil* and M.A. Yıldız, Ankara University, Agricultural Faculty, Animal Science Department, Biometry and Genetics Major, 06110, Diskapi-Ankara/Turkey

Bovine leukocyte adhesion deficiency (BLAD) and deficiency of uridine mono phosphate synthase (DUMPS) are genetic disorders, with a great economic importance in dairy industry. In this study, 150 candidate Holstein bulls which are reared in Turkey were tested in order to identify BLAD carrier (or affected) and DUMPS-carrier. DNA extractions were obtained from fresh blood and frozen semen. The amplification of BLAD and DUMPS were obtained by using Polymerase Chain Reaction. PCR products were digested with *Taq*I and *Hae*III restriction enzymes for BLAD genotypes. PCR products were digested with *Ava*I restriction enzyme for DUMPS genotypes. Results showed that only one BLAD-carrier and no DUMPS-carrier were found for 150 candidate Holstein bulls in Turkey.

Effect of β- and κ-casein genotypes on milk coagulation properties, milk production and content, and milk quality traits in Italian Holstein cows

A. Comin[1], M. Cassandro[1], M. Ojala[2], G. Bittante[1], [1]Padova University - Department of Animal Science, 35020 Legnaro (PD), Italy, [2]Helsinki University - Department of Animal Science, 00014 Helsinki, Finland

The goal of this study was to investigate the effect of β-κ-casein genotypes on milk coagulation properties (MCP), production traits (daily milk, fat, and protein yield), content traits (fat, protein, and casein percentage) and quality traits (titratable acidity and somatic cell score). Because of the close linkage, the two genotypes were considered as a composite genotype. Data consisted of individual milk samples from 1,071 Italian Holstein cows of 54 sires and reared in 34 herds in Northern Italy. Single-trait animal model included: herd, days in milk, age within parity, β-κ-casein genotype as fixed effects and animal as a random effect. The most frequent β-κ-casein genotype was A_2A_2AA (20.1%), and the least frequent ones were those involving κ-casein as homozygote BB. The effect of the β-κ-casein genotype was statistically significant for MCP and for milk and protein yield, but non-significant for milk content and quality traits. The most favourable genotypes for MCP contained at least one B allele in both loci: A_1BAB and A_2BBB; the genotypes unfavourable for MCP were A_2A_2AA and A_1A_2AE. The genotypes A_2A_2BB and A_1A_1AB were associated with the highest milk and protein yields, whereas the lowest values were related with A_1A_2BB and A_1BAB ones.

Sire effect on variability and heritability of body measurements in performance tested Simmental bulls

V. Bogdanovic*, R. Djedovic and P. Perisic, Institute of Animal Sciences, Faculty of Agriculture, University of Belgrade, Nemanjina 6, 11080 Zemun – Belgrade, Serbia

Beside growth traits, in performance testing of dual-purpose bulls target traits are also body development charactertics. In order to estimate sire effect on variability and heritability of various body measurements, data on 371 Simmental performance tested bulls was used. Analysed traits were height at withers, circumference of chest, depth of chest, width of round and body length. Sire effect and heritability of traits were obtained by restricted maximum likelihood (REML) methodology applied to sire model. Model with effects of sire, year and month of birth, farm and group in test was used. Effect of sire was significant but non-linear source of variation (P<0.001) for all body measurements beside circumference of chest at the end of test. In sum of total variation in body measurements, sire component was in range from 19.72-23.65%, 15.36-7.43%, 25.19-21.30%, 44.08-31.48% and 24.60-14.08% for test-on and test-off height at withers, circumference of chest, depth of chest, width of round and body length, respectively. Obtained results show that sire has significant influence on body measurements but its component of variation for almost all traits declines as the end of test approach. Heritability estimates for linear body measurements were 0.63-0.43, 0.45-0.30, 0.49-0.33 and 0.30-0.29 for test-on and test-off height at withers, circumference of chest, depth of chest and body length, respectively.

Estimates of heritability of meat quality traits in Piemontese cattle

A. Boukha[*1], M. De marchi[1], A. Albera[2], M. Cassandro[1], L. Gallo[1], P. Carnier[1] and G. Bittante[1], [1]Department of Animal Science, University of Padova, 35020 Legnaro (PD), Italy, [2]ANABORAPI, 12061 Carrù (CN), Italy

The study aimed to investigate variation and to estimate heritability of meat quality traits of Piemontese young bulls. A total of 988 young bulls, offspring of 127 AI sires, were sampled from 153 fattening farms. Animals were slaughtered in different days at the same commercial abattoir. Carcass were weighed and scored for fleshiness using the European grading system. Bulls were 536 ± 73 d old and average carcass weight was 425 ± 42 kg. Individual samples of Longissimus Dorsi were collected 24 h after slaughter and held refrigerated at 4 °C for 8 d. Measured traits were pH, meat colour (L, a, b, hue and chroma), shear force, drip and cooking loss. Meat colour, pH and shear force were measured repeatedly. Three measures of pH were taken at collection of muscle samples. Colour, pH and shear force were measured five times on samples after ageing. Pedigree information was provided by the Italian Association of Piemontese Cattle. Estimation of variance components was based on a linear animal model and was accomplished using Bayesian methodology and the Gibbs sampler. Results are presented.

Estimates of genetic parameters and breeding values for calving interval, body condition score, production and linear type traits in Italian Brown Swiss cattle
R. Dal Zotto, M. De Marchi, L. Gallo, P. Carnier, M. Cassandro and G. Bittante, Department of Animal Science, University of Padova, 35020 Legnaro, Italy*

Aims of this study were the estimation of heritabilities, genetic correlations and estimated breeding values for calving interval (CI), body condition score (BCS), linear type traits (LTT), and yield traits (YT) using data recorded from 2002 to 2004. A total of 32,359 records of Italian Brown Swiss (IBS) primiparous cows reared in 4885 herds, were used. Pedigree information was based on 96,661 animals. A multi-traits animal model REML was used to estimate (co)variance components, without repeated observations. Heritabilities of BCS and test day milk yield were 0.15 and 0.14 respectively, while LTT and CI showed respectively higher values (from 0.23 to 0.31) and lower value (0.05). Genetic correlations between CI and MY and between CI and LTT were positive and unfavourable, while between CI and BCS was negative and favourable. The selection for MY and LTT seems to have an important and negative effect on reproduction ability of IBS cattle. A perspective to improve cows' fertility might be to include BCS in selection index as an indirect way to define a reliable fertility genetic index. Information of BCS and LTT could be predictable for fertility index of animals that are culled during first lactation or are unable to calve for second time.

Genetic diversity and relationships of Algerian camels in Tamanrasset region (south of Algeria)
H. Derradji and R. Bouhadad, Houari Boumedien Algiiers University of sciences & tecknology, Animal genetics and ecology, BP 32 El Alia Bab Ezzouar, Bab Ezzouar, 16111, Algeria

The famils of Camelidae comprises the old world (Camelini) and the new world Lamini Mbuu &al.(2002). The old world tribe, represented by *Camelus dromedary* commonly known as the Arabian Camel or one- humpred Camel and C. bactriunus or two-humpred camel. The Arabian Camel is chractrised by a large geographical distribution covering north western India region, Afghanistan, Arabian Peninsula, African dry areas including Australia. Fifty two (52) different breeds were reported even though the studies were based solely on morphological characters. After domestication 3000 years ago BC, the camels reached in their first of migration North Africa where they play an important role in human life history. This situation induced for a large fringes of the population in Algeria (in dry aereas) a substantial contribution in protein (milk, meat), and provide transport to the indigenous population. The number of camels gathered in Algeria is about 240 000, this state of fact brought them to occupy an economical important position in rural development. This study is a contribution to chracteruze the phenotypic variability of camels in Algeria using multivariate methods., even though this part of livestock is poorly studied.

Relationship of GH, GHR and PRL genes polymorphisms with milk production traits in Lithuanian Black and White cows

R. Skinkyte[1], N. Peciulaitiene[1], S. Kerziene[1], L. Zwierzchowski[2] and I. Miceikiene[1], [1]K. Janusauskas Laboratory of Animal Genetics, Lithuanian Veterinary Academy, Tilzes 18, LT 47181 Kaunas, Lithuania, [2]Polish Academy of Sciences Institute of Genetics and Animal Breeding, Jastrzebiec, 05-552 Wolka Kosowska, Poland

Lithuanian Black & White (LBW) cows belong to the modern cattle breed and form 66 % of the total Lithuanian cattle population. The purpose of this study was to estimate associations of milk production traits between five different polymorphic sites within GHR, GH and PRL genes of LBW cattle. DNA samples extracted from blood were genotyped using PCR-RFLP method. Milk yield (MY), fat yield (FY), protein yield (PY), fat percentage (F%) and protein percentage (P%) was analyzed using ANOVA general linear model (MINITAB Release 14, 2005) to estimate associations between different polymorphic sites and production traits. GHR RFLP-Fnu4HI $locus$ had no effect on milk composition traits. GHR RFLP-AluI had significant influence to fat %: GHR$^{(AluI+/+)}$ +0.63%*, GHR$^{(AluI+/-)}$ +0.55%** and GHR$^{(AluI+/+)}$ genotype to protein yield ($p \leq 0.04$). GHR-AccI $locus$ was clearly favourable for fat content: GHR$^{(AccI+/+)}$ +0.84 %*, GHR$^{(AccI+/-)}$ +1.05 %***. GHR$^{(AccI+/-)}$ genotype had positive effect to milk yield ($p \leq 0.04$). Homozygote PRL^{AA} genotype significantly influenced fat content -0.58 %*. Heterozygous GH^{LV} was superior for fat % +0.51 %**. Identified genotypes having relationship to milk production traits can be used as selection criteria for Lithuanian dairy cattle population improving.

DNA polymorphism of bovine pituitary-specific transcription factor and leptin gene in Iranian Bos indicus and Bos taurus cattle using PCR-RFLP

*Nader Asadzadeh[*1], Adam Torkamanzehi[2], Arash Javanmard[3], Mohammad Hossein Banabazi[4] and Javad Tavakolian[4], [1]Department of Animal Production and Management, Animal Science Research Institute of Iran (ASRI), Karaj, Iran. [2]Department of Animal Science, University of Zabol, Zabol, Iran. [3]Department of Genomics, West and North-West Agriculture Biotechnology Research Institute (ABRII-T), Tabriz. Iran. [4]Department of Biotechnology, Animal Science Research Institute of Iran (ASRI), Karaj, Iran*

Variations at DNA level contribute to the genetic characterization of livestock populations and this may help to identify possible hybridization events as well as past evolutionary trends. The leptin and pit-1 are attractive candidate genes for production and reproduction traits in cattle. A total of 247 Animals from four Iranian cattle populations in two groups include Bos taurus (Sarabi, Golpayegani) and Bos indicus (Sistani, Taleshi) were genotyped for the Pit-1 HinfI and leptin Sau3AI polymorphisms by the polymerase chain reaction and restriction fragment length polymorphism (PCR-RFLP). The genotype and gene frequencies for each group were determined and shown to be quite variable among the breeds. The highest frequencies of allele B for the leptin gene and allele A for the Pit-1 gene were found in Bos indicus group. Our result supported the previously proposed long evolutionary separation of these cattle subrace. Candidate gene approach may be a useful method to measure of genetic distance for cross breeding program between taurin and indicine cattle.

Genetic characterisation of busha in Croatia based on mtDNA
M. Konjacic, P. Caput, A. Ivankovic, J. Ramljak, Z. Lukovic, M.Bradic, University of Zagreb, Faculty of Agriculture, 10 000 Zagreb, Croatia

Busha is one of the autochthonous Croatian cattle breeds. It belongs to a group of primitive short horned cattle (*Bos brachyceros europeus*). Busha formed the basis of cattle production in Croatia during past centuries, while today it is almost extinct. The awareness of decrease of the genetic diversity on the national and global level stimulated the question of Busha preservation. Having visited the areas where it has been traditionally bred, it has been realised that only a few dozens of Busha remained. Mitochondrial DNA (mtDNA) was used for the purpose of phylogenetic research of the remaining population. Eight sequences of mtDNA D-loop region of Busha have been researched out of the total of 44 individuals which are the subject of the whole analysis. A proximal part of mtDNA D-loop regions (from 15900 to 16225 nt) has been sequenced. By analysis of sequenced regions, ten polymorphic sites have been noticed. The transition has been noticed on eight positions, while transfersion occurred on two positions. Transition T→C makes 75% of the total transitions. In the individual comparison of sequences, the number of polymorphic sites was between 1 to 5. The variety of researched sequences was between 0.31% and 1.53% and the genetic distance level from 0.0062 to 0.0156.

The sustainability of autochthonous cattle breeds in Croatia
A. Ivanković[1], P. Caput[1], P. Mijić[2], J. Ramljak[1], M. Konjačić[1] and F. Poljak[3], [1]Faculty of Agriculture, Svetošimunska 25, Zagreb, Croatia, [2]Faculty of Agriculture, Trg. Sv. Trojstva 3, Osijek, Croatia, [3]Croatian Livestock Centre, Ilica 101, 10000 Zagreb

Three autochthonous cattle breeds in Croatia are in the endangered status. State with constant financial stimulations from the budget encourages the preservation of endangered breeds. Necessary is seems the routine overview over the breeding work on autochthonous breeds in order to notice inadequacies in view of genetic and economic sustainability. Has been analyzed genetic structure and sustainability of three autochthonous cattle breeds which are in centre scientific interest. The Istrian cattle has been partly consolidated in the genetic sight, beside still is unclear breeding strategy with regard to the accessible lineages. The estimated level of inbreeding of the Istrian cattle on the basis of pedigree data is moderate, but the number and distribution of lineages suggest the needs of more careful implementation breeding program. Level of inbreeding Slavonian syrmian podolian cattle (0,0186) suggest on the systematic selection work, but numerical situation of populations and distribution of herds suggest on the existing seriousness situations. Completed inventarization of the Busha breeds and great interest of breeders for this breed encourage hope in her protection. The height of the annual financial support and undeveloped programs of economic exploitation, call in question too physical sustainability populations in the 'in situ' protection programmes.

The diversity in Yaroslavl, Kholmogor and Black and White breeds of cattle

A.F. Yakovlev*, N.V. Dementeva and V.P. Terletzky, Russian Institute of Farm Animal Genetics and Breeding, 196601, St.Petersbug, Pushkin, Russia

For definition of diversity and heterozygosity of Yaroslavl, Kholmogor and Black and White breeds of cattle have been used a method DNA-fingerprints which allows to estimate objectively heterogeneity of a population and to calculate genetic distances between researched groups of animals. DNA- fingerprinting used with digoxigenated oligonucleotide probe (GTG)5. For treatment of polymorphic DNA fragments used enzyme HaeIII. For study unrelated animals have been selected. At animals of the Yaroslavl breed is present a marker fragment 12,1 kb (frequency 0,9), 7,28 kb(frequency 0,70) which do not meet in Kholmogor and Black and White breeds. The number of bands on 1 path at animals Yaroslavl, Kholmogor and Black fnd White breeds made accordingly:13.5± 0.65,12.38 ±1.95, 9.2±1.38; number of loci: 8,05, 6,89, 5,11; number alleles on 1 locus: 6,21, 6,68, 7,24.; heterozygosity 0,68, 0,80, 0,80, index of genetic similarity inside breeds: 0,31, 0,30, 0,32. Index of genetic similarity between Yaroslavl and Kholmogor breeds has made 0,29 (genetic distances (D) - 0,015); between the Yaroslavl and Black and White breeds - 0,27 (D - 0,045), between Kholmogor and Black and White breeds - 0,32 (D - 0,02). The carried out estimation of heterozygosity has revealed a smaller genetic variety in a population of cattle of the Yaroslavl breed. Probably consolidation of breed is under influence of use of low number of bulls.

Estimation of genetic trends for milk production traits in Iranian Holsteins

H. Farhangfar*, M.R. Asghari and H. Naeemipour, Animal Science Department, Agriculture Faculty, Birjand University, Birjand, P.O. Box 97175-331, Iran

In order to estimate genetic trends for milk production traits in Iranian Holsteins a total of 18,989 adjusted 305d-2X-ME first lactation records collected from 18,989 heifers calved between 1995 and 2001 and distributed in 202 herds over the country were used. The traits were milk, fat and protein yields as well as fat and protein percentages. A multiple trait animal model was utilized. The effects included in the animal model were fixed contemporary group of province-herd-year-season of calving, linear covariate of Holstein genes and random additive genetic effect. Restricted maximum likelihood estimates of variance and covariance components and best linear unbiased prediction of breeding value of individual cows for each trait were obtained by using MTC and MTJAAM programmes respectively. Genetic trends for the traits under consideration were estimated based upon simple linear regression analysis of BLUP solutions on year of first calving. The results obtained in this study revealed that the average breeding value of cows had a significant increase of +33.178 Kg per year ($P<0.01$) for milk yield. At the same time, fat and protein yields had positive genetic trends of +245 g ($P>0.05$) and +789 g per year ($P<0.05$) respectively. In contrast to yield traits, the genetic trends calculated for fat and protein percentage showed an annual decrease of -0.011 % ($P<0.001$) and -0.003 % ($P<0.05$) respectively which were significant statistically.

Estimation of lactation yield by partial milk yield of Black and White Cattle raised at private farm in Bolu provinces of Turkey

M.I. Soysal[1], F. Mutlu[2] and E.K. Gurcan[1], [1]Trakya University, Agricultural Faculty of Tekirdağ 59030 Tekirdağ, Turkey, [2]Trakya University, Graduate School of Natural and Applied Sciences, Edirne, Turkey

In this study was aimed to reveal the relationship between several lactation yield estimation methods and also relationship of partial milk yield and whole milk which is obtained daily milk control recording. The lactation records of 48 animals were used in the study. Using daily obtained lactation record as references for real lactation yield; it is searched the accuracy of estimating lactation yield by 15-30-60 day interval control yields according to the real lactation yield obtained by daily record. Using Sweden, Holland and Wogel methods of estimation 305 day lactation yields by 15-30-60 days interval control yields; comparison between these result and real lactation yield obtained by daily recorded data were done in data of Bolu province. Using first 15-30-60 day partial lactation yield for estimating whole lactation yield by means of linear, logarithmic, compound and exponential regression equations. Best estimation among to equations were determined according to the coefficient of determination (R^2).

Expression of milk protein genes in cows of different κ-casein genotypes and fed wheat gluten or soybean meal supplements

M. Pawelska, T. Szulc and P. Nowakowski, Institute of Animal Breeding, Agricultural University of Wroclaw, Chelmonskiego 38C, 51-630 Wroclaw, Poland*

The aim of research was to analyse changes in the yield of protein classes in cows of AA, AB and BB κ-casein genotypes which were supplemented daily with 300 g of total protein as wheat gluten (WG) or soybean meal (SM). Protein concentrations were determined using polyacryloamide gel at 4°C temperature, 280 V voltage and current intensity 10mA for 3 hours in the pH 8.3 buffer. WG supplement caused significant increase ($P<0.05$) in milk total protein (+0.18%) and casein (+0.14%). Content of β-casein increased by +0.5% ($P<0.01$) and κ-casein by + 0.3% ($P<0.01$). The highest rise ($P<0.01$) in the levels of protein (+0,29%), casein (+0.22%), α-casein (+0.9 g/l), β- casein (+0.9 g/l) and κ-casein (+0.4 g/l) was found in cows of AB κ-casein genotype. SM caused significant increase ($P<0.05$) in milk total protein (+0.20%) and casein (+0.15%) as well as α-casein (+1g/l) and β-casein (+0.5 g/l). Cows of AA and AB κ-casein genotypes characterised with increased ($P<0.05$) levels of protein (+0.22%), casein (+0.17%), α-casein (+1.1 g/l) and β-casein (+ 0.5 and 0.6 g/l) while no significant changes were stated in the milk proteins of BB genotype cows. Expression of milk protein genes in cows supplemented with WG or SM varied due to milk protein genotype.

Productive and reproductive characteristics of Holstein cows raised in intensive farming system in Egypt

S. Abou-Bakr[1], H.O.A. Alhammad[2], Y.Y. Sadek[1] and A.A. Nigm[1], [1]Department of Animal Production, Faculty of Agriculture, University of Cairo, Giza, Egypt, [2]Department of Animal Production, Faculty of Agriculture, Allepo University, Syria

A total of 4382 lactation records for 1868 Holstein cows sired by 182 bulls in a herd belongs to The Modern Agricultural Development Company (Dina). The data were analyzed using SAS (1998) and DF-REML (Meyer, 1998) for estimating the phenotypic and genetic parameters of some productive and reproductive traits.

Means of total milk yield (TMY), 305-day milk yield (305-dMY, lactation period (LP), annualized milk yield (AMY), daily milk yield (DMY), dry period (DP), days open (DO) and calving interval (CI) were 13172 kg, 10847 kg, 370 day, 10899 kg, 35.5 kg, 63.47 day, 154 day and 430 day, respectively.

All the studied traits, except AMY, were significantly affected by parity of the cow. Also, year of calving showed a significant effect on all traits. Season of calving showed any significant effect on all traits except TMY and DP.

Estimates of heritability ranged between 0.0003 for DP and CI to 0.13 for 305-dMY, and DMY. All estimates of the phenotypic and genetic correlation coefficients among these traits were positive. The results indicated that Holstein cattle in Egypt could produce high amount of milk if kept under reasonable management standards.

The influence of Red Holstein gene share on reproduction and production traits of first-calf heifers

P. Perisić[1], Z. Skalicki[1], M.M. Petrović[2], V. Bogdanović[1], G. Trifunović[1] and R. Đedović[1], [1]Faculty of Agriculture, Belgrade University, [2]Institute for Animal Husbandry, Belgrade-Zemun, Serbia and Montenegro*

Certain reproduction and performance traits of first calving Simmental cows, as well as the demonstration of those traits depending on the participations of different proportions of Red Holstein-Friesian genes (HF), were investigated. The investigation included 185 first calving cows, divided according to percentage of HF genes, in 5 groups (0%, 12,5%, 25%, 50%, 75%). The effect of age at first fertilization varied highly significantly between groups of first-calf heifers and the youngest was in the fifth group (the group with 75% of HF genes). The average age at first conception for all examined groups was 536,7 days.

Results showed that different proportion of HF genes was not significant (P>0.05) body weight of calves. The least body weight of calves was in group of first calving cows with 75% HF genes. Gestation period was 283.3 days and the difference between the groups was not significant (P>0.05). The average period of lactation was 321.32 days and the longest (348.5 days) was established in the fifth group of cows with 75% HF genes. Average milk yield was 4224.34 kg with 3.83% milk fat or 4116.6kg 4%FCM. Cows with 75% HF genes produced more milk compared to the cows in other groups.

Genetic parameters of clinical mastitis for Dutch Holstein cattle based on farm management software data

E.P.C. Koenen and W. Steeneveld, NRS, P.O. Box 454, 6800 AL Arnhem, The Netherlands*

Recent ICT developments have facilitated the use of farm management software by dairy farmers. The disease recording modules might be a source of information for breeding organisations to reduce the incidence of clinical mastitis (CM) by genetic selection. Hence, the aim of this study was to evaluate the suitability of CM observations recorded by dairy farmers for selection purposes. A group of 202 Dutch dairy farmers routinely recorded CM treatments in 30,202 lactations of 18,758 Holstein Friesian cows in their farm management software between 1999 and 2005. Incidences of CM during the first 210 days of lactation increased from 12% in first parity to 19% in third parity with most CM cases in early lactation. Genetic parameters of CM (0/1) were estimated within and across parities using a linear animal model. Heritabilities of CM were 0.03 in first parity and 0.05 in second and third parity. Additive genetic standard deviations for CM ranged from 6 to 9%. Genetic correlations across parities were high (0.81 to 0.92), but clearly deviated from unity. The results of this study indicate that CM data recorded by dairy farmers can be useful for selection purposes. With the increasing numbers of farmers recording disease data in their management software, the inclusion of direct observations on CM in selection decisions by breeding organisations has become more attractive.

Frequency of hereditary anomalies in populations of offspring of Holstein-Friesian sires

Radica Djedovic, V. Bogdanovic and P. Perisic. Institute of Animal Sciences Faculty of Agriculture, Universiyu of Belgrade, Nemanjina6, 11080 Zemun-Belgrade, Serbia*

The frequency of hereditary anomalies in calves, offspring of 51 sires of the Holstein-Friesian breed, was investigated during a six year research period, on 14 farms in the Republic of Serbia. Of a total of 7036 investigated calves, hereditary anomalies were present in 51. A total of 22 carious types of lethal and semilethal anomalies were registered. The most frequent anomaly was atresia ani with 16 cases (30.77%), followed by disorders of locomotory organs: ancylosis (3 cases), acroter iasis (2), arthrogryposis (1), ataxia (2), syndactilia (1). Following anomalies, with one case each, were also found: two-head, achondroplasia, brachygnatia inferior, hydrocephalus, theratoma, hernia umbilicalis, hernia inguinalis, tailless, and others. All congenital anomalies were described in detail. Their frequency in the total number of investigated calves was established, as well as the incidence by sires. Simultaneously, pedigrees of parents were analyzed in order to establish the potential existance of any mutual ancestors.

Frequency of BLAD factor in the population of calves and young bulls of the Holstein-Friesian breed

Radica Djedovic, V. Bogdanovic and P. Perisic, Institute of Animal Sciences Faculty of Agriculture, Universiyu of Belgrade, Nemanjina6, 11080 Zemun-Belgrade, Serbia*

BLAD *(Bovine leucocyte adhesion deficiency)* is a hereditary, autosomal, recessive disease which appears in populations of Holstein-Friesian cattle. The disease is caused by a spot mutation of the CD18 gene bonded to the exterior glycoprotein receptor of leukocytes.

The presence of the BLAD allele mutation in calves and young bulls of the Holstein-Friesian breed was investigated using the DNA (PCR) test during 4 years, on a total of 103 individuals. In the 1st year of testing, a total of 4 heterozygous indivduals (10%) were discovered, which were also carriers of the BLAD factor. No affected, i.e. homozygous recessive (bb) individuals were found.

Molecular and genetic analyses carried out during the 2nd and the 3rd year showed that all tested calves were homozygous dominant i.e. free from BLAD. One heterozygous individual was identified during the 4th year of research. This indicates that a total of 5 carriers of the BLAD factor were found, i.e. that the frequency for carriers of the mutated CD18 gene in the population of calves and young bulls in the process of testing was 4.85%. The frequency of the dominant B allele is 0.97, while the frequency of the mutated b allele was low, at 0.03. Carriers of this undesirable gene were excluded from further testing, and from bull testing programs.

The effect of leptin (LEP), of growth hormone (GH) and PIT-1 genotypes on carcass traits in cattle

J. Oprządek, E. Dymnicki and L. Zwierzchowski, Institute of Genetics and Animal Breeding Polish Academy of Sciences, Jastrzębiec 05-552 Wólka Kosowska, Poland*

Dynamic development of molecular biology in the last years made possible to identify genes affecting animal production traits. Identified quantitative trait markers can be used in marker-assisted selection (MAS). The objective of this study was to examine the effect of polymorphism in the leptin (LEP), growth hormone (GH) and PIT-1 gene on the traits related to meat production in the cattle. Genotypes of leptin (LEP) growth hormone (GH) and PIT-1 were analysed using the PCR-RFLP technique. Crude DNA was prepared from blood samples according to Kasai. The digestion fragments were identified on 2% agarose gel. The allele frequencies at the loci studied were 0.60/0.30 for LEP C/T variants and 0.77/0.23 for LEP A/B variants, 0.89/0.11 for GH L/V variants and 0.25/0.75 for PIT-1 A/B variants. The statistical model fitted for year/season, genotypes of LEP, GH, PIT-1, two-way interactions between the genotypes at the studied loci and regression on cold carcass weight. The differences between genotypes were tested by Duncan's test. The effect of interaction LEP/*Kpn*2I x GH was found significant on valuable cuts and fat in the carcass. LEP/*Kpn*2I x GH interaction significantly affected the live body weight at the age of 8 moths. Amount of fat in the carcass was affected by GH x LEP/*Hph*I interaction.

Estimation of genetic parameters for type traits in beef cattle in the Czech Republic
Z. Veselá[1], J. Přibyl[1], P. Šafus[1]*, L. Vostrý[1], L. Štolc[2] and K. Šeba[3], [1] Research Institute of Animal Production, P.O. Box 1, 10401, Prague 114, Czech Republic, [2] Czech University of Agriculture, Prague, Czech Republic, [3] Czech Beef Cattle Association, Prague, Czech Republic

The type was evaluated in 8,316 young animals of twelve beef breeds. Ten traits were evaluated: height at sacrum (HS), body length (BL), live weight (LW), front chest width (CW), chest depth (CD), pelvis (P), shoulder muscling (SM), back muscling (BM), rump muscling (RM) and production type (PT). The traits represent two groups: 1. traits scoring body measurements and body capacity (HS, BL, LW, CW, CD, P) and 2. scoring muscling (SM, BM, RM, PT). These effects were included in the model: breed, sex, HYS, mother's age and Legendre polynomial on age at evaluation and linear regression on average gain. The highest heritability coefficient were estimated for HS ($h^2 = 0.57$), SM ($h^2 = 0.51$) and LW ($h^2 = 0.50$). BL had lower heritability coefficient ($h^2 = 0.35$). The values $h^2 = 0.28 - 0.46$ were calculated for the traits scoring body capacity (CW, CD, P). The range of values for muscling traits was $h^2 = 0.46 - 0.51$. The heritability coefficient for PT was $h^2 = 0.34$. All traits scoring muscling and PT showed high genetic correlations ($r_g > 0.94$). The traits scoring body capacity were highly genetically correlated with muscling $r_g > 0.82$.

Genetic parameters and breeding values of weaning results of Hungarian Fleckvieh beef calves
Sz. Bene[1], I. Füller[2], Z. Lengyel[1], B. Nagy[1], F. Szabó[1], [1]University of Veszprém, Georgikon Faculty, Hungary, H-8360 Keszthely Deák F. str. 16, [2]Association of the Breeders of Hungarian Simmental Cattle, H-7150 Bonyhád, Zrínyi str. 3

Weaning weight, preweaning daily gain and 205-day weight of 8929 Hungarian Fleckvieh calves (4539 male and 4390 female) born from 232 sires between 1980-2003 were examined in two farms. Variance, covariance components and heritabitity values and correlation coefficients were estimated. The effect of the maternal permanent environment on genetic parameters and breeding values were examined. Two animal models were used for breeding value estimation. The direct heritability (h^2_d) of weaning weight, preweaning daily gain and 205-day weight was between 0.37 and 0.42. The maternal heritability (h^2_m) of these traits was 0.06 and 0.07. The direct-maternal correlations (r_{dm}) were medium and negative -0.52 and -0.74. Contribution of the maternal heritability and maternal permanent environment to phenotype is smaller than that of direct heritabilities ($h^2_m + c^2 < h^2_d$). The proportion of the variance of maternal permanent environment in the phenotypic variance (c^2) changed from 3 to 6 %. Estimated breeding values changed whether the permanent environmental effect of dam wasn't taken into consideration but the rank of the animals has not modified.
The genetic value for weaning results of Hungarian Fleckvieh population has increased since 1997.

Steer production from progeny of Friesian cows and Belgian Blue or Charolais bulls
M.G. Keane, Teagasc, Grange Beef Research Centre, Dunsany, Co. Meath, Ireland

Belgian Blue bulls are preferred to Charolais for crossing on Friesian cows in Ireland. This study compared Belgian Blue x Friesian (BB) and Charolais x Friesian (CH) spring-born steers for growth, feed intake and carcass traits. After calf rearing indoors, the animals were turned out to pasture on May 31 for 182 days. They were castrated on September 20. During the first winter of 120 days they were offered grass silage plus 1 kg concentrates per head daily. The duration of the second grazing season was 224 days after which the animals were finished on silage plus 5 kg concentrates per head daily for 126 days. Silage and grass intakes were measured. Carcasses were measured and the 6 to 10th rib joint was dissected. There was no difference between the breed types in silage or grass intake. CH calves had higher live weight gains during the first grazing season (P<0.05). Otherwise there was no difference between the breeds in growth rate, slaughter weight or carcass weight per day of age. Neither was there any difference in kill-out proportion but CH had better conformation (P<0.05) while BB had a lower fat score (P<0.05). There were no differences in carcass measurements or in ribs joint composition. Substitution of Belgian Blue for Charolais sires in dairy herds would not impair the productivity of the male progeny for beef.

Prevention of butcher's fraud in meat sale by bovine carcass sex determination
Arash Javanmard[1]*, Ahmad Rezban Haghighe[2], Ali Javadmansh[3], [1,2]West and North- West of Agriculture Biotechnology Research Institute (ABRII-T), Tabriz. I.R. Iran [3]Department of Animal Science, Ferdowsi University of Mashhad, P.O. Box: 91775-1163, Mashhad, I,R, Iran

Fraudulent of butchers for mature and old female beef sale has increased in recent years, several analytical methods have been developed in order to protect consumers from these fraud. The objective of this study to develop a technique that would permit the sexing of boned, packaged and chilled, ready for sale beef. one part of second intron of leptin gene(international control) and one 307 bp fragment from male-specific DNA region (BRY1 locus) were amplified by polymerase chain reaction We analyzed 50 samples identified as belonging to males, 15 of which (30%) presented a discordant result, i.e., they actually belonged to females. This method was a cheap and efficient technique for the determination of cattle sex from the carcass. It is technically simple and rapid, permitting its utilization in the prevention of fraud in the commercialization of beef

Zoonoses in the rural environment: Risks and factors for emergence and re-emergence

B.B. Chomel, WHO/PAHO Collaborating Center on New and Emerging Zoonoses, Department of Population Health and Reproduction, School of Veterinary Medicine, University of California, Davis. Davis, CA 95616, USA.

Farm animals, especially in traditional farming systems, can be a source of zoonoses for farmers, veterinarians and animal health workers. Similarly, infections may occur in slaughter house workers and food processing plants. Emergence of new zoonoses, such as *E. coli* O157:H7 or BSE/vCJD, underlines the importance of production animals as a source of foodborne zoonoses. New concerns have also emerged in developed countries with the presence of petting farms in various fairs, leading to pediatric infections by zoonotic agents (*Salmonella, E.coli, Coxiella*). Consumption of trendy foods and culinary preferences has also been associated with outbreaks of trichinellosis and more recently hepatitis E. Development of open farming and contact with wildlife is also another source of domestic animal infection and contamination of humans, as illustrated by the re-emergence of bovine tuberculosis from the badger reservoir in the UK and the emergence of swine brucellosis in pig herds in Western Europe.

It is surprising that only very limited surveys on prevalence of zoonotic infections in rural populations have been conducted worldwide. A discussion on available data will be discussed. It appears essential that surveillance networks be established by veterinary and human health agencies to determine risk levels of zoonoses in the rural environment and that comprehensive control measures be set.

Prevalence of Brucella antibodies in blood serum of Holstein cattle in the Western Highlands of Cameroon

P.H. Bayemi[1], E.C. Webb[2], H. Njakoi[3], M.V. Nsongka[1], V. Tanya[4], C. Ndi[1], H. Unger[5] and B.M.A.O. Perera[6], [1]IRAD Bambui P O Box 51 Bamenda Cameroon, [2]Head: department of animal and wildlife science, university Pretoria, South Africa, [3]Heifer International Cameroon, [4]Ministry of Scientific Research and Innovation, Cameroon, [5]IAEA, Vienna, Austria, [6]University of Peradeniya, Sri Lanka

Holstein cattle were screened for Brucella in Cameroon. The laboratory test involved ELISA assays. Results show a general seroprevalence of 8.4 %. Females (75%) were more infected than males (P= 0.0143). There was no evidence (P=0.11) of difference in the seroprevalence of age groups. Brucella infection varied with location with 3 over 21 locations accounting for 64% of infected animals: Kutaba (32%), Bamendankwe (16%) and Finge (16%) (P=0.0147). A specific control programme should be organized in these locations and care should be given to determine causes of the spread of the infection. Owing to the fact that animals screened are from the Holstein high yielding breed, measures should be taken to ensure eradication of the disease within the population and a sound control adopted as the animals are within a larger cattle population in the region. Infected animals should systematically be slaughtered. All farmers should be advised to boil milk before consumption. Vaccination against Brucella should be instituted. This test should be regularly carried out as is distinguishes vaccinated from infected animals.

Shedding routes of Coxiella burnetii in dairy cows: Implications for detection and infection control

R. Guatteo, F. Beaudeau, C. Fourichon* and H. Seegers, Veterinary School of Nantes – INRA, BP 40706, 44307 Nantes Cedex 3, France

Q fever is a widespread zoonosis caused by a bacteria, *Coxiella burnetii (Cb)*. The disease is endemic worldwide. Ruminants are the main sources for human infection. Reliable detection of *Cb* shedders is a critical point for the control of *Cb* transmission between animals and from animals to humans. *Cb* is assumed to be shed by ruminants in vaginal mucus, milk and faeces. However, the informative value of these biological samples to identify shedders under field conditions is unknown. Our aim was to describe the distribution of *Cb* shedding routes in dairy cows. Vaginal swab, raw milk and faeces were taken on the same day in 262 dairy cows in 5 commercial herds, 6 times over a 1.5-month period. Each sample was tested using the LSI Taqvet *Coxiella burnetii*® (Laboratoire Service International, Lissieu, France) real-time PCR assay. *Cb* was detected in 14, 22 and 3% of vaginal mucus, milk and faecal samples, respectively. Sampling days with only one (out of 3) sample found positive accounted for more than 85% of positive samples. By contrast, days with positive results for all samples were scarce (2%). Detection of *Cb*-shedding cows cannot rely on only one type of biological sample. This study is the first showing *Cb* shedding in faeces of dairy cows. Consecutive risk for human infection is discussed.

The relationship between serum acute phase proteins, health and performance in finishing pigs

K. Scott[1]*, F.M. Campbell[2], D. Chennells[3], B. Hunt[4], D. Armstrong[5], L. Taylor[5], B.P. Gill[5] and S.A. Edwards[1], [1]University of Newcastle, NE1 7RU, UK; [2]Glasgow Veterinary School; [3]Acorn House Surgery; [4]VLA; [5]MLC

Acute Phase Proteins (APP) are elevated by infection, inflammation and stress, and may provide an index of health status. From a multidisciplinary study on different housing (fully-slatted or straw-based), a database was developed with information on a sample of 384 finishing pigs over a 3-year period. This included information on growth, health records, serology and serum APP levels of C-Reactive Protein (CRP) and Haptoglobin (Hp). Hp, but not CRP, was higher in the fully-slatted system at slaughter ($P<0.01$). At entry, mid-point and slaughter, there was a significant correlation between levels of Hp and CRP ($P<0.001$). Pigs which received veterinary treatment for clinical illness showed no differences in APP levels compared with untreated pigs. Pigs which showed seroconversion for PRRS had lower levels of Hp, but not CRP, than pigs whose PRRS status remained unchanged ($P<0.01$). Conversely, seroconversion for *Actinobacillus pleuropneumoniae* tended to increase Hp levels ($P=0.09$). Daily gain in the grower ($r=-0.209$; $P<0.001$) and finisher phase ($r=-0.187$; $P<0.01$) was correlated with Hp level. In conclusion, levels of CRP and Hp did not reflect observed clinical symptoms of ill health in finishing pigs. However, the significant weak negative relationship between Hp and growth suggests that this APP might be a useful indicator of sub-clinical infection affecting performance.

Evaluation of the microbial environment of rabbit cages with and without enrichment strategies

P.A. Martino[1], C. Lazzaroni[2]*, F.M.G. Luzi[3], P. Panagakis[4] and M. Verga[3], [1]DIPAV, Microbiology and Immunology Unit, University of Milan, Italy, [2]Dept. Animal Science, University of Turin, Italy, [3]Institute of Zootechnic, University of Milan, Italy, [4]Agricultural Engineering Dept., Agricultural University of Athens, Athens, Greece

A high microbial charge in the environment could compromise both the animal welfare and the human health; so it's necessary to explore the presence of bacteria, viruses, fungi (environmental and dermathophytes) and parasites. Aim of this work is the evaluation of the environmental charge of microorganisms in a rabbit farm. We investigated, during wintertime, the total bacterial and fungal charge in rabbit nests and in cages with or without environmental enrichment. We used the SAS System (PBI, Italy) with plates filled with media for bacteria and fungi. In nests and in environment we found that the bacterial and the fungal charges were similar (50-100 UFC/m^3, <50 UFC/m^3 respectively). In control cages and in cages with enrichment we always found Bacillus, Alternaria, Aspergillus niger and Microsporum gypseum with different percentage of isolation (ranging from 8.3% to 100%). The value of the total microbial charge was low with regards to the environmental risk according to the parameters supplied by the SAS System's producer. The widespread presence of the geophilic fungus M. gypseum could suggest that the coat of these rabbits would be colonised by this dermathophyte. Moreover, there is no zoonosic risk for worker's health.

Evaluation of passive immunity in calves due to antigen pressure from the environment

A. Zachwieja[1], P. Nowakowski[1]*, J. Twardoń[2], T. Szulc[1], J. Molenda[3], A. Dobicki[1] and A. Tomaszewski[1], Agricultural University of Wroclaw, [1]Institute of Animal Breeding, [2]Department and Clinic of Obsterics, Ruminanat Diseases and Animal Health, [3]Department of Pathological Anatomy, Pathophysiology, Microbiology and Forensic Veterinary Medicine, Chelmonskiego 38C, 51-630 Wroclaw, Poland

The aim of research was to study the effect of bacterial contamination of the environment on volume and composition of Charolaise cows' colostrum, its technological properties, immunoglobulins absorption efficiency in calves. 57 cow – calf pairs were analysed - calves born from March to May. Bacteria were cultured from the animal building for denomination to genera and bacteria colonies were counted. There was trend in increasing number of bacteria in colostrum in consecutive months, from March to May. The highest number of somatic cells in colostrum was stated in April and May and it coincided with longer coagulation time and lower level of thermal stability. These observed changes were not related significantly to immunoglobulin concentrations in cows' colostrum. There were no differences stated in total protein and immunoglobulin content of calves blood serum except of Ig of class A, which the highest level was observed in March (p≤0,05). Environmental studies showed significant variability in bacteria numbers with the decreasing tendency towards May. Performed research did not allow confirm hypothesis that the level of environment antigenicity may affect immunological properties of colostrum and the level of passive immunity of calves.

Developed system of animal product quality control in Hungary
Sz. Simai, D. Mezőszentgyörgyi and F. Flink, National Institute for Agricultural Quality Control, 24. Keleti K. u. H-1024, Budapest, Hungary

Due to recent and previous food scandals, the Europian consumers have lost their trust in certain animal products. The consumption is decreasing dramatically in poultry meat and beef meat, as well. Over the food scandals, there are new focuses and aspects of keeping animals in wellfare conditions, adequate housing and in safe environment.

Besides – or instead – of different, additional control of certain authorities, the quality assurance systems are beginning to become part of industrial animal production systems. There is huge demand from consumers about the reliability of tracing animal products "from stable to table".

Serving these demands, the Hungarian National Intstitute for Agricultural Qualty Control has established – as a pioneer in this area – a trademark called "Hungarian Quality Product from Controlled Animal Production".

At first step, the trademark was available for pig breeders, but from 2005., it has been expanded to beef cattle and broiler chicken.

The updated trademarked and acredited quality system serves as a proof for all producers taking voluntarly its obtainment that the product made complying with the animal breeding, animal health, animal wellfare, feeding and environmental protection rules in the controlled production is appropiate for food safety.

Trademarked farms meet the expectations of slaughterhouses and food processing companies.

Assessment of human-animal relationship in cattle kept on slatted floor
F. Napolitano[1], C. Tripaldi[2], A. Braghieri[1], F. Saltalamacchia[2], F. Grasso[3] and G. De Rosa[3], [1]Dipartimento di Scienze delle Produzioni Animali, Università della Basilicata, 85100 Potenza,Italy, [2]Istituto sperimentale per la zootecnia, 00016 Monterotondo (Rome), Italy, [3]Dipartimento di Scienze Zootecniche e Ispezione degli Alimenti, Università di Napoli, 80055 Portici (NA), Italy

The nature of human contacts toward cattle can affect their subsequent behavioural response. The aim of this study was to evaluate the test-retest reliability of some variables used to assess the quality of human–animal relationship. These variables were: time needed to isolate a subject from the group in a corner of the pen (TI), isolation time (IT) and number of flight attempts (FA). They were recorded in 1 farm on 30 Italian Freisian males aged 10 months. Observations were conducted in the home pens by two trained people: one always recording the variables (observer), the other always performing the test (handler). Three recordings were performed at 15 day intervals. Reliability was tested using the Kendall's W coefficient. The three observations were significantly similar (P<0.05, P<0.001 and P<0.01 for TI, IT and FA, respectively), although the W coefficient was high for IT (0.81) but lower for TI (0.50) and FA (0.67). In addition, the Spearman coefficient showed a significant correlation between IT and FA (r_s=-0.40, P<0.05). These results indicate that the time of isolation may be a reliable indicator of the animal responsiveness to the stockperson.

Identification of Eimeria spp in industrial poultry flocks in Tehran province of Iran

A.R. Mohammadi[1]* and G.Motamedi[2], [1]Research Center of Agriculture and National resources, Tehran, Iran, [2]Department of parasitology Razi Institute, Karaj, Iran

Coccidiosis is one of the most important parasitic disesases in poultry.There are clinical and subclinical forms. It is a cosmopolitant disease and depend on host distribution.This disease is mainly seen in extensive poultry rearing systems.Great economical losses is believed to be caused by sub-clinical forms,leading to the increasing of food conversion rate and decrease of meat as well as to high cost of anticoccidial drugs and vaccines. Nowadays although the disease is controlled by use of anticoccidial drugs(including coccidiocidal and coccidostat agent)and vaccines. It seems that the existance of drug resistant starins and unsuccessful vaccination.The need for more research on poultry Eimeria spp is essential. The present work was conducted to record the prevalence of Eimeria in poultry.A total of 2910 litter fecal samples of 354 broiler farms and 242 poulet rearing farms during four seasons were examined. In broiler farms the Eimeria species were E.tenella. E.maxima and E.acervulina.in poulet rearing farms in addition to 3 above spcies the E.necatrix was found positive.for identification of Eimeria spp their morphological characteristics were measured. In broiler farms the E.t,E.m and E.a were high in summer while decreased in spring. In poulet farms E.t and E.m were low in winter while were high in autumn and summer respectively. The E.a and E.nec were low in spring while were high in autumn and summer respectively.

Impact of reproduction technology on horse breeding programmes

E. Bruns[1] and E.P.C. Koenen[2]*, [1]Institute of Animal Breeding and Genetics, Georg-August-University Göttingen, Albrecht-Thaer-Weg 3, D 37075 Göttingen, Germany, [2]NRS, P.O. Box 454, 6800 AL Arnhem, The Netherlands

Horse breeding programmes aim at the improvement of performance, conformation and health traits by genetic selection. The efficiency of breeding programmes depends on the systematic use of testing, evaluation and selection procedures. Advanced reproduction techniques such as artificial insemination (AI) and embryo transfer (ET) are increasingly applied in horse breeding breeding to increase selection efficiency. Using chilled or frozen semen, AI has now become standard technology (>80%) in many warmblood riding horse populations. The number of inseminated mares per stallion and season varies substantially and can go up to 800 inseminations. Because less stallions are needed to produce the next generation, the selection intensity in the male path increases. Furthermore, the use of frozen semen has greatly facilitated the international availability of good stallions.

The ET technique with fresh or frozen embryos, increases the selection intensity in the female path. For example, it allows a mare to produce about six foals within three years compared with two without ET. The effect of ET in terms of genetic progress is less compared to AI. ET enables mares to reproduce while performing in sport competition. If applied to genetically superior breeding horses, both AI and ET offer great opportunities to substantially increase the genetic progress in horse populations.

Session H6

Theatre 2

Cloning and embryo transfer in selection plans in horses
A. Ricard and C. Dubois, SGQA, INRA 78352 Jouy en josas, France

Genetic trend was calculated from $ir\sigma_a/T$ with i=selection intensity, r=reliability, T = generation interval σ_a genetic standard deviation in different situations involving cloning and embryo transfer. The first situation was use of somatic cloning of geldings with own performances to become stallions when they were 3. In current sport situation, comparing to selection among males only, the genetic progress increased +4% with 40% of geldings among horses with performances. With 90% of geldings among horses with performances, cloning increased genetic progress 52% compared to selection on males with performances but only 6% comparing to selection of young males aged 3 on parental information (females and geldings as half sibs), before gelding. The second situation was the use of cloning of embryo as multiplication of own performances for the horse subject to selection. In the male way, the increase of genetic progress was very low, about 1% with 5 horses per clone and 40% of candidates with clones in competition (which supposed a very high technology). On the female way, the increase of reliability on candidates due to the existence of clones is also following by an increase of number of progeny per female. The late increase may also be obtained by embryo transfer (ET). The genetic progress increased 98% with 20% of female with ET among whole population and only 10% with cloning only but 204% with cloning and ET.

Session Ph7

Theatre I

Gene expression and pig mammary gland function
M.T. Sørensen* and P.K. Theil, Danish Institute of Agricultural Sciences, P.O. Box 50, DK-8830 Tjele, Denmark

In mammalians, the mammary gland is a very important organ since the offspring are dependent on milk for their survival and development. Much effort has been devoted to manipulate mammary output in a desired direction, i.e. mainly to increase milk volume or to change milk composition. Changes in milk volume and composition are functions of changes in gene expression of the mammary gland. With the emergence of techniques to measure gene expression, we are now able to understand a bit more of the complex mechanisms that govern mammary output. In an individual animal, changes in gene expression are due to the physiological state (e.g. pregnant or lactating) or to stimuli from the environment (e.g. diet or climate), while in individuals of the next generation, changes in gene expression may also be due to changes in the genome whether they are a result of selection programmes or whether they are a result of insertion of extra copies of a gene by gene manipulation. This presentation will discuss the contribution of measures of gene expression in understanding mammary gland function in the pig.

Research into the importance of antimicrobial peptides in the bovine mammary gland
J.Reinhardt[1], P.Regenhard[1], K. Knappstein[2], C. Looft[1] and E. Kalm[1], [1]Institut für Tierzucht und Tierhaltung der Christian-Albrechts-Universität zu Kiel, D-24098 Kiel, [2]Bundes-forschungsanstalt für Ernährung und Lebensmittel, Standort Kiel, Germany

β-defensin genes are coding for multifunctional peptides, having a broad-range of anti-microbial activity. Basis of this project is the hypothesis that β-defensin genes may be candidate genes for the resistance to mastitis. Several β-defensin genes in the bovine mammary gland and their localisation on chromosome 27, a genomic region with influence on the somatic cell count were identified. Analysis of mammary gland tissue derived cDNA from 15 cows with different clinical findings showed a constitutive and induced expression of β-defensin genes. Closer analysis tends to show differences in the amount of the expression of individual β-defensin genes and their inducing period relating to different mastitis causing pathogens. Next to the quantitative and qualitative analysis of the udder tissue, a MAC-T-Cell-Line is used to analyse different dose- and time-dependent stimulation experiments in order to show differences in the amount of the expression of individual β-defensin genes and their inducing period relating to different mastitis causing pathogens and the amount of pathogens. Further, the production of recombinant bovine psoriasin is done in order to show the specific activity of this protein and to generate polyclonal antibodies to do some immuno-histochemical research. Recombinant production of other β-defensins is planned.

Rapid and sensitive identification of buffalo's, cattle's and sheep's milk using species-specific PCR and PCR-RFLP techniques
S.M. Abdel-Rahman* and M.M.M. Ahmed, Department of Nucleic Acid Research, Genetic Engineering Institute, Mubarak City for Scientific Research and Technology Applications, Post Code 21934, Alexandria, Egypt

For the rapid, specific and sensitive identification of buffalo's, cattle's and sheep's milk, species-specific PCR and PCR-RFLP techniques were developed. DNA from small amount of fresh milk (100 μl) was extracted to amplify the gene encoding species-specific repeat (SSR) region and the mitochondrial DNA segment (cytochrome *b* gene). PCR amplification size of the gene encoding SSR region was 603 bp in both buffalo's and cattle's milk, while in sheep's milk it was 374 bp. Polymerase chain reaction-restriction fragment length polymorphism (PCR-RFLP) technique was used to discriminate between buffalo's and cattle's milk. Restriction analysis of PCR-RFLP of the mitochondrial cytochrome *b* segment (359 bp) analysis showed a difference between buffalo's and cattle's milk. Where in buffalo's the fragment length (bp) generated by *Taq*I PCR-RFLP were 191 and 169, no fragments were obtained in cattle's milk for cytochrome *b* gene (359 bp). The proposed PCR and PCR-RFLP assays represent a rapid and sensitive method applicable to the detection and authentication of milk species-specific.

Gene expression profiling in bovine liver and mammary gland during the production cycle using cattle-specific microarrays

J.J. Loor, Mammalian NutriPhysio Genomics, Department of Animal Sciences, University of Illinois, Urbana, 61801, USA

The development of cattle microarrays has allowed high-throughput gene expression analysis in tissues of economic importance such as mammary gland and liver. cDNA and oligonucleotide microarray platforms developed at the University of Illinois have been used extensively to study the effects of physiological state and/or nutrition on global gene expression patterns. Initial studies utilized a 7,872 cattle cDNA microarray, containing approximately 6,300 unique genes, to examine temporal liver gene expression profiles in multiparous Holstein cows biopsied at -65, -30, -14, 1, 14, 28, and 49 days relative to parturition. More than 4,500 unique genes were expressed in liver tissue. In addition, most drastic changes in expression occurred during the periparturient period (i.e., day -30 through day +28) and were associated with inflammation, acute-phase response, oxidative stress, and fatty acid oxidation among others. A recently-developed 13,257-gene oligonucleotide microarray, containing approximately 12,400 unique genes, was used to study global mammary gene expression patterns in multiparous Holstein cows biopsied at -30, -15, 1, 15, 30, 60, 120, 240, and 300 days relative to parturition. More than 10,000 sequences were expressed in mammary tissue. Changes in expression patterns throughout lactation included genes associated with metabolism, transcription regulation, apoptosis, and cell proliferation. Gene expression data have been used to compliment measurements of whole-animal tissue function and animal performance to develop integrative models.

Gene expression associated with beef sensory qualities

C. Bernard[1], I. Cassar-Malek[1], H. Dubroeucq[1], G. Renand[2] and J.F. Hocquette[1], [1]INRA, Herbivore Research Unit, Theix, France, [2]INRA, Quantitative and Applied Genetics Unit, Jouy-en-Josas, France

Beef quality and more particularly its sensory traits (tenderness, flavour, juiciness and colour) depend on many factors including muscle characteristics, which explain, however, less than 35% of the variability in sensory qualities. They are under the control of many genes. So to get a better understanding of the physiological processes influencing meat quality, a transcriptomic analysis was conducted. *Longissimus thoracis* muscles from fourteen 15- and 19-month-old Charolais bulls among 25 were analysed using microarray technology. Gene expression profiles were compared between high and low meat quality scores of tenderness, flavour and juiciness. Several genes (n=215) were differentially expressed (Fold Change>1.4; FDR<1‰). They are mostly involved in different metabolic pathways and their regulation including nucleic acid and protein metabolism, developmental processes, signal transduction, cell structure and muscle contraction. Among these genes, 19 were significantly related simultaneously with the three quality attributes (tenderness, flavour and juiciness): 15 were up-regulated and 4 down-regulated in the most tender, juiciest and tastiest meats. One of the 4 genes had a strong negative correlation with tenderness and 14 of 15 genes were highly correlated with both meat flavour and juiciness. In conclusion, the expression of some genes was associated with the sensory qualities of beef and could account for their variability.

MicroRNA: A recently discovered class of small RNAs involved in gene regulation
F. Le Provost, INRA, Laboratoire de Génétique biochimique et Cytogénétique, batiment 440, Jouy-en-Josas 78 350, France

RNA silencing mechanisms were first recognized as a defence mechanism against RNA viruses or random integration of transposable elements. But the general role of RNA in the regulation of gene expression only became apparent when it was realized that specific genes, which encode short forms of single-strand RNAs, could silence the expression of target messenger RNA (mRNA). Many of these small (19-22 nt) non-coding RNA, called microRNAs (miRNAs), are evolutionarily conserved and may account for 2 to 3 % of the total number of genes in human. In animals, most miRNAs are imprecisely complementary to their target mRNA and inhibit protein synthesis. In some few cases, this complementary is total leading to the cleavage and the destruction of the target mRNA. Although the biological functions of most miRNAs are unknown, their importance for development, cell proliferation, cell death, and morphogenesis has been demonstrated in several species. The expression of some miRNAs is regulated in a developmental and spatial manner. The genetic analysis of miRNA genes in model organisms is beginning to put into place the piece of a mosaic that will eventually show us the range of functions that miRNAs have in the control of animal development and physiology.

Quantitative trait loci for osteochondrosis in Hanoverian warmblood horses
C. Dierks and O. Distl*, University of Veterinary Medicine Hannover, Institute for Animal Breeding and Genetics, 30559 Hannover, Germany

The objectives of this work were to perform a whole genome scan for quantitative trait loci (QTL) of equine osteochondrosis/osteochondrosis dissecans (OC/OCD) and to refine their positions in order to develop marker tests based on intragenic single nucleotide polymorphisms (SNPs). The whole genome scan included 218 microsatellite markers, genotyped for 104 progeny of 14 paternal half-sib families of Hanoverian Warmblood horses, their dams and eight sires. Traits analysed were OC and OCD in fetlock and hock joints. Chromosome-wide significant QTL were located on equine chromosomes (ECA) 2, 3, 4, 5, 15, 16 and 19. A genome-wide QTL was found on ECA2p for fetlock OC and fetlock OCD. Furthermore, 32 SNP-markers located on ECA2p and ECA4q were developed using BAC end and equine EST sequences. Regarding ECA2, the test statistic peaked in a 3 cM interval close to the candidate gene *COL9A2*. On ECA4, an intragenic SNP marker was significantly associated with hock osteochondrosis. The additive genetic effects of the SNP marker on ECA4 were -0.85 for the occurrence of hock OC and -0.83 for the occurrence of hock OCD. The significant dominance effects were +0.47 for hock OC and +0.53 for hock OCD. This is the first report where QTL for equine osteochondrosis and the development of a marker test for application in horse breeding were shown.

Accelerated expansion of group IID like phospholipase A2 genes (PLA2s) in Bos taurus

M. Golik[1]*, M. Cohen-Zinder[1], J.J. Loor[2], J.K. Drackley[2], M.R. Band[2], H.A. Lewin[2], J.I. Weller[1], M. Ron[1] and E. Seroussi[1], [1]Agricultural Research Organization,Institute of Animal Sciences, Bet Dagan 50250, Israel, [2]University of Illinois, Department of Animal Sciences, Urbana, Illinois 61801, USA

Secreted low-molecular-weight PLA2s are clustered within a syntenic group on human 1p35-36 and mouse 4qD3. We reassembled trace files available from the whole genome sequencing project obtaining a 86 kb contig with 3 tandem PLA2G2D duplications in the Hereford strain, with a coverage exceeding 5-fold. The genomic structure indicated that most of the PLA2G2D transcripts are formed by four exons. Two alternative first exons were present in all duplications. Linkage and comparative mapping placed the bovine PLA2G2 locus on BTA2, indicating that it evolved from an ancestral PLA2G2D locus common to human, cattle and rodents. Bovine PLA2G2D variants were capable of encoding 147 amino acid polypeptides that consisted of putative signal peptide and metal-binding domains. Cysteine residues were conserved in positions analogous to those forming the seven disulfide bonds characteristic to PLA2G2 genes. Quantitative PCR analysis of bovine PLA2G2D transcripts indicated that their expression levels varied between the dry period and lactation in the mammary gland samples and that their expression was polymorphic in liver tissue. The recent burst of duplication and divergence of the bovine PLA2G2D genes and their polymorphic nature are typical of innate immune response genes.

Evaluation of hierarchical clustering methodologies applied to proteomic data mining

B. Meunier[1], E. Dumas[2], I. Piec[1], M. Hébraud[2] and J.F. Hocquette[1], INRA, [1]Herbivore Research Unit, [2]Microbiology Unit, Theix, France

Hierarchical clustering of proteomic data resulting from 2D-gel electrophoresis is the technique of choice for classifying samples and proteins. To this end, improved methods and user-friendly bioinformatics tools are needed. Unlike transcriptomic studies in which these techniques have already been standardized, in proteomics we can observe as many clustering approaches as papers. Therefore, we evaluated the major clustering methods using the F-Measure, a powerful clustering result quality metric (Bioinformatics, 21:3201-3212). We analysed two different proteomic data sets under PermutMatrix. This freely available software developed for microarray analyses offers the entire panel of algorithms needed for the study: three distance measures (Pearson correlation, Euclidean, and Jaccard) and three aggregation procedures (Complete, UPGMA, Ward). This software also allows the visualization of the clustering results: dendrograms of the samples and of the proteins and heat map representation of the data matrix. Additionally, we introduced several data pre-processing procedures improving the clustering results and its exploitation e.g. a suitable data normalization consisting of a logged ratio transformation of the raw protein volume was beneficial for both data sets. The Pearson correlation and the Ward aggregation method gave the best clustering results based on the F-Measure. In conclusion, hierarchical clustering is very intuitive and if tuned to proteomic data it would serve as a useful, visual data exploring technique.

Molecular markers used in genetic characterization of Romanian Black Spotted Cattle

R. Vatasescu, S. Georgescu, E. Ionica, A. Dinischiotu and M. Costache*, University of Bucharest, Molecular Biology Center, phone/fax 0040213181575, 91-95 Spl. Independentei, 050095, Bucharest5, Romania

Romanian Black Spotted cattle are specialized in milk production and were homologated as a breed in 1987. Genetic characterization of Romanian Black Spotted population was made using eleven microsatellites, three genes (bovine k-casein, β-lactoglobulin and α-lactalbumin) were used as molecular markers to characterise the milk properties and three genes implicated in three important recessive hereditary disorders in cattle (Bovine Leukocyte Adhesion Deficiency, Deficiency of Uridine 5-Monophosphate Synthase and Bovine Citrullinaemia). The microsatellites analysed were chosen based upon the polymorphism detected in other breeds. High levels of polymorphism were observed over the population. This technology has great potential for use in cattle breeding and provides a more efficient method for parentage and individual identification. The importance of polymorphism of milk proteins in the selection of dairy cows has been shown, since milk protein genetic variants have a significant effect on the composition of milk. Using PCR amplification and RFLP technique we analyzed the genetic polymorphism of bovine k-casein, β-lactoglobulin and α-lactalbumin loci. In cattle breeding, the genetic diseases produce significant effects that lead to a decrease in or loss of performance. Using the RFLP-PCR technique, we can develop easy and efficient methods that can be used to correctly identify the normal and the affected animals for these genetic disorders.

Impact of nutrition on mammary transcriptome and its interaction with the CSN1S1 genotype in lactating goats

S. Ollier[1], C. Robert-Granié[2], S. Bes[1], M. Goutte[1], Y. Faulconnier[1], Y. Chilliard[1] and C. Leroux[1]*, [1]INRA, URH, 63122 Theix, [2]INRA, SAGA, 31326 Castanet-Tolosan, France

Genetic and nutritional factors significantly affect milk composition, influencing its nutritional quality. Milk lipid synthesis and secretion by the mammary gland involve the expression of a large number of genes which regulation remains poorly defined. To investigate the pathways regulated by nutrition and its interaction with genetic factor, the mammary transcriptomes comparison of 12 lactating goats under extreme nutritional conditions (standard diet *vs.* 48-h fasting) and presenting different alpha-s1-casein (CSN1S1) genotype ("High" *vs.* "Low") was performed. Therefore, a bovine oligochip containing 8,400 different genes was used. Then, to complete the analysis, mRNA quantification of transcripts involved in milk fat synthesis and secretion was done (qRT-PCR). Something like 200 differentially expressed genes between the two nutritional conditions were observed, 84% of which were downregulated by fasting. Feed deprivation altered several metabolic pathways in mammary gland. In particular, the decrease in mRNA abundance of genes encoding major milk proteins and lipogenic genes was in agreement with milk protein and fat secretion decrease. Moreover, among these 200 genes, 69 were common between the two genotypes whereas the fasting effect on the expression of 9 genes was different between the genotypes. This experiment shows an interaction between nutritional and genetic factors, which is currently under investigation (*Work funded by the LIPGENE EU-FP6 Project, www.lipgene.tcd.ie*).

L. *lactis* upregulates immune gene expression in the bovine mammary gland

M. Daly[1*], K. Klostermann[1], W.J. Meaney[2], F. Crispie[1], C. Hill[3], L. Giblin[1] and P.R. Ross[1], Moorepark Food Research Centre, Teagasc, Fermoy, Co. Cork, Ireland. [2]Dairy Production Research Centre, Teagasc, Fermoy, Co. Cork, Ireland. [3]Department of Microbiology and Alimentary Pharmabiotic Centre, University College Cork, Cork, Ireland

Lactococcus lactis has been shown to be an effective live culture treatment of mastitis. This novel treatment results in an immunomodulatory effect with an enhanced recruitment of polymorphonuclear neutrophils. The intricacies of this immunomodulation are not clearly understood. The objective of this study was to monitor immune gene expression (IL-8, TNF-α, NF-κB, TLR2, TLR4 and Lactoferrin) in response to *L.lactis*. Cows were selected based on their low somatic cell counts. The left hind quarter was infused with *L.lactis* and the localised gene response was monitored over various time points (0 hrs, 3 hrs, 8 hrs, 24 hrs, 3 days, 7 days). RNA was isolated from leukocytes from blood, somatic cells from milk of the infused quarter and from milk of the adjacent quarter. Real time PCR was performed using Lightcyler and SyBr green technology with gene specific primer pairs. Copy number was determined by absolute quantification. Infusion increased the copy numbers of all six genes several hundred-fold, over the levels detected in the adjacent uninfused quarter and pre-infusion. The magnitude of the IL-8 burst was noteworthy with peak increases at 3-8hrs. Induction of these immune genes is not systemic but is confined to a localised response in the infused quarter.

Influence of form and concentration of Cu, Zn and Fe on swine production

Gretchen Myers Hill, Michigan State University, USA

The physiological need and interaction of Cu, Zn and Fe have been known for many years, but new laboratory techniques and leaner genetics may have altered the dietary requirements. The mode of action of pharmacological copper as sulfate has been hypothesized, and it now appears that some organic forms (Bioplex Cu, Mintrex Cu, Cu lysine) and copper tribasic copper chloride may be able to produce the same results with lower dietary concentrations. Feeding pharmacological Zn as oxide (2,000 ppm Zn) will result in improved growth, but if fed for only 10 or 11 days, it does not result in excessive Zn excretion. Our work indicates that villus height and crypt depth may be improved and proteins associated with antioxidant activity may be increased with Zn oxide. Recent work from our lab demonstrates that 100 ppm Fe should be added in nursery diets by using a highly available source such as ferrous sulfate. The ability of organic forms of Fe to reduce this dietary concentration has not been investigated with the newer Fe biomarkers such as iron regulatory proteins and transferrin. In collaboration with researchers at North Carolina State University, we have shown that feeding a 50:50 mixture of organic and inorganic minerals at the recommendations of the National Research Council from weaning through farrowing was superior to inorganic minerals fed at higher concentrations.

Trace mineral nutrition of pigs: Meeting production and environmental objectives
W.H. Close, Close Consultancy, Wokingham, Berkshire, RG41 2RS, UK

Trace mineral nutrition has been a neglected area of pig nutrition. There is little recent information on the trace mineral requirements of modern pig genotypes and it has become customary to provide levels in the diet much higher than those recommended. Some minerals, such as copper and zinc, are added at 'pharmacological' levels to increase growth, to enhance immunity and to reduce diarrhoea in piglets. There is, however, concern about the large quantities of undigested elements being excreted and causing environmental pollution.

Although inorganic sources of trace minerals have been widely used, there are questions about their availability to the animal and this has created interest in proteinated or chelated (organic) trace elements. These are better absorbed and are more available to the animal. As a consequence, inclusion levels can be reduced while maintaining, or even enhancing, performance. The results of studies comparing inorganic and organic sources of copper, zinc, iron and selenium and combinations of these, are discussed. Indeed, providing a balance of inorganic and organic minerals may be the most effective way to meet the animal's need and studies with sows have shown improvements in reproductive performance.

In the future, the source of mineral may therefore be of increasing importance in attempting to satisfy both production and environmental demands.

Magnesium balance in lactating dairy cows fed different potassium levels
C. Kronqvist, E. Briland, R. Spörndly and K. Holtenius*, Department of Animal Nutrition and Management, Swedish University of Agricultural Sciences, SE-753 23 Uppsala, Sweden

High levels of dietary potassium (K) may negatively interact with ruminal magnesium (Mg) absorption. Most studies of interactions between K and Mg have been performed in small ruminants or dry cows. Hence the objective of the present study was to investigate the effects of K intake on Mg balance in lactating dairy cows. The experimental set up was a Latin square design with six lactating cows in a 2 x 3 factorial arrangement of treatments. The cows were fed two levels of magnesium (1.9 and 3.7 g/kg DM) and three levels of potassium (19, 28 and 37 g/kg DM). The average DM intake during the 6 periods was 16.1 kg/day (range 15.3-16.9). The experiment was performed during 12 weeks in mid-late lactation. The K level in the diet did not negatively affect apparent Mg digestibility ($P > 0.1$) or excretion via milk and urine ($P > 0.1$). Neither was the concentration of Mg in rumen fluid or blood plasma affected ($P > 0.1$).

The higher dietary Mg level gave rise to an increased Mg concentration in rumen fluid ($P<0.0001$) and in plasma ($P<0.001$). The excretion of Mg in faeces and in urine was increased ($P<0.0001$). The calculated apparent digestibility was reduced in diets with high Mg content.

It is concluded that an increased K intake did not reduce Mg uptake in these lactating dairy cows.

Session P8   Theatre 4

Status of copper deficiency of blood serum of sheep breeds in Iran

A.A. Javan and A. Ezzi, Animal Science Research Institute of Iran, P.O. Box: 31585-1483,Karaj, Iran

The present study was led to determined copper concentration over 129 serum from different Breeds in sheep flocks. In addition soil, forages and liver (dead lambs) were examined in The Research Center of Animal Science. Three suspected dead cases were autopsied. Spinal cord and brain were removed and examined under microscope.The results showed that the copper concentration was marginal or deficient.By using three statistical tests (T,Odds Ratio and Duncan) the mean copper values were not differ in serum. However these values were significantin breeds and cities (P<0.01).

Session C9   Theatre 1

New methods for grading beef and sheep carcasses

P. Allen, Teagasc, Ashtown Food Research Centre, Ashtown, Dublin 15, Ireland

Carcass grading is important as it gives information to producers about the quality of their animals in relation to the requirements of the market. These quality signals are most powerful when they are linked to price differentials that reflect the worth of different quality categories in carcass. There is resistance to such price differentials when the grading is considered by producers to be inaccurate or inconsistent. In Europe beef and sheep carcases have been graded visually by trained graders who judge their conformation and fat cover according to the SEUROP grading scheme and assign each carcass to a conformation (S,E,U,R,O,P) and fat (1,2,3,4,5) class. Since the 1980's, mechanical grading systems that use video image analysis (VIA) to assign carcases to the grades have been developed. In 2003, the EC changed the rules to allow their use for official beef carcass grading and Ireland was the first country to carry out an authorisation trial. VIA systems can also estimate the saleable yield of carcases and this would be a better basis for producer payments. This paper reviews the development and introduction of these new grading systems and outlines how they might be use din the future.

Traditional and new methods to assess meat quality

J.F. Hocquette[1], G. Renand[1], E. Dufour[2], J. Lepetit[1], G.R. Nute[3], INRA, [1]Theix and [2]Jouy-en-Josas, [2]ENITAC, Lempdes France, [3]University of Bristol, UK

Quality can be defined as "The properties of a product that contribute to and satisfy the needs of the end-user". Therefore, extrinsic quality attributes (including production characteristics) need to be considered in relation to how they impact on intrinsic quality attributes. The latter are focused on muscle characteristics for rearing and genetic purposes or on the product itself for sale. In muscle biology, the recent advent of genomics has the potential to identify new biological predictors of meat quality. Genomics will yield new information that will increase our knowledge of the genes affected by production factors. To assess meat quality closer to the point of sale, different indicators (pH, colour, mechanical properties, ultrasound, electrical impedance and conductivity, etc) and emerging technologies (NMR, image analysis, fluorescence and NIR spectroscopy, etc) can be used. Final assessments by analytical tests with trained panels or consumer tests are methods that are chosen to evaluate meat sensory quality. Nutritional quality is assessed by the analysis of meat composition (proteins, fatty acids, etc). In all cases, relevance with regard to the initial objectives, repeatability, accuracy, speed, on-line application and cost are essential criteria in selecting the most appropriate methods. Large scale programmes are in progress to assess and explore these criteria and the interrelationship between quality indicators. This approach may define the best combination of the indicators of quality which together with modelling approaches may produce a labelling system for meat based on quality that complies with the existing legislative framework.

Plasma leptin and insulin-like growth factor I (IGF-I) as potentially phenotypic markers for carcass composition and growth rate in lambs

E. von Borell[1]*, H. Sauerwein[2] and M. Altmann[1], [1]Institute of Animal Breeding and Husbandry with Veterinary Clinic, Martin-Luther-University, Halle, Germany, [2] Institute of Physiology, Biochemistry and Animal Hygiene, Bonn University, Bonn, Germany

The selection for IGF-I concentrations at an early developmental stage is a well documented and patented method to improve growth, feed efficiency and carcass quality of livestock species such as pigs and cattle. However, studies on the relationship between the adipocyte derived hormone leptin measured at an early growth phase and final performance traits are scarce. Therefore, 30 ad libitum fed intact male lambs were blood sampled at 20, 25, 30, 35, and 40 kg live weight. After slaughtering at 40 kg, lean, visceral, subcutaneous and intermuscular fat were assessed by dissection. Average daily gain from birth to slaughter correlated to leptin at 30 and 35 live weight (-0.56 and -0.61, $P < 0.01$). Subcutaneous and intermuscular fat correlated to leptin only at 40 kg live weight ($r = 0.45$ and 0.36, $P < 0.05$), while visceral fat correlated to leptin concentrations analyzed at 25 to 40 kg live weight ($r = 0.41$, $P < 0.05$ to $r = 0.58$, $P < 0.01$). This study indicates that leptin concentrations measured at a time when lambs reach their slaughter weight have the greatest potential to assess body fat content, whereas an earlier prediction does not seem to be feasible. Measurements of IGF-I concentrations confirmed that this hormone has the potential to serve as an early marker for average growth rate ($r = 0.52$ to 0.59, $P < 0.01$). However, correlations to carcass composition were not significant in most cases.

Relationship of carcass grades with carcass composition and value of steers

M.J. Drennan, M.G. Keane and M. Nolan, Teagasc, Grange Beef Research Centre, Dunsany, Co. Meath, Ireland

The objectives were to study the relationship of carcass conformation (CS) and carcass fat (FS) scores with carcass meat, fat and bone proportions and value of steers. A total of 134 steers of 619 (s.e. 62) kg liveweight, 319 (s.e. 36) kg carcass weight, 5.7 (s.e. 2.2, scale 1 to 15) CS and 8.7 (s.e. 1.8 scale 1 to 15) FS were used. Carcasses were mechanically graded according to the EU Beef Carcass Classification Scheme. Carcass meat, fat and bone proportions were obtained following dissection of the right side of each carcass. Regression analysis showed the following relationships: meat (g/kg) = 652 +13.7 (s.e. 0.84) CS –6.6 (s.e. 1.00) FS; fat (g/kg) = 78 –5.9 (s.e. 0.76) CS +8.7 (s.e. 0.91) FS; bone (g/kg) = 270 -7.8 (s.e. 0.45) CS –2.0 (s.e. 0.54) FS. Carcass value was calculated as the sum of the commercial values of the fat trimmed boneless meat cuts with a deduction for bone. Value (c/kg) = 234 +6.2 (s.e. 0.36) CS –2.5 (s.e. 0.43) FS. Correlations of CS with meat proportion and carcass value were 0.78*** and 0.80***, respectively. Correlations between FS with fat proportion and carcass value were 0.57*** and -0.28**, respectively. In conclusion, increasing CS increased meat yield and carcass value. The effects of changes in FS were less than half as important as changes in CS.

Use of ultrasound technology in selecting meat quality in fat-tailed sheep

S. Bedhiaf-Romdhani[1] and M. Djemali[2], [1]Laboratoire des Productions Animales et Fourragères, INRAT, Rue Hédi Karray, 2049 Ariana, Tunisia, [2]Laboratoire des Ressources Animales et Alimentaires, INAT, 1082 Cité Mahrajène, Tunisia*

The objective of this work was to estimate genetic parameters of rib muscle and fat depth measured on live lambs by ultrasound. A total of 745 lambs from 559 ewes and 97 rams of a fat tailed breed known as the "Barbarine" was used in this study. Ultrasound measurements of external fat thickness (UFD) and muscle (UMD) taken at the 12-13th rib and palpation of body conditions were made from 150 d till 520 days of lamb ages. Variance components estimation of fat and muscle depth at 180 d, 200 d, 240d and 520 days of age were computed based on REML method using an additive model. Main results showed that after 200 days of age UMD, UFD, loin and tail scores increased as live weight increases. The highest muscle depth for males was reached at 180 d and the lowest fat depth was recorded between 180 and 200 days. Average loin scores became greater for males than females from 240 days of age. Male lambs had greater tail scores at all ages. Heritabilities were higher for fat and muscle depths at 240 d of age but relatively low in other ages. Genetic correlations were relatively low between fat and muscle depths with a negative sign.

Quality classification of the *M. rectus abdominis* of Charolais Heifers

M.P. Oury[1]*, J. Agabriel[2], C. Agabriel[3], J. Blanquet[1], D. Micol[2], B. Picard[2], M. Roux[1], R. Dumont[1], [1]ENESAD, BP87999, 21079 Dijon, France, [2]INRA, URH, 63122 St-Genès-Champanelle, France, [3]ENITAC, 63370 Lempdes, France

Ninety-nine Charolais heifers were used to study the variation of meat quality attributes. The *M. rectus abdominis* was excised 24 hours post-slaughter and sensory attributes analysed at 14 days *post mortem*. Meat quality was evaluated by sensory evaluation, using four descriptors : initial and general tenderness, juiciness and flavour intensity. Biochemical characteristics analysed were : shear force, intramuscular fat content, collagen content and solubility, lactate deshydrogenase, isocitrate deshydrogenase and cytochrome c oxydase activities, fibre size, MyHC isoforms, haem pigment content, CIELAB colour measurements.

Samples of meat were organised in types using a principal component analysis and a hierarchical cluster analysis. Five meat types were classified by increasing tenderness, from the type I (low tenderness, low juiciness, medium flavour intensity) to the type V (high tenderness, medium juiciness, high flavour intensity).

The types IV and V leads to both tender and juicy meat and content the most intramuscular fat. The lowest level of tenderness (type II) was explained by a high shear force and more total and unsoluble collagen.

Lactate deshydrogenase, isocitrate deshydrogenase and cytochrome c oxydase activities, fibre size, MyHC isoforms, haem pigment concentration and CIELAB colour measurements of the *rectus abdominis* muscles were no different between the five meat types.

Use of carcass weight, community scale for carcass classification and carcass ultrasound measurements to predict carcass composition of young beef bulls

R. Delfa*, G. Ripoll, M. Joy and P. Albertí, Unidad de Tecnología en Producción Animal. Centro de Investigación y Tecnología Agroalimentaria de Aragón.Avda. Montañana, 930. 50016-Zaragoza, Spain

Sixty nine carcass of young bulls of four different Spanish local breeds with a hot carcass weight ranging from 199.9 to 449.6 Kg were classified by the Community Scale Carcass Classification and scanned by ultrasound to determine *M. Longissimus Thoracis depth* (BCAR) and subcutaneous fat thickness (CCAR). Carcasses were halved and cooled at 4° C for 24 h. The left side was divided in standardised commercial joints. Each joint was then dissected into muscle, subcutaneous and inter-muscular fat and bone.

The hot carcass weight, carcass classification and ultrasound measurements were used to predict carcass tissue composition by Stepwise regression analysis.

The developed model with hot carcass weight and carcass classification explained 97, 60 and 85 % of the muscle, fat and bone weight variation, respectively. When ultrasound measurements were included into the above model, it was observed an increase of percentage of variation explained for a further 1 and 23 percent units for muscle and fat weight, respectively. For bone weight, the RSD was reduced by 8.8 %.

Relationship of visual muscular and skeletal scores and scanned muscle measurements with carcass grades, composition and value of steers

M.J. Drennan and M.McGee, Teagasc, Grange Beef Research Centre, Dunsany, Co. Meath, Ireland

The objectives were to examine the relationships of live animal scores and scanned muscle measurements with carcass grades, composition and value. A total of 134 steers of 619 (s.e. 62) kg liveweight and 319 (s.e. 36) kg carcass weight were used. Pre-slaughter visual muscular and skeletal scores were recorded at 6 and 3 locations respectively. Eye muscle depth at the 3rd lumbar vertebra and fat depth at the 13th rib and 3rd lumbar vertebra were recorded using a Dynamic Imaging Scanner. Post-slaughter records included carcass classification and meat, fat and bone contents following dissection of one side of each carcass. Carcass value was estimated as the sum of the commercial values of each fat-trimmed, boneless cut with a deduction for bone. Correlations between average muscular score and carcass conformation, meat proportion and value were 0.87***, 0.69*** and 0.69***, respectively. Corresponding values for the 3 skeletal scores varied from -0.39*** to -0.51***, -0.34*** to -0.42*** and -0.33*** to -0.41*** and with scanned muscle depth 0.79***, 0.66*** and 0.69***. Correlations between scanned fat depth and carcass fat score, fat proportion and value were 0.54***, 0.30*** and 0.07, respectively. In conclusion, muscular scores and scanned muscle depth were good indicators of carcass conformation, meat yield and carcass value, while skeletal scores were poor indicators.

Dependence between the area of MLLT and the muscling score of selected body areas in live animals of certain cattle breeds

J. Dvorakova[1], L. Stadnik[1], A. Jezkova[1], V. Kuprova[1] and F. Louda[2]*, [1]Czech University of Agriculture Prague, Czech Republic,[2]Research Institute of Cattle Breeding, Rapotín, Czech Republic*

During rearing, at the age of 120, 210 and 365 days an area (cm^2) of musculus longissimus lumborum et thoracis (AMLLT) in the area of 1st and 6th lumbar vertebrae in 30 heads of Blonde d'aquitaine cattle and 30 heads of cross-breds (BAxC) was measured via ultrasound. At the same age the muscling of shoulder, back and rump was estimated (using 10 point scale). Further, basic body measurements were evaluated. At the age of 120, 210 and 365 days in the animals, where reached shoulder muscling score was higher than 7-8 points, the dependence (P=0,05) with the AMLLT was detected. The AMLLT in the animals with the muscling score higher than 7-8 points detected at the age of 120, 210 and 364 day it was from 38,9 to 50,8 cm^2, from 44,6 to 62,6 cm^2 and from 49,9 to 78,8 cm^2, respectively.Between the point evaluation of muscling of back and rump higher than 7-8 points and with the AMLLT at the age of 120, 210 and 365 days significant dependence (P≥0,05) was detected.In relation to the meatiness of shoulder, back and rump on a live animal and ultrasound measurement of the AMLLT it is possible to specify the grade of meatiness and slaughter yield.

The use of ultrasound measurements in estimation of Valuable cut content in fattened bulls
P. Polák*, J. Daňo, M. Oravcová and E. Krupa, Slovak Agricultural Research Authority, Research Institute for Animal Production, Hlohovská 2, 949 92 Nitra, Slovakia

The aim of our investigation was to analyse possibilities of using ultrasound as a method for objectification of price of fattened bulls via estimation of valuable cut content. There were 483 bulls of seven breeds analysed. Results of detail dissection were used to evaluate the dependence of weight of valuable cut, weight of hot carcass and EUROP grading for carcass conformation obtained by dissection and assessment by ultrasound. The average class for carcass composition of all bulls was 2.706 (R⁻). Weight of valuable cats was assessed by linear regression model with weight before slaughter and muscle thickness on the last thoracic vertebra as independent variables. No statistical differences were found out between weights of valuable cut obtained using both methods. Relative value difference was 0.088% i.e. 123 g of meat. For modelling the dependence of response variable weight of valuable cut on carcass composition, coefficients of non-linear function ($y = 19.5963 + 18.961x - 2.38542 x^2$; Iyx = 0.497995) were estimated. Maximum of this function for carcass composition was $x_{max} = 3.974$ and for weight of valuable cut assessed by ultrasound $y_{max} = 95.74$ kg. Calculated value for extremes of function for classification of carcass composition (3.974) reflects the present structure of the bull's population in Slovakia.

Comparison of two methods for longissimus muscle area measurements
D. Karolyi*, A. Džidić, K. Salajpal, M. Đikić and I. Jurić, Faculty of Agriculture University of Zagreb, Svetošimunska cesta 25, 10 000 Zagreb, Croatia

The aim of the study was to determine the variance of longissimus muscle area (LMA) measurements associated with the standard use of polar planimeter (REISS Precision 3005, Germany) and Image tool® software program. Measurements were conducted on six dorsal samples taken between 7th and 9th rib of the right halves of beef carcasses. Firstly, LMA was traced on acetate paper and measured using a planimeter. Secondly, the images of longissimus muscles for computer analysis were obtained by scanning (UMAX PowerLook II) dorsal samples in ratio 1:1. LMA size was determined with Image tool® software. Triplicate measurements of LMA were performed by six measurers for each sample and method. Variance components for LMA measurement method associated with measurer, sample and measurement replicates were estimated by REML. The proportion of the total variance in LMA measurement beetwen samples was greater with Image tool® (99.79%) than with planimeter (96.49%). The proportion of variance between measurers was 0.12% with Image tool® and 0.34% with planimeter. The proportion of error variance due to measurement replicates associated with the use of Image tool® program was smaller (0.09 %) than with the planimeter (3.17%). The application of Image tool® showed greater precision in measuring LMA than standard planimeter method.

In vivo ultrasonic measurements and live weight for predicting carcass quality in Churra Tensina mountain breed lambs

R. Delfa[1]*, M. Joy[1], A. Sanz[1], B. Panea[1] and A. Teixeira[2], [1]Unidad de Tecnología en Producción Animal. Centro de Investigación y Tecnología Agroalimentaria de Aragón. Avda. Montañana, 930. 50016-Zaragoza, Spain, [2]Departamento de Zootecnia. Escuela Superior Agraria. Apto. 727. 5301-855 Bragança. Portugal

Forty six single male lambs of Churra Tensina mountain breed with a live weight ranging from 19.9 to 24.4 Kg producers of carcass type *Ternasco*, were scanned by ultrasound (ALOKA model SSD-900, with a 7.5 MHz sounder) to determine *M. Longissimus dorsi* depth and subcutaneous fat thickness between the 10th–11th, 12th-13th dorsal vertebra and 1st-2nd, 3rd-4th lumbar vertebra. Lambs were slaughtered after 24 h. fasting. Carcasses were cooled at 4° C for 24 h. and halved. The left side was divided according to a standardised jointing procedure, based on six anatomically regions: shoulder, long leg, anterior ribs, ribs, flank and neck. Each joint was then dissected into muscle, bone plus remainder and subcutaneous, inter-muscular, kidney and pelvic fat.

The *in vivo* ultrasound measurements plus slaughter weight were fitted to predict carcass tissue composition by Stepwise regression analysis.

All the developed models were highly significant ($P<0.001$) and explained 70, 51, 82, 56, 59 and 41 % of the muscle, bone plus remainder, subcutaneous, inter-muscular, kidney and pelvic fat variation respectively. The models residual standard deviations were lower than 124.3 g.

Slaughter value evaluation of large weight Ile de France and Hungarian Merino lambs by CT and traditional slaughter cutting

Gy. Toldi[1], A. Molnár[2], T. Németh[2], S. Kukovics[2], [1]Kaposvár University Faculty of Animal Science, Kaposvár, Hungary, [2]Research Institute for Animal Breeding and Nutrition, Herceghalom, Hungary

The slaughter value of two sheep breeds, Hungarian Merino (HM) and Ile de France (IdF) has been compared, by investigation of ten lambs of each genotype and sex on the day before slaughter. CT-images were taken from individuals from occiput to knee.

The individuals were slaughtered at ~30 kg live weight and categorized according to the S/EUROP carcass qualification method. After 24h cooling the cut halves were divided according to the slaughterhouse practice applied in Hungary, then boned. The divided parts were classified to 1st class (roast meat) and 2nd class (non-roast meat) quality categories.

Close correlations were stated in both genotypes between the tissue (fat, muscle, bone) area results calculated from the CT-images and the weight data measured after slaughter. There was also a favourable correlation between the numbered „body conformation" values of the S/EUROP cut body, and the measured slaughterhouse cutting and boning values.

Cut halves of the IdF lambs were superior to the HM lambs mainly in the pieces of roast meat category. The muscle tissue areas of the IdF lambs were always superior as compared to HM lambs of the same live weight. A bit higher fat incorporation was verified in the IdF individuals by both the CT-images and slaughterhouse evaluation.

Longissimus thoracis et lumborum muscle volume calculation using in vivo real time ultrasonography

S. Silva, A.L.G Lourenço*, C. Guedes, V. Santos, J. Azevedo and A.A. Dias-da-Silva. CECAV-UTAD Department of Animal Science, PO Box 1013, 5001–801 Vila Real, Portugal

The determination of three-dimensional structure from serial sections is a common problem in animal corporal composition studies. Volume measured *in vivo* by computer tomography or magnetic resonance image accurately estimate the volume of the dissected tissues ($r^2>0.8$). Little information is available on muscle volume measurements using real time ultrasonography (RTU). The purpose of the present study was to study the in vivo RTU to estimate the *Longissimus thoracis et lumborum muscle* (LM) volume. Thirteen Île-de-France mature ewes were used. RTU images were taken over the 7th, 9th, 11th and 13th thoracic vertebrae and over the 2nd and 4th lumbar vertebrae. These images were acquired using an Aloka 500V ultrasound scanner with a 5 MHz probe, and were analysed by Image J software for the determination of LM areas. The distance between 6th thoracic vertebra and 5th lumbar vertebra was determined and it was divided by six to obtain the length of each slice. The LM volume was calculated by multiplying the areas and slice lengths. For the equivalent carcass measurements a joint from 6th thoracic vertebra to 5th lumbar vertebra were used. LM volume measured in carcass and in vivo by RTU was highly correlated ($r^2=0.92$, 0.95 and 0.97 for lumbar, thoracic and all vertebrae, respectively). These results strongly support evidence that in vivo RTU accurately estimate LM volume.

The effect of ultrasound probe on accuracy of intramuscular fat content and marbling prediction in beef longissimus dorsi muscle

J. Tomka, P. Polák*, E. N. Blanco Roa, E. Krupa, J. Huba, M. Oravcová and D. Peškovičová, Slovak Agricultural Research Authority, 94992 Nitra, Slovakia

A frequency of ultrasound probe is one of the main factors influencing ultrasound images. The aim of this paper was to evaluate the accuracy of intramuscular fat content prediction using two ultrasound probes (5 MHz/ 64 mm and 3,5 MHz/ 172 mm). The study was carried out on 142 bulls of which 74 Slovak Spotted, 6 Pinzgauer, 22 Holstein, 20 Beef Simmental and 20 Charollais. Ultrasound images were obtained between 12th and 13th rib. Then analyses using special software for computer image analysis and SAS software for statistical evaluation were performed. No differences in level of marbling and intramuscular fat content were found between 5 MHz and 3,5 MHz probe. Correlation coefficients between level of marbling and ultrasound image values were calculated; r = 0,263 (5 MHz) and r = 0,248 (3,5 MHz). The optimal performance for ultrasound machine ALOKA SSD – 500 was proposed; 65% (5 MHz probe) and 80% (3.5 MHz probe).

Ethanol specimen of beef muscles samples for NIRS is rapid but efficient as the freeze-dried

G. Masoero, M. Iacurto, S. Gigli and G. Sala, CRA - Istituto Sperimentale per la Zootecnia, Meat department, Via Salaria 31, Monterotondo, Italy*

Rapid muscle preparation and conservation for NIRS by ethanol was studied. In a 3x3 trial (N=36) two homogeneous Semitendinosus bovine muscles were split in slices, (20mm*Ø20mm) and plugged in number of 1-2-3 into a 50ml tube containing 30ml ethanol 95.0, 63.3 or 31.7%. At 0h and 24h the samples were weighted, tested by a penetrometer, then scanned by NIRS. The low ethanol concentration gave erratic results while the higher concentrations gave homogeneous weight losses (35.0 and 33.6%) and tender losses (64 and 59%), the two traits being correlated (r=0.80). Not the weight of the samples nor their numbers were perceived by NIRS, while the two high ethanol concentrations were clearly distinguished (R^2_{val}=0.65). So it could be recommend putting into a tube a maximum of 15g/30ml 95% ethanol, indifferently as single or multiple specimen, even from different tissues. In a discrimination trial muscle samples from 54 Piemontese and 16 Italian Friesian were immersed in ethanol or freeze-dried. In Semitendinosus m. the classification ability of the NIRS for the two breeds reached 97% in ethanol vs. 94% in freeze-dried preparations. In Sternomandibularis m. ethanol preparation breed identification was 91%. It was concluded that this very simple and rapid method get good efficacy in separation of (onto)genetic factors at a very high level of efficiency. Further studies are necessary to link this specimen to traits and treatment effects.

Genetic trends for reproduction and maternal abilities in French Large White pigs

L. Canario, T. Tribout and J.P. Bidanel, INRA, Station de Génétique Quantitative et Appliquée, 78352 Jouy-en-Josas, France*

An experiment was set up at INRA to estimate genetic trends in French Large White (LW) pigs from 1977 to 1998 using frozen semen. LW sows were inseminated with semen of LW boars born either in 1977 (G77) or 1998 (G98). Their female progeny was tested for reproduction and maternal ability traits. Selection significantly increased ovulation rate (+3.9±1.2 corpora lutea) and total number of piglets born (+2.8±0.8 piglets/litter), but also resulted in a higher number of stillbirths (+0.7±0.3 piglets/litter). No significant trend was observed for farrowing kinetics traits. Though heavier (+182±63 g when adjusted for litter size) and more heterogeneous, G98 piglets were less mature at birth than G77 piglets. They also suffered from more respiratory difficulties at birth, had a lower viability and a higher risk of dying from starvation during the first day after birth, although colostrum production did not significantly change. G98 sows were more attentive towards their piglets at farrowing and had a larger number of functional teats (+1.8±0.4). G98 piglets had a similar or higher growth rate than their G77 counterparts, but both G77 and G98 piglets nursed by G98 sows had a lower weight gain up to 21 days of age than those nursed by G77 sows, thus suggesting a decrease in milk production from 1977 to 1998. Colostrum and milk composition remained unchanged.

Session G10

Theatre 2

Reliable computing in estimation of variance components

I. Misztal, University of Georgia, Athens, GA 30602, USA

Estimates of genetic parameters can be derived by a number of statistical and computing methodologies. For medium size problems, all are expected to result in similar estimates. For many problems, AI REML is the best choice. When it fails to coverge, EM REML may, although it is typically much slower and does not provide estimates of standard errors. It is also less reliable with random regression models particularly when starting values are small. When the number of traits is large, general REML approaches fail. A series of bivariate analyzes often results in a non-positive definite matrices. REML via canonical transformation is inexpensive and reliable but applicable to relatively simple models. Optimized MCMC can support a large number of traits with complicated models. Priors and heuristics are important for stability especially in small data sets. Determining the convergence is not obvious. A very slow mixing indicates a need for a different parameterization or for a more parsimonius model. Many problems in estimation with REML or MCMC are due to a limited data. For example, in maternal models few MGS could be connected to sires. A pedigree that appears complete may be sparse after pruning. The amount of information is usually much lower in categorical than in continuous traits. One way to evaluate estimability for a given model and data is to simulate observations using the structure from the field data.

Session G10

Theatre 3

Genetic parameters for lactation efficiency and its components in sows

R. Bergsma and E.F. Knol, IPG, Institute for Pig Genetics, P.O. Box 43, 6640 AA Beuningen, The Netherlands

Genetic and management changes during the last decades have increased litter size and milk production of sows. Voluntary feed intake of lactating sows has not increased in proportion with the higher energy requirements. A loss of body reserves is the result, with potentially negative consequences for sow performance in the subsequent cycle. An obvious solution is to increase feed intake capacity of sows. An alternative solution could be to improve the efficiency of using feed and body reserves during lactation.

The aim of the current study was to estimate the heritability of lactation efficiency and its underlying components: Fat mass at start of the lactation (FM), Feed intake capacity (FI), estimated milk yield (MY). Lactation efficiency (LE) was defined as the ratio of output (basically MY) over input (FI and body tissues) both expressed as MJ/d. The traits were estimated using weight-, backfat- and feed intake recordings of sows and weight recordings of piglets. The lactation feed intake dataset involved ad libitum fed gilts and gilts and multiparous sows fed according to a scheme.

Heritabilities for LE and its components were: 0.56 (0.60), 0.38 (0.15), 0.25 (0.18) and 0.05 (0.05) for FM, FI, MY and LE respectively, estimated under ad libitum feeding conditions (combined dataset between brackets). Genetic differences in lactation efficiency exist, but are smaller than those of the components.

Genetic analysis of mothering ability in sows and the relationship to reproduction and piglet mortality

B. Hellbrügge[1],K.-H. Tölle[2], C. Henze[1], J. Bennewitz[1], U. Presuhn[3] and J. Krieter[1*], [1]Institute of Animal Breeding and Husbandry,Christian-Albrechts-University, 24118 Kiel, Germany, [2]Chamber of Agriculutre Schleswig Holstein, 24327 Bleckendorf, Germany, [3]farm concepts, 23812 Wahlstedt, Germany

Records were available from 13,971 piglets and 1,538 purebred Landrace litters of the year 2004. Sows were kept in one nucleus herd. To characterise mothering abilities the sow's reaction to the separation from her litter during the first 24 h after birth, the reaction towards the playback of a piglets distress call and the reaction towards an unknown noise (music) were used. The heritabilities were estimated with a multiple ordered threshold model and rank from $h^2 = 0.09$ to $h^2 = 0.14$. For the number of piglets born alive, stillborn piglets and piglets born in total the heritabilities were estimated with a linear model. Different causes of piglet losses were evaluated as binary traits of the sow with survival rate (84.3 %), different definitions for crushing by the sow, underweight and stunting. The variance components were estimated with a threshold model and rank from $h^2 = 0.03$ to $h^2 = 0.05$. The genetic correlations were analysed bivariate and showed that more responsivness sows had fewer piglet losses. The analysis of the different behaviour-tests suggested that they were responsive to different patterns of behaviour.

Genetic variation for MMA treatment in sows

J. Krieter[1*] and U.Presuhn[2], [1]Institute of Animal Breeding and Husbandry, Christian-Albrechts-University, 24118 Kiel, Germany, [2]farm concepts, 23812 Wahlstedt, Germany

The objective of the present study was to analyse the genetic variation of the treatment for MMA (Mastitis, Metritis, Agalactia) syndrome in sows. 2,597 purebred litters from 544 German Landrace sows kept in one nucleus herd were available. Total number born, number of piglets born alive and piglets weaned were 10.6, 10.1 and 9.1, piglet lossses from birth to weaning were 10.2%. If the rectal temperature increased above 39.4°C sows were treated for MMA over three days. 38.5% of sows were treated for MMA, first parity sows showed a higher incidence (54.1%) compared to sows in higher parities (30.2%). The treatment for MMA had no effect on the number of piglets weaned, but sows treated for MMA showed higher piglet losses ($p<0.05$). Mixed linear models were used to analyse variance and covariance components for the different traits. The models included fixed effects (parity, year and season) and the random effects of permanent environment and the genetic effect. The heritability for MMA treatment was 0.13 (±0.08) and for piglets born alive 0.08 (±0.04). The genetic correlation between treatment and number of piglets born alive was negative

Between-breed variability of stillbirth and its relationship with sows and piglet characteristics

L.Canario[1]*, E.Cantoni[2], E. Le Bihan[3], Y. Billon[4], J.P. Bidanel[1] and J.L. Foulley[1], [1]INRA, Station de Génétique Quantitative et Appliquée, 78352 Jouy-en-Josas, France, [2]Université de Genève, Département d'économétrie, 1211 Genève, Suisse, [3]University of Luxembourg FLSHASE, 7201 Walferdange, Luxembourg, [4]INRA Unité de Génétique Expérimentale en Productions Animales, 17700 Surgères, France

Characteristics at birth were recorded from 969 litters of F1 Duroc x Large White (82), Large White (651), Meishan (63) and Laconie (173) sows. The data were analysed using models including the random effects of sow and/or litter, the fixed effects of sow genetic type, parity, birth assistance, piglet sex, as well as gestation length, piglet birth weight and litter size as covariates. Model selection was performed based on DIC criteria from hierarchical models fitted to the data: the random litter effect and a binomial distribution of stillbirth (as compared to a Poisson distribution) were selected. Then, statistical analyses were performed with standard Generalized Estimating Equations (GEE). The three most important factors involved in stillbirth were difference in piglet birth weight from the litter mean (2.36%), piglet sex (1.01%), and sow genetic type (0.94%). Stillbirth probability was higher for lighter and male piglets. Piglets born from Meishan sows had a lower risk (p<0.0001) and were little affected by the different factors of variation as compared to the three other sow genetic types. Standard and robust GEE approaches gave similar results despite some disequilibrium in the data set structure highlighted with the robust GEE approach.

A study of the reproductive traits of large white, Landras and Danube white breeds of swine and specialized maternal line "Srebrena", bred in Bulgaria

T. Yablanski[1], A. Stoikov[2], E. Jeliazkov[1], T. Kunev[3] and S. Nikolov[3], [1]Trakia University Department of Genetics, Animal Breeding and Reproduction, 6000 Stara zagora, [2]Institute of Agriculture, 9700 Shumen, [3]Silistra-hibrid Ltd, 7500, Bulgaria

On the basis of genetic parameters the reproductive traits of the major swine breeds used of pig production in Bulgaria were investigated. Additionally the very new but going to popularity specialized maternal line "Srebrena" were compared with the other breeds. Born, live born piglets, number of piglets aged 21 days; weight of piglets in birth and weight of piglets aged 21 days were included in the study. Danube White sows have better weight of piglets in birth and 21 days then the other breeds, but smaller number of live born and 21 aged pigs in a litter. The "Srebrena" line is in a good competition with the investigated breeds.

The results between the first and second and the next litters of sows were compared. The genetic biodiversity in the population were analyzed.

Genetic correlations between body condition scores and fertility using bivariate random regression models

Y. de Haas and H.N. Kadarmideen, ETH Zurich, Institute of Animal Sciences (UNS D8), CH-8092 Zurich, Switzerland

Body condition score (BCS) was recorded by the Swiss Holstein Association on 22,075 lactating heifers. Fertility data during first lactation were extracted from 40,736 cows. Heritabilities and genetic correlations were estimated using a sire model in ASREML. A bivariate random regression model was developed to estimate genetic correlations between BCS as a longitudinal trait and daughter's fertility at sire level as a lactation-based score. Heritability of BCS was 0.17, and heritabilities for fertility traits were low (0.01-0.08). Random regression analyses showed that genetic correlations between BCS and fertility changed over the lactation; genetic correlations of BCS with fertility traits varied from: -0.45 to -0.14 for days to first service, -0.75 to 0.03 for days between first and last insemination, -0.59 to -0.02 for calving interval, -0.47 to 0.33 for number of inseminations, and 0.08 to 0.82 for conception rate to first insemination. These results show (genetic) interactions between fat reserves and reproduction along the lactation trajectory of modern dairy cows, which can be useful in both genetic selection and management. It is suggested that maximum genetic gain in fertility from indirect selection on BCS should be based on measurements taken in midlactation when the genetic variance for BCS is largest, and the genetic correlations between BCS and fertility is strongest.

Estimation of performance and genetic parameters of reproductive traits in Lori-Bakhtiari sheep using linear and threshold models

M. Vatankhah[1], A. Poortahmasb[2] and H.R. Merzaei[2], [1]Department of Animal Science, Agriculture and Natural Resources Research Center, Shahrekord, [2]Department of Animal Science, Collage of Agriculture, University of Zabul

The performance and genetic parameters of reproductive traits were estimated using linear and threshold univariate animal model on the data of 5374 records of 1696 ewes, collected during 1989 to 2004 in rearing and improvement of Lori-Bakhtiari sheep Station (Sholi) in Shahrekord. The model was included fixed effects (year and age of ewe) and random effects additive genetic of ewe, permanent environmental of ewe and residual. The overall mean (s.d) of traits were as 0.90 (0.30) for conception rate, 1.05 (0.50) for number of lambs born per ewe joined, 1.01 (0.55) for number of lambs born alive per ewe joined, 0.94 (0.55) for number of lambs weaned per ewe joined and 1.05 (0.48) for number of lambs weaned per ewe conceived. The heritability coefficient traits were estimated as 0.01, 0.05, 0.04, 0.02 and 0.06 respectively, resulted from linear analysis and 0.08, 0.10, 0.08, 0.10 and 0.23 respectively, resulted from threshold analysis. The estimation of repeatability coefficient of traits was as 0.10, 0.13, 0.13, 0.09 and 0.09 respectively, for linear analysis and 0.67, 0.56, 0.67, 0.56 and 0.24 respectively for threshold analysis. The results of this study showed that, using threshold models for analyzing reproductive traits in genetic evaluations, comparison of linear models, caused relatively increasing genetic parameters and accuracy of evaluations.

PCR-sexing and genotyping of bovine preimplantation embryos
R. Nainiene, V. Juskiene and J. Kutra, Institute of Animal Science of LVA, R. Zebenkos 12, Baisogala, LT-82317 Radviliskis distr. Lithuania

The objectives of the study were to determine the effects of amplified DNA fragments, magnitude and method of biopsy on the efficiency of sex determination and kappa-casein gene loci amplification of bovine embryos. Embryo sex determination was achieved by two methods: 1) detection of male-specific repetitive Y-chromosomal DNA sequences after PCR amplification, 2) amplification the ZFX/ZFY loci. The kappa-casein gene loci were PCR amplified with primers KCN1 and KCN2. Twenty-seven assays of biopsies were used for sex determination by the first method. The biopsy of 1 to 6 and 7 to 30 blastomeres resulted in sex determination efficiency of 84.6% and 100% respectively. The sex determination efficiency was dependent on the method of biopsy, i.e. biopsy by suction resulted in 75.5% and that by cutting in 100% successful sex determination of embryos. Seventy-three biopsy assays were used for sex determination of embryos by PCR-amplified ZFX/ZFY loci. It was determined that the number of blastomeres had no influence on the efficiency of embryo sexing. The biopsy of 1 to 2 and 5 to 10 cells resulted in sex determination efficiency of 80% and 76.9%, respectively. The method of biopsy had no effect on the formation of the PCR product. After suction and after cutting, sex for 82.5% and 81.8% of embryos has been determined. The kappa-casein gene loci were identified in 20 samples of the analysed 27 (74.1%).

Building and process technique requirements in horse husbandry systems: Investigations and trends in regard to animal welfare and environment protection
F.-J. Bockisch[1], P. Kreimeier[1], G. Hoffmann[1] T. Hohmann[1], W. Bohnet[2] and U. Brehme[3], [1]Institute of Production Engineering and Building Research, Federal Agricultural Research Centre, Braunschweig, [2]Centre for Animal Welfare, University of Veterinary Medicine, Hannover, [3]Institute for Agricultural Engineering Bornim, Potsdam, Germany

This overview deals with designing and performance of different horse husbandry systems and their effects to welfare parameters. Actual investigations at our new research facilities, existing of a group housing system and 6 single boxes, both directly connected with paddocks - together 12 warm blooded mares, which are changed after every project period of 6 month, are focused. For additional locomotion a computer-operated free walking unit is used. Monitored Parameters are: Quantitative behaviour criteria by continuous video monitoring and analyzing, online measuring of heart rates as well as of activity and resting by ALT-pedometers, time and frequency of feed intake by computerized feeding systems, continuous monitoring of climate parameters, analyzing the contamination of water and soil in paddocks, etc. Examples for results: In group housing systems it is possible to reduce the ratio of feeding place to horses to 1:3; there is a worse heart frequency variability, if the time span between every single concentrate feed application is growing; different offers of space and movement-possibilities in paddock-stables are showing, that it is necessary to have more then 45 m^2 per horse; 80 to 90 % of the urinating activities of the horses are taking place within the stable and not on the outside paddock area.

Session H11

Theatre 2

Impact of nutrition and feeding practices on equine behaviour and welfare

P.A. Harris,Equine Studies Group, WALTHAM Centre for Pet Nutrition, Melton Mowbray, Leics LE14 4RT, UK.

In the wild, horses spend most of the day roaming and foraging in an externally variable environment as part of a herd. As non-ruminant herbivores they are well suited to a high fibre, low starch diet. They rarely fast voluntarily for more than 2–4 hrs at a time and would naturally forage for 16-18hrs a day. Modern horse management often brings horses into small-enclosed, isolated environments and limits the feeding occasions. What and when they are able to eat, is now predominantly determined by ourselves and we therefore have to take responsibility for the effects that our choice of managemental practices has on their health as well as their behaviour and welfare. Many managemental practices have the potential to be less than optimal for horse welfare. In addition, many people, in developed countries, have horses in order to ride them or to watch them being ridden. Being naturally fear and flight animals, their behaviour becomes increasingly important when being ridden in environments beyond the rider's control or under competitive circumstances. Currently much of the interest around diet and nutrition therefore is focused on the areas of temperament and tractability. This paper will highlight the impact that diet and the way we feed can have, both positively and negatively, on the behaviour of horses especially the young growing animal.

Session H11

Theatre 3

Can behavioural observations give an evaluation of welfare? A study of time budget and social behaviour of mares in paddock

H. Ben Haj Ali[1], M. Leroux[1], M.A. Richard-Yris[1], M. Ezzaouia[2], F. Charfi[3] and M. Hausberger[1], [1]Rennes I University, 35042 Rennes, France, [2]FNARC, Sidi Thabet, 2020, Tunisia, [3]Tunis University, Tunis, Tunisia*

A herd of 44 Arab mares were observed in a 2500m^2 paddock in summer in Tunisia. Fifteen-minute focal animal samples and scan sampling were used to determine the time budget of the mares during the period from 9am to 3pm and study their social behaviour. A total of 42 hours of observation were performed. Chi square test and spearman correlation were used to analyse data. Locomotion was the most frequent activity with 27, 9 ± 19, 47%. Feeding and stand-resting were the following activities. The mares spent 14, 88 ± 18, 83% of their time alert-standing, 1, 03 ± 3, 51% self-grooming, 4, 75 ± 7, 19% drinking and 0, 83% ± 2, 64% interacting socially. Rolling was observed only once and mares were never been observed allogrooming, lying down, urinating, defecating nor vocalizing. Social behaviour was restricted to agonistic interactions and preferential associations that could be determined were very rare. The data obtained reveals restricted behavioural repertoires, unusual time budgets with a high frequency of active walk that constitutes the most frequent activity and a low level of social interactions. By modifying some management practices, we have tested, in a second part, the effect of some environmental factors on mares' time budgets and social behaviour. These results are discussed in relation to other similar studies on domestic horses and questions of welfare assessment.

Session H11

Theatre 4

Importance of social contact for housing of horses

E. Søndergaard, Danish Institute of Agricultural Sciences, P.O.Box 50, 8830 Tjele, Denmark

Horses are by nature social animals but are often housed singly with limited access to social contact with conspecifics. Previous work suggests that allowing horses to sniff but not touch each other may not fulfil their need for social contact. Other studies have shown that group housed horses are easier to train than single housed horses and show a more appropriate behaviour towards humans. It may be possible, however, to provide horses with sufficient opportunities to fulfil their need for social contact in single housing systems, but this requires more knowledge of which aspects of social contact is most important to the horse. This was investigated in an ongoing operant conditioning experiment aiming to assess the motivation for social contact in horses. Each horse worked individually for rewards of 3 min access to a known companion horse under 3 different types of social contact. These were full contact, head contact or muzzle contact. Preliminary results from this experiment indicate that the demand for social contact is elastic i.e. the higher the price the lower the number of rewards taken by the horses. There was high variation between horses in the number of rewards taken but, so far, no apparent effect of the type of social contact. Results on various demand parameters (elasticity, area under curve etc) will be presented as well as recordings of individual horse behaviour during reward periods.

Session H11

Theatre 5

The behaviour of horses in different paddock sizes, with and without exercise

G.H.M. Jørgensen and K.E. Bøe, Department of Animal and Aquacultural Sciences, Norwegian University of Life Sciences, N-1432 Ås, Norway

In a 2 x 3 factorial experiment with exercise (no exercise / daily exercise) and paddock size (small: 150 m^2, medium: 300 m^2 and large: 450 m^2) as main factors we examined the effect on behaviour of horses. Nine (3 cold blood and 6 warm blood) adult horses were individually exposed to all the three paddock size treatments for one week at a time. A pilot study revealed the same horses' social activity and behaviour when kept together in a large enclosure (2700 m^2).

In the non-exercise treatment period the horses walked significantly more, they travelled a longer distance, explored more and stood more alert in the paddock than in the period with exercise. The horses showed higher activity in the large paddock compared to the medium and the small paddocks), and they also travelled a longer distance. When kept in a social group the nine horses were generally more active, spending less time standing passively and eating more grass than in individual paddocks. Aggressive social behaviours were rarely observed.

Based on the activity in the individual paddocks we conclude that daily turnout is important for horses. The size of the paddocks seems to be of less importance when comparing paddocks sizes with relatively restricted space allowance. Turnout in social groups may be a good alternative to individual paddocks.

Resting behaviour of horses depending on the offer of exercise and climate
C. Gille[1], E. Moors[1], F.-J. Bockisch[2], P. Kreimeier[2] and M. Gauly[1], [1]Institute of Animal Breeding and Genetics, Albrecht Thaer Weg 3, University of Goettingen, 37075 Goettingen, [2]Institute of Production Engineering and Building Research, Federal Agricultural Research Centre, 38116 Braunschweig, Germany

The present study was conducted on 24 Standardbred Hannoverian mares in three different housing systems – each with a outdoor exercise area, investigating the resting behaviour of horses. Mares of group 1 and 2 were kept in group housing systems, additionally mares of group 1 moved in a horse walker 1h/day. Mares of group 3 were kept in single boxes, although with a outdoor exercise area. Resting behaviour, defined as dozing while standing and lying, was analysed by 24h-video recordings.

There were no significant differences between the three groups neither in the overall resting time ($p = 0.278$) nor in the duration of lying ($p = 0.543$). However mares kept singly dozed significantly longer per doze action (16.3 ± 12.5 min, $p \leq 0.05$) than mares of group 1 (14.1 ± 12.4 min) and 2 (15.2 ± 12.4 min). Durations of lying actions were not significantly different between the three groups ($p = 0.704$).

Outside temperature had no significant influence on the mean resting duration of the mares ($p = 0.713$).

Investigations on the activity of trotters and the relationship to growth
L. Voswinkel[1]*, K.-H. Tölle[2], C. Henze[1], K. Blobel[3] and J. Krieter[1], [1]Institute of Animal Breeding and Husbandry, Christian-Albrechts-University, 24118 Kiel, Germany, [2]Chamber of Agriculture Schleswig Holstein, 24327 Blekendorf, Germany, [3]HIPPO-Blobel, 23812 Ahrensburg, Germany

The aim of this study was to determine factors affecting the activity of juvenile horses as well as to analyse the relationship between activity and growth parameters. Data of 60 horses of one stud were available. After weaning two foals were housed in one loose-box (pair) with daily exercise. Pedometers and videotaping were used to measure the activity. Data were recorded 15 times for periods of two weeks ($n=210$ per animal). Climatic data were taken each test day. For intervals of ten weeks, the development in height at withers, cannon bone diameter and circumference, length of foreleg, weight, and diameter of flexortendons were recorded. Data were analysed using a linear mixed model with fixed (temperature, wind speed) and random effects (pair, animal, test day). Higher temperature increased the activity values ($p<0.001$) and a stronger wind speed tended to decrease the activity. Individual horses showed different activities despite the same environment. Over the investigation period the individual activity remained constant in comparison to the average of the herd. Pairs of foals showed a more similar activity pattern ($p<0.001$). Positive phenotypic correlations were found between activity and diameter of superficial digital flexortendon ($r_p=0.24$) and deep digital flexortendon ($r_p=0.27$).

Three-dimensional design of a horse stud like better toll for technical choices of housing and welfare

N. Miraglia and A. Simoni, Molise University, Dept. Animals, Vegetables and Environmental Sciences, Via De Sanctis, 86100 Campobasso, Italy

It is well known that captivity and the following upsetting of the horse natural habits and physiological activities have made the horse defenceless towards the numerous dangers and have exposed it to a wide range of risks. As a consequence, the safety of the infrastructures where the athlete spends the majority of its life time represent a fundamental aspect strictly linked to the prevention modalities of many pathologies found within equestrian veterinary practice. In recent years the three-dimensional designs are more and more utilised to have a complete vision of the structures. In the case of horse studs, the three-dimensional design represents the best solution to verify, in virtual imagines, the technical choices not only for housing but also for improving welfare conditions. In fact it is possible to hypothesize and to virtually test the operative conditions needed both in terms of structures and welfare. The result of this kind of representation is optimal to research the best welfare conditions. This study refers to a virtual three-dimensional design of an equestrian centre destined to show jumping horses. This facility includes the realization of different structures such as studs and wasting areas, covered round, open air field, open field, rings, club services, competition field with tribunes, jury, services for public, etc.

Drinking behaviour of horses depending on the type of trough, concentrate feeding and climate condition

B. Niemann[1], E. Moors[1], F.-J. Bockisch[2], P. Kreimeier[2] and M. Gauly[1], [1]Institute of Animal Breeding and Genetics, University of Goettingen, 37075 Goettingen, [2]Institute of Production Engineering and Building Research, Federal Agricultural Research Centre, 38116 Braunschweig, Germany

Water demand of horses is influenced by many factors, i.e. dry matter intake, exercise, age, breed, lactation and climate. In this study the impact of drinker systems (low pressure valve vs. float valve), the frequency of concentrate feeding (3 vs 10 – total amount 1.5 kg) and climate (temperature) on drinking behaviour of 24 two years old Hannoverian mares was evaluated. Frequencies and total time of water intake were recorded. Each animal had both drinker systems available. Concentrate was given in different frequency to each animal for a special period of time. No significant differences were observed in both parameters between the watering systems. But in tendency the animals preferred the float valve system. No significant differences in the average drinking frequencies were found between animals fed three times (1.72 ± 1.01) or ten times a day (1.67 ± 1.17). However the temperature had a significant impact on the drinking duration and frequency. On days with average temperatures below 0 °C animals visited the troughs 1.32 ($\pm$ 0.7) times, whereas on temperatures between 0 and 14 °C the frequency was 1.55 ($\pm$ 0.97), between 15 and 24 °C 1.90 ($\pm$ 1.22) and > 24 °C 2.5 ($\pm$ 1.49).

Session Ph12

Theatre 1

Maximising piglet survival

J. Le Dividich[1], Y. Le Cozler[2] and D. Causeur[2], [1]INRA UMR-SENAH, 35590 St-Gilles, France, [2]Agrocampus ENSA-INSA, 35000 Rennes, France

Currently in the EU one out of five piglets born (i.e., ± 20%) is stillborn or does not survive to weaning at about 3-5 weeks of age. After a brief survey of peri-natal mortality, this paper provides new insights on the importance of early nutrition (colostrum) for piglet's survival. Emphasis is given to the nutritional and immunological roles of colostrum. Colostrum production of the sow, consumption of the piglet and factors of variation are presented. Colostrum production is a characteristic of the sow. Within the litter, colostrum consumption of the piglet is independent on birth order, but positively and negatively related to birth weight and litter size, respectively. Other factors, including cold stress and hypoxia on colostrum consumption, are also examined. Colostrum IgG concentrations vary widely between individual sows both in initial concentration and in the rate of decline after the onset of parturition. Main factors controlling the acquisition of maternal antibodies in the piglet, including amount of colostrum intake, IgG concentration of colostrum and birth order are discussed. Finally, the effect of colostrum consumption on neonatal survival is discussed. Colostrum consumption in amounts sufficient to meet the energy requirement of the piglet is a major determinant for neonatal survival while the acquisition of a high immune protection is important in later resistance to diseases. Colostrum production appears to be a good marker for the maternal quality.

Session Ph12

Theatre 2

Regulation of growth and development of the gastrointestinal tract and adipose tissue from birth to weaning in pigs: Influence of birth weight

A. Morise, I. Louveau and I. Le Huërou-Luron, INRA/Agrocampus Rennes, UMR Systèmes d'Elevage, Nutrition Animale et Humaine, 35590 Saint Gilles, France

Pig birth weight is an important risk factor for preweaning mortality. Low birth weight pigs exhibit lower postnatal growth rates and feed efficiency, which may be explained by an inadequate digestion and/or nutrient use as a consequence of prenatal under-nutrition. It is now documented that factors like under-nutrition during prenatal development may influence the development of organs, tissues and the endocrine system and the long-term physiology of the individual. During the neonatal period, the rapid somatic growth is accompanied by tremendous anatomical, physiological and chemical composition changes. The present review focuses primarily on the gastrointestinal tract and adipose tissue during the suckling and around weaning periods in relation to birth weight. Hormone levels in plasma as well as receptors expression in target tissues are also considered. The ability of the neonate to express its growth potential is related to milk intake and milk quality. The relationship between nutrition, endocrine parameters and growth is presented.

Commensal microbial influences on intestinal health and development
A.G. Van Kessel and B.P. Willing, Department of Animal and Poultry Science, University of Saskatchewan, Saskatoon, SK, Canada, S7N 5A8*

Microbial colonization of the gastrointestinal tract occurs rapidly after birth, increases in complexity with age and is affected in composition by various environmental factors including diet. To investigate the effect of microbial colonization and succession in the neonate we established a germ-free (GF) pig model, which allowed control over gastrointestinal colonization. Initially, we compared intestinal responses in GF and conventionalized (CV) pigs versus pigs monoassociated with one of two model early-colonizing commensal bacteria; namely, Gram-positive *Lactobacillus fermentum* and Gram-negative non pathogenic *Escherichia coli*. In the proximal small intestine (SI), morphology was remarkably similar among all 4 groups suggesting that diet constituents and/or endogenous intestinal secretions contribute significantly to proximal intestinal responses. In the distal SI, GF and *L. fermentum*-associated pigs showed an acellular *lamina propria*, shallow crypts and villi 3-4 fold longer than CV pigs. *E. coli*-associated pigs presented a morphology intermediate between GF and CV pigs. Intestinal morphological responses were associated with significant changes in local gene and protein expression including toll-like receptors, proinflammatory cytokines (IL-1, 1L-6, TNFalpha) and indicators of apoptotic activity (TNFalpha, Fas ligand, active caspase-3) and cell proliferation (proliferating cell nuclear antigen). Functionally, brush border enzyme activity was lower in CV pigs despite increased enzyme gene expression. We conclude that the pattern of early microbial succession may significantly affect postnatal intestinal development.

Porcine muscle satellite cell growth in response to estrogens and isoflavones
M. Mau[1], C. Kalbe[1], T. Viergutz[2], K. Wollenhaupt[2] and C. Rehfeldt[1], Research Institute for the Biology of Farm Animals (FBN), [1]Muscle Biology and Growth Research Unit, [2]Reproductive Biology Research Unit, Wilhelm-Stahl-Allee 2, D-18196 Dummerstorf, Germany*

The study aimed at the *in vitro* effects of estrogens and isoflavones on porcine muscle cell growth. The receptors for estrogens (ERα and ERß) and epidermal growth factor (EGF) as possible targets of isoflavones were shown to be expressed in satellite cells, derived from *semimembranosus* muscle of newborn piglets. The effects of various concentrations of 17ß-estradiol (E2), estrone (E1), genistein, and daidzein on DNA synthesis were measured as 6 h-[^{3}H]thymidine incorporation. After 7 h exposure, E2 (1nM; 1µM), E1 (1nM; 1µM), genistein (1; 10µM) and daidzein (1; 10; 100µM) slightly decreased, while 100µM genistein substantially lowered DNA synthesis. Declines in cell number (DNA) were observed with genistein (0.1; 1; 100µM) and daidzein (0.1; 100µM). After 26 h exposure, 100µM genistein caused a cell cycle arrest in G_2/M and S phase. Additionally, DNA synthesis was markedly reduced by 100µM genistein, whereas it was significantly increased by 10µM genistein, 10 and 100µM daidzein. Cell number decreased with 10 and 100µM genistein and 100µM daidzein. In part, this was associated with elevated lactate-dehydrogenase (LDH) activity in cell culture supernatants, indicative of cell death. The results suggest that both estrogens and isoflavones may directly affect porcine muscle cell growth in a dose- and time-dependent manner.

Evaluation of the influence of dietary fat content and fatty acid composition in four diets based on different fat sources on loins (*M. longissimus dorsi*) of newborn piglets

N. Panella[1]* and J. Pickova[2], [1]IRTA-Centre de Tecnologia de la Carn, 17121 Monells, Spain, [2]Department of Food Science, Swedish University of Agricultural Sciences, PO Box 7051, S-750 07 Uppsala, Sweden

The aim of this study was to explore if a change in dietary fatty acid content in sow feed had impact on the loin (*M. longissimus dorsi*) fatty acid (FA) composition of newborn piglets. Four cereal based diets were used: a conventional low fat (3%) sow diet (LF) and three high fat (6%) diets: Saturated diet (HFSat), Omega-6 diet (40% high fat oats, HFn-6) and Omega-3 diet (20% high fat oats and Linseed oil, HFn-3). Loins from newborn piglets (four per treatment) coming from 39 selected litters (mean parity=3.9) were used to analyse FA composition in triacylglycerols (Tg) and phospholipids (Pl) fractions by GC. Statistical analysis was performed using proc GLM (SAS v. 8.02). Both Tg and Pl fractions presented significant changes among treatments. In Tg fraction, HFn-3 group showed a higher C18:3 n-3 content (p<0.001) than the other groups, and the same results were found for C20:5 n-3 (p<0.001), C22:5 n-3 (p<0.001) and C22:6 n-3. Similarly, in Pl fraction, HFn-3 diet showed a higher content of n-3 (p<0.001) and a lower n-6/n-3 ratio (p<0.01) than the other diets. The results indicated than loins from newborn piglets were affected by sow diet.

Choice of a probiotic Bifidobacterium and of the best dietary dose by a selection protocol for the highest ability to multiply in the gut of weaning pigs

S. De Filippi, M. Modesto, P. Trevisi, L. Casini, M. Mazzoni, B. Biavati and P. Bosi, University of Bologna, Bologna, Italy

A protocol was set up to select the Bifidobacterium strain with the best ability to develop in the pig intestine, and a prebiotic with possible stimulating effect on Bifidobacteria growth. Then the best probiotic dose x prebiotic was investigated to improve weaning pig growth. In a series of 2 trials, 12 bifidobacteria were tested on 60 weaning pigs. Two strains of *Bifidobacterium animalis* and one of *Bifidobacterium choerinum* showed the highest number of bifidobacteria in cecum. Then 3 prebiotics (Fructo-oligosaccharides (FOS) from chicory inulin or from sugar beet; Galacto-oligosaccharides from milk whey) were tested on 60 pigets with 2 doses (1% and 4% of the diet). A trend for increased number of bifidobacteria was observed for FOS from sugar beet compared with the other supplements. One of the two *Bifidobacterium animalis* and the *Bifidobacterium choerinum* were finally tested at different doses (0; 10^7; 10^9; 10^{11} per pig per day) crossed with 0% or 2% sugar beet FOS in 128 piglets. The first bifidobacteria improved linearly the live weight growth, while the second did not. FOS supplementation did not interact with the probiotics for growth performance, nor affected it directly, except for an increase of bifidobacteria count in cecum contents in the 1st trial.

Session M13

Welfare aspects of dairy cows in Western European countries in the context of zero grazing systems

H. Hopster[1]*, C.J.C. Phillips[2], U. Knierim[3], S. Waiblinger[4], L. Lidfors[5], C.C. Krohn[6], E. Canali[7], H. Valk[1], I. Veissier[8] and B. Beerda[1], [1]Animal Sciences Group of Wageningen UR, Animal Production Division, PO-Box 65, 8200 AB, Lelystad, The Netherlands, [2]University of Queensland, Australia, [3]University of Kassel, Germany, [4]University of Veterinary Medicine, Austria, [5]Swedish University of Agricultural Sciences, Sweden, [6]Danish Institute of Agricultural Sciences, Denmark, [7]Universita degli Studi di Milano, Italy, [8]INRA URH-ACS Centre de Clermont-Ferrand, France

In some areas of Europe, "zero-grazing" of dairy cows increases, due to the ability to lower labour inputs, to better control nutritional inputs and to increase the dairy output potential of the cows. This silent intensification may have irreversible consequences for the welfare of dairy cows, and in particular for their behaviour and health. In a literature study these issues are explored in relation to the different regions of Western Europe. The reasons for practising zero-grazing may vary with the different types of farming. Access to pasture is associated with low levels of disease, e.g. reduced mastitis and lameness, and it facilitates the cows to express their natural behaviour. In that sense, grazing is also an important quality aspect of dairying. Despite that many agree on the positive aspects of grazing, a growing number of farms with larger and more productive dairy herds are likely to drastically reduce grazing across Europe. Possible strategies to maintain a high level of dairy cow welfare in a competetive market are discussed.

Session M13

Dairy cows with access to both indoor and outdoor environments - experiences from automatically milked cows on pasture in temperate regions

E. Spörndly, Swedish University of Agricultural Sciences, Department of Animal Nutrition and Management, Kungsängen Research Centre, SE-753 23 Uppsala, Sweden

Pasture based milk production offers many advantages such as high quality feed at a low cost and the opportunity for the cows to exercise, graze and move more freely in a paddock outdoors. However, large variations in herbage mass and quality can lead to an uneven nutrient supply and high temperatures and other unfavourable environmental conditions can sometimes occur which may be stressful for the animals. A model with combined grazing and indoor feeding has in some cases proved to be advantageous.

Experiments have shown that automatic milking (AM) can successfully be combined with grazing. During grazing periods cows are offered a choice between indoor and outdoor environments. Various factors affect that choice and farmers with AM face the new management challenge of stimulating the cows to move regularly between the barn and the pasture area throughout the grazing season. Grazing experiments have been performed to study how factors such as of distance between barn and pasture, location of drinking water and amount of supplements offered in the barn affect cow traffic and milk yield. Results have in some cases differed substantially depending on conditions, thus illustrating the importance of a detailed analysis of the indoor and outdoor environments actually offered to the cows in each situation.

Session M13

Theatre 3

Dairy cows behaviour under free stall and no stall fully roofed housing systems
E. Maltz, A. Antler and N.Livshin, A.R.O. The Volcani Center, P.O. Box 6, Bet dagan, 50250, Israel

Rest and activity are fundamental and complementary indices of animal behaviour. Behaviour of individual dairy cows can indicate animal's comfort under different housing conditions. We studie diurnal behaviour of dairy cows under normal commercial management routine, and compare thed behaviour under two housing systems.

A leg-mounted sensor monitored lying time and a commercial sensor (S.A.E. Afikim®) activity. Data were downloaded during milking times.

In one trial 12 multiparous cows under comfortable thermal conditions in a roofed no-stalls barn, lay for 8.8 ± 1.6 h per day. Lying periods ranged from 3.7 ± 1.3 h, between 20:00 and 05:00, and 2.3 ± 0.8 h, between 13:00 and 20:00. In a second trial, 8 first-calving cows were monitored in each of two adjacent completely roofed barns: one no-stall and the other free-stall; the third trial repeated the second trial, with 4 cows from each group interchanged. In both trials, cows of both groups demonstrated diurnal lying patterns similar to that of trial 1, except that those in the no-stall barn lay for 2 h more then those in the free-stall barn. The free-stall cows were more active; there was a significant negative correlation between activity and lying time in the free-stall barn, and no correlation in the no-stall barn.

A sensor that monitors behaviour parameters such as activity, lying time, standing time and lying bouts was developed and can be used to assess animal comfort and suitability of housing conditions and management routines.

Session M13

Theatre 4

High-tech and Low cost farming: An indoor versus outdoor farming system
M.H.A. de Haan, C.J. Hollander, S. Bokma and Kees de Koning, Animal Sciences Group of Wageningen University and Research Centre, P.O. Box 65, 8200 AB Lelystad, The Netherlands

The Low cost farm and High-tech farm are two experimental farms in The Netherlands established to study the perspectives of these farming systems. The main objective for both farms is to realize a cost price of € 0,34 or less per kg milk.

The High-tech farm is an indoor housing system with: 75 high yielding Holstein Friesian dairy cows, an automatic milking system, an automatic feeding system and 800.000 kg milk quota. Special attention has been paid to barn design with respect to ventilation, climate, animal well being and animal health. The low cost farm is a typical Dutch outdoor system with milk quota and herd size equal to the Dutch average (about 450.000 kg quota). Grazing is essential in this outdoor system. Montbeliarde and "low producing" HF-cows are kept on this farm. The milk yield is 7800 kg, in spite of low concentrate feeding (50% lower than Dutch average). Both farms are managed by a herdsman in less than 50 working hours per week, which is 40% less than comparable farms.

Results of these two farms representing an indoor and an outdoor system will be presented and discussed. Special attention will be paid to housing conditions and management strategies including grassland, feeding and animal health management and control.

Environmental management of feedlot cattle

F.M. Mitloehner[1]*, J.L. Morrow[2], M.L. Galyean[3], J. Stockstill[3] and J.J. McGlone[3], [1]Animal Science, University of California, Davis;USA, [2]USDA-ARS Lubbock; USA, [3]Animal Science, Texas Tech University, Lubbock, USA

Effects of shade, sprinklers, misters, and hormone implant strategies on heat stress in feedlot cattle were investigated. In the first two studies, water (sprinkling and misting) was largely ineffective, but shade reduced heat stress. Effects of shade on feedlot heifer performance, physiology, and carcass traits were tested in a third study. Six pens were shaded (SHADE) and six pens the unshaded control (CONT). Heifers in SHADE compared with CONT had higher DMI (P < 0.01) and ADG (P < 0.05). Carcass traits were similar between treatments, but more carcasses of heifers in SHADE graded Choice, (P < 0.01) and the incidence of dark cutters (P < 0.05) was decreased in SHADE. Respiration rates were decreased among SHADE compared with CONT heifers (P < 0.01). Most behaviors were not different between treatments, but cattle in SHADE showed less agonistic behavior than CONT heifers (P < 0.05). Finally, in the fourth study, we studied the effects of shade and implant strategy during heat stress conditions. Results suggest that feedlot cattle performance was maximized when shade was provided to heifers that received an implant at initial processing; however, shade did not benefit heifers implanted after 56 d on feed. In conclusion, heat stress in feedlot cattle can be minimized by use of proper pen design, housing, and management.

Characteristics of barn design for dairy cows under harsh conditions derived by a new stress model

E. Shoshani[1]*, A. Hetzroni[2], A. Levi[2] and R. Brikman[2], [1]Extension Service, P.O. Box 28 Bet dagan 50250, Israel; [2]A.R.O. The Volcani Center, P.O. Box 6, Bet dgan, 50250, Israel

Careful barn planning is required to minimize heat stress conditons. During summer of 2004 and 2005, ambient conditions in 39 barns were recorded. Two meteorological stations were installed inside the barn, downstream the dominant wind and a third- out of the barn. Ambient temperature (AT), relative humidity (RH), wind velocity and wind direction were measured and recorded every 10 min 3 to 5 days during summer. A heat-stress model simulating the threshold temperature (TT) in which a cow begins to increase her respiratory rate (RR) was used. This model takes into account wind velocity, RH, AT and physiological characters of a cow: milk yield 45 Kg, 3.5% fat and 3 mm fur depth. SAS-GLM procedure was used for statistical analyses. TT was used as the dependent variable. Independent variables were: orientation, barn type, roof slope, roof ridge, margin heights, roof type (close vs. mobile), and barn width. Results show that optimal barn for high milking cows is loosing house (vs. free stalls), oriented perpendicular to the dominant wind, with opened roof, opened ridge, with roof margins of around 5 m., roof slope 19%-22%, and width between 30 to 51 m. Barn orientation parallel to dominant wind requires higher roof margins and narrower width.

Seasonal performance of different breeds of feedlot beef cattle grown under the Mediterranean conditions

Y. Bozkurt, Suleyman Demirel University, Faculty of Agriculture, Department of Animal Science, Isparta, Turkey, 32260

In this study, data from Holstein (11), Brown Swiss (27), Simmental (8) cattle as European type (ET) and Boz (12) and Gak (48) as Indigenous type (IT) grown under feedlot conditions were used to evaluate and compare performance differences in the Mediterranean type of climate, covering summer, autumn and winter seasons. Initial average weights of cattle were 202, 194, 210, 203 and 220 kg for Holstein, Brown Swiss, Simmental, Boz and Gak repsectively. There were statistically significant (P< 0.05) differences in daily live weight gains (DLWG) of both type of cattle. ET cattle were performed better than IT cattle for all seasons. There were no statistically significant differences in performance between Holsteins, Brown Swiss and Simmental cattle and between Boz and Gak cattle themselves. However, Simmentals tended to perform better than the rest for all seasons, following Holsteins, Brown Swiss, Boz and Gak respectively. There was no significant (P> 0.05) interaction between seasons and breed types. Overall DLWGs of animals in winter (0.80 kg/day) was statistically higher (P< 0.05) than those of both summer and autumn (0.68 and 072 kg/day respectively) which was not statistically significant. The results showed that under the Mediterranean conditions the ET cattle were better suited to the feedlot beef systems than IT cattle. The higher overall performance of cattle in winter indicated that animals may suffer from heat stress during summer, causing a decrease in performance in the Mediterranean conditions.

Effect of early weaning and calf supplementation on cow and calf performance in dry mountain areas

M. Blanco*, A. Sanz, A. Bernués, R. Revilla and I. Casasús. CITA-Aragón. Apdo 727. 50080 Zaragoza. Spain

In order to evaluate strategies for maximising feed resources use in dry mountain areas, 27 cow-calf pairs were distributed in four managements: early weaning (d90) with (EWS) or without calf supplementation during lactation (EWNS), traditional weaning (d150) supplemented (TWS) or non-supplemented (TWNS). After weaning calves received an intensive diet until slaughter (450 kg). EW cows were turned out to forest pastures from d90 to d150. Thereafter, they grazed high mountain pastures with their TW counterparts. Cows only presented different weight gains between d90 and d150, with EW cows losing weight and TW maintaining it. In the same period, TWS calves (1.442 kg/d) had the highest growth rates followed by EWS (1.287 kg/d) and EWNS (1.154 kg/d), and TWNS had the lowest (0.794 kg/d). Besides, EWS calves growth differed from the other calves from d150 to slaughter, showing the lowest ADG. Despite growth differences, slaughter age was similar. EW calves had higher concentrate consumption, fatter carcasses and tended to present higher carcass yield. TWNS calves had the worst conformation and lowest income. EW calves and TW cows had the higher feed costs. Thus TWNS cow-calf pair reported the lowest extra benefit (289€), followed by TWS (323€) and EWS (335€) and the highest EWNS (361€). Moreover, EW cows had shorter calving-interval than TW cows (374 vs. 403 days, respectively).

Effect of rearing systems on growth performance and carcass composition of Podolian young bulls

A. Braghieri[1], C. Pacelli[1], A. Sevi[2], A Girolami[1], A. Muscio[2], R. Marino[2], [1]Dipartimento di Scienze delle Produzioni Animali, Università degli Studi della Basilicata. Via dell'Ateneo Lucano, 10 - 85100 Potenza, Italy, [2]Dipartimento PRIME, Università degli Studi di Foggia. Via Napoli, 25 - 71100 Foggia, Italy

Twenty-four Podolian young bulls (305 ± 13.43 kg of body weight) were grouped according to different production systems: indoor (IND); outdoor, at pasture, with lower (12% D.M; PLP) or higher (16% D.M.; PHP) protein supplementation. Average daily gains, feed conversion index and final body weight at about 18 months of age showed similar values (469.50 ± 27.58 kg, 464.25 ± 27.58 kg and 470.75 ± 27.58 kg for IND, PLP and PHP groups, respectively). No significant differences emerged for carcass composition and dressing percentage ($56.18\pm 0.67\%$, $56.787\pm 0.67\%$, $57.09\pm 0.67\%$ for IND, PLP and PHP groups respectively). Carcasses were graded R for muscular conformation and 2+, 2- and 2+ (for IND, PLP and PHP groups, respectively) for fatness. Plasma leptin content, predictor of carcass composition, was not affected by different production system. On the economic point of view, the absence of differences between the two rearing systems leads to prefer extensive practices, with the lower protein integration. This is particularly appropriate for this indigenous breed because it imply an animal welfare friendly production and the complete safety of products as well as the acquisition of peculiarities closely related to the typical rearing environment.

Dispersion of large beef cattle herd in free-range system of grazing in the National Park "Warta Mouth"

P. Nowakowski[1], A. Dobicki[1], S. Coimbra Ribeiro[1], K. Wypychowski[2], [1]Agricultural University of Wroclaw, Chelmonskiego 38b, 51-630 Wroclaw, Poland, [2]National Park „ Warta Mouth", Chyrzyno 1, 69-113 Gorzyca, Poland*

Beef cattle herd (n = 725; Hereford, Limousine, Simental, Charolaise, Salers and crossbreeds) as cows (C) with calves (Cf), heifers (H) and bulls (B) were kept all year round on pastures. Observations were taken 3 times a day for 3 hours: after sunrise, in midday and before sunset in May – July 2005. The subgroup structure and distance between animals were estimated and area occupied per 1 animal (Animal Unit = AU) was calculated. Within the sub-group cattle were grazing in distances less than 50 m apart. Between subgroups distances were exceeding 200 m. There were 4 categories of subgroups: C+Cf (animals per group: average $\pm$ sd = 34.0 ± 27.5), C+Cf +H (50.6 ± 35.9), C+Cf +B (45.5 ± 26.2) and C+Cf+H+B (90.0 ± 35.4 AU). The structure and number of animals in subgroups were changing in time. Dispersion of animals was related to the functional region of pasture: the largest was connected with open pasture (grazing – 1283 m^2/AU; standing – 1332 m^2 and lying – 859 m^2/AU), the moderate dispersion was observed during movements (921 m^2/AU) and the lowest next to watering places (343 m^2/AU). The clear-cut picture of dispersion and sub-grouping was observed on open pasture, while next to watering places it was the least clear.

Behaviour of beef cattle herd on free range of National Park "Warta Mouth" in relation to weather pattern

A. Dobicki[1]*, P. Nowakowski[1], S. Coimbra Ribeiro[1], R. Mordak[1], J. Glowacki[2], [1]Agricultural University of Wroclaw, Chelmonskiego 38b, 51-630 Wroclaw, Poland, [2]National Park „Warta Mouth", Chyrzyno 1, 69-113 Gorzyca, Poland

Beef cattle herd consisting of cows with calves (340 dam-offspring pairs), single cows and heifers (27 animals) and bulls (18) were kept on natural pastures. Observations were taken 3 times a day: after sunrise, in midday and before sunset in May - July of 2005. There were 150 3-hour sessions performed, where type of weather (fine, variable, rainy), region of pasture (open pasture, next to water-course, trees and shrubs) and forms of behaviour (grazing, laying, standing, moving, drinking) were recorded. Cattle most of the time spent on open pasture when it was raining (75.0% of total time). When weather was variable herd stayed on open pasture 301.2 minutes = 55.8%. During fine weather herd spent 42.4% of time next to watering sites. It was a significantly longer stay next to watercourse during fine weather when compared to time spent there during variable weather conditions (150.1 minutes) and rainy weather (84.4 minutes = 12.5% of time). Time periods cattle used to stay next to trees and shrubs during fine and rainy weather were similar (consequently 12.0% and 12.5% of total time) while during variable weather this stay was longer (16.4%). Weather conditions significantly influence behaviour of cattle in free-range conditions in temperate climate.

Awassi sheep as a genetic resource and efforts for their genetic improvement

Salah Galal[1]*, Oktay Gürsoy[2] and Ihab Shaat[3], [1]Animal Production Department, Faculty of Agriculture, Ain Shams University, P.O. Box 68, Hadaeq Shubra 11241, Cairo, Egypt, [2]Department of Animal Sciences, University of Çukurova, Faculty of Agriculture, Adana, Turkey, [3]Animal Production Research Institute, Nadi Elsaid Street, Dokki, Cairo, Egypt

Awassi is the most widespread sheep breed of non-European origin. The breed adapts to a wide range of environmental conditions from the steppe to the highly intensive system. Performance of the breed varies according to production environment and strain, the Israeli Improved Awassi being the heaviest and producing the highest amount of milk. Efforts to genetically improve milk production yielded positive results. In Israel the phenotypic average of lactation milk production increased from 297 kg in the 1940's to over 500 kg in the 1990's, while in Syria a selection program succeeded to increase it by 13% in eight years. In Turkey, the mean milked yield of ewes increased from 67 to 152 kg in a selection/outcrossing program that lasted for seven years. Although Awassi is best known for its high milk production, it is often used as a triple purpose sheep in most of the countries of its origin in the Middle East. Heritabilty estimates in different traits are within those for other populations of sheep but recent estimates for milk yield in the Improved Awassi indicated lower heritabilty and higher contribution of non-additive genetic effects.

Session S14

The Awassi and Assaf breeds in Spain and Portugal

L.F de la Fuente[1], D. Gabiña[2]*, N. Carolino[3] and E. Ugarte[4], [1]Producción Animal I, Universidad de Leon, 24071 Leon, Spain, [2]Mediterranean Agronomic Institute of Zaragoza, Apartado 202, 50080 Zaragoza, Spain, [3]Estação Zootecnica Nacional, 2005-048 Vale de Santarém (Portugal), [4]NEIKER, 01080 Vitoria-Gasteiz, Spain

The Awassi was introduced into Spain in 1971, reaching a population size of between 150,000 and 200,000 practical purebreds, besides a large number of crossbreds. Today the population has greatly decreased to 6,000 purebred ewes and a further 80,000 ewes crossed with Assaf, Churra or Castellana. Production is high (approximately 300 l of milk per lactation) however it is being substituted by the Assaf breed.

The Assaf was introduced in Spain in 1991 and has achieved great success with 600,000-700,000 ewes more than 80 % pure and approximately 500,000 crossed ewes. The reasons for success lie in a higher milk production than local breeds (278 l vs. 127 l in the Churra), although it is not so rich in fat and protein. It's prolificacy size is good, it is well adapted to intensive systems and it is easy to obtain purebred animals.

The Assaf was first introduced in Portugal in 1991 and nowadays there are around 15,000 purebred Assaf and some 15,000 F_1 Assaf crossed with local breeds. Average milk yield is 359 l in 220 days of lactation, with milk of 7.2% fat and 5.5% protein.

Session S14

Exploitation of Awassi sheep breed in the Central-, Eastern and South European countries

S. Kukovics[1]*, D. Dimov[2], K. Kume[3] and N. Pacinovski[4], [1]Research Institute for Animal Breeding and Nutrition, Gesztenyés u. 1. Herceghalom, 2053, Hungary, [2]Agricultural University of Plovdiv, 12 Mendeleev str, Plovdiv, 4000, Bulgaria, [3]Institut de Recherches Zootechniques, Rr; "Abdyl Frasheri" Pall 3/3 Ap. 5, Tirana, Albania, [4]Institute for Animal Sciences, str. Ile Ilievcki 92A, Skopje 1000, Macedonia

The Awassi sheep arrived to Central-, Eastern and South European countries in more well distinguishable waves. The first one was in 1969-70 when 450 lambs were imported to Former Yugoslavia, and some 2,000 female lambs to Macedonia, separately. Romania imported in 1973 80 female and 20 ram lams, as next one. In Bulgaria the Awassi sheep appeared in 1977-79 importing 1,786 lambs. Between 1989 and 1998 in five separate importation 540 females and 34 rams arrived to Hungary. From Hungary tow other countries (Albania, Greece) imported some Awassi breeding stock during the last couple of years.

The imported sheep started to be the bases of purebred Awassi breeding and they were also used in crossbreeding to increase the milk production ability of the local milking sheep breeds. State and cooperative farms were dealing with Awassi breeding and crossbreeding.

In order to summarise the knowledge about the Awassi sheep in the region survey was made including number of animals imported, the history of using this breed in the given countries, and the present of the breed (number of animals, milk and meat production, reproduction, price of products) in the countries.

Comparison of East Friesian and Lacaune dairy sheep breeds in the USA

J. Casellas[1], D.L. Thomas[2]* and Y.M.Berger[3], [1]Departament de Ciència Animal i dels Aliments, Universitat Autònoma de Barcelona, 08193 Bellaterra, Spain, [2]Department of Animal Sciences, [3]Spooner Agricultural Research Station, [2,3]University of Wisconsin-Madison, 53706 Madison, Wisconsin, USA

A total of 2,554 lactation records of 1,068 ewes with variable contributions from East Friesian and Lacaune dairy breeds and the meat-type breeds of Dorset and seven others (e.g. Rambouillet, Polypay, Targhee) were analyzed to estimate breed differences for milk, fat and protein yield, fat and protein percentage, somatic cell count, lactation length and litter size. Data were collected at the Spooner Agricultural Research Station between 1996 and 2005. The analysis was performed with a multivariate animal model solved through Bayesian methodologies, and breed differences were modeled following the standard approach of genetic groups. East Friesian and Lacaune breeds were not significantly different for milk, fat and protein yield, whereas the Lacaune breed, relative to the East Friesian, had increased ($P < 0.05$) fat percentage (+0.16 %), protein percentage (+0.14 %) and log somatic cell count (+0.19), and reduced ($P < 0.05$) lactation length (-14.4 days) and litter size (-0.13 lambs). Relative to the East Friesian breed, the Dorset breed had reduced ($P < 0.05$) milk (-113.6 kg), fat (-3.8 kg) and protein yield (-4.5 kg) and lactation length (-43.8 days) and increased ($P < 0.05$) fat (+1.08 %) and protein percentage (+0.65 %). Awassi sheep are currently not available in North America.

Structure and performance of Awassi and Assaf dairy sheep farms in the NW of Spain

G. Caja[1], M.J. Milán[1], R. González[2] and A.M. Fernández[2], [1]Grup de Recerca en Remugants, Universitat Autònoma de Barcelona, Bellaterra, Spain, [2]Cargill-España, Benavente, Zamora, Spain

Awassi and Assaf (East Friesian × Awassi) dairy sheep totalize near 900,000 sheep in the NW of Spain. Data of 30 farms (30% Awassi and crossbred; 70% Assaf), in León, Valladolid and Zamora (Castilla-León, Spain), were used to study their current structure and performances. On average, they had 74.5 ha (pastures, 40%; cereals, 33%; forages, 22%; other, 5%), 2.19 annual work units (familiar, 92%; hired, 8%), 489 ewes, and yielded 157.500 L/yr. Farmers were tenant (93%), young (<45 yr, 77%), had new houses, and were grouped in cooperatives (73%). All farms had modern loose stalls, machine milking (100%), and 97% did planned matting (summer to fall; hormonal treatments, 55%; artificial insemination, 17%). Annual sales averaged 322 L/ewe (fat, 6.52%; protein, 5.30%; SCC, 524×10^3) and 1.38 lambs/ewe ('lechal'; 10.7 kg/lamb). Artificial rearing was done in 60% of farms (automatic, 83%; manual, 17%). Total mixed rations were used in 30% of farms, and the rest used forage (dehydrated, 83%; hay, 77%; fresh, 37%; silage, 17%) and separated concentrate (482 kg/ewe). Half of the farms bought >50% forages, and 93% of them bought >50% concentrates. Annual income was 301 €/ewe (milk, 83%; lambs, 17%), being 139 €/ewe the feeds bought (concentrates, 76%; forages, 24%). Estimated Annual gross margin, including the European subsidy, was 155 €/ewe.

Awassi sheep production and the development of breeding program options in Syria
M.Y. Amin, K. J. Peters, Humboldt University Berlin, Dept. of Animal Breeding in the Tropics and Subtropics, Germany

The fat-tailed Awassi sheep is the only sheep breed in Syria adapted to the harsh environmental conditions and it is distributed all over the country. There are about 13.5 million heads of Awassi sheep in Syria contributing 78%, 30% and 100% of the total red meat, milk and wool production, respectively (FAO, 2004).These animals are raised under three husbandry systems (extensive, semi-intensive, intensive). To improve Awassi sheep the General Commission for Scientific Agricultural Research (GCSAR) in Syria established some stations. The commission intends to produce specialized lines of animals for milk and meat production. Their target is to distribute those animals to farmers and test the on-farm performance of the sheep. Despite this effort, Awassi sheep in Syria produce on average 200–300 kg milk per year compared to Israel, which was able to double milk production (over 500 kg) of its strain through intensive selection. This study aims to analyze breeding objectives and effective breeding programs. During the first step of the study productions systems characterized, breeding objectives determined and local as well as regional breeding activities implemented by sheep producer identified. In a second step the effective of government breeding stations with regard to their impact will be evaluated and alternative breeding programs including a "young ram" scheme analyzed in relation to genetic progress and the operational challenges.

Awassi productivity under different production systems in Lebanon
G. Srour[1,2], S. Abi Saab[2] and M. Marie[1], [1]URAFPA, ENSAIA-INPL-Nancy, B.P. 172, 54505 Vandœuvre lès Nancy, France, [2]Université Libanaise, Faculté d'Agronomie, B.P. 5368/13, Horch Tabet, Beyrouth, Liban*

Studies have been conducted on 11222 Awassi sheep from 39 farms in order to investigate the management practices, productive performance and economical return of different production systems in Lebanon. Ten percent of the surveyed flocks were specialized in fattening lambs and 67% owned goats together with sheep. Sheep farmers were relatively old (the average age being 54 years, with a range from 27 to 75 years). Fattening of lambs for 6-7 months results in an average weight of 55 kg/head in fattening flocks, while this weight did not exceed the mean of 48 kg/head in milk and meat flocks after 12-15 months. Fertility and prolificacy rate were respectively 0.94 and 1.28. The yield of milk per year per ewe during all lactation period was 112 kg, and varied between production systems from 30 to 170 kg. Suckled milk per lamb for a period of 2-3 months ranged from 1 to 93%, the mean being 23%. The average fat and protein content were respectively 6.9 and 5.4 percent. Economical return was better in farms where cheese processing and direct sale of dairy product is important in Vertical Transhumance and Sedentary systems. Rent of rangelands and crop residues constitutes a significant cost in Semi-Nomadic and Horizontal Transhumance systems.

The Awassi sheep in Australia

Rafat Al Jassim[1]*, Herman Raadsma[2], Roberta Bencini[3], and George Johnston[4], [1]School of Animal Studies, The University of Queensland, Gatton QLD 4343 Australia, [2]Faculty of Veterinary Science, The University of Sydney, NSW 2570 Australia, [3]School of Animal Biology, The University of Western Australia, WA 6009 Australia, [4]Yathroo, PO Box 96 Dandaragan WA 6507 Australia

The Awassi is the most popular breed of sheep in the Near and Middle Eastern region. The Awassi is well adapted to semi-arid and arid conditions, raised mainly for its high quality meat and high milk yield. By contrast, in Australia the sheep industry is based on the production of wool from Merinos. For prime lamb production surplus Merino ewes are crossed with Border Leicester rams and the resultant F1 ewes are mated with rams from meat breeds to produce high quality prime lambs. Australia is a major producer and exporter of lamb and mutton. However, due to the need to meet the demand of the lucrative market of the Middle East for live fat tail sheep, Australia imported the Awassi sheep in 1987. At present there are two Awassis operations in Western Australia for the production of prime lamb for live export. There is a pure Awassi stud in Wickepin and prime lamb production operation at Dandaragan. This paper reviews the genetic, milk production, wool production and nutrition work that have been carried out on the Awassi sheep in Australia since its release on the early 1990s.

Traditional dairy products of Turkey manufactured from awassi sheep's milk

B. Özer[1]*, C. Koçak[2], M. Güven[3], [1]Harran University, Faculty of Agriculture, Department of Food Engineering, Şunlıurfa, Turkey, [2]Ankara University, Faculty of Agriculture, Department of Dairy Technology, Ankara, Turkey, [3]Çukurova University, Faculty of Agriculture, Department of Food Engineering, Adana, Turkey

In Turkey, total milk production is around 10.5 million tonnes per annum and roughly 12% of this volume is produced from sheeps. Awassi sheep breed is one of the high yielding milking breeds with 90-155 kg milk production per lactation period. Awassi sheep breeding is fairly intensified in the Southeastern Anatolia region where Turkey's largest development project (namely GAP) is ongoing. Awassi sheep breed occupies roughly 2.3% of Turkey's total sheep population. Awassi sheeps milk is largely reserved for home and/or small scale production of some traditional dairy products including Urfa cheese, Tulum cheese, Örgü (braided) cheese, yoghurt and butterfat (sade yağ). Among the traditional sheep's milk products, Urfa cheese variety has a unique place since, in recent years, this product has enjoyed national and, to some extent, international recognition. Recently, efforts are being made to characterize the geographical origin of Urfa cheese. Tulum cheese is another major cheese variety produced from Awassi sheep's milk in Southeastern Anatolia region of Turkey. Total amount of Tulum cheese production is around 600 tonnes per year and almost 80% of this amount is exported. This paper reviewes the manufacturing technologies of traditional milk products produced only from milk of Awassi sheeps.

Increasing prolificacy of the Awassi and the Assaf breeds using the FecB (Booroola) gene
E. Gootwine, Institute of Animal Science, The Volcani Center, PO Box 6, Bet Dagan, 50250, Israel

Mostly, Awassi sheep are managed under extensive conditions relying on pasture utilization in transhumant systems. Low prolificacy, about 1.2 lambs born/lambing (LB/L) interfere with transition of the Awassi into intensive production systems. Crossbreeding programs was initiated in 1986 aimed at improving the prolificacy of the Awassi by introducing the *B* allele of the *FecB* locus to the Improved Awassi strain. A marker assisted selection approach and monitoring induced ovulation rate were applied. To this end, prolificacy (2nd parity) of *B+* Awassi ewes is 1.9 LB/L. Average milk yield in the 2nd lactation of *B+* and non-carrier Awassi ewes managed under intensive conditions is 482 and 533 l, respectively. A breeding program has been lunched where *BB* Improved Awassi rams are introduced to native Awassi flocks to improve lamb production while keeping the desirable Awassi type. Performance of native Awassi *B+* ewes under semi-extensive conditions is under investigation. The *B* allele has also been introduced to the intensively managed Assaf breed aiming to further improve its profitability. Average prolificacy of *BB* and *B+* Assaf ewes in the 2nd parity is 2.5 and 2.4 LB/L, respectively, as compared to 1.6 LB/L of non carriers Assaf ewes. It was found in the Assaf that the *B* allele has an adverse effect on birthweight of lambs and mature body size of ewes.

The effect of different body condition on the reproductive performance of Awassi ewes
Mona Abboud[1], K.J.Peters[2], [1]Min. Agric.-Small Stock Livestock Rehabilitation Project, Lebanon, [2]Humboldt University Berlin, Dept. of Animal Breeding in the Tropics and Subtropics

The effect of accumulation and mobilization of fat deposit on the reproductive performance has been studied seventy two Awassi ewes. Experimental animals were grouped according to age (ewe lamb A, primiparous P, multiparous M) and weight (S,L) and kept under traditional and improved conditions (barley supplement). The reproduction period has been divided into four phases: early season (P1: mid July- mid August), mid season (P2: mid August- mid September), decline phase (P3: mid September- mid October) and late (P4: mid October- mid November). Body fat conditions were assessed using BCS and tail measurements. BCS, tail measuares and body weight were highly correlated (p<0.01). The percentage of females detected at the fourth week post introduction of males was higher for the (ML) than for other groups MS, PL, PS (93% vs 89%, 87% and 80%) respectively. The percentage of pregnancy and parturition were higher for the LW than SW and in the improved system than the traditional system.
From September till October, the poor pasture offered in traditional system is insufficient for attaining optimal body condition and reproductive performance. The adoption of a supplementioan strategy can signifivcantly improve the reproductive efficiency.

Physiological (hormonal and spermatological) development of testes in Awassi lambs

Kadir Kirk[1], Tuncay Ozgunen[2] and Oktay Gürsoy[3], [1]Faculty of Agriculture, University of Yuzuncu Yil, Department of Anim.Sci. Van, Turkey [2]Faculty of Medicine, University of Çukurova, Department of Physiology, Adana, Turkey, [3]Faculty of Agriculture, University of Çukurova Department of Anim. Sci. Adana, Turkey

Physiological development (hormonal and spermatological) of testes was investigated in Awassi lambs in the 3-9 month age period. Blood sera were taken from 12 lambs monthly to determine the hormonal development. Radioimmunoassay technique (RIA) was used for FSH, LH, DHEA-S and Testosterone hormones. The FSH, LH, Testosterone and DHEA-S levels in the third, sixth and the ninth months of the lambs were 0.1 nmol/ml, 0.32 nmol/ml, 0.19 ng/ml and 2.93 ng/ml; 0.19 nmol/ml, 0.32 nmol/ml, 1.78 ng/ml and 2.79 ng/ml; 0.25 nmol/ml, 0.32 nmol/ml, 1.41 ng/ml and 1.71 ng/ml respectively. It was seen that in the sixth month both FSH and Testosterone levels boomed up, but the DHEA-S decreased while LH remaing the same. In the ninth month the FSH increased significantly while Testosterone and DHEA-S dropped appreciably. The spermatological development of 9 month old lambs was determined by collecting and evaluating one ejeculate daily for 5 consecutive days. The semen volume, motility, mass activity, sperm concentration, dead/live sperm ratio, abnormal spermatozoa ratio and pH were found to be 1.02 ml, 91.3 %, 4.04, 1.59×10^9, 6.3%, 11.6% and 7.24 respectively.

Hormonal levels and the spermatological values indicated that the lambs reached sexual maturity and could be used in the natural mating and AI successfully.

Effect of flushing treatment on fertility of oestrus synchronized Awassi ewes

Ü. Yavuzer and A. Can, Department of Animal Science, Faculty of Agriculture, Harran University, Sanlıurfa, Turkey

The aim of this research was to determine the effect of flushing on fertility of oestrus synchronized Awassi ewes. The research was carried out with 40 Awassi ewes at the different ages. The ewes were randomly allocated into two treatment groups with equal numbers. Ewes were fed according to guideline of NCR, (1985) for flushing and control groups. 1 ml of prostaglandin F_2 alpha ($PGF_{2\alpha}$) was two times injected with 10 day intervals for oestrus synchronisation during their breeding season for both groups in order to compare their pregnancy rates.

Oestrus rates of 94% and 87% were obtained from flushing and control group, respectively. Real-time transrectal ultrasonography was performed on all ewes on day 35 of pregnancy. Pregnancy rates were obtained 94% and 68% for flushing and control groups, respectively. Number of pregnant ewes per ewe exposed to the ram at the lambing was significantly higher in ewes of the flushing treatment group ($P < 0.01$) than in those of the control group.

Effect of fibrolytic enzyme inclusion in fattening diet on growth performance of Awassi lamb

M.M. Muwalla*, S.G. Haddad and M.A. Hijazeen, Jordan University of Science and Technology, P. O. Box 3030 22110 Irbid, Jordan

The objective of this study was to investigate the effect of supplementing fibrolytic enzyme in high concentrate fattening diet on growth performance of Awassi lamb. Thirty Awassi lambs (initial weight 20.4 ± 0.3 kg) housed in individual pens were randomly divided into two groups (15 lambs/ group) with or without enzyme supplementation. Lambs were fed the diet for 60 days including a 14-day adaptation period in a completely randomized design. Feed intake was recorded daily and weight gain weekly. The diet contained 58, 17, 15, 7, 3% barley grain, wheat straw, soybean, wheat bran, and salt, limestone, bicarbonate, respectively. Dry matter, OM, CP, NDF and ADF intake were similar in lambs receiving the finishing diet with or without the enzyme supplementation and averaged 1078, 941.5, 171.5, 432, and 223.5 $g.d^{-1}$, respectively. Dry matter, OM, CP, and apparent ADF digestibility were all not affected ($P > 0.05$) by the enzyme supplementation and averaged 64.8, 66.9, 65.7 and 50.4%, respectively. Body weight changes, average daily gain and final body weight were all unaffected by the enzyme supplementation and averaged 13.25 Kg, 222 $g.d^{-1}$, 33.8 Kg, respectively. In conclusion the use of fibrolytic enzyme for fattening Awassi lambs fed on high concentrate diet did not improve nutrient intake or growth performance.

Light and electron microscopic study of testes development in Awassi lambs

Kadir Kirk[1], Sait Polat[2], Oktay Gürsoy[3], [1]Faculty of Agriculture, University of Yuzuncu Yil, Department of Anim.Sci. Van, Turkey [2]Faculty of Medicine, University of Çukurova, Department of Histology, Adana, Turkey, [3]Faculty of Agriculture, University of Çukurova Department of Anim. Sci. Adana,Turkey

A total of 28 testes samples of 3-9 month old Awassi lambs were analyzed under light and electron microscope to investigate testicular development. Light microscope analyses indicated the presence of vast number of Sertoli cells (SC) and few primary spermatocytes (PS) in the seminiferous tubules (ST) in the third and fourth months, followed by the presence of SC in the ST and Leydig cells (LC) in the intertitial tissues (IT). In the sixth month spermatids and very few spermatozoa were seen in addition to increased numbers of cells of earlier stages of spermatogenesis. In the seventh month further increases in the numbers of cells were noted. In the eighth and ninth months greater numbers of spermatozoa, buried in nursing SC as well as mature spermatozoa were observed implying satisfactory functioning of testes.

Electron microscopic analyses showed spermatogonia, SC and a few PS in the third month followed by the appearance of LC and increased numbers of PS in the fourth month. In the fifth month there was further increases in the previously seen cells. However in the sixth month spermatids and spermatozoa were observed for the first time. The seventh month samples showed further increases of PS, SC, spermatids, LC and spermatozoa. Finally in the eighth and ninth month samples indicated fully functioning testes incorporating vast numbers of all cells in various stages of maturity.

The effect of using urea-based feed blocks on Awassi ewe's nutrition under farm conditions
A. AL-Jamal and S.G. Haddad*, Department of Animal Production, Jordan University of Science and Technology, PO Box 3030, Irbid, Jordan

The aim of the experiment was to determine the effect of using urea-based feed blocks as a supplement to Awassi ewes grazing cereal stubble. One hundred seventy ewes in three different farms were used in this experiment using a complete randomized block design. On each farm ewes were divided into two groups, the control (A) group, and the feed block (B) group. The three farms contained 95, 36, and 39 ewes, respectively. Ewes in farms 1 and 2 grazed 40 hectares and ewes on farm 3 grazed 8 hectares of cereal stubble. All animals grazed twice a day from 04:00 till 11:00 and from 16:00 till 21:00. Feed blocks were offered at night from May until late August of 2004. The supplemented group gained weight more than that in the unsupplemented group. The increase in body weight in farms 1 and 2 was higher (P<0.05) than farm 3. The stocking rate in both farms 1 and 2 was 3.3 (131 ewes/40 hectares) whereas in farm 3 it was 4.9 (39 ewes/8 hectares). Pregnancy rates were not significantly different (P<0.05) in the three farms. Feed blocks had no effect on lambs weight (P<0.05). In conclusion, feed blocks improved live weight gain of ewes grazing cereal stubble but had no effect on reproductive performance.

Partial replacement of barley grain for corn grain: Associative effects on lambs' growth performance
S.G. Haddad* R.E. Nasr and R.W. Sweidan, Department of Animal Production, Jordan University of Science and Technology, PO Box 3030, Irbid, Jordan

To study the positive effects of partial replacement of barley grain for corn on growth performance of growing lambs, thirty-three male Awassi lambs were divided into 3 groups and offered three isonitrogenous diets. The control diet (B) contained 81 and 14 % barley grain and wheat straw, respectively all on dry matter (DM) basis. Corn grain replaced barley grain by 10 and 20 % of the diet DM in the low corn (LC) and high corn (HC) diets, respectively. The digestion rate for the B diet was higher (P < 0.05) than the HC diet whereas the LC diet was intermediate. Lambs fed the LC and HC diets consumed more DM compared with lambs fed the B diet. DM and CP digestibility were similar in all the experimental diets. Final body weights for lambs fed the HC diet were higher than the B and LC diets. Feed to gain ratio was lower for lambs fed the HC diet than the B and LC diets. In summary, positive associated effects of partial replacement of corn for barley in high concentrate diets for fattening sheep were detected. However, a minium of 20% replacement of the diet DM was needed to positively improve both performance and efficiency.

Effect of yeast culture supplementation on nutrient digestibility of Awassi Sheep
A. Can, N. Denek, M. Seker and H. Ipek, Department of Animal Science, Faculty of Agriculture, University of Harran, Sanliurfa, Turkey

Two apparent digestion trials were conducted to determine the effect of yeast culture (YC) supplementation on nutrient digestibility of Awassi lambs. In trial 1, nine 3 years old Awassi rams (65.8 ± 2.3 kg) were used as experimental animals in apparent digestion trial. They were allocated for three treatments (3 per treatment) at random within live weight. Treatments were 0, 5, and 10 g/d of a YC with basal diet containing 70% wheat straw (WS) and 30% commercial concentrate pellet feed (CCPF) with low CP level. The diet was offered 2.5% of body weight as fed basis. In trial 2, the same rams in trial 1 were used as experimental animals. They were allocated for three treatments which were 0, 10, and 20 g/d of a YC with a basal diet that contains 30%WS and 70% CCPF with high CP level. The basal diet was offered 2.1% of body weight as fed basis. Yeast culture supplementation did not change DM, OM, and ash digestibility in both in vivo digestion trials (P>0.05). The YC supplementation of 20g/d in trial 2 increased ruminal pH and NH_3-N. As a result, there was no advantage of supplementing YC on DM, OM, and ash digestibility of diets.

Effect of feeding cost-efficient diets on the meat quality of fattened Awassi lambs
M. Zaklouta[1], W. Knaus[2], L. Iñiguez[1], M. Hilali[1], B. Hartwell[3] and M. Wurzinger[3], [1]International Center for Agricultural Research in Dry Areas, POB 5466, Aleppo, Syria, [2]University of Natural Resources and Applied Life Sciences, Gregor-Mendel-Strasse 33, A – 1180 Vienna, Austria, [3]The Royal Veterinary and Agricultural University, Grønnegårdsvej 2, 1870 Frederiksberg C.,Denmark*

The meat of Awassi lambs fed and fattened in two experiments designed to reduce feeding costs in fattening production systems in Syria, was evaluated using sensory tests to assess the possible effect of diets on meat quality. The assessed meat derived from two sets of experiments: on-station (E1) and on-farm (E2). In E1, two low-cost diets were tested in comparison with the traditional diet used by farmers. In E2 the on-station most promising diet was compared also to the traditional diet on three farms at Khanasser, near Aleppo. Lambs were slaughtered at the age of 6.5-8.5 months. The carcass pH was measured at different *post mortem* times and samples of the *Longissimus dorsi* muscle were obtained. Sensory panels assessed the quality of cooked meat samples from E1 and E2, according to local preferences. The pH was unaffected by diets (P>0.05). The sensory evaluation did not revealed differences in quality among diets in E1 and between diets within farms in E2 (P>0.05). The results suggest that the alternative diets that promoted important growth and were less costly in the fattening experiment, did not affect the quality of the meat.

Performance of Awassi sheep and Shami goats fed cereal-legume hay supplemented with potato chips

B. Jammal[1], A. Larbi[2] S. Haj[1], B. Jammal[1], Kh. Houshaimi[3], M. Jerdi[1], L. Hnoud[1], D. Zaidouni[1], A. Mansour[1], E. Haj[1] and A. Nassar[2], [1]Lebanese University-Faculty of Agricultural Sciences, P. O. Box 13-5368, Beirut, Lebanon, [2]International Center for Agricultural Research in the Dry Areas, P. O. Box 5466, Aleppo, Syria, [3]Lebanese Agricultural Research Institute, Tal Amara, Rayak, P. O. Box 287, Zahle, Lebanon*

Although cereal-legume hay have potential to improve availability of quality feed in Lebanon, information on performance of small ruminants fed cereal-legume hay as sole diet or basal with protein or energy supplement is limited. Two trials compared the growth rates of weaned male Awassi lambs (Experiment 1) and female Shami kids (Experiment 2) fed sun-dried barley-vetch (B-V) or oat-vetch (O-V) hay supplemented with 0, 10 and 20% potato chips. A 2 x 3 factorial treatment arrangements in a randomized complete block with 6 animals per treatment were used in both trials. In Experiment 1, lambs fed O-V hay gained more (P<0.05) weight than those fed on B-V (183 vs. 143 g/head/day). Similarly, average daily gain of kids fed O-V was relatively higher than those fed B-V (79 vs. 71 g/head/day) in Experiment 2. Supplementation increased daily weight gain in both trials. Oat-vetch hay supplemented with 10-20% potato chips has higher potential as fattening diets for Awassi sheep ad Shami goats than B-V supplemented with potato chips.

Growth rate of Awassi lambs grazing promising vetch lines in spring

A. Larbi[1], A.A. El-Moneim[1], L. Iniguez[1] and S. Rihawi[1], [1]International Center for Agricultural Research in the Dry Areas (ICARDA), P. O. Box 5466, Aleppo, Syria*

Lack of quality feed during the early spring is a major constraint to small ruminant production in the dry areas of Central and West Asia and North Africa (CWANA). Grazing common vetch (*Vicia sativa*) is an option for filling the early spring feed gap. ICARDA's Forage Legume Germplasm Improvement Project has developed elite lines of common vetch and bitter vetch (*Vicia ervilia*) that could fill the early spring feed gap, but, their potential to fill the early spring feed gap has not been evaluated. A 2-year trial was conducted in northeastern Syria to compare growth rates of Awassi lambs grazing common vetch line 2566 and bitter vetch line 3330 with a released common vetch cultivar (Baraka) between March and May. Daily gain ranged from 160 – 181g/head or 381 – 420 kg/ha. Forage-on-offer and average daily gain of the promising bitter vetch line were similar (P>0.05) to the released cultivar. The results show that bitter vetch line 3330 could be integrated into small-scale lamb fattening systems to reduce the early spring feed gap in the dry areas and also help farmers benefit from the ever increasing market for livestock products as a result of population growth.

Screening for allele frequency at the PrP locus in Awassi and Assaf populations in Israel and in the Palestinian Territories

E. Gootwine[1], A. Abdulkhaliq[2] and A. Valle Zárate[3], [1]Institute of Animal Science, The Volcani Center. PO Box 6, Bet Dagan, 50250, Israel, [2]Agriculture Engineers Home Society, Ramallah, Palestinian Authority, [3]Institute of Animal Production in the Tropics and Subtropics, University of Hohenheim, Garbenst. 17, 70593 Stuttgart, Germany

Screening for polymorphism at the prion protein locus was carried out in 17 flocks of local Awassi (LA), one flock of Improved Awassi (IA) and 11 flocks of Assaf (As) sheep located in Israel and in the Palestinian Territories. All the rams present in the flocks at the time of the survey: 218, 29 and 306 rams from the LA, IA and As breeds, respectively, were genotyped. The ARQ allele found to be the predominant allele in all three populations with frequencies of 0.75, 0.69 and 0.88 for the LA, IA and the As breeds, respectively. Frequencies of the desirable ARR allele were 0.09, 0.03 and 0.12, respectively. No VRQ carriers were found in this survey. ARH, AHQ and ARK alleles were identified at low frequencies. Considerable variation was found in PrP allele frequencies between the different Awassi and the Assaf flocks, reflecting them being managed as closed flocks. Few Scrapie outbreaks were documented in the past in the region. To increase ARR allele frequency in the Awassi and the Assaf populations, distribution of rams from flocks with high ARR frequency is recommended.

Lamb and milk production in improved Awassi crosses with the Kazak Fat Tail and the Kazakh Fine Wool breeds

N. Malmakov[1], K. Kanapin[1], V.A. Spivako[1], K. Seitpan[1] and E. Gootwine[2], [1]Kazakh Research Technological Institute of Sheep Breeding Mynbaevo Village, Djambul District, Almaty Region, 483174, Republic of Kazakhstan, [2]Institute of Animal Science, ARO, The Volcani Center. POB 6 Bet Dagan, 50250, Israel

Sheep population in Kazakhstan was reduced dramatically due to the fall in prices of lambs and wool. To test the option that milk production may improve flocks' profitability, breeding nucleus of the Improved-Awassi - a fat tail breed known for its high milk production and hardiness was exported from Israel to Kazakhstan in 1998 and used for crossbreeding with the Kazakh fine wool (KFW) and the Kazakh fat rump (KFR) non-dairy breeds. F1 Awassi x KFW crossbreds (n=214) and F1 Awassi x KFR (n=180) crossbreds raised under extensive management conditions were similar to their contemporary KFW and KFR purebreds in survival rate, growth, body size and prolificacy. Milk yield in 105 days of Improved Awassi (n=22), KFR (n=16) and Awassi-KFR F1 ewes (n=23) was 152.4, 42.6 and 119.2 liters, respectively. During the lactation period lambs were separated from their mothers on the day of lambing and ewes were hand-milked twice a day. No difference was found between genotypes in concentration of fat or protein in the milk. Our results support the assumption that the Improved Awassi can be used to increase milk production ability of Kazakh sheep by crossbreeding.

Possibilities for prediction of the test day milk yield based on only one individual test per day in Awassi sheep

M. Gievski[1], G. Dimov[2], N. Pacinovski[3] and B. Palasevski[3], [1]Awassi Mediterranean Farm - Selection Center, Ivan Milutinovic, 6, 1300, Kumanovo, Macedonia, [2]AgroBioInstitute, 8 Dragan Tsankov Blvd, Sofia, Bulgaria, [3]Institute of Animal science, 1000, Skopje, R Macedonia, P.O. Box 207

400 test day records were used for testing the accuracy of measurement of the test day yield of three time a day milking Awassi ewes. The actual test day yield was approximated by only one individual measurement and its weighting by the ratio of total yield in the farm in that milking and of the whole day.

Coefficients for separate test days and period of the day wary from 1.9 to 3.4. The average difference for the morning, midday and evening milkings three practically zero, and the corresponding extremes were from -0.67 l to 0.81 l, from -1.05 l to 0.79 l and from -1.14 l to 1.22 l. Only about 30% of the deviations in predicted and actual yields were in the range ±5%.

Differences were not affected by the age, lactation stage, consequence and level of the test day.

It was considered that the differences from the actual yield are too big and the method is of use for herds in the initial stages of recording practice and with limited importance for the breed improvement.

Crossbreeding scheme of imported Mutton and Prolific sheep breeds with local Awassi sheep and the development of training and extension center in Jordan

Mohamed Momani Shaker[1], Abdullah Y. Abdullah[2], Rami T. Kridli[2], Ivan Šáda[1], [1]Czech University of Agriculture in Prague, Institute of Tropics and Subtropics, Czech Republic, [2]Jordan University of Science and Technology Irbid, Faculty of Agriculture, Irbid, Jordan*

Jordan belongs to an important farm animal in many countries. Jordan is one of them, where sheep breeding holds the most significant part in Animal Husbandry. The studies of many materials and resources from Jordan show, the demand for high quality lamb meat is arising gradually. The local breeders of sheep are not able to meet this demand at present. Two alternative methods of production development exist: using knowledge of genetics (selection of local breed and crossbreeding with exotic breeds) and improvement of environmental conditions. The objective of the project is to recommend and implement a set of methods and technologies leading to the increment and improvement of mutton production in Jordan through a planned crossbreeding scheme between the local Awassi sheep and imported mutton and prolific sheep breeds (Charollais [Ch], Suffolk (S) and Romanov [R]). The second component deals with establishing a Center for research, training and extension. The Center basic goals are: workplace for implementation and to continue with the crossbreeding and also and to produce purebred animals of the breeds used in this project. Educate farmers and students about the theoretical and practical knowledge of animal breeding and presentations of project crossing acquired results.

*The project is supporting by the Czech Republic of Development Cooperation.

Evaluation of Awassi genotypes for milk production improvement
L. Iñiguez, M. Hilali* and G.Jessr, International Center for Agricultural Research in Dry Areas (ICARDA), P.O. Box 5466, Aleppo, Syria

More farmers in Syria claim for access to sources of improved animals. The reason: a demand for animals that could produce better under conditions where farmers invest in improving their feeding systems. A small flock of selected Turkish Awassi sheep (T) was introduced at ICARDA experimental station, Aleppo, Syria in the early 1990's and their performance evaluated. Furthermore, animals of this genotype were crossed to ICARDA's stock of Syrian sheep (S) to produce S×T females (ST). The milk production and milk composition of S, T and ST ewes produced were evaluated at ICARDA experimental station. Results from three years of milk production after weaning (at day 56) show that milk production of T and S genotypes amounted to 123.54 kg and 109.41 kg, respectively (P=0.019), while the production of the ST ewes amounted 119.18 kg, was no different than the average of the parental genotypes (116.48 kg) (P=0.94). Milk composition among the T, S and ST ewes, although significantly different, varied little in amount in relation to fat (6.51 %, 6.22 %, and 6.24 %, respectively; P<0.0001), protein (5.09 %, 5.33 % and 5.43 %, respectively; P<0.0001) and Solids non fat (11.11 %, 11.43% and 11.39 %, respectively; P<0.0001). The superiority of T ewes to produce milk was consistent through the years of performance and suggests a potential avenue to improve milk production in more intensive dairy sheep systems in Syria.

Lamb fattening strategies and growth of different Awassi lamb genotypes
S. Rihawi*[1], L. Iñiguez[1], M. Wurzinger[2], W. Knaus[2], A. Larbi[1], M. Hilali[1] and M. Zaklouta[1], [1]International Center for Agricultural Research in Dry Areas, POB 5466, Aleppo, Syria, [2]University of Natural Resources and Applied Life Sciences, Gregor-Mendel-Strasse 33, 1180 Vienna, Austria

A test of lamb fattening cost-efficient diets was conducted with a farmer in El Bab area, northern Syria. The growth performance assessment of pure Syrian Awassi (S) lambs (n=36) was also compared to F1 crossbred lambs (ST) with a Turkish Awassi genotype (T) (n=31). After weaning, all equally raised lambs were allocated to three diets reflecting intensiveness (on concentrates) (n=24), semi-intensiveness (grazing vetch, minor supplementation) (n=22) and traditional fattening practices (n=21). For the semi-intensive group, vetch (*Vicia sativa*) was used for grazing in April-May 2005. Overall, ST lambs tend to gain more weight and grow faster (15.5 kg/lamb; 315 g/day) than S lambs (14.9 kg/lamb; 304 g/day) (P>0.05). The combination of vetch grazing and minor concentrate supplementation did not promote different gain weights and average daily gains (15.9 kg/lamb; 323 g/day) than the concentrate feeding (15.3 kg/lamb; 313 g/day) and the traditional feeding system (14.5 kg/lamb; 294 g/day) (P=0.264). It was apparent that the contribution of a different genotype to growth performance was minimal. Furthermore, vetch grazing resulted in an interesting option with the potential not only to reduce feeding costs providing green fodder for about 60 days, but also to promote a high growth response.

Milk production characteristics of ewes belonging to various Awassi genotypes

S. Nagy[1], T. Németh[2], I. Tisler[1], A. Molnár[2], S. Nagy[1], A. Jávor[3] and S. Kukovics[2], [1]Bakonszegi Awassi Corporation, Hunyadi u. 83. Bakonszeg 4164 Hungary, [2]Research Institute for Animal Breeding and Nutrition, Gesztenyés u. 1. Herceghalom, 2053 Hungary, [3]University of Debrecen Centre of Agricultural Sciences, Böszörményi út 138. Debrecen 4032 Hungary*

Milk production characteristics of 2,000 ewes belonging to various Awassi genotypes (purebred, (F_1, R_1, R_2, R_3, R_4) were studied between 2000 and 2005. The effects of genotype and year on the milk yield, daily milk yield, number of milking days, fat%, protein%, SCC were examined. Each genotype of ewes was kept on the same unit under intensive production system. The lambs were weaned at the day of birth from their dams, and the ewes were milked for 180-240 days. Milk yield of them was individually measured based on test milkings (twice a day) four weekly started two weeks after lambing, when the milk of each ewe was sampled. Milk was examined in official raw milk laboratory using Milkoscan 600 and Fossomatic equipments.
Data received were processed using Microsoft Excel 7.0 and SPSS 9.0 for Windows programmes.
According to the results the milk yield data was increasing up to R_2 genotype, while the benefit was small after it, however, the year had a strong effect on the data. There were significant differences among the various genotypes concerning the milk composition as well.

Artificial rearing of Awassi lambs

I. Tisler[1], S. Nagy[1], T. Németh[2], S. Nagy[1] and S. Kukovics[2], [1]Bakonszegi Awassi Corporation, Hunyadi u. 83. Bakonszeg 4164 Hungary, [2]Research Institute for Animal Breeding and Nutrition, Gesztenyés u. 1. Herceghalom, 2053 Hungary*

The authors examined the artificial lamb rearing system (after 0 day of weaning from ewes) utilised in the Bakonszegi Awassi Corporation during the period between 2000 and 2005. The effects of genotype, sex and year were studied on the weaning results (age, bodyweight), followed by the results of first mating and lambing concerning the artificially reared lambs. There are about 2,000 heads of purebred and crossbred (F_1, R_1, R_2, R_3, R_4) Awassi sheep bred in the farm, along with the some 3,000 heads of Transylvanian Racka (indigenous breed). The Awassi sheep are kept in intensive while the Racka sheep in extensive production system. Concerning the latter one the possible use of artificial lamb rearing system were also studied. Some 1,500-1,800 lambs were artificially reared in each year from the abovementioned genotypes.
Data received were processed using Microsoft Excel 7.0 and SPSS 9.0 for Windows programmes.
According to the results the sex, genotype and the year had strong effect on the rearing and the weaning results. The lambs of indigenous breed could not tolerate well the artificial rearing but the purebred Awassi could reach the best results in the system.

The markets of Jameed in Jordan an export window for Syrian dairy sheep farmers
A. Fukuki* and L. Iñiguez, International Center for Agricultural Research in Dry Areas (ICARDA), P.O. Box 5466, Aleppo, Syria

A market niche for Jameed was identified in Jordan with the potential to add to income generating opportunities for Syrian sheep milk producers. Jameed is a sheep milk yogurt byproduct, often unused in Syria but with a high and growing demand in Jordan. By using survey data both from Syria and Jordan, this poster examines the feasibility for Syrian sheep producers to export Jameed to Jordan and thus expanding their income generation options. Actually Syrian Jameed is already being exported to Jordan to fill the widening supply/demand gap and represents 93% of the Jordan's imported Jameed. The demand is expected to increase in view of population growth (3% in 1996-2003) and higher purchasing power. Furthermore, the local supply is not likely to keep up with the population growth pace due to constraints including limited skilled labor, suitable facilities, and limited milk production volumes. Intermediaries indicate that currently there is a market niche for cheaper Jameed, usually produced in and imported from Syria (US$3.5-5.6/kg), and a more expensive high quality Jameed produced mostly in Jordan (US $4.9-11.3/kg). Interest in targeting the Jordan market was expressed by non-exporter Syrian producers if this shows clear profitability. This poster also discusses policies issues involved, such as export/import regulations, price competitiveness and opportunity niches for Syrian and Jordanian sheep milk producers.

Markets and market opportunities for dairy small ruminant products in Syria
L. Iñiguez and A. Fukuki*, International Center for Agricultural Research in Dry Areas (ICARDA), P.O. Box 5466, Aleppo, Syria

Sheep milk derivatives (SMD) contribute significantly to smallholders' income in Mediterranean West Asia. In Syria the demand for cheese and yogurt has increased; however, the benefits of these enhanced opportunities are not yet fully captured by producers. Based on a survey in El Bab region in northeastern Syria, a traditional dairy sheep production area, this poster examines what are the SMD market channels, constraints for production and marketing, and opportunities for better marketing. According to producers, feed shortages and feed costs were main production constraints to produce milk whereas product quality a main constraint for SMD marketing. Thus, technologies to reduce feeding costs and improve the SMD quality are pertinent. Selling of milk directly in the market is rare. Producers sell only processed products (cheese and yogurt), mainly processed by women, particularly to dealers due to lack of time for marketing, unavailability of vehicles, labor shortage and lack of knowledge in marketing. An assessment of consumers' preferences showed that 61% of consumers will pay for quality while producers will improve quality if asked by intermediaries. However, intermediaries are not getting the message and consumers' feedback does not reach efficiently producers and intermediaries to prompt for improvements. In Aleppo, 55% of consumers reported that some SMD were missing in the market, a condition that shows a potential window for product diversification.

Investigations on carcass and meat quality characteristics of Awassi and Turkish Merino x Awassi (F1) lambs
R. Çelik and A. Yilmaz*, Istanbul University, Faculty of Veterinary Medicine, Department of Animal Breeding and Husbandry, Avcilar, Istanbul, Turkey

The study was planned to determine the carcass and meat quality characteristics of two-way crossbred lambs obtained by the mating of Turkish Merino (Karacabey Merino) as a sire line with Awassi ewes, in comparison with purebred Awassi lambs. For Awassi purebred and Turkish Merino x Awassi (F_1) crossbred lambs the live-weights before slaughter were 30.32 kg and 33.71 kg; hot carcass weights were 12.99 kg and 14.84 kg; cold carcass weights were 12.61 kg and 14.50 kg; dressing percentages were 41.52% and 42.83%; musculus longissimus dorsi (MLD) muscle areas were 12.65 cm^2 and 13.42 cm^2; values of pH at 24th hour were 5.84 and 5.80; percentage of drip losses were 2.35% and 2.69%; percentage of cooking losses were 31.41% and 30.94%; tenderness of meat were 3.42 kg and 2.63 kg; P:S ratios were 0.14 and 0.16; (n-6/n-3) ratios were 3.70 and 4.66, respectively. As the result of the present study, Turkish Merino x Awassi (F_1) lambs had higher performance for live-weight before slaughter, hot carcass weight, cold carcass weight, dressing percentage and tenderness of meat than Awassi lambs and these results showed that Turkish Merino genotype have positive influence on obtaining quality slaughter lambs.

Comparison of estimation methods used in milk yield on Awassi sheep
Ü. Yavuzer, Department of Animal Science, Faculty of Agriculture, Harran University, Sanliurfa, Turkey

The main purpose of present study was to determine the best and reliable method to estimate lactation milk yields the closest to those produced by Awassi ewes, by comparing the most common methods applied in this purpose. The material was the 50 real lactation records collected from 10 ewes in a sheep farm. Lactations was divided in to 8 different lengts of periods: of 7-day, 14 day, 21-day, 28-day, 35-day, 42-day, 49-day, and 56-day. It was pretended as if the test day milk records were collected at those periods, and the test day records were used in calculation of estimated lactation milk yields by methods of Holland I, Holland II, Sweden, Trapez and Vogel. The analyses were revialed that different methods estimated the lactation milk yield closer the real values depending on the parity, first control day after lambing, length of control periods. In general, the best method estimating the real milk yield was Trapez method. On the other hand the worst method estimating the real milk yield was Vogel's method. It was also determined that the period of 7-day was the best period resulted in taking test day milk records to estimate real lactation milk yields.

Factors affecting the fibre yield and quality of Awassi sheep
Ü. Yavuzer, Department of Animal Science, Faculty of Agriculture, Harran University, Şanlıurfa, Turkey

The aim of this research was to determine factors affecting the fibre yield and quality of Awassi sheep. The animals used in these studies were 60 male and female Awassi sheep, ranging 10 months to 8 years and in body weight from 25 and 90 kg. The fleece is mostly carpet type with a varying degree of hair. Awassi sheep produce coarse mixed wool. These fibers are used wieldly in carpet production. Mean fibre diameter of the undercoat increased with age up to years from 31.5μm up to 35.4. There was little difference in fibre production of male and female animals. The range of mean fibre diameter between animals within each age group was high. There was a relationship between body weight and fleece weight (r = 0.945). Grease fleece weight of Awassi sheep are increased with age up to years from 1.9 kg up to 5.3 kg. In Awassi, grease fleece weight and wool yield are markedly influenced by nutrition and such effects have been shown to vary according to season and age. The fleece weight for male Awassi was slightly than that for females and was probably a reflection of greater body size. The overall correlation between grease fleece weigth and live body weight was high (r = 0.875).

Recent advances in decision support systems for the development of sustainable livestock production systems
M. Herrero[1,2], P.K. Thornton[1,2], F. Bousquet[3], B. Eickhout[4], P. Titonnell[5,6], C.F. Nicholson[7] and D. Peden[1], [1]International Livestock Research Institute, Nairobi, Kenya and Addis Ababa, Ethiopia, [2]Institute of Atmospheric and Environmental Sciences, University of Edinburgh, West Mains Road EH9 3JG Scotland, [3]Centre de Coopération Internationale en Recherche Agronomique pour le Développement.(CIRAD), Campus International de Baillarguet 34398 Montpellier Cedex 5, France, [4]Netherlands Environmental Assessment Agency, PO Box 303, 3720 AN Bilthoven, The Netherlands, [5]Plant Production Systems, Department of Plant Sciences, Wageningen University, P.O. Box 430, 6700 AK Wageningen, The Netherlands, [6]Cornell University, Applied Economics and Management Department, 316 Warren Hall, Ithaca, New York, US
Livestock systems throughout the world are undergoing significant changes. They are under considerable pressure to generate enough food to support consumer demands for livestock products while increasing incomes for farmers and ensuring the maintenance of ecosystems goods and services. The present paper reviews some of these aspects and how their associated trade-offs have been studied using a range of decision-support systems. It discusses the importance of considering spatial and temporal scales, the different modelling methods available, how to account for the interactions between stakeholders, and how to consider the impacts on different natural resources (land, water, nutrients) and on people dependent from these systems. The paper also discusses how recent methodological advances could help move the R&D agenda to improve our understanding of options to balance livelihoods opportunities while increasing ecosystems resilience. The paper concludes by mentioning some aspects related to the adoption of DSS beyond the scientific community.

Immaterial resources mobilized by farmers and flexibility of farming systems: Two concepts linked to decision making and sustainabilty

S. Ingrand, M.A. Magne, E. Chia, C.H. Moulin and M. Cerf, INRA, SAD, UMR Metafort, equipe TSE, 63122 Saint-Genes Champanelle, France

This study falls within the continuity of a project dealing with the evolutions of livestock farming systems, considering on one hand the decision making process of farmers for technical aspects ogf herd management (the information they need and/or use), and on the other hand on the adaptative ability of livestock farming systems face to uncertainty (especially climate and market).

Concerning decision making process, we propose a new framework for the analysis of livestock systems to formalize the information mobilized by the farmers to design, and monitor their farming activity by taking into account the meaning they assign to it. Such an approach has to be link with the social sciences, i.e; the framework of the activity analysis and the notion of immaterial resources.

Concerning the adaptative properties of livestock farming systems, we propose the concept of flexibility, resulting also from management sciences. Two types of studies have been carried out: i) one oriented on "internal flexibility" (within the farming system), targetting to explain what "to change" means for farmers and the consequences on the technical management practices and workforce organization. Ii) the other one oriented on relational flexibility, more especially the comercial policy of farmers. We analysed the links between internal flexibility (playing on the product type and the production process) and external flexibility (the type and the number of purchasers, as well as the relation established with them).

Evaluation of animal production sustainability with IDEA in the Mediterranean context

M. Marie[1]*, G. Srour[1,2] and N. Bekhouche[1,3], [1]ENSAIA-INPL-Nancy, B.P. 172, 54505 Vandœuvre cedex, France, [2]Université Libanaise, Horch Tabet, Beyrut, Lebanon, [3]INA-El Harrach, 16200 Algiers, Algeria

The IDEA (Indicateurs de Durabilité des Exploitations Agricoles, or Farm Sustainability Indicators) method has been implemented for sustainability assessment of small ruminant production systems in Lebanon, and dairy cows farming systems in Algeria. This method is based on 40 indicators representing the agro-ecological, socio-territorial and economic dimensions of the farming system's sustainability. In the context of the Lebanese small ruminant pastoral or agro-pastoral systems, 3 indicators have been omitted (stocking density, N balance, financial autonomy) and 9 have been adapted by modification of the scores (1 for diversity, 2 for space organisation, 3 for farming practices, 1 for quality of products, 1 for employment and services, and economical efficiency). For zero-grazing dairy cows in Mitidja, where conditions are closer to those observed in systems for which IDEA has been designed, modifications concern 8 indicators. Difficulties to compute indicators arose from: i) unavailability or uncertainty of information such as those related to fertilisation, pesticide use, or economical data, ii) different scales for local references such as land surface, or iii) land use in pastoral systems.

This paper discusses the approach used for adapting the method to these contexts, while keeping its ability to fit the criteria of assessment methods, namely sensibility, specificity, robustness. A typology of systems according to their sustainability is presented.

A dynamic model to assess the efficiency of grazing strategies for biodiversity conservation

M. Tichit[1]*, L. Doyen[2] and J.Y. Lemel[1], [1]INRA SAD, 16 rue Claude Bernard, 75231 Paris, France, [2]CERESP, MNHN-CNRS-UPMC, 55 rue Buffon 75005 Paris, France

Improving the environmental performance of grazing is a key issue for sustainable livestock systems in Europe. In grasslands, grazing is a major driver of habitat quality which is crucial to the survival of threatened ground nesting bird species. Our objective is to investigate how cattle grazing may be used to sustain bird biodiversity by creating suitable habitats without penalizing farmers. We develop a model that described the temporal dynamics of (1) a grass sward controlled by grazing where the efficiency of grazing depends on technical and economical constraints linked to feeding costs for farmers and (2) a community of three bird species formalised using an age-structured matrix model that incorporates the impact of habitat quality, measured by sward height, on the vital rates of each species. The results show that grazing is a key component to prevent extinction of bird community. In absence of grazing the growth of sward during spring does not allow the development of a favourable sward structure. According to habitat quality targeted (optimal or sub optimal) different grazing strategies emerge and some are more cost-effective than others. Our results are discussed in the light of agri-environment schemes aimed at bird conservation in Europe and highlight the need to develop integrated modelling approaches linking agricultural and conservation issues.

Fenced grazing systems based on knowledge of shepherding practices

E. Lécrivain* and J. Lasseur, National Institute for Agronomical Research, 84914 Avignon, France

Shrub-invaded Mediterranean hills are opened up to sheep grazing for environmental purposes as a contribution to biodiversity management. Traditionally, these lands used to be grazed by shepherded flocks. Their present environmental value results from this past type of management based on the daily steering by the shepherd of the animal plant relationships. In the last decade, following the renewed use of these areas, the need to improve the sheep farmers' work productivity has led to generalizing the paddocking. However, current paddock management indicators are derived from pasture situations (stocking rate, grass height) and do not fit the management requirements for these environments. Our work aims to identify new indicators by taking into account shepherd know-how. It is based on surveys of some twenty shepherds and field observations on the behaviour of paddocked flocks in the Luberon regional Park. Environmental components which were considered to be significant as attractive areas or obstacles are topographical (high points, slope ruptures, depression) and linked to vegetation structure (open patches, corridors). Identified behavioural aptitudes are linked to the animals' learning capacities (regarding food and foraging) and positioning abilities (in space and time). The structural features will serve to design paddocks (shape and organisation) while the functional ones will serve for flock and resource management (animal learning, feeding transitions, food diversity). The sustainability objective for sheep systems is to inject shepherding knowledge into new fenced system.

Supporting livestock-farming contribution to sustainable management of natural resources and the landscape: A case study in the Davantaygue valley (Pyrenees, France)

A. Gibon[1], A.Mottet[1], S. Ladet[1] and M. Fily[2], [1]UMR 1201 Dynafor INRA – INPT/ENSAT, BP 52627, 31326 Castanet-Tolosan cedex, France, [2]Direction Départementale Agriculture & Forêt, Cité administrative Reffye - rue Amiral Courbet, 65017 TARBES cedex 9, France

Past change in agricultural systems over the past mid-century brought out important change in agricultural land use in the Mediterranean mountains. Agricultural abandonment and resulting landscape encroachment and reforestation impact negatively natural resources and landscape sustainability. Sustainable development of livestock husbandry in such contexts calls for assessing the possiblity to integrate further sustainable use of natural resources and landscape quality in livestock farming system development strategies. In this poster we present a participatory research approach aimed at providing local farmers and land policy-makers with decision-support tools allowing exploring the perspectives for enhancing land use sustainability at the village level. The approach developed relies on (i) the assessment of the spatial partterns of farmland organisation and the spatio-temporal dynamics of land use at livestock farms, and (ii) the participatory design and assessment of alternative scenarios for change using GIS modelling and 3D visualisation technology.

Local primary and secondary school collaboration in a meadow's study balancing livestock production and biodiversity conservation

J. Aguirre and F. Fillat, Pyrenean Institute of Ecology. Avda. Regto. Galicia s/n E-22700 Jaca (Huesca), Spain*

A research and demonstration project was carried out in Broto Valley (Central Spanish Pyrenees) during 2000-2002 trying to find a correct intermediate management way considering simultaneously the traditional and improved meadow's techniques and their impact on the herbage diversity preservation. The participation of young people as future interested farmers was planned in a intergenerational transfer, discussing very defined subjects. Twelve local primary and secondary scholars collaborated in the project following different familiar calendars of these four techniques: fertilization, irrigation, cutting and grazing. Every correspondent described the management technique filling a specialized questionnaire on every topic and marking the position of animals or the considered event on a detailed map. All of this information was collected directly in the family house at the end of every season, leading to an interesting participation of the different family group members and allowing to present them in a digital map representation considering the temporal sequences developed in the field. A general discussion with professors and collaborators is planned that takes place in Broto School during the course.

Impact of livestock production on the environment: View from the United States
Henry F. Tyrrell (Retired) United States Department of Agriculture

Legislation has been in place in the United States since the early 1970's which would allow regulation of pollution from concentrated animal feeding operations (CAFO). A court ruling in the mid 90's required the U.S. Environmental Protection Agency (EPA) to implement new regulations limiting pollution of water by CAFO's. Also, in the mid 90's, EPA reduced particulate matter standards from PM10 to PM2.5 in response to research implicating fine particulate matter as detrimental to human health. The PM2.5 regulation is significant for animal production because ammonia reacts with other gases in the atmosphere to form fine particles. Animal agriculture is the largest source of ammonia release into the atmosphere. The new water quality regulations have greatest impact on phosphorus and then nitrogen. Each CAFO is required to develop a comprehensive nutrient management plan for the total facility. Implementation of air quality standards is more complex because of regulations other than PM2.5 which determine permissible release into the atmosphere and includes odor as a public nuisance. EPA has entered into an agreement with several livestock industries to delay strict enforcement of ammonia emission violations pending development of acceptable methods to measure ammonia release and database development of factors related to ammonia release from CAFO's. This research is currently in progress funded in part by livestock industries. Technologies required to assist livestock producers to comply with environmental regulations will be discussed.

Session N16   Theatre 1

The importance of autochthonous Bakan goat in sustainable farming in Serbia
S. Jovanović[1]*, M. Savic[1], A. Vranes[2] and S. Bobos[2] [1]Faculty of Veterinary Medicine, University of Belgrade [2], Faculty of Agriculture, University of Novi Sad

Program of the rural development of the Eastern Serbian mountain regions has been designed, providing favourable models of sustainable farming systems and food processing technologies. The project should also provide conditions for the protection of local natural resources and the rational utilization of capacities in production of food of high quality and safety in the best way. The breeding strategies are focused on utilization of autochthonous breeds. Balkan goat, as a good locally adapted breed, is a good basis for sustainable livestock production development.
The Balkan goat is an indigenous breed traditionally reared in mountain regions of the Balkan Peninsula, under very modest conditions and poor pasture. The obtained results of 92 milking Balkan goat showed its average body weight of 40 kg, withers height of 65 cm and milk yield of 160 kg, with 3,4 % milkfat and 3,3 % protein, litter size was 1,6. Milk is used for traditional manufacturing of cheese, certified for organic production and registered as an autochthonous product in the list of protected food names (PDO, PGI). According the our data, mastitis, ketosis and locomotory disorders have not been diagnosed. The genetic resistance to diseases is the most prominent characteristic which gives advantages in sustainable farming. Also, the Balkan goat genetic characterization based upon protein polymorphism has been done.

Possibilities for reduction of excretion (N, P, CH4, trace elements) from food producing animals from the European view

G. Flachowsky, Federal Agricultural Research Centre (FAL), Institute of Animal Nutrition, Bundesallee 50, D-38116 Braunschweig, Germany

High portions of consumed energy, protein and minerals are excreted with feces, urine or gases depending on substance type, level of feeding and production and animal species/-categories. To reduce the excretion of trace elements, upper limits in the feedingstuffs are introduced by the EU for Cu and Zn. For nitrogen and phosphorus the maximum level is related to the agricultural area. Animal nutritionists have some possibilities available to reduce the excretion from ruminants and nonruminants. It is well established that an increase of performance increases the excretion per animal. However when related to the unit food of animal origin the excretion decreases. Furthermore there exist specific possibilities to decrease the excretion. Examples are given in the following:

Nitrogen: - N-excess in the ration should be avoided,
           - amino acid supply should meet the requirements of the animals,
Phosphorus: - to meet the requirements of animals
           - supplementation of phytase to improve the P-availability from phytat in nonruminants
Methane: - consider ration composition and include methane reducing substances
Trace elements - consider the native content of trace elements in the feeds for requirements
(e.g. Cu, Zn): - use trace elements with a high bioavailability.

The presentation deals with details to reduce the excretion of food producing animals under consideration of European production systems and the EU regulations.

Nitrogen balance and ammonia emission from slurries of heavy pigs fed diets with high fibre contents

G. Galassi, L. Malagutti and G.M. Crovetto*, Istituto di Zootecnia Generale, Università degli Studi, via Celoria 2, 20133 Milano, Italy

Aim of the experiment was to investigate the influence of two different kind of non-starch polysaccharides (NSP) on N balance and ammonia emissions from slurries in the Italian heavy pig. Thirty Landrace x Large White barrows (153 kg BW) were divided into 5 groups fed the following diets: C (control: 51,5% maize, 36% barley, 8% soybean meal), and 4 high fibre diets where 12 or 24% barley was replaced by equal amounts of wheat bran (WB) or dried beet pulp (BP). Each pig was placed in individual metabolic cage for 7 days of adaptation and 7 days collection of excreta. Ammonia emission was registered, for 14 days and for each pig, on 2 kg slurries (Derikx and Aarnink, 1993). Faecal N excretion (% of intake N) was significantly increased ($P<0.05$) by NSP: 10.0, 12.3, 16.3, 16.3, 19.3 for C, WB12, WB24, BP12, BP24. On the contrary, urine N was numerically decreased: 46.2, 41.3, 44.2, 38.7, 41.9% of IN, respectively. Ammonia emission from the 2 kg slurries in the 14 days was not reduced by WB, but it was numerically lowered by BP: 613, 603, 647, 511, 458 mmol equivalent to 10.4, 10.3, 11.0, 8.7, 7.8 g for pigs fed C, WB12, WB24, BP12, BP24, respectively. For BP12 and BP24 this represents a reduction of 16.6 and 25.3% in comparison with C.

Effect of diet manipulation on manure characteristics: I - Agronomical features and atmospheric emissions

O.C. Moreira[1]*, S.A. Sousa[2], M[a].A. Castelo-Branco[1], J.R. Ribeiro[1] and F. Calouro[1], [1]INIAP - Estação Zootécnica Nacional. 2005-048 Vale de Santarém. Portugal, [2]Environmental Technologies Center -ISQ. TagusPark, Apartado 012, 2780-994 Porto Salvo, Portugal

The aim of this study was to evaluate the effect of nitrogen and phosphorus dietary manipulation on the agronomical features and gaseous emissions of manure.

Three experimental diets: T (18% crude protein), N (15% crude protein and amino acids balanced) and P (18% crude protein, monocalcic phosphate) were studied in efficacy trials with pigs (35-100 kg). The originated manures were evaluated in fertilization studies, at four levels of application (0, 42.5, 85, 170 kg ha^{-1} of organic N). Sorghum was used as plant test. Air emissions were monitored at six times after application in the soil. The leachates from each treatment were collected from lysimeters. Vegetal biomass production was quantified.

Manure from pigs fed diet N presented the lowest levels of organic matter, nitrogen (about 35% less total nitrogen), ammonia, calcium, phosphorous, magnesium, copper, and zinc. Also the lowest ammonia and carbon dioxide emissions were observed for this manure. Leachate nitrates were superior where the highest levels of manure were applied and for P derived manure, followed by N and T. The maximum level of fertilisation presented the highest vegetal biomass productions, which corresponded in the 1st cut to the manures from diets N and P and in the 2nd cut to those from diet T.

Effect of diet manipulation on manure characteristics: II - Energy recovery of biogas

S.A. Sousa[1], M[a].A. Castelo-Branco[2], F. Calouro[2], J.R. Ribeiro[2] and O.C.Moreira[2]*, [1]Environmental Technologies Center -ISQ. TagusPark, Apartado 012,2780-994 Porto Salvo. Portugal, [2]INIAP - Estação Zootécnica Nacional -.2005-048 Vale de Santarém, Portugal

Minimizing livestock manure's impact on environment is nowadays one of the major agricultural challenges. The aim of this study was to evaluate the effect of the dietary nitrogen and phosphorus manipulation on the optimisation of biogas production.

Three experimental diets: T (18% crude protein), N (15% crude protein and amino acids balanced) and P (18% crude protein and bicalcic phosphate replaced by monocalcic phosphate) were studied in efficacy trials with pigs (35-100 kg).The originated manures were anaerobically incubated in experimental reactors with 3 liters capacity, at 35°C, for periods of 35 days. Media pH and fermentation gases: oxygen (O_2), carbon dioxide (CO_2) and methane (CH_4) were monitored.

Manure from pigs fed diet P presented the best profile of CH_4 production, with a peak of 33% of total gas productions in the 15th day of test, decreasing to near an half on the 24th day of incubation. Also for CO_2, this manure presented the highest production and a peak at the 15th day. After this date, all manures presented an increase of O_2 values indicator of a degradation of anaerobic conditions. Manure from pigs fed diet N showed the lowest gas productions, associated with higher pH values. It was conclude that N manure presents limited efficiency in energy recovery as biogas.

Development of technique for manufacturing activated carbon from livestock manure and its characteristics

H.C. Choi[1], J.I. Song[1], D.J. Kwon[1], J.H. Kwag[1], Y.H. Yoo[1], C.B. Yang[1], Y.T. Park[2], D.K. Park[3], K.S. Park[3] and Y.K. Kim[4], [1]National Livestock Research Institute, Suwon, Korea, [2]DongYang Carbon Co. [3]National Horticulture Research Institute, [4]Choongnam National University

This study was carried out to develop the technique for manufacturing activated carbon and to know its different characteristics from livestock manure and litter. Layer manure(LM), litter from broiler house(BL) and dairy barn(DL), compost from layer manure(LC) and pig manure(PC), and coconut shell(CS) were used. Dried livestock manure or litter was ground, coal tar was added as a binder, pelletized, dried, heated with N_2 gas at 400 °C for 1 hour, and activated with steam at the temperature of 750 °C for 1 hour. Volatile matter of LM, BL, DL, LC and PC was 18.8%, 31.0%, 49.8%, 22.3% and 11.6% respectively. Surface area(BET) of LM, BL, DL, LC, PC and CS was 259.8, 209.8, 63.5, 442.3, 812.9 and 1,040 m^2/g respectively. With scanning electron microscope (SEM) examination micropore appeared as a sponge like particles honeycombed with chambers. Pore size of activated carbon was ranged from 0.39 to 5.02 Å. Whereas coconut shell size was 0.30 Å. Iodine absorptiveness of activated carbon from livestock manure and coconut shell was 530-580 and 1000 mg/g repectively. Absorptiveness of activated carbon from layer manure for hydrogen sulfide and trimethyl amine was 74.5 % and 73.9 % at the accumulated flux of 60,000 ml. Whereas ammonia showed 15.2 % of absorptiveness at the accumulated flux of 10,000 ml.

Analysis of combining ability for double cocoon characters in silkworm

A.R. Seidavi[1]*, A.R. Bizhannia[2], S.Z. Mirhoseini[2], M. Mavvajpour[2] and M. Ghanipoor[2], [1]Islamic Azad University, Science and Research Branch, Iran, [2]Iran Silkworm Research Center

In Sericulture, the parents with high general combining ability are used generally for improvement of the silkworm population performance. Double cocoons are one important section of production by silkworm. Analysis of combining ability for double cocoon character is beneficial for understanding of its genetic specifications and improvement of production level. Heterosis, general combining ability and special combining ability were estimated for double cocoon character in silkworm lines including Japanese lines of 31, 103 and 107 and Chinese lines of 32, 104 and 110 and their hybrids. Heterosis effects were significant for all characters (P<0.001). Heterosis percentage for number of double cocoons at six hybrids of 31×32, 32×31, 103×104, 104×103, 107×110 and 110×107 were -32.673, -16.831, -60.000, -55.555, 24.066 and -9.991% respectively. Furthermore heterosis values for double cocoon percentage at these six hybrids were -54.034, -42.488, -39.161, -33.663, 32.456 and -4.854. It was distinguished that general combining ability for hybrids is accorded with their heterosis and studied lines is affected from additional effects and also their application resulted to production improvement. In crossbreeding programs combining ability and heterosis must be put under attention jointly together. The obtained results indicated that the traits related to double cocoon production are strongly under heterotic effects. This criterion can be applied to improve the efficiency of silkworm egg and cocoon production.

Optimization on salting out method for DNA extraction from animal and poultry blood cells

A. Javanrouh, M.H. Banabazi, S. Esmaeilkhanian, C. Amirinia, H.R. Seyedabadi and H. Emrani, Dept. of Biotechnology, Animal Science Research Institution, Iran

High quality and adequate DNA is a principle component in any molecular genetic study. This will support successful next steps. Genomic DNA can be extracted from different tissues, including hair, sperm and blood cells. There are several protocols for DNA extraction. Salting out is a choice method due to using high saturated sodium chloride instead of phenol in order to denaturizing and precipitating of proteins attached to DNA. We optimized salting-out procedure (Miller et al., 1989) in order to DNA extraction from animal blood cells. Optimization includes utilization of separate buffer instead of Buffy coat isolation, chloroform for DNA phase isolation and achievement to purified DNA and sodium acetate for more concentrated DNA. The optimized protocol would be more safe, simple, cheap and rapid. We used this protocol for DNA extraction from whole blood collected from sheep, goat, cattle, buffalo, horse and poultry in our lab and DNA samples had high quality and quantity. In contrast to animals, red blood cells in poultry are nucleated. Therefore, very low blood is enough to DNA extraction from poultry blood cells.

Bottleneck effect in four Japanese quail based on assign test using ten microsatellite loci

H. Emranii[1], F. Radjaee Arbab[3], C. Amirinia[1], M. H. Banabazi[1], A. Nejati Javaremi[2], A. Javanrouh[1], H. R. Seyedabadi[1], A. Abbasi[4] and A. Masoudi[1,4], [1]Dept. of Biotechnology, Animal Science Research Institution, Iran, [2]Dept. of Animal Science, Faculty of Agriculture, University of Tehran, Iran, [3]Dept. of Animal Science, Faculty of Agriculture, Tarbiat modares University, Iran, [4]School of Natural Science and Technology, Okayama University, Japan

The Japanese quail is valued for its egg and meat. Advantages of small body size, rapid generation turnover, and high egg production make it particularly suited for laboratory research and it has been recommended as a pilot animal for poultry.Fore strains of Japanese quail including Pharach, Panda, Tuxedo and Golden were first introduced to Iran at 1993. Understanding the population genetic factors that shape genome variability is pivotal to the design and interpretation of studies using large-scale polymorphism data. Bottleneck is important effect because it can increase demographic stochasticity, rates of inbreeding; lose genetic variation and fixation of mildly deleterious alleles. We study bottleneck effect for these populations based on assign test using 10 microsatllite loci. The results showed that Tuxedo population was bottleneck (P >0.05). Initial low size population can be a possible reason for this bottleneck. The introduction of new individuals into this population is suggested to resolve this problem.

Chicken DNA fingerprinting with molecular markers designed on interspersed repeats

M. Soattin[1]*, G. Barcaccia[2], M. Cassandro[1] and G. Bittante[1], [1]Department of Animal Science and [2]Department of Environmental Agronomy and Crop Science, Agriculture Faculty, University of Padova. Agripolis, Via dell'Università 16, 35020 Legnaro, Italy

The chicken genome contains 15% of repetitive DNA organized as short tandem repeats and several families of longer interspersed elements. This research deals with the development of a forensic genomics assay for the chicken DNA fingerprinting based on the analysis of micro- and minisatellite polymorphisms. The identification of breed-specific markers was based on the S-SAP and M-AFLP systems derived from the AFLP technology. Genomic DNA fingerprints were generated in 84 individuals belonging to six local breeds and one commercial line. A number of variation statistics were computed: the effective number of alleles per locus (ne=1.570), total and single-breed genetic diversity (Ht=0.334 and Hs=0.162, respectively) and the fixation index (G_{st}=0.515). The mean genetic similarity coefficients within and between local breeds were 0.799 and 0.557, respectively. Markers useful for the genetic traceability of breeds revealed significant sequence similarities with either genic or intergenic regions of known chromosome position. Sequence tagged site primers were designed for the most discriminant markers in order to develop multiplex non-radioactive genomic PCR assays. The identification of single local breeds according to multilocus genomic haplotypes is currently under evaluation. In conclusion, the setting up of a molecular reference system seems to be feasible for the precise identification of the chicken breeds considered.

Genetic polymorphism of two Egyptian buffalo breeds in comparison with Italian buffalo using two ovine and bovine-derived microsatellite multiplexes

A.R. Elbeltagi*[1],Salah Galal[2],F. ElKeraby[1], A.Z. AbdElSalam[2]; M. Blasi[3] and Mahasen Mohamed[1], [1]Animal Production Research Institute, Dokki,Giza,Egypt, [2]Faculty of Agriculture, Ain Shams Univ.,Cairo,Egypt, [3]LGS Genetic Service Lab, Cremona, Italy

The study was carried out to investigate genetic diversity in Nile-Delta and Southern-Egypt buffalo populations in comparison with the Italian buffalo using 15 polymorphic microsatellite loci, in two multiplexes. Animals used were 28, 38 and 38 representing the Delta, Southern-Egypt and Italian buffalo, respectively. Studied microsatellites were CSSM38, CSSM70, CYP21, CSSM42, CSSM60, MAF65, BM0922, CSSM19, INRA006, D5S2, BM1706, BMC1013, CSSM47, INRA026 and RM4. All microsatellites showed allelic polymorphism. Chi-square test for Delta and Soutrhern-Egypt showed significant differences in allelic distribution at CSSM70, CSSM38, BM0922, ETH02 and BM1706. Testing HW equilibrium showed significant deviations for both Italian and Delta populations. F_{IS} estimates indicated that Italian buffalo is relatively the most inbred population while Southern-Egypt is the only outbred. High level of genetic differentiation (F_{ST}) between the Italian group and each of the Delta and Southern-Egypt groups (0.083 and 0.076, respectively) were observed while Southern-Egypt group showed a lower level of genetic differentiation with the Delta group (0.014). Italian buffalo had the highest distance values with the two Egyptian groups (0.25 and 0.23) while much lower values between the two Egyptian groups (0.06) was observed. In conclusion, there was a reasonable genetic variation between the Italian and the Egyptian buffalos and a lower level of it between Southern-Egypt and Delta buffalo. Southern-Egypt buffalo could be considered as population distinct from the Nile-Delta buffalo.

Optimisation of the sampling strategy for establishing a gene bank: Storing PrP alleles following a scrapie eradication plan as a case study

J. Fernández[1]*, T. Roughsedge[2], J.A. Woolliams[3] and B. Villanueva[2] [1]INIA, Carretera Coruña Km 7, 28040 Madrid, Spain, [2]Scottish Agricultural College, West Mains Road, Edinburgh, EH9 3JG, UK, [3]Roslin Institute, Roslin, Midlothian, EH25 9PS, UK

We present an algorithm to determine the optimum number of animals contributing to a gene bank and their contributions for storing alleles of a locus at specific frequencies, while maintaining the genetic variability for other unlinked loci. Its application focuses on gene banks linked to eradication programmes of particular deleterious alleles. Besides being reservoirs of general genetic diversity, gene banks also would allow future reintroduction of the removed alleles and the associated diversity. The efficiency of the algorithm is tested using the case of scrapie eradication programmes. Different restrictions on the average coancestry of candidates are considered. The total costs incurred in the creation of the semen bank are also accounted for. Results show that the algorithm is able to find the combinations of candidate contributions fulfilling different objectives. The most important factors determining the optimal contributions are the genetic structure of the population (i.e., the allele frequencies) and the levels of diversity (coancestry) present in the population. Restrictions on the budget for the establishment of the bank limit the feasible solutions very much. Heterozygous are favoured over homozygous individuals as, for a given number of animals contributing to the bank, the use of heterozygotes leads to lower levels of coancestry.

Effect of erroneous pedigree data on efficiency of methods to conserve of genetic diversity in small animal populations

P.A. Oliehoek* and P. Bijma, Animal Breeding and Genetics Group, Wageningen University, PO Box 338, 6700 AH Wageningen, The Netherlands

Small populations are at risk because genetic drift lowers genetic diversity. An important strategy in conservation genetics is to minimise average coancestry, by breeding and/or conserving animals which are genetically important. An efficient method of minimising coancestry is using optimal contributions. Optimal contributions rely on information of coancestry among individuals, such as pedigree or molecular marker data. In practise, however, pedigree-data is often erroneous. Effects of pedigree errors on preservation of genetic diversity were investigated using Monte-Carlo simulation of random mating populations. Pedigree data of these populations were made erroneous in two ways. First parents were made unknown. Even when as little as 5% of parents were unknown, application of optimal contributions decreased diversity because offspring of missing parents appeared more diverse than they actually are. In the second scenario fathers were replaced with other males. Benefit of using optimal contributions decreased strongly and linearly with the proportion incorrect pedigree. Optimal contributions only minimised half of average coancestry when 10% of fathers were replaced compared to when pedigrees were completely correct. Results indicate that use of optimal contributions is very sensitive to quality of information on coancestry. Accurate pedigree administration has a strong influence on the possibilities to maximise diversity in small populations without aid of molecular markers.

Improving test day model genetic evaluation for the Holstein breed in Italy
F. Canavesi, S. Biffani, F. Biscarini, M. Marusi and G. Civati, ANAFI, Via Bergamo 292, 26100, Cremona, Italy

Genetic evaluation for production traits in the Holstein breed in Italy is based on a Random Regression Test Day Model (RRTDM) since November 2004. More specifically the model is a multiple lactation, multiple traits RRTDM, similar to the model used in Canada for official genetic evaluation. Fixed regression curve effect include time, region, age at calving, parity and season of calving. Last changes in the model included a new definition of the proof scale and of the genetic base. Production and somatic cells data from the February 2005 official evaluation were used to test a different definition of fixed regression curves in the model including year effect, days open effect and a combination of the two. Number of fixed curves increased from 456 to 17024 levels. Simple correlations among proofs ranged from 0.99 to 0.98. Rank correlation among proofs varied from 0.99 to 0.97. The effect of the year of production on the fixed curves showed that there was an increase in production level from 1988 to 1998 and then a slow decrease. The effect of days-open was generally toward the end of lactation. Mean absolute difference between observed and predicted valued was analyzed. Research is still ongoing in order to determine which of the four models better predict breeding values.

Effect of the threshold nature of traits on heritability estimates: Comparison between linear and threshold models
N.G. Hossein-Zadeh[*], A. Nejati-Javaremi, S.R. Miraei-Ashtiani and H. Mehrabani-Yeganeh, Animal Science Department, University of Tehran, Karaj, Iran

Two general classes of phenotypes are measured in animal breeding data, continuous and discrete. Many traits of importance, such as litter size, calving ease, disease resistance and livability are measured on a discrete scale. Among reproductive traits, litter size has most often been used as a selection criterion. Litter size is relatively easy to measure and report, and heritability estimates for litter size are generally higher than those of other reproductive traits such as fertility or lamb survival. In this study we used stochastic method to simulate litter size as a categorical trait. For genetic analysis with linear animal model, DFREML program and for genetic analysis with non linear animal model, MATVEC program were used and run for ten different replicates for each threshold and for each heritability within threshold.
Results show that when comparing heritability estimates obtained by MATVEC with those obtained using DFREML, the linear model has always underestimated true heritabilities but threshold model of MATVEC did not show a consistent trend. While threshold model may be trusted at higher levels of true heritability, it does not behave consistently at lower levels of heritabilities. Estimation of heritability by threshold model is completely depends on the threshold point and real heritability.

Heterogeneity of variance for milk traits at climatical regions in Holstein dairy cattle in Iran and the best method(s) for data transformation

Sh. Varkoohi[1]*, S.R. Miraei-Ashtian[1] and H. Mehrabani-Yeganeh[1], [1]Animal Sci. Dept, College of Agriculture, University of Tehran, Karaj, Iran

In this research, Heterogeneity of variance components in cattle populations at climatical regions in the first three lactations was studied. The data set included the following: 161328 records of first, 123369 records of second and 81013 records of third lactations, which was collected by Animal Breeding Center of Iran from 1983 to 2004. Records of three lactations were divided in the base of Domarten method. Bartlett test for heterogeneity of variance components was significant among all subgroups. In order to decrease the heterogeneity of variance components, we used several data transformation methods including Logarithmic, Square root and Arc sin transformations. Logarithmic transformation decreased the heterogeneity of variance components in the three lactations and other methods had not effect for removing the heterogeneity in any group. Genetic parameters and heritability were estimated for three lactations by MATVEC Microsoft, using animal model. Results showed that the heritability estimates of milk yield were decreased from the first lactation to the third; also the heritability estimates of transformed data were slightly higher than the original data. Comparison between estimated parameters in single trait and two traits analysis, before and after data transformation, showed that there were not significant differences between derived results.

For a new paradigm in farm animals' taxonomy

C. Drăgănescu, Agicultural University-Bd. Mărăşti 59 Bucharest Romania

Mason noticed (1951..1999) that "confusion can arise by the same breed having entirely different names in different part of the world", and "because comparable information about all breeds was hard to obtain" and to put order in Farm Animals Taxonomy, his "work remained at the stage of a dictionary" The dictionary have a great scientific importance, but did not solved a fond the problem. The present FAT paradigm is at the empirical level of 19-th century. The confusions persists and have the tendency, by some subjective raisons, to increase.. So the name Tsigai is used for three sheep breeds, from two different phyletic group (Tsigai, Ruda). The Buffon and Darwin Valachian sheep have now some four name (Ratska, Racka, Zackel, Valaska vitoroga) and is included in a wrong phyletic group ("Zackel", a translation of his latin name-O.a.strepsiceros, but given to a phylogenetically different group of valachian breeds). The Saracatsaniko (the sheep o Saracaciani Vlah tribe), is consideret a different breed by Bulgars and Turks and named Karakchan etc As the sine qua non basis of international cooperation in the management of AnGR is a "common language"; a standard international breeds nomenclature, based on a standardized description of each breed and a correct classification of them, it seem necessary that EAAP and FAO acted, for the development of a modern paradigm, of a matur science of Farm Animal Taxonomy, similar to the system and paradigm existing in zoology.

Native horse breeds diversity in Lithuania as assessed by blood markers
*R. Šveistienė and V. Jatkauskienė, Institute of Animal Science of LVA, R Žebenkos 12, LT-82317
Baisogala, Radviliškis distr., Lithuania*

The objective of our study was to examine the genetic differences and genetic variation of Lithuanian
native horse breeds. The main study was conducted in 1996-2005 at LVA Institute of Animal
Science. The pedigree, gene frequencies of blood group and serum protein markers, homozygosity,
and genetic similarities was analysed from 427 horses.

The genetic analysis of alleles indicated that Al^{AB} genotype was dominant of large type Zemaitukai
and Lithuanian heavy draught. The Al^{AA} genotype was dominant of Zemaitukai breed. Genotype
Tf $^{FF, FO, DF}$ in all three horse breeds were also frequent. The frequent of alleles EAD^{dk} of large
type Zemaitukai and EAD^{ad} of Lithuanian heavy draught breeds was distinguishing feature and
statistical reliable.

The effective population size $Ne=58$ for large-type Zemaitukai and $Ne=78$ for Zemaitukai are also
driftless reproduction and, therefore, the survival of the population in uncertain, but degree of
homozygosity is low. The Lithuanian Heavy-Draught horses have effective population size $Ne=157$
of vulnerable risk status but degree of homozygosity was higher for other breeds. The higher genetic
similarities, was detected between large-type Zemaitukai and Lithuanian heavy draught horses
breeds and the lowest between Zemaitukai and Lithuanian heavy draught horse breeds.

**Differentiation and relationship of three Iranian horse breeds (Turkoman Sahra, Turkoman
Jergelan and Caspian miniature) by using microsatellite markers**
*S. Behroozinia and S.Z. Mirhoseini, Fars Research center for Agriculture and Natural resoueces.
P.O.Box 71555-617, Shiraz, Iran*

Genetic diversity of two Iranian horse breeds (Turkoman Sahra, Turkoman Jergelan) have been
studied using microsatellite markers (HMS1, HMS3, HMS6, ASB2, AHT4). 30 individuals selected
from each population individually and randomly and their blood samples were taken. DNA samples
extracted by using phenol-chloroform method. Number of alleles varied from 1 (AHT4) to 5 (HMS3
and HMS6). AHT4 locus was monomorph in all populations. The highest averge hetrozigosity and the
lowest average of heterozigosity belonged to Turkomansahra (0.549) and Turkoman jergelan (0.534)
respectively. Polymorphic information content (PIC) estimated in all populations. HMS3 locus in
Turkomansahra indicated the highest PIC (0.7033) and the lowest this cretria (0.5517) was belonged
to HMS1 locus in Turkoman Jergelan population. Highest and lowest Shannon index were belonged
to HMS3 locus in Turkomansahra (1.43) and HMS1 locus in Turkoman Jergelan (1.037) respectively.
Genetic distance between populations estimated by using Nei method.The low genetic distance
between Turkomansahra and Turkoman Jergelan indicates that these populations have same origin.
UPGM dendrogram drawed for two studied populations by NTsys 2.02. Turkomansahra population
included three groups of horses (Akhal, Yamud and Chenarani) whereas, Jergelan population can cause
the only one group (Akhal). Therefor this population has less divers. The low number of Jergelan
population is a result of their lower genetic distance between these two populations.

Genetic diversity of the Czech and Slovak Thoroughbreds using 17 microsatellite loci
M. Burócziová, J. Říha,P,Humpolíček* Mendel University of Agriculture and Forestry, Zemedelska 1, 613 00 Brno, Czech Republic

We compared the genetic diversity between two horse breeds to Czech and Slovak Thoroughbred by means of 17 microsatellite loci /AHT4, AHT5, ASB2, HMS3, HMS6, HMS7, HTG4, HTG10, VHL20, HTG6, HMS2, HTG7, ASB17, ASB23, CA425, HMS1, LEX3 - recomended by ISAG. In same population size /234 individuals of each breed/, the allele frequencies, observed and expected heterozygosity, test for deviations from Hardy-Weinberg equilibrium and Polymorphism information content have been calculated for each breed. In conclusion, the main objective of study was to show the level of genetic distance among the Czech and Slovak Thoroughbred horse breeds with very short history of breeding. We analyzed genetic distance and diversity between them on the base of the dataset of highly polymorphic set of microsatellites representing all autozomes using set of PowerMarker v3.28 analysis tools and three ML algorithms /IB1, k-means clustering and Naïve Bayes classifier/ for results comparison. *This work was supported by Czech Science Foundation (Project No. 523/03/H076)*

The genetic structure of the autochthonous horse breeds in Croatia
A. Ivanković[1], P. Caput[1], P. Mijić[2], J. Ramljak[1], M. Konjačić[1], B. Mioč[1], [1]Faculty of Agriculture, Svetošimunska 25, Zagreb, Croatia, [2]Faculty of Agriculture, Trg. Sv. Trojstva 3, Osijek, Croatia

Autochthonous horse breeds make the most numerous group of horses in Croatia. They were bred on the then horse population "busak" which, depending on the climate, was differently conformational and genetically profiled. With the aim of improving the conformation of the then "busak", during the 20th century, quality stallions of heavy European horse breeds were imported and used for "improving" the existing population. Posavina horse and Croatian coldblood remained in the marginal pasture areas in a small number which demands a constant monitoring. Murinsulaner horse population is reduced to thirty heads. Genetic variation of the autochthonous horse breeds in Croatia was analysed using eight microsatellite loci (43 Posavina horses, 40 Croatian coldblood and 18 Murinsulaner horses). The frequency of observed alleles in Croatian coldblood and Murinsulaner horses were more similar compared to the Posavina horses. The number of alleles were higher in Posavina horse. The Croatian coldblood and Murinsulaner horses are closely related, whereas their estimated genetic distances to the Posavina horses are significantly higher. Genetic distance between breeds is important for breeding strategies and preservation of originality of autochthonous horse breeds.

Pedigree analysis of the endangered Amiata donkey breed (Tuscany, Italy)

F. Cecchi[1], R. Ciampolini[1], E. Ciani[1], E Mazzanti[1], M. Tancredi[2], C. Dominici[2], B. Matteoli[2], S. Presciuttini[3] and A. Rosati[4], [1]Dipartimento di Produzioni Animali, Viale delle Piagge 2, 56124 Pisa, Italy, [2]Collaboratore esterno. [3]Centro di Genetica Statistica, S.S. Abetone e Brennero 2, 56127 Pisa, Italy. [4]EAAP, Via Nomentana 134, 00162 Roma, Italy*

The genetic-demographic structure of the Amiata donkey, an endangered breed from Tuscany (Italy), was investigated using information from pedigrees. Genealogical data of 608 donkeys were recorded in a database and analysed. Population size increased from 89 subjects in 1995 to 503 (129 males and 374 females) in 2005. Animals were distributed among 152 herds, though the effective number of herds was 21; this suggests that a small number of herds provided stallions for the entire breed. Mean generation lengths were 7.0 ± 3.7 in females and 5.9 ± 2.6 in males. Number of animals in the reference population was 286, whereas number of animals in the base population was 232. As 83 of these animals had one parent unknown, the number of half founders was 190.5. Among 24 animals with 4-generation history, 3 (12.5%) were 25% inbred. The present work represents a first step towards an efficient management of the breed to control the level of inbreeding with an appropriate breeding scheme and to improve the desired population traits.

Assessment of Gene flow level among nucleus of feral Losina horse using molecular coancestry information

P.J. Azor[1], M. Valera[2], J. Martínez-Sáiz[3], M.D. Gómez[1] and A. Molina[1], [1]Department of Genetic. University of Cordoba. Campus de rabanales. 14071 Cordoba. Spain, [2]Department of Agro-forestry Science. EUITA. University of Seville. Spain. [3]Consejería de Sanidad y Bienestar Animal. Juntad e Castilla y León. Burgos. Spain*

Losino horse is a endangered local breed of Spain with only 200 animals left. There are two isolated subpopulations into the breed (a feral nucleus: Pancorbo and the other stabling nucleus: Quincoces). The aim of this work is determine the gene flow level among both subpopulation using molecular data. 120 individuals were genotyped using 20 microsatellite. Genetic variability parameters, within populations molecular coancestry coefficients (f_{ij}) and between subpopulations kinship distance (D_k) were computed using the program MolKin v2.0. Recent migration rates among populations (m) were estimated using the BayesAss+ program using a burning and a data collection period of 3×10^6 iterations. The average number of alleles per locus was 7.5 for Pancorbo and 6.3 for Quincoces nucleus. F-statistics, F_{IS}, F_{ST}, and F_{IT} for the whole analyzed population were, respectively, -0.019, 0.022 and 0.002. Within subpopulations molecular coancestry was 0.267 for both nucleus and Dk among nucleus was 0.408. Pancorbo nucleus has not received a significant proportion of migrants. Significant recent introgression rates (approximately 10 %) were found from the Pancorbo to Quincoces nucleus. Average standard deviations of all distributions of m for each source population ranged from 0.006 to 0.039.

Genetic relationships among Spanish goat breeds using molecular, nuclear and mitochondrial, information: Preliminary results

P.J. Azor[1], M. Luque[2], M. Valera[3], M. Herrera[4], A. Membrillo[1], E. Rodero[4], I. Cervantes[1]* and A. Molina [1], [1]Department of Genetic. University of Cordoba.E-14071. Córdoba. Spain, [2]Agriculture Department. FAO. Rome. Italy, [3]Department of Agroforestry Science. EUITA. University of Seville. Spain, [4]Department of Animal Production. University of Cordoba. Spain

In this work we have studied the genetic relationships among six Spanish goat breeds of meat aptitude using molecular coancestry information by means of a set of 20 microsatellite. We have sampled 210 unrelated goats (30 per breed) belonging to : Blanca Celtibética, Blanca Andaluza, Pirenaica, Azpi-Gorri, Moncaína and Negra Serrana and Malagueña breed of milk aptitude as out group. The within breeds molecular coancestry *(fij)* coeffient varied from 0.314 for Malagueña breed to 0.450 for Negra Serrana breed. The between breeds molecular coancestry values varied from 0.235 (for the Blanca Andaluza-Malagueña breeds pair) to 0.419 (for the Moncaina-Pirenaica pair). The average molecular coancestry for the whole population was of 0.347 and the mean self-coancestry was 0.745. The between breeds kinship distance, (Dk) varied from 0.342 (for the Pirenaica-Moncaína pair) to 0.491 (for Blanca Andaluza-Blanca Celtibérica pair). F-statistics, F_{IS}, F_{ST}, and F_{IT} for the whole analysed population were, respectively, of 0.130, 0.103 and 0.220. F_{IS} value ranged from 0.106 for Negra Serran breed to 0.172 for Azpi Gorri breed. Malagueña breed presented the lowest value for this parameter (0.089). Moncaína and Pirenaica breed (from the north of Spain) showed the lowest F_{ST} distance (0.021).

Genetic substructure of the Spanish Guadarrama goat breed

J.H. Calvo[1]*, M. Martínez[2], F.J. Cuevas[2], P. Díez de Tejada[2], A. Marcos-Carcavilla[3] and M. Serrano[3], [1]CITA, apdo 729, 50080-Zaragoza, Spain, [2]Complejo Agropecuario de Colmenar Viejo, Colmenar Viejo, Spain, [3]INIA, 28040-Madrid, Spain

The Guadarrama goat breed is an autochthonous dairy breed of Spain, which is managed under a special protected way. The objective of the present work was to develop a genetic characterization of this breed to evaluate its genetic variability and population structure. Genetic variability at 9 microsatellite loci was analysed at two different hierarchic levels in the data of the Guadarrama Spanish goat breed: (1) the breed itself, i.e. the whole data set without division into flocks; and (2) eight flocks. The 25% of the total population (2100 animals from eight flocks) were analysed. Allele frequencies and heterozygosity revealed high genetic variation. Noticeably, private alleles at a significant frequency (p>0.1) were found for some flocks. The degree of population subdivision calculated between flocks from FST indices was around 4% of the total variability. A Bayesian model-based clustering analysis assuming from k=1 to k=8 clusters were performed using STRUCTURE 2.1 software. Results showed that the model assuming k=8 clusters (subpopulations) was the most probable grouping. Values of DReynolds genetic distance indicated a clear differentiation between the flocks. DReynolds distance between flock pairs revealed that Population1/Population2 and Population7/ Population8 populations are very close. In conclusion, Guadarrama goat breed show good diversity measures that can permit a higher selection pressure without increasing inbreeding related to the whole population (FIT = 0.051) and within population (FIS = 0.007).

Genetic variability in twelve microsatellites and their utility for parentage analysis in Portuguese Serrana goats

M.F. Santos-Silva[*], A.I. Lopes, M.I. Carolino, M.C.O. Sousa and L.T. Gama, Estação Zootécnica Nacional, Vale de Santarém 2005-048, Portugal*

The objective of this work was to analyze genetic variability for twelve microsatellite *loci* (OarFCB0304, ADCYC, HSC, OarCP0049, McM527, CSRD247, OarFCB0020, OarAE129, MAF214, MAF65, SRCRSP0008, BM1818) in Serrana goats, as a preliminary activity to be enlarged to other Portuguese goat breeds. Samples of 31 unrelated individuals were obtained from 19 flocks. Laboratory techniques were developed to coamplify all *loci* in a single PCR reaction, and 10 *loci* were successfully amplified in a multiplex system. To reduce costs and time, BM1818 and OarCP49 were excluded from the analysis, because they could not be amplified in the same reaction. Amplified fragments were separated by capillary electrophoresis in an ABI 310 automatic sequencer, and analysed with appropriate software.

The effective number of alleles ranged from 2,02 to 5,77, and polymorphism information content from 0,47 to 0,81, for the ADCYC and HSC locus, respectively. Overall, the average observed and expected heterozygosity was 0,65 and 0,73, respectively. The probability of parentage exclusion was calculated for each locus (0,29 to 0,67), and for all 10 *loci* taken together (0,999552).

In Serrana goats, HSC is the most, and ADYCD the least informative among the microsatellites analyzed. The set of ten *loci* studied should be enough to give a satisfactory paternity identification.

Ancient autochthonous genetic type (AAGT) 'Casertana' pig: Genetic characterization by microsatellite analysis

D. Matassino[1,2], M. Occidente[1], N. Castellano[1], D. Fornataro[1] and C. Incoronato[1], [1]ConSDABI - NFP.I.- FAO.- Centro di Scienza Omica per la Qualità e l'Eccellenza nutrizionali, Piano Cappelle 82100 Benevento, Italy, [2]DSBA- Università degli Studi del Sannio - Via Port'Arsa,11-82100 Benevento, Italy*

Casertana pig AAGT, known ever since antiquity and defined the 'Italian pig pride' by Hoesch (early half of 10th century), is currently, mostly present in Campania region (Italy). A wide safeguard programme promoted by ConSDABI is in progress for this AAGT. The aim of this study is to contribute to the knowledge of genetic variability degree of this AAGT using potential of microsatellite markers. The research was carried out on 30 subjects reared at ConSDABI experimental Farm, examining 20 microsatellite *loci*. All considered *loci* were polymorphic. The most polymorphic *loci* (SW 2038, SW 1370, SW 1035, SO017, SW 1823, SW1556, Sw 1873, SW 240), with five alleles, had a 51,3 % incidence upon the total allele number up to now typified. The average allele number per *locus* was 3,9 + 1,04 (*c.v.*, 27%); the average expected and observed heterozigosity were 0, 59 + 0, 13 (*c.v.*, 22%) and 0,66 + 0, 17 (*c.v.*, % 22), respectively.

The results evidenced that, within the limits of the observation field, the population preserves an amount of genetic variability though it was constituted by a reduced number of *'founders'*.

Ancient autochthonous genetic type (AAGT) 'Nero Lucano' pig: Genetic characterization by microsatellite analysis. Preliminary results

C. Incoronato[1], M. Blasi[2], C. Cosentino[3], M. Occidente[1], A. Perna[3] and D. Matassino*[1,4], [1]ConSDABI-NFP.I.- FAO.- Centro di Scienza Omica per la Qualità e l'Eccellenza nutrizionali, Località Piano Cappelle 82100 Benevento, Italy, [2]LGS - Via Bergamo, 292 – 26100 Cremona, Italy, [3]DSPA– Università degli studi della Basilicata – viale dell'Ateneo Lucano, 10 - 85100 Potenza, [4]DBSA- Università degli Studi del Sannio - Via Port'Arsa, 11-82100 Benevento, Italy

The recovery of AAGTs represents an example of integration between environmental and productive goals. The aim of this study was to provide, by means of typification of 41 subjects to 19 microsatellite *loci*, a first contribution to the knowledge of the genetic variability of '*Nero Lucano*', whose origins could go back to the ancient pig present in Lucania region (Italy). The results, within the limits of the observation field, evidenced an average allele number per *locus* equal to $3,2\pm0,89$ (*c.v.* = 28 %) ranging from 5.0 for SW 2038 *locus* and 1 for SO 386 *locus*; on the contrary, the latter *locus* is polymorphic in other 12 AAGTs ($2\div9$ alleles).
The mean observed and expected heterozygosity were $49,13 \pm 22,44$ (*c.v.* = 46 %) and $47,86 \pm 18,02$ (*c.v.* = 38 %), respectively. These preliminary results must be confirmed by increasing sampling. Indeed, monomorphism of SO 386 *locus* might be used for genetic discrimination of meat and its products.

Microsatellites applicable for determination of genetic diversity in Sea eagle (Haliaeetus pelagicus) population from Sachalin

L. Putnova[1], A. Hovorkova[2], B. Kral[3], V. Masterov[4], M. Hovorka[3] and P. Humpolicek[1*], [1]Department of Animal Morphology, Physiology and Genetics, MZLU, Zemedelska 1, 61300 Brno, Czech Republic, [2]Laboratory of Molecular Genetics, Genservice Ltd., Palackeho 1/3, 61242 Brno, Czech Republic, [3]The Zoological garden of the Brno City, U zoologicke zahrady 46, 63500 Brno, Czech Republic, [4]Department of Vertebrate Zoology, Biological faculty of Moscow State University, Leninske gory 1-12, Russia

No microsatellite markers have been described for Sea eagle (*Haliaeetus pelagicus*). Application of such markers is essential to the management and conservation of this species. From this reason we used 16 markers developed for Spanish imperial eagle and 3 markers developed for *aquila* and *haliaeetus* eagles to assess the individual identification, parentage analysis and genetic variability of Sea eagle population from Sachalin. A total number of 36 alleles were obtained. The number of alleles at individual loci ranged from 1 to 5. The highest heterozygosity and polymorphism information content was observed for locus *Aa39* (75%), *Aa57* (63%), *IEAAAG12* (47%), *Aa15* (34%), *Aa27* (31%), *Aa36* (28%) and *IEAAAG04* (28%). The probabilities of paternity exclusion/ one parental genotype unavailable/and parentage exclusion were for this panel of 19 microsatellites 90.62%/67.37%/98.16%, respectively. Microsatellites were tested in three other eagle species: the white-tailed eagle (*Haliaeetus albicilla*), the golden eagle (*Aquila chrysaetos*) and the imperial eagle (*Aquila heliaca*). Supported by Eurasian Regional Association of Zoos and Aquariums (EARAZA) and by Czech Science Foundation (no. 523/03/H076).

Formation process study of Andalusian bovine local breeds in danger of extinction

E. Rodero[1], P. Azor[2], M, Luque[1], M. Valera[2], A. González[1], A. Molina[2] and I. Cervantes[2], [1]Departament of Animal Production, Universidad de Córdoba. Spain, [2]Genetic Departament. Universidad de Córdoba. Spain*

A paper on breed characterization and genetic relationships among Berrenda en Colorado(BC), Berrenda en Negro (BN), Pajuna (P) and Cardena (C) cattle breeds was presented at a previous EAAP meeting. The present paper involves 168 BN animals (23 farms), 262 BC animals (25 farms), 14 C animals (1 farm) and 51 P animals (3 farms); 32 DNA markers were used for the genetic studies. Gene flow among the four breeds and farms within breeds was estimated: values for among breeds are low, (3,78 and 3,29) among C and Berrenda (BN and BC); little higher among P and Berrenda (BN and BC) (6,71 and 6,89); lower limits were found (2,91)among P and Berrenda (BN and BC) and the highest between BN and BC (18,42). The values obtained are not determinants for farms with the breed, especially for Berrenda cattle breeds (BN and BC). The results agree with previous ethnologic studies about common breed origins and mating systems in populations. The divergence times among the populations studied were obtained, resulting 13,22 generations for Berrenda cattle breeds (BN and BC) and 403,75 generations for P and C. These values were obtained without taking into account gene flow among breeds, when previous data show that it is high, at least for Berrenda breeds.

Crossbreeding effects of temperament traits in beef cattle

C. Plachta[1], H. Brandt[1], M. Gauly[2] and G. Erhardt[1], [1]University of Giessen, Institute for Animal Breeding and Genetics, 35390 Giessen, [2]University of Göttingen, Institute for Animal Breeding and Genetics, 37075 Göttingen, Germany*

In order to estimate crossbreeding effects of temperament traits in beef cattle, German Angus (GA, n=44) and German Simmental (GS, n=97) cattle and their reciprocal crosses (n=320) were tested in two consecutive years using a tethering test at an age of 5 weeks and a separation and restraint test 2 weeks after weaning at an age of 233 ± 12 days, respectively. GS calves received higher scores in both tests (tethering score, handling score, before handling score), showed a higher locomotory activity (time in locomotion during tethering, time spent running with/without handler, time until animal reached the corner, time spent running in restraint yard) and could be restrained in the corner for a shorter time than GA calves. Estimated heterosis effects ranged from -14 % (time until animal reached the corner) to 13 % (time in locomotion during tethering). However, none of these heterotic effects were significant ($p > 0.05$). The reciprocal crosses showed significant ($p<0.05$) differences in the behavioural scores during the restraint test (before handling score, handling score); crossbred calves received similar scores as their pure bred counterparts. Heterotic effects show no consistent tendency in the parameters taken, which underlines the complexity of behavioural genetics and the need for more research in this field.

Grinding level and extrusion effects on the nutritional value of lupin seed for ruminants

E. Froidmont[1]*, M. Bonnet[2], V. Decruyenaere[3], P. Rondia[1], N. Bartiaux-Thill[1], [1]CRA-W, Animal Production and Nutrition Department, rue de Liroux 8, 5030 Gembloux, Belgium [2]FUSAGx, Unité de Zootechnie, Passage des Déportés 2, 5030 Gembloux, Belgium, [3]CRA-W, Farming systems, rue de Serpont, 6800 Libramont, Belgium

Four grinding levels (average particle size: 1.0 mm, 2.5 mm, 4.0 mm and 5.5 mm) and extrusion (180°C, 30 sec.) were applied to lupin seeds (*Lupinus albus*, var. Lublanc) for measuring the effects of these treatments on the supply of digestible proteins in the small intestine (DP) of Belgian Blue bulls. The experiment was led on 5 fistulated animals according a 5*5 Latin square design. DP from microbial and dietary origins were reduced by 1.0 mm grinding. Extrusion and 4.0 mm grinding optimized dietary DP. However, the 5.5 mm grinding induced a higher ruminal degradability of lupin proteins due to a more intense rumination. The 4.0 mm grinding reduced the rumen degradability compared to 1.0 mm grinding (74 *vs* 86%). Total DP supplies by lupin seeds were determined to 99, 149, 151, 134 and 176 g/kg for the 4 increasing grinding levels and extrusion. The difference amongst 1.0 mm grinding (raw lupin) and extruded lupin is confirmed by most of feeding standards. However, the coarse grindings (closer to practical using forms) supplied 45% more DP than 1.0 mm grinding. This difference is not considered in feeding standards which underestimate the true nutritional value of lupin seeds for ruminants. These results were completed by an *in vitro* trial.

Effect of carvacrol and cinnamaldehyde on performance of growing lambs

A.V. Chaves[1],*, K. Stanford[2], L. Gibson[3], T.A. McAllister[1] and C. Benchaar[4], [1]Agriculture and Agri-Food Canada, Lethbridge, AB, Canada, [2]Alberta Agriculture Food and Rural Development, Agriculture Centre, Lethbridge, AB, Canada, [3]Agriculture and Agri-Food Canada, Lacombe, AB, Canada, [4]Agriculture and Agri-Food Canada, Lennoxville, QC, Canada

Effects of plant extracts on lamb performance and carcass characteristics were determined using 60 lambs (24.6 ± 0.77 kg initial live weight, LW) receiving 0 (control) or 260 mg/d of plant extracts (carvacrol or cinnamaldehyde) in iso-nitrogenous and iso-energetic barley- or corn-based diets. Diets were fed ad libitum in a 2×3 random block design over 11-week period. Saleable meat yield (as proximal cuts) from the carcasses were assessed. Feeding plant extracts did not affect dry matter intake (DMI), but they increased the average daily gain (ADG) of ewe lambs fed barley, as compared to the control diet (278.2 vs. 214.8 g/d; $P < 0.05$). Male lambs exhibited greater ($P < 0.01$) DMI and ADG (1402.5 g/d and 346.1 g, respectively) than did the female lambs (1222.3 g/d and 257.8 g). The males had heavier final liveweights, but saleable meat yield was similar ($P > 0.05$) between sexes (22.3 ± 0.40 kg in males vs. 21.9 ± 0.55 kg in females). No effects of plant extracts on ruminal pH or concentrations of ammonia, total VFA or individual VFA were observed, but acetate:propionate ratio was reduced by plant extracts when barley was fed (average 1.16 compared with 1.34 without supplements).

Effect of sheep breed on hay intake, in vivo digestibility and hay ruminal kinetics as affected by soybean supplementation
A.L.G. Lourenço*, P. Fontes and A.A. Dias-da-Silva, CECAV-UTAD Department of Animal Science, P.O. Box 1013, 5001–801 Vila Real, Portugal

The effects of sheep breed and soybean supplementation of a low quality grass hay on intake, in vivo digestibility and ruminal kinetics were evaluated. Four rumen cannulated Churra-da-Terra-Quente autochthonous ewes (CTQ, 39.2 kg BW) and four Ile-de-France exotic ewes (IF, 71.4 kg BW) were used. All animals were fed hay ad libitum and, within each breed, half of the ewes were supplemented with soybean meal (150 g/kg hay DM intake). After two weeks of adaptation and four weeks of measurements, diets within breeds were switched and all procedures and measurements were repeated. The intake and in vivo digestibility data are consistent with results of previous work Soybean supplementation didn't improved hay intake (P<0.55) or its in vivo digestibility (P<0.70) but improved rumen N-NH3 (P<0.001) and the ruminal rate of NDF disappearance (P<0.061). Dry matter intake, expressed per kg of BW, was higher (P = 0.03) and in vivo digestibility was lower (P<0.051) in CTQ than in IF ewes. No difference in intake between breeds was observed per kg metabolic BW ($BW^{0.75}$). Breed didn't affect ruminal pH (P=0.36) or N-NH3 (mg/L, P=0.88). There were no differences between breeds in rumen fermentation patterns of the hay (P>0.11). The lower effective rumen disappearance of hay DM and NDF in CTQ and its lower in vivo digestibility is a consequence of its higher rumen passage rate.

Ruminal protein degradation characteristics of roasted soybean meal using nylon bags and SDS-PAGE techniques
A.A. Sadeghi[1]* and P. Shawrang[2], [1]Department of Animal Science, Faculty of Agriculture, Science and Research Branch, Islamic Azad University, Tehran, Iran, [2]Department of Animal Science, Tehran University, Karaj, Iran

This study was carried out to determine effect of roasting at 125°C for 10, 20 or 30 min on protein degradability and intestinal digestibility of soybean meal (SBM). Four non-lactating Holstein cows fitted with rumen fistulas and duodenal cannulae were used for *in situ* incubations. Ruminal disappearance of CP was measured for up to 48 h. Fitting degradability data of SBM to exponential model showed that roasting decreased linearly (P<0.001) the water soluble fraction, rate of degradation and increased linearly (P<0.001) the potentially degradable fraction of CP. Roasting for 10, 20 and 30 min decreased linearly (P<0.001) the effective degradability of CP by 10, 17 and 20%, respectively. Roasting for 30 min decreased (P<0.05) intestinal mobile bag digestibility of ruminally undegraded CP, compared with 20 min. Electrophoresis results showed that β-conglycinin subunits of untreated SBM disappeared completely within the shortest incubation period, whereas glycinin subunits were degraded in the middle of incubation period. In roasted SBM, β-conglycinin subunits were degraded in the middle of incubation and glycinin subunits were degraded in the longest incubation period. It was concluded that roasting at 125°C for 20 min had the greatest potential to increase rumen undegradable protein of SBM.

Comparison of ensiled grass-white clover and grass-red clover mixture for dairy cattle
S. De Campeneere*, J.L. De Boever and D.L. De Brabander, ILVO, Animal Science Unit, Scheldeweg 68, 9090 Melle, Belgium

Recent legislation, including financial stimuli, promotes the use of clover in Flanders. However, pure culture of clover has certain disadvantages, which may disappear by growing a mixed culture of grass and clover (red or white). At the same time, the advantages of these two species can be combined. To examine the potential use of these crops in dairy cattle nutrition, two diets were compared using 14 lactating Holstein cows in a cross-over design with 2 periods of 4 weeks. The diets consisted of grass-clover (diet 1: white and diet 2: red clover) and maize silage (60/40 on DM base). Before and after the trial a three-week control period was added to compare the clover diets with a traditional diet of maize silage/prewilted grass silage concerning intake and performance. During the trial roughages were fed ad libitum and completed with concentrates to provide approx. 105% of their requirements. The net energy of both crops was calculated from in vivo digestibility determined with wethers, whereas the protein value was based on rumen degradability characteristics determined in fistulated cows.

Dietary potassium affects fecal consistency in dairy cows
J. Sehested[1]*, P. Lund[1] and H.B. Bligaard[2], [1]Danish Institute of Agricultural Sciences, P.O Box 50, DK-8830 Tjele, Denmark, [2]Danish Agricultural Advisory Service, Udkaersvej 15, DK-8200 Aarhus N, Denmark

Liquid feces has negative impact on hoof and udder hygiene and health. We hypothesised that fecal consistency is affected by potassium(K) and sodium(Na) intake through electrolyte homeostatic mechanisms in the hindgut, and by indigestible dietary fibers through their water binding capacity in the hindgut.
A 4*4 Latin square experiment with 20 lactating dairy cows (mean ECM=28.2 kg), 4 periods of 7 days and 4 dietary treatments was conducted. A total mixed ration based on maize silage was used on all treatments. Dietary treatments were: 1) Low-Na/Low-K; 2) Low-Na/High-K; 3) High-Na/High-K; 4) High-Na/High-K-high fiber. Treatment levels were 12 or 35 g K and 1 or 10 g Na per kg dry matter by addition of chloride salts, and low or high indigestible fiber content by iso-energetic substitution of rolled oats by alfalfa pellets (appr. 4 kg dry matter cow^{-1} day^{-1}).
Fresh fecal consistency was scored by a 9 step score-system (1=liquid, 5=firm), and fecal dry matter content and water-binding capacity were measured. Fecal score was linearly related to fecal dry matter content. Fecal score decreased from 3.9±0.1 with low Na and K to 2.8±0.1 with low Na and high K. Fiber did not influence fecal score or water-binding capacity.
In conclusion, dietary K is a so far unrecognised factor with high impact on fecal consistency.

The role of cactus pear (Opuntia ficus-indica) for ruminant feeding systems in dry areas
Firew Tegegne, K.J.Peters* and C.Kijora, Animal Sciences Institute, Humboldt University of Berlin

Cactus pear (*Opuntia ficus-indica*) has multiple importances in dry areas including the Mediterranean Regions of Europe. Two 90-day experiments in a randomised complete block design (8 sheep/treatment) were conducted to determine the optimum cactus pear (C) supplementation level and assess its contribution as water source and complementarity with urea-treated straw. In experiment I, C replaced pasture hay at 0, 20, 40, 60 and 80% (C0, C20, C40, C60, and C80, respectively), on DM basis. Experiment II consisted of untreated wheat straw(S), S+C(SC), S+C+ wheat bran(B)(SCB), urea-treated straw(TS)(5% urea), TS+C(TSC) and TS+C+B(TSCB), with supplement(C and/or B) rate of 40%. In experiment I, cactus pear supplementation affected total DM, nutrients and water intakes and live weight gain (LWG)(P<0.001). The highest DMI (92 g/kgW$^{0.75}$) was recorded in C60 while comparatively high (+33.0 g/day) LWG was recorded in C20. Sheep on C40, C60 and C80 drunk negligible amount of water (8, 17, and 6 ml/day, respectively). In experiment II, cactus pear and urea-treatment significantly (p<0.001) affected total DM and nutrients intakes and LWG. Sheep on TSCB had the highest LWG (75.5 g/day). In conclusion, cactus pear could optimally substitute pasture hay up to 60% and satisfy the water requirement of sheep. Cactus pear and urea-treatment appeared complementary, improved straw-based diet intake and sheep performance. Diet six (TSCB) appears to be a promising package for dry season feeding systems.

Stable C and N isotope analyses in muscle to discriminate dietary background in lambs
S. De Smet[1], K. Raes[1], E. Claeys[1], P. Boeckx[2] and S. Kelly[3], [1]Laboratory for Animal Nutrition and Animal Product Quality, Ghent University, Proefhoevestraat 10, 9090 Melle, Belgium, [2]Laboratory of Applied Physical Chemistry-ISOFYS, Coupure Links 653, 9000 Gent, Belgium, [3]Institute of Food Research, Norwich Research Park, Colney, Norwich NR4 7UA, UK

The potential of stable C and N isotope analyses in muscle for tracing back diets was examined in two lamb studies. In trial 1, lambs (n=4 per group) were fed indoors on a concentrate/hay diet for three months. The diets contained 0 or 18% maize, and concentrates and hay were produced in conventional or organic systems. Four pastured lambs were sampled at the onset of the trial for control. In trial 2, lambs (n=7 per group) were exclusively pastured for three months on an intensive ryegrass, a botanically diverse or a leguminosa rich pasture. In trial 1, the muscle δ^{13}C value allowed discrimination between grass feeding and indoor feeding, and 0 vs. 18% maize in the diet. However, the mean δ^{13}C and δ^{15}N values did not significantly differ between groups reared according to organic or conventional feeding regimes. In trial 2, discrimination between the three pasture groups was possible on basis of the muscle δ^{15}N but not the δ^{13}C values, with the highest and lowest mean δ^{15}N value for the intensive ryegrass and the leguminosa rich pasture group respectively. Combined stable C and N isotope analyses did not allow discrimination between organic and conventional feeds, but the use of concentrates vs. pasturing and the type of pasture could be distinguished.

The determination of in situ protein degradability characteristics of some feedstuffs and comparing to AFRC standard Tables in the feeding of lactating cows

N. Karimi, M.D. Mesgaran and A.G. Golian, Department of Animal Science, Islamic Azad University- Varamin branch, P.O.Box: 33817,Varamin, Iran

Absence of true information about native feedstuffs is a prime problem for diet formulation in Iran. this study attempted for determination of protein degradability coefficients of some native feedstuffs and compare them with AFRC (1995) data. Experiment I) protein degradability coefficients of five feedstuffs (barley grain, wheat bran, soybean meal, beet pulp and cottonseed meal) determined with *in situ* technique. Experiment II) 16 lactating Holstein cows used in a complete block design.The diets were formulated to meet the nutritional requirements of cows according to either AFRC (1995) or the experimental data. Two diets (1 and 2) were formulated according to AFRC coefficients. Diet 1 contained forage (lucerne hay and maize silage). Diet 2 contained forage (lucerne silage, lucerne hay and maize silage) and the same concentrate. Diets 3 and 4 were formulated according to experimental data and the same concentrate.The results indicated that the cows fed with diets based on the degradability coefficients of experiment I had higher milk yield (kg/d), milk protein (g/d), milk NPN(g/l) and milk dry matter (g/d). The results showed that current coefficients are more capable to supply requirements of animals for milk yield persistency than AFRC coefficients. Use of diets formulated with experimental data did not have significant effect on milk composition

---

Supplementary feeding of breeding ewes consuming grass based diets

A.S. Chaudhry and C.J. Lister, School of Agriculture, Food & Rural Development, University of Newcastle upon Tyne, NE1 7RU, UK.

This study examined the effect of feeding a supplement as feed blocks on the performance of breeding ewes. Around 160 Mule ewes and four Texel rams were separately divided into Control and Treatment groups. The ewe and ram groups were flushed on separate fields of perennial ryegrass. A tub containing supplement for Treatment rams and four tubs for Treatment ewes were evenly placed in respective fields for *ad libitum* access by Treatment rams from six weeks and Treatment ewes from 2 weeks pre-joining onwards. Measurements were made to check grass quality and quantity, supplement intakes by the Treatment ewes, ewe liveweight (LW) and ewe cover and pregnancy. The data were statistically analysed to test significance at P<0.05. Both ewe groups maintained their LW despite decline in grass availability over time. While adequate quantity of grass was available to both ewe groups, the Treatment ewes grew (g/ewe/day) faster (25) than the Control ewes (19) (P>0.05) even when grass availability to Treatment ewes declined more over time. This faster growth is attributed to the consumption of nutrionious supplement causing greater grass digestion and utilisation by the Treatment ewes. More Treatment ewes had ram covers within 17 days post-joining than the Control ewes. Although the scanning followed by lambing data did not differ between Treatment and Control groups, it appeared that the Treatment ewes consuming the supplement grew faster and had a better condition and better performance than the Control ewes.

Effect of Eucalyptus leaves supplementation on cattle calf performance

N.E. El-Bordeny[1]*, M.A. El-Ashry[1] and G. Hekal[2], [1]Animal production Dept. Fac. of Agric. Ain Shams Univ. Egypt, [2]Egytech company, Egypt

Forty four calves with an average initial body weight 149 kg in complete random design were used in fattening trial. The aim was to study the effect of *Eucalyptus* leaves supplementation on calves' performance. The animals were divided into 2 similar groups of 22 animals each. The experimental period was divided into two stages, the first extended for 143 days and the second extended for 79 days. Animals of each group were fed total mixed ration containing concentrate feed mixture contained 14% CP during the first period and 10.57 % CP during the second period. The *Eucalyptus* leaves (22 g / h / d) were added to ration of group G2. Feed intake DM, TDN and DCP per kg $^{0.75}$ were higher (P<0.05) for G1 than that G2 during all experimental periods. The animals of G2 presented higher nutrients digestibility (P≤0.05) than G1. Also G2 had higher serum total protein and albumin concentrations than G1. Animals of G2 showed higher (P<0.05) mean daily gain than those of G1 during the second stage as well as the overall period. Feed conversion g or kg /kg gain of DM, TDN and DCP, were better (P<0.05) than G1 during the second stage and the over all period.

Using *Eucalyptus glopulus* leaves as feed additives by the recommended dose enhance digestion and improve ADG and feed conversion.

Selenium increases immunoglobulin G absorption by the intestinal pinocytosis of newborn calves

H. Kamada[1], I. Nonaka[2], Y. Ueda[1] and M. Murai[1], [1]National Agriculture Research Center for Hokkaido region, Department of Animal Production, Sapporo, 062-8555 Japan, [2]National institute of Livestock and Grassland Sciences, Tsukuba, 305, Japan

Newborn animals are very sensitive to pathogens because of their immature immune system. Therefore, for healthy growth during infancy, the immunoglobulin supplied from the maternal colostrum is an important factor. There have been many discussions about the quantity, quality and timing of feeding of colostrum to calves; however, techniques to increase the efficiency of IgG transfer from the colostrum to calves have not been found. Here, we discovered that the addition of inorganic selenium to colostrum increased the IgG transfer to calf blood plasma. It is known that the absorption of colostrum IgG is mediated by intestinal pinocytosis, which continues for only 24 hrs after birth. The addition of selenium to colostrum may directly activate this pinocytosis of intestinal epithelial cells, as indicated by the rapidity of the reaction. This effect is not nutritional, but rather pharmacological, and can not to be explained by any of the known effects of selenium.

Effect of ruminally protected amino acids on milk nitrogen fractions of Holestein dairy cows fed processed cottonseed

A.R. Foroughi[1], A.A. Naserian[2], R. Valizadeh[2] and M. Danesh mesgaran[2], [1]Education centre of khorasan Jihad-Agriculture, Animal Science Department, Mashhad, Iran, [2]Ferdowsi University of Mashhad, Animal Science Department, Mashhad, Iran

The experiment was conducted to investigate the effect of processing (grinding and moist heat) of whole cottonseed (WCS) and ruminally protected lysine(Lys) and methionine(Met) on milk nitrogen(N) fractions of Holestein lactating cows during early lactation. Multiparious cows (n=12) were used in a 4×4 Latin square design. Cows were fed: 1) WCS; 2) WCS + 16gr Met&20gr Lys(WCS2); 3) ground cottonseed (GCS) heated in 140°C and steeped for 20 minute (GHCS1); or 4) GCS heated in 140°C and steeped for 20 minute +20gr Met&30gr Lys(GHCS2). MY was significantly (P<0.01) affected by the diets and was greatest for HGCS2 (35.78 kg/d) and the lowest for WCS (33.07kg/d). Milk protein percent was progressively increased, averaging 3.21%, 3.30%, 3.28% and 3.48% for 1,2,3 and 4 treatments, respectively. Met and Lys supplementation of HGCS2 resulted in increased (p<0.05) in casein N and in treatments of 1,2,3 and 4 were 0.37%, 0.39%, 0.39% and 0.42%, respectively. Total and whey N showed the same pattern of response as observed for casein N. There isn't significant difference between milk NPN treatments. Physical processing of WCS and amino acid supplement can affect milk protein and N fractions.

Effect of feeding ration on sizes and density of rumen mucous membrane papillae in calves before weaning.

B. Hucko[1]*, Z. Mudrik[1], A. Kodes[1], J. Havlik[1] and V. Christodoulou[2], [1]Czech University of Agriculture Prague, 16521 Prague, Czech Republic, [2]NAGREF, 58100 Giannitsa, Greece

Differences in the production of volatile fatty have been previously shown to affect the development of rumen. The physiological development of rumen of calves fed by different rations was evaluated. The experimental group was fed with milk replacer and starter, whereas the control group with milk replacer, hay and mixture.

The size and density of rumen papillae was evaluated. The test was performed on 24 calves in the experimental group and 18 calves in the control group, both being slaughtered at 8 weeks of age. Sampling was performed from the ventral portion of the cranial sac of rumen and papillae were measured using a stereomicroscope.

Length of papillae in the control group averaged at 2588 (±117) μm, whereas their length in the test group was 3087 (±81) μm. Widths of papillae were 1113 (±51) μm in the control group vs. 1405 (±98) μm in the test group. Density of papillae in rumen mucous membrane ranged between 59.7 papillae / cm^2 in the control groups and 74.2 papillae / cm^2 in the test group. The results confirmed that higher microbial production of butyrate and propionate together with lower pH in the rumen leads to increased sizes and density of rumen papillae.

This work was supported by grants: VZ MSM 6046070901 and NAZV 1B 44035.

The effect of crude protein level of starter on growth rate in Iranian Holstein calves

A. Towhidi[1]*, A.H. Ahadi[2] and M.R. Sanjabi[2], [1]Department of Animal Science, University of Tehran, P.O.Box 4111, Karaj,Iran, [2]Department of Biotechnology and Agricultural Science, IROST, Tehran, Iran

It has been well established that the starter feeding increase growth rate in calf. The goal of this study was to determine the effect of crud protein (CP) level on growth rate and feed conversion ratio (FCR) in Iranian-Holstein calf. Sixteen female calf at birth day, were randomly assigned to four group. Group I (control) were fed whole milk from birth day to 93 days of age. Group II, III and IV had been fed whole milk and the rations as starters, that provided 17, 19 and 21% CP ad lib. since 3 to 30 days of age, respectively. In 31st day, these calves were weaned, then fed with the ration based on NRC (2001). Body weight (BW) was measured once in each 5 days. The data of BW were analyzed by a repeated measured using the MIXED procedure of SAS. In group IV, average daily gain was significantly ($P<0.01$) higher than the other groups. Contrary to, in group I, FCR (total feed intake (kg of DM)/total gain (kg)) was significantly ($P<0.01$) higher than the other groups.

The effect of different yeast strains on milk yield, fatty acids profile and physiological parameters in dairy cows

V. Kudrna[1], K. Poláková[1], P. Lang[1]*, J. Doležal[2], [1]Res. Inst. of Anim. Prod., Přátelství 815, 10401, Praha10, Uhříněves, The Czech Republic, [2]MZLU, Zemědělská 1, 61300, Brno, The Czech Republic

24 dairy cows (Holstein & Czech Pied breeds) were allocated into 3 well-balanced groups for 84 days to find out an effect of two different yeast strains (*Saccharomices cerevisiae)* supplemented into TMR (groups „L" a „B"). The control group „K" was without the yeast supplement. TMR was fed *ad libitum*. Yeasts didn't influence DMI positively: the highest (20.867 kg/head) was determined in the K (B-20.009, L-19.841 kg/head). An average daily milk yield of the L and B was significantly higher than that in the K (by 1.12 and 0.73 kg, resp.). The highest contents of basic milk components were found in the K. The lowest main milk components contents, and fat and FCM produces were found in the B. The highest differences in milk fatty acids proportions were found in the stearic acid (the highest in the K), whereas contents of linolic and α-linoleic acids were higher in the L a B. The lowest level of NH_3 in rumen liquid was detected in the L. The yeast supplement also considerably affected activity of celulase in the rumen. It could be concluded, that, statistically, the supplement of two strains of yeasts significantly improved milk yield and celulolytic activity into rumen liquid. Simultaneously, the contents of milk fat and also (the B-group) milk protein were reduced. The proportions some of milk fatty acids were also influenced.

The effect of artificial dietary on growth rate and biological indices Caspian Sea white fish larvae (*Rutilus frisii Kutum* K.)
S. Shafiei Sabet*, R.M. Nazari and H. Noveyrian, Guilan Natural Resource University, Fisheries department, Iran

An experiment was conducted for determination on the artificial dietary relationship with supply optimum growth of rutilus fish larvae for increasing of growth efficiency until reaching to4gr weight in controlled condition aquariums. Rutilus fish larvae divided into 4 experimental groups in aquariums with controlled condition laboratory that each group consisted of 30 larvae with three replication. Artificial diet that fed on the every groups of rutilus fish larvae have same fixed food supplement.A,B,C and D groups fed on 3,7,11 and 15 % body weight respectively. Feeding process doing three times in everyday. In our experiment at the end were calculated total biomass of each aquarium and natural mortality rate percentage. Biological parameters in rutilus fish larvae were determined by using standard technique in Guilan fisheries research center Bandar-e- anzali.The level of feeding which leads to illegally difference growth of rutilus fish larvae weight in aquariums for rutilus fish larvae that average weighting between 1.9 gr-3.3 gr. But by comparison biological characteristics and optimum growing rate in (B) treatment group rather better than other groups. From obtained results, it was showed that biological characteristics and parameters relationship measurement such as, water contaminate and natural mortality in A, B, C and D treatment groups were found.

Strategies to capture energy from agro-industrial residues of vegetable origin for sustainable animal production
Mohammad Imran Khan, National Dairy Research Institute, Dairy Cattle Nutrition, #714, Pocket 13, Phase I, Dwarka, New Delhi, 110045, India

Developing countries need to find alternative feed resources, which should not compete with the resources being utilized for human population. The vegetable residues after industrial processing are available in bulk and cause environmental pollution due to decomposition if left as such.
The residues of vegetables such as cabbage, cauliflower, pumpkin, okra, peas, and carrots were good source of nutrients i.e., CP (6.78 to 22.40%), fibre, essential fatty acids, essential and nonessential amino acids and minerals. Carrot residues were mixed with wheat straw and treated with propionic acid or urea + common salt and analyzed for different chemical and microbiological properties. Buffering capacity of substrate treated with propionic acid or urea & common salt was better than untreated substrate, which reduces the losses of nutrients and improves the hygienic quality. *In vivo* studies on 3 groups of growing kids and lactating goats were conducted. The study revealed the vegetable residues when fed to the goats (growing & lactating) at 40 and 60 % of total DM requirement were found equally palatable and high in nutrient quality which resulted in better growth, milk production, milk composition, feed utilization, nitrogen and mineral balances. These residues could be a choice to replace high quality conventional roughages like oat and berseem hay upto 60% of total DM requirement. This type of approach could shorten a partial gap between demand and supply of quality roughages.

The effect of yeast (*Saccharomyces cerevisiae*) on nutritive value and digestibility of sunflower heads
S. Safari, F. Kafilzadeh and M. Moeini, Razi University, Egypt

Five sanjabi sheep (63±2.31 kg and 18 m. age) were used in a completely randomized design to determine the addition of yeast culture (*saccharomyces cerevisiae*) on Sunflower head digestibility. As a feedstuff the Sunflower head by-product is poorly utilized by animals. The aim of this study was to determine the chemical composition, in vivo and in vitro digestibility of Sunflower heads in sheep nutrition. Sunflower head contained: 92.5 %DM, 87.4 %OM, 9.07% CP, 18.02 %CF and 4.03 %EE. The digestibility of DM, OM, CP and EE of sunflower head were 66.82, 66.45, 48.77 and 70.03 %, respectively. The result of in vitro experiment indicated that the digestibility of DM and OM were 83.88% and 80.37%, respectively and DOMD (digestible OM in DM) was 71.29%. Yeast culture had no significant effect on digestibly of DM, OM, CF, CP, and GE of sunflower head, but improved apparent digestibility of EE significantly ($p<0.005$). Animal fed sunflower head supplemented with yeast culture had a lower ($p<0.001$) serum sodium concentration but no significant effect on the serum potassium concentration was found. It can be concluded that sunflower head can be used as a part of ruminant feed but the addition of Yeast culture had no significant effect on its digestibility.

Evaluation of the inactivation of heat sensitive antinutritive factors in fullfat soybean
J. Csapó[1,2], É. Varga-Visi[1], Cs. Albert[2], K. Lóki[1], Zs. Csapó-Kiss[1], [1]University of Kaposvár, Faculty of Animal Science, Deperment of Chemistry-Biochemistry, Guba S. u. 40., H-7400 Kaposvár, [2]Sapientia – Hungarian University of Transylvania, Csíkszereda Campus, Department of Food Science, Szabadság tér 1., Csíkszereda, Romania

The regular quality control on the adequacy of heat treatment of fullfat soybeans requires the application of rapid chemical methods. In the present work the trypsin inhibitor activity test and the urease test were applied on fullfat soya samples that were cooked in a pressured steam (toasted) or extruded at different temperatures and speed rates. In the case of toasting both of the results of the laboratory examinations proved that the heating was adequate, while in the case of the extruded samples the two tests gave different results. In the case of certain temperature and time combinations the more rapid and less accurate urease test claimed that the heat treatment reached the aim, while the results of the trypsin inhibitor activity test showed that the level of the inhibitors is still high and the fullfat soya is underheated.

Studies on Rare Earth Elements (REE) in animal nutrition
Hartwig Böhme, Doris Förster, Ingrid Halle, Ulrich Meyer and Gerhard Flachowsky, Federal Agricultural Research Centre (FAL), Institute of Animal Nutrition, Bundesallee 50, D-38116 Braunschweig;

In China REE (e.g. Lathan, Cer or other elements) are used for decades in plant production and also in livestock feeding as growth promoter (increase of daily weight gain of pigs between 4 and 32%). Even under European production conditions some studies show effects of dietary REE-supplements on animal performance. The objective of our experiments was not only to quantify the growth-promoting effect but also to contribute to the mode of action (influence on feed digestibility and thyroid hormone status) and to the REE transfer in edible tissues.
The experiments, which were carried out with rearing piglets, growing pigs, broilers and calves, comprised supplements of various REE-compounds (La, Ce, Pr, Nd) in concentrations between 100 and 800 mg per kg feed, which were compared in each case to un-supplemented controls
In piglets the effect of REE on live weight gain was found to be between –10 and+6%, in growing pigs –3%, in broilers in the range between + 2 and + 7% and in calves + 11% as compared to the un-supplemented controls respectively. REE did not significantly influence the digestibility of organic matter in pigs. Thyroid hormone levels increased with REE supplementation in piglets The. highest REE-concentrations in the carcass of broilers were found in the kidney and liver (up to 194 µg kg^{-1} DM). Further research is necessary for understanding the mode of action of REE-supplements in food producing animals.

Determination of phthalic acid esters in premixtures and feeding staffs
J. Harazim[1]*, A. Jarošová*[2] *and J. Mylyszová*[2]*, [1]Central Institute for Supervising and Testing, Brno, [2]Mendel University of Agriculture and Forestry in Brno, Czech Republic*

Samples of feed stuffs were taken from the industrial feed stuff producers (P1 – P5) in the Silesian-Moravian Region of the Czech Republic in accordance with the National Law of Feeding Stuffs. The samples were being taken within the official inspections.
The samples of complete feeding stuffs (CF), mineral complementary feeding stuff (MCF), premixtures (M) and all of their components (C), which include feed supplements and feed stuffs, were taken from CF given for laying hens (P1, C = 11) and for piglets (P2, C = 14), from MCF for pigs (P3, C = 7), from premixture for dairy cattle (P4, C = 13) and premixture for pigs (P5, C = 19). Concentration of phthalic acid esters was also measured.
The highest concentration as a sum of di-n-butyl phthalate (DBP) and di-2-ethylhexyl phthalate (DEHP) was detected in feeding material – soy oil (131,42 mg. kg –1) as a component of complete feeding stuff and in feeding supplement – vitamin E (9,71 mg. kg –1) as premixture components. Concentrations of phthalic acid esters as a sum of DBP and DEHP (mg. kg –1) were detected (n=8) in CF (P1) 1,32 ; CF (P2) 1,55; CMF (P3) 0,45; M (P4) 0,54 and M (5) 1,04. There were evident statistical differences in all feeding stuffs and premixtures (P < 0,05).
The work was supported by the Czech National Agency for Agricultural Research. Project No.: QG60066/2005.

Effect of Pleurotus florida on nutritive value of wheat stubble
F. Kafilzadeh[2], M. Kabirifard[* 1] and H. Fazaeli[3], [1]Animal Science Division, Research Center of Agriculture and Natural Resources, Boushehr, Iran, [2]Department of Animal Science, Faculty of Agriculture, Razi University, Kermanshah, Iran, [3]Research Centre of Animal Science, Karaj, Iran*

This study was carried out to investigate the effect of *Pleurotus florida* on the voluntary feed intake and digestibility of wheat stubble (WS). A complete randomised design was used with eight shall male sheep (four replicates/ treatment). The treatments were: 1) Untreated wheat stubble (UTWS); 2) Mycelial treated wheat stubble (MTWS); 3) Fungal treated wheat stubble after the first harvesting of mushroom (FTWS1); and 4) Fungal treated wheat stubble after the second harvesting of mushroom (FTWS2). The OMI and DOMI (34 and 13g/kg $W^{0.75}$ respectively) of the MTWS were significantly (P<0.05) higher than UTWS (23 and 7.5g/kg$W^{0.75}$ respectively). DMI and OMI of FTWS1 were not significantly (P>0.05) from those found with UTWS or MTWS. The OM digestibility of the MTWS (38.4%) was higher (P<0.05) compared to the UTWS (32.7%) or FTWS2 (30.9% The DM and OM digestibilities of the FTWS1 were higher than those from FTWS2 but not different from those found with MTWS. The results indicated that the voluntary feed intake, and organic matter digestibility of wheat straw improved as a result of mycelial treatment but this advantage diminished after harvesting the mushroom.

Relationship between INDF and ADL lignin values of grass silages, haylage, hay, straw, dried grass and dried alfalfa
P. Nørgaard[1], L. Raff[1], U.P. Jørgensen[1], L.F. Kornfelt[1], M.R. Weisbjerg[2], [1]The Royal Veterinary and Agricultural University, Department of Basic Animal- and Veterinary Sciences, 1870 Frederiksberg, Denmark, [2]The Danish Institute of Agricultural Sciences, Department of Animal Heath, Welfare and Nutrition, P.O. Box 50, DK-8830 Tjele, Denmark*

The aim of the experiment was to study the relationship between lignin (ADL) and INDF values from different forages. Forage samples of dried alfalfa (n=1), grass silage (n=3), haylage (n=1), dried grass (n=2), grass hay (n=2), grass seed straw (n=1) and barley straw (n=2) were analysed for the NDF and ADL concentration by use of fibre bag technique, Foss Electric, Denmark. INDF was analysed by in situ incubation in the rumen for 508 hours in nylon bags with 37 µm pore size by triplicate determinations and corrected for particle loss. The mean and the coefficient of variation on twelve forages were (in % of DM) 4.2±1.9, 1.7±0.9 for ADL, and 13.2±6.6 and 6.4±4.1 for INDF, respectively. The INDF values were linearly related to the ADL values as: INDF= 3.4 x ADL - 0.9, n=12, R^2=0.96, P<0.001, S_{yx}=1.33, S_a=1.1, S_b=0.22 on DM basis and INDF= 3.0 x ADL + 0.4, n=12, R^2=0.88, P<0.001, S_{yx}=2.3, S_a=2.8, S_b=0.37 on NDF basis. In conclusions, the ADL analysis appears to have a high potential as a fast and cheap method for prediction of forage INDF values.

Stable carbon isotope fractionation may be diet dependent

S. De Smet[1], K. Raes[1], E. Claeys[1] and P. Boeckx[2], [1]Laboratory for Animal Nutrition and Animal Product Quality, Ghent University, Proefhoevestraat 10, 9090 Melle, Belgium, [2]Laboratory of Applied Physical Chemistry-ISOFYS, Coupure Links 653, 9000 Gent, Belgium

Stable isotope analysis of animal tissues has some potential for discriminating between diets, e.g. the $\delta^{13}C$ value is known to reflect the proportion of C_3 and C_4 plants in the diet. Due to depletion of ^{13}C during metabolism (fractionation), which is higher during lipid synthesis than protein synthesis, the $\delta^{13}C$ value is generally higher in animal samples compared to the diet. When back-calculating diets, it is often assumed that this trophic shift is constant for a given tissue. We verified this hypothesis by re-examining plasma and dietary $\delta^{13}C$ values from four trials with wethers, bulls, lambs and pigs, and a proportion of C_4 plants in the diet between 0 and 46% (n=9 treatment mean values). Plasma samples had been taken after at least two months following a dietary change, so that equilibrium could be assumed. Treatment mean plasma $\delta^{13}C$ fractionation was on average 1.4‰ ranging from -1.3‰ to 3.8‰. In two trials, there was no indication for fractionation depending on the diet, whereas in two others there was. Across trials, there was a significant linear decrease in fractionation with increasing (less negative) $\delta^{13}C$ values and increasing proportions of C_4 plant material in the diet.

The nutritional characteristics of raw and treated protein feeds for ruminants

M. Chrenková*, Z. Čerešňáková and A. Sommer, Slovak Agricultural Research Centre, Institute of Animal Nutrition, Hlohovská 2, 949 92 Nitra, Slovak Republic

The aim of our experiment was to determine crude protein and soluble protein. The protein escaping rumen degradation and nitrogen disappearance after intestinal passage for raw soybean, soybean meal, extruded full fat soybean, roasted soybean, raw pea, extruded pea, rapeseed meal, expanded rapeseed meal, and rape cake were determined using in sacco and mobile bag techniques. The proportion of soluble N fraction in the total N content ranged from 19% for soybean meal to 68% for rape cake which had the highest CP degradability (>80 %). Heat treatment decreased ruminal CP degradability at pea from 87% (raw) to 68% (extruded), at soybean 70% to 43% and rape seed meal from 65% to 35% (at 130° C). The reduction of CP degradability provide increasing the protein flow to the intestinum. We determined in treated feeds higher intestinal digestibility of by pass protein than in untreated (for soybean 91% and/or 98%, for rape 65 and/or 75%, etc.). Decreasing of effective degradability and increasing of intestinal digestibility by thermal treating had positive effect on PDIE and PDIN values of feeds, too. Results of this study demonstrate positive effect of optimal treatment of feeds on utilization of nutrients. The effect of treatment depends on height of temperature and length of its action.

Influence of absorbents addition on the fermentation quality of brewers' grains silage
P. Dolezal, L. Zeman, J. Dolezal, I. Vyskocil, V. Pyrochta, Department of Animal Nutrition and Forage Production, Mendel University of Agriculture and Forestry Brno, Zemedelska 1, 613 00 Brno, Czech Republic

In the experiment was evaluated the effect of absorbents supplementation on the fermentation quality of brewers' grains silage by comparing with the untreated control. As effective substance of experimental groups were used barley groats and malt sprouts. The addition of malt sprouts and barley groats in our experimental conditions increased statistically significantly ($P<0.01$) the content of DM in silage. The addition of malt sprouts decreased pH value in experimental silage (4.29 ± 0.007) in comparison with control silage (4.43 ± 0.049). The malt sprouts increased significantly ($P<0.01$) the content of lactic acid (67.15 ± 2.796 g/kg DM) and sum of acids (84.30 ± 2.97 g/kg DM) and decreased ($P<0.01$) ethanol content (0.51 ± 0.102 g/kg DM) and acetic acid content (17.15 ± 0.227 g/kg DM). Silage with malt sprouts has the highest ($P<0.01$) ammonia content from all silages in trial ($966,67\pm33,33$ mg/kg DM). The use of absorbents inhibited significantly ($P<0.01$) in comparison with control silage (without absorbents) the content of propionic and butyric acid production. Brewers' grain silage with malt sprouts and barley groats addition were free of butyric and propionic acid, but have had higher lactic acid content. This study was supported Project QF 4027.

Nutritive effect of enrichment mulberry leaves with L-Alanine amino acid on the economic traits and biological indices of silkworm *Bombyx mori* L.
R. Rajabi* and R. Ebadi, Esfahan University of technology, Entomology department, Najaf abad street, Esfahan, Iran

An experiment was conducted for investigation on effects of supplementation of mulberry leaves for increasing of silkworm yields. Larvae divided into 5 experimental groups including control (normal leaf). Each group consisted of 100 larvae with tree replication. L-alanine amino acid dissolved in distilled water and diluted to 0.1, 0.2, 0.5 and 1 % doses. Daily fresh mulberry leaves were soaked in each concentration for 15 min and then were dried in air for 20 min. Silkworm larvae fed using supplementary leaves from 1st to 5th instar, once a day. Biological and economic parameters were determined by using standard technique in sericulture. 0.5% dose of L-alanine increased silk index as well as larval and male pupal weight indices, but Female pupal weight index showed irregular variation compared with other indices with high value in 0.1% concentration. Cocoon characteristics were different between male and female due to different food consumption and absorption between two sexes ($P<0.05$). Female cocoon had higher amounts in higher dose. In contrast, male cocoon showed high amount in low concentration. From obtained results, it was showed that treatment with L-alanine resulted to significant increase for biological and economical parameters such as larval weight, pupal weight, cocoon weight and cocoon shell weight($P<0.05$) although fortification with L-alanine decreased cocoon shell ratio percentage ($P>0.05$).

Utilization of lupin seed as the main protein source in broiler chicken feeding: Influence of the variety and interest of protease addition

E. Froidmont[1]*, Y. Beckers[2],P. Rondia [1], P. Saive[1], N. Bartiaux-Thill[1], [1]CRA-W, Animal Production and Nutrition Department, rue de Liroux 8, 5030 Gembloux, Belgium [2]FUSAGx, Unité de Zootechnie, Passage des Déportés 2, 5030 Gembloux, Belgium

360 chicks received 3 iso-energy and iso-first limiting amino acids (Met, Lys, Thr and Trp) diets composed of 34% of wheat and 51% of lupin seeds in a randomised blocks design. All diets contained β-glucanase, cellulase, xylanase, hemicellulase and pectinase activities. They differed by the variety of lupin used (*Lupinus albus*, cv Lublanc and Amiga, *Lupinus angustifolius*, cv Boltensia) and were supplied with or without a supplement of protease (Biofeed Pro CT, Novozymes, Denmark). Zootechnical performances (ADG and FCR) were lower with Boltensia diets, inducing a higher duodenal (6.1 *vs* 3.0 cP) and ileal (13.7 *vs* 4.5 cP) chyme vicosity and also an increase in the relative size of duodenum and jejunum length (0.097 *vs* 0.087 cm/g). The protease addition did not influence the performances of chicks fed with *Lupinus albus* varieties but tended to improve the ADG for those receiving the *Lupinus angustifolius* diets. This effect was due to a better ingestion with protease. The fatty acid profile of leg meat reflected the composition of lupin fat: *Lupinus angustifolius* diets increased C18:2 to the detriment of C18:1 and C18:3. More information concerning the nutrient digestibility will be available soon.

The effect of organic acids in broiler nutrition

H. Ghahri[1], M. Shivazad[2] and J. Eghbal[1], [1]Urmia Islamic Azad university. Urmia, Iran, [2]Department of animal science, university of Tehran, Iran

This experiment was conducted to evaluate different levels (0.0, 0.5, 1.0 and 1.5%) of organic acids (Biotronic supplement) on Performance of day-old chickens, in a completely randomized design for 42 days. Birds were divided to 12 experimental groups with 16 birds in each. Results showed that using of organic acid had significant effect on feed intake and wet litter. Supplementation diets with organic acids had significant effect on body weight (BW) in starter Period (P<0.05); although at the end of the experimental Period, BW was not affected by treatments. Organic acid improved, BW compared with control group. Diets containing organic acids decreased feed conversion ratio (FCR) significantly in both periods (P<0.05). Intestinal pH decreased significantly by using organic acid at the end of 42 days (P<0.05). The control group had the highest pH while the treatment containing 1.5% of organic acid had the lowest pH. The effect of organic acid on the number of cecum salmonella was not significant but the number of cecum salmonella decreased

The effect of canola meal (hula) on the performance of broiler chicks
M. F. Harighi[1] and F. Kheiri[2], [1]Islamic Azad University of Krmanshah, Iran, [2]Islamic Azad University of Shahrekord, Iran

Canola meal is one of the most widely used protein sources in animal feed. Hula is a new variety of canola meal. In order to determine the effect of replacing different levels of canola meals to soybean meals (0, 5, 10, 15 and 20 percent) on broilers performance with 3 treatmentas. A randomized complete design were used. Average daily gain and feed conversion were measured. At the end of the trail one male and one female of each pen were selected and killed. Dressing percentage, abdominal fat, intestine, liver, gall bladder and pancreas weights were determined. Data from this experiment show feed conversion and average daily gain of the group with 5 percent canola meal had better performance than the other groups. The best body weight was also related to group 5. But the highest levels of abdominal fat were related to groups 15 and 20 percent canola.

Use of microbial phytase in decreasing environmental phosphorus pollution due to poultry excreta
A. Musapuor[1], M. Afsharmanesh[1], H. Moradi Shahrbabak[2],M. Soflaei[3], [1]Department of Animal Science, Faculty of Agriculture, Shahid Bahonar University of Kerman, Iran, [2]Faculty of agriculture, University of Tehran, [3]Kerman's Jihad-e-Agriculture Educational center, Iran

This experiment was conducted to study the effects of different levels of microbial phytase (0, 500 and 1000 FTU/kg diet), calcium (2.275 and 3.25 percent) and available phosphorus (0.175 and 0.25 percent) on phytate phosphorus utilization in laying hens and benefits of phytase supplementation on decreasing environmental phosphorus pollution. One hundred ninthy two 30-week aged White Leghorn (Hy-line W-36) laying hens were randomly allocated in cages for 12 dietary treatments in a 3*2*2 factorial experiment with four replicates, four hens per replicate. The experimental period lasted 90 days, when the age of hens was 42 weeks. Dietary phytase caused a significant ($P<0.05$) improvement in feed intake, feed conversion ratio, tibia ash weight, tibia ash percentage, tibia phosphorus, plasma phosphorus and phosphorus digestibility. Available phosphorus levels had significant effect ($P<0.05$) on tibia ash weight and tibia ash percentage. Reduction dietary available phosphorus caused a significant ($P<0.05$) decrease in feed consumption. Effect of dietary calcium were significant ($P<0.05$) on tibia ash weight, and feed consumption. Interaction between phytase and available phosphorus on tibia phosphorus were significant ($P<0.05$). Overall, it could be concluded that in low phosphorus diet where feed consumption is low, phytase would increase feed consumption as well as retention of phosphorus in tibia bone. Also, one of greatest benefits of phytase supplementation appeared to be maintaining lower excreta of phosphorus, that could decrease environmental pollution.

The using of natural fullerenes (shungites) in feeding of poultry of meat and eggs production

A.E. Bolgov[1], N.A. Leri[1], N.N. Tyutyunnik[2] and N.V. Grishina[1], [1]Petrozavodsk State University, 185910, Petrozavodsk, Russia, [2]Institute of Biology, Karelian Scientific Center, 185000, Petrozavodsk, Russia

In submitted work we investigated possibility and efficiency of using the natural mineral Karelian shungite in poultry feeding for meat and eggs production, which has 30% of similar-fullerene carbon in contents, from unique large deposit at Northern coast of the Onego lake (Republic Karelia, Russia).

The additives of shungites in a ration of broilers in amount 1% and 1,5 % to weight of the basic feeds had positive influence on birds safety, the gain of alive weight is from 8,8 -9,0 % up to 5,0-5,3 %. The weight of drawn and half-drawn carcasses of tested broilers were higher on 12,0-13,1%, and the expenses of feeds for 1 kg of gaining weight was lower on 10,8%. Under influence of shungites the eggs productivity and production of hens have increased from 4% to 16%, the quality of eggs and shell, the expenses of feeds for 1 kg of eggs on 6,7-13,5 %. Shungite did not made negative influence on the physiology of broiler-chickens and hens, but it was the reason for vitamins A and E increasing and for common protein in blood, improving of energy and mineral exchanges.

Therefore, the additive of Karelian shungite in a ration of poultry increases its production, stress-stability and improvement of quality of meat and eggs.

Effect of different levels of lysine in starter period on broiler chicken performance, carcass characteristics, and N excretion

M. Rezaei, Dept. of Animal Science, College of Animal Science and Fisheries, Mazandaran University, P.O.Box 578, Sari, Iran

An experiment was carried out whit male Ross broiler chicks from o to 42 days of age to test the effect of different levels of lysine in starter period on performance, carcass characteristics, and N excretion. Experimental diets consisted of six levels of lysine (0.87, 0.97, 1.07, 1.17, 1.27, and 1.37 %) in starter period with 18.6 % CP and 2900 Kcal ME/kg. For adjusting lysine level in diets L-lysine HCl were used. Chicks were fed with a commercial diet containing 18.2 % CP and 2900 Kcal ME/kg diet in grower period. The results showed that with increasing lysine levels, feed intake increased significantly in starter, grower and whole period of the experiment ($P<0.05$). Increasing lysine level more than requirement (0.97 %) hadn't significant effect on weight gain and gain to feed ratio, but increased breast meat percentage ($P<0.05$). With increasing lysine level abdominal fat percentage increased significantly ($P < 0.05$). Lysine level hadn't significant effect on N excretion and mortality percentage. The results of present experiment showed that lysine requirement in starter phase established by the NRC 1994 are considered to be safe estimate for Ross broiler chicks.

The effect of diferent crude protein level and metabolic energy on performance and blood characteristic of broilers

Z. Steiner*, M. Šperanda, M. Domaćinović, Z. Steiner, Z Antunović and Đ. Senčić, Faculty of agriculture, Trg sv. Trojstva 3, 31 000 Osijek, Croatia

A study was conducted to evaluate crude protein level and low energy supply on broiler production parameters. Three- hundred Ross 308-day-old broiler chicks were divided and placed in 3 cages (100 chicks/cage). Treatments consisted of a control diet (220 g/kg CP) formulated to meet NRC (National Research Council, 1994. Nutrient Requirements of Poultry. 9th ed(revised). NationalAcademy Press, Washington), a low crude protein diet (205 g/kg CP), and the third low crude protein diet (190 g/kg CP) with the same ratio 1:135-138 ME and crude protein in starter (first 21 days). Finisher (22 day till end) control diet consisted 200 g/kg CP, the second diet with low crude protein consisted 185 g/kg CP, and the third diet consisted 170 g/kg CP, with the ratio 1: 158 ME and crude protein. Feed intake tended to decrease with increasing crude protein and energy. Feed conversion (g gain per g feed) improved as crude protein and energy increased. Concentration of total protein, albumin, and creatine in blood serum were significantly lower (P<0,05) in second and third group. The values of research parameters have been statistically processed with a computer program for analysis of the variance (Statistica Stat Soft Inc., 2001).

Use as additive of *Lycopodium clavatum* on control of ascitic syndrome on broiler chicken

D. Camacho-Morfin*, L. Morfin-Loyden, J. Bruno-Valencia, G. Franco-González and J. Garibay Bermúdez, Department of Animal Sci. MVZ. Faculty of Superior Studies Cuautitlán. UNAM. Campo 4. km 2.5 Carretera Cuautitlán-Teoloyucan, Edo. de Méx. México

The efficacy of *Lycopodium clavatum* as homeopathic preparation was tested for seven weeks on broiler chicken for ascitic syndrome control. The test was conducted on 256 one day old *Arbor Acress* chickens, randomly allocated on two treatments with four repetitions. The treatments were alcohol 72^0 like control (A) and the other one *L. clavatum* prepared how homeopathic dilution to 200c on alcohol 72^0 (L). All animals were weighted every seven days and every day was register dry matter feed intake. Average of weekly body gain (WBG) and feed conversion (FC) were calculated, general mortality and for ascites were register too. WBG and FC dates were analyzed with the t student test for sample's independents and X^2 test for the mortalities. Differences of WBG and FC were not significant between treatments. The final body weight were 1910 and 1880 g, and FC were 1.47 and 1.46, for A and L respectively. The general mortality did not depend on the treatment, 5 and 4 % respectively, but mortality for ascites depend on the treatment, 5 and 0.8 % respectively. These results suggest that *L. clavatum* could contribute to control the ascitic syndrome and do not affect weekly body gain nor feed conversion.

Effects of higher amounts of triticale with multienzyme composition on pig growth and meat quality
J. Norvilienė, R. Leikus and V. Juskiene, Institute of Animal Science of LVA, R. Zebenkos 12, LT-82317, Baisogala, Radviliskis distr., Lithuania

The feeding trial with fattening pigs was carried out to determine the effects of higher amounts of triticale with multienzyme composition in the diet (α-amylase-70 U/g, ß-glucanase-700 U/g, xylanase-1800U/g, protease-0.8 U/g) on pig growth and meat quality. The trials indicated that the highest weight gains for pigs were reached when the compound feed containing 60 to 70 % triticale was supplemented with 0.05 % of multienzyme composition. During the whole treatment, the average daily gains of these pigs were 11 % (P<0.05) higher, feed consumption per kg gain was 7 % lower and preslaughter weight was reached 9 days earlier in comparison with the control pigs. Daily gains of pigs in the period between 30 to 40 kg weight were 18 % (P<0.05) higher and feed consumption was 3.2 % lower. Supplementation of the triticale based diets with 0.05 % multienzyme composition resulted in improved growth of pigs in subsequent periods, too. In the period from 40 to 60 kg weight, the average daily gains were 8 % (P>0.4) higher and feed consumption per kg gain was 2.7 % lower. From 60 kg weight to slaughter, the daily gains were 9.4 % (P>0.2) higher and feed consumption per kg gain was 6.7 % lower. Lower amount (0.035 %) of multienzyme composition in the diets with triticale had no significant influence on the growth of pigs. Higher amount (60-70 %) of triticale supplemented with 0.035 and 0.05 % of multienzyme composition in the diet had no negatyve effects on carcasses and physico-chemical indicators of meat.

Effect of partial dehulling of barley on metabolizable energy determined in pigs
U. Hennig, S. Kuhla, A. Tuchscherer and C.C. Metges, Research Institute for the Biology of Farm Animals, 18196 Dummerstorf, Germany

Two experiments (E1, E2) were carried out to determine the effect of partial dehulling of two-row and six-row barley varieties (TRB and SRB) on ME using digestible nutrients determined by a standardized difference method using quantitative feces collection. In E1 and E2 8 and 5 barrows (92 and 91 to 111 kg BW, respectively) were used to estimate nutrient digestibilities of diet supplements and of the assay barleys, respectively. In E2 diets were provided containing intact TRB + wheat gluten (WG), dehulled TRB + WG, intact SRB + WG, and dehulled SRB + WG. Pigs were fed according to a repeated group-period design. Diets were supplied at daily rates of 37 and 32 g DM $\times$ kg BW $^{-0.75}$ in E1 and E2, respectively.
The intact TRB was characterized by higher digestibilities of OM, CP, EE, CF, and N free extracts (all $P < 0.01$), and by higher content of ME ($P = 0.001$) compared to intact SRB. Over all, dehulling increased the digestibility of OM ($P = 0.034$), N free extracts ($P = 0.021$), N free residuals ($P = 0.001$). The ME content (MJ/kg DM) was lightly improved due to dehulling from 15.8 to 16.1 and 15.1 to 15.4 in TRB and SRB (overall $P = 0.033$).

Fat coated butyrate induces mucosal hypertrophy in the gut of pigs
R. Claus[*], *H. Letzguß and D. Günthner, University of Hohenheim, Institute Animal Husbandry and Animal Breeding, 70599 Stuttgart, Germany*

Components of normal feed may modify the equilibrium between mitosis and apoptosis of enterocytes thus adapting the functional status of gut mucosa to varying demands. High starch contents initiate mucosal hypertrophy in the jejunum by stimulating mitosis via the IGF-I system. In the colon butyrate out of dietary fibres inhibits apoptosis thus stimulating crypt depth.
Fat coated butyrate can be targeted into the small intestine so that we investigated whether its function can be combined with mitosis stimulation by starch. Three groups of pigs (mean weight 75 kg) were fed for 5 days 3.6 kg/day of either a low energy ration (6.6 MJ ME/kg), a high energy ration (13.7 MJ ME/kg) or an isocaloric high energy ration where starch was partly substituted by 29 g calcium butyrate/kg. Whereas villus size was not influenced, the number of villi and thus the size of plicae increased. Plica area in the medial jejunum was 2.3 mm^2 in the lower energy group, 2.7 in the high energy group and 4.2 in the butyrate group. Activity of the brush border enzyme sucrase in the ileum content due to shedded cells was about 4-fold compared to low energy pigs and 2-fold compared to high energy pigs. The effects of butyrate are explained mainly by the induction of crypt fission.

Effect of Fusarium mycotoxin on serological and immunological parameters in weaned piglets
I. Taranu[1*], *D. E. Marin*[1], *F. Pascale*[2], *P. Pinton*[3], *J.D. Bailly*[4] *and I. P. Oswald*[3], [1]*Institute of Biology and Animal Nutrition, Balotesti, Romania,* [2]*Institute Pasteur, Bucarest, Romania,* [3]*Laboratoire de Pharmacologie-Toxicologie, UR 66, Institute National de Recherche Agronomique, Toulouse, France,* [4]*Ecole Nationale Vétérinaire, Toulouse, France*

Among mycotoxins produced by *Fusarium*, fumonisin B$_1$ (FB$_1$) is the most frequent contaminant of cereals. Due to their maize-rich diet, pigs potentially are exposed to the intoxication with fumonisins. In the present study we investigated *in vitro* and *in vivo* the effect of low doses of FB$_1$ on some immunological and serological parameters in piglets. *In vitro* results show that FB$_1$ induced a decrease of interleukin 4 (Th2 cytokine) and an increase of interferon-γ (Th1 cytokine) produced by peripheral blood mononuclear cells (PBMC). This imbalanced was validated *in vivo* exposure of weaned piglets to feed contaminated with 8 μg/g of FB$_1$ for 28 days. In fact, the food contamination with this mycotoxin decreased the level of mRNA encoding for IL-4, IL-6 and IL-10 synthesis (minus 42.1%, 31.5% and 61.5 % respectively) by whole blood cells upon mitogenic stimulation and led to a disruption of vaccinal immunity by significantly diminished the specific immune response after vaccination of weaned piglets. The presence of FB$_1$ in the diet significantly increased the creatinine and phosphorus concentrations and decreased that of triglyceride in the serum of treated animals.

Seasonal variation in concentrations of fibre, crude protein, condensed tannins and in vitro digestibility of eight browse species: Nutritional implications for goats
Aziza Gasmi-Boubaker[1] and H. Abdouli, [1]cité Mahrajène INAT, Tunis, Tunisia, [2]Ecole Supérieure d'Agriculture Mateur, Tunisia*

Leaves and twigs from 8 shrub species (*Arbutus unedo, Cistus salvifolius,Cytisus triflorus, Erica arborea,Myrtus communis, Pistacia lentiscus, Smilax aspera* and *Viburnum tinus*) consumed by range goats were evaluated for condensed tannins, proximate composition and in vitro organic matter digestibility. Samples were collected in winter (February), spring (April) and summer (July) in a Tunisian forest land (Sejnane). Mean content (g catechin kg^{-1} DM) of condensed tannins as determined by the vanillin-HCl method ranged from 3.1 in *Cytisus triflorus* to 147.3 in *Pistacia lentiscus*. Shrubs contained low crude protein (range 55, 100) except *Cytisus triflorus* with 227 g kg^{-1} DM crude protein in winter. The content of phosphorus was below recommended level required by goats for growth and productivity. The *in vitro* OM digestibility ranged from 34 for *Erica arborea* to 64.7 for *Myrtus communis*. Most shrubs had moderate Net Energy contents in spring but low in winter and summer. Therefore, supplementation should be provided to goats ranging during winter and summer.

Selenium status of sheep on some areas in Bosnia and Herzegovina and Croatia
S. Muratovic[1], Z. Steiner[2], E. Dzomba[1], Z. Antunovic[2], Z. Steiner[2], S. Cengic-Dzomba[1], M. Šperanda[2] and M. Drinic[3], [1]Faculty of Agriculture, Zmaja od Bosne 8, 71000 Sarajevo, B&H, [2]Faculty of Agriculture, Trg sv. Trojstva 3, 31000 Osijek, Croatia, [3]Faculty of Agriculture, Banjaluka, B&H*

The selenium status of sheep on some area in Bosnia and Herzegovina and Croatia was assessed by surveys of selenium content in soil-plant-animal continuum. Total 12 localities were selected and during three consecutive years samples of soil, plant, blood and wool have been taken. A very wide variation in levels of selenium in all substrates has been observed. Generally, selenium content in serum (0.443-1.520 µmol l^{-1}), plants (0.006-0.057 mg kg^{-1} DM), soil (0.065-0.975 mg kg^{-1} DM) and wool (0.003-0.059 mg kg^{-1} DM) in Craotia was lower compared to selenium in serum, plants, soil and wool in B&H containing 0.861 to 2.591 µmol l^{-1}; 0.032-0.784 mg kg^{-1} DM; 0.396-1.134 mg kg^{-1} DM; and 0.022-0.499 mg kg^{-1} DM respectively. The results suggest that sheep in almost all areas in Croatia have marginal Se status. Serum vs wool; soil vs plants and plants vs serum regression analysis had low coefficient determination although linear regression equation for wool Se concentration on serum Se in samples from Croatia had R^2=0.90 indicating that wool also may be used as selenium status criteria. Further study is required to examine the soil factors affecting of selenium availability as well its seasonal variation.

Effect of Cistus ladanifer L. tannins on digestion, ruminal fermentation and microbial protein synthesis in sheep
Maria T.P. Dentinho*, Ana T.C.Belo and Rui J.B. Bess, Estação Zootécnica Nacional, 2005-048 Vale de Santarém, Portugal

Three rumen cannulated rams were used in a 3 x 3 Latin square design experiment to measure the effect of a purified extract of tannins of *Cistus ladanifer* L. on the ruminal degradability, intestinal and apparent digestibility and rumen microbial protein synthesis. Animals were daily fed with a basal diet consisting of oat straw (600g) and manioc (300g). Dietary treatments were the basal diet and soybean meal (100g) treated with 0, 1.5 and 3% of tannin extract. Apparent digestibility of DM and CP of the diets and intestinal digestibility of CP were not affected by the incorporation of tannins extract. Tannin inclusion did not affect the effective ruminal degradability (P) of DM and CP of soybean meal. However, the addition of 3% of tannin extract induces a decrease (P< 0.05) in the rapidly degradable fraction "a" of CP and an increase in the slowly degradable fraction "b". There are no significant differences in the rate of degradation "c" (P> 0.05). The microbial nitrogen yield was significantly lower (P<0.05) in rams fed diets containing soybean meal treated with 3% of tannins (3.7 g/day) than in rams fed diets with soybean meal 0% or 1.5% of tannins (6.1 and 5.4 g/day respectively). Rumen pH, NH3-N ad VFA concentrations were unaffected by treatments.

Effects of frequency of GnRH injection on the ovulation and lambing rates during the ovine breeding season
A.A. Sabetan-e-Shirazi, Animal science of Islamic Azad university branch of Marvdasht

Ewes were run with rams and were ascertained to be cyclic. Estrus was synchronized by 2 i.m. injections of a PGF_2a 1 ml lutalyse, 9 days apart. The ewes were then run with fertile rams. On the day of the second injection of lutalyse, the ewes were randomly divided into 4 groups of 20 ewes each. The ewes in group 1 received 3 injections of GnRH at 12, 36 and 60 h after the second injections of lutalyse. The ewes in group 2 received 6 injections of GnRH at 12,24,36,48,60 and 68 h after the second injection of GnRH at 12,20,28,36,44,52,60 and 68 h after the second injection of lutalyse. The ewes in group 4 (control) were injected with normal saline according to the protocol of GnRH injection in group 3. Five ewes from each group were randomly salughtered on 12-13 days after matingThe results showed that total corpora lutes number and weight in group 1 were significantly greater as compared with group 3 and control. Mean corpora lutes and ovarian weights were not effected by the treatment. Fecundity in in group 2 was lower than that of control. Fertility in group 3 was higher than other groups. But this differents was not significantly fertility in group one was not significantly different from control, but fertility in group 2 was significantly lower than the control group.

Mycotoxins in poultry diets and mycotoxins controling
S.A. Abd EL-Latif, Department of Animal Production, Faculty of Agriculture, Minia University, Minia, Egypt

Mycotoxins are produced as secondary metabolites by moulds or fungi growing on agricultural products before or after harvest during transport and storage. Mycotoxins are metabolised in the liver and the kidneys and also by micro organisms in the digestive tract. Harmful effects of mycotoxins on human and animal health have been known for about 80 years. However, studies on mycotoxins and mycotoxin induced diseases began only recently when in the 1960s a toxic molecule was extracted from *Aspergillus flavus*.

The problem of mycotoxins do not end in feed or reduced animal performance. Many are transferred into meat, visceral organs, milk and eggs. Their concentration in food is usually lower than the levels present in the feed consumed by the animals and unlikely to cause acute toxicity in humans. Various dietary manipulation techniques have been documented to alleviate the adverse effects of mycotoxins. Higher levels of methionine, selenium, carotenoids and vitamin supplementation have been found to be beneficial although not necessarily cost effective.

Chemical detoxification is a possibility using products such as ammonia, sodium bisulphite, peroxide acids etc but most of these chemical methods are not practical and raise further questions about safety of reaction products and palatability. Mineral clay products such as bentonites, zeolites and aluminosilicates have been found to be effective in binding/adsorbing mycotoxins. The aim of this article is to evaluate the effect of mycotoxine in poultry diets on poultry production and haw to reduce this toxic effect.

The effect of sample grinding on gas production profiles and end-products formation in expander processed barley and peas
A. Azarfar*, A.F.B. van der Poel and S. Tamminga, Wageningen University, Animal Nutrition Group, P.O. Box 338, 6700 AA Wageningen, The Netherlands

Grinding is a technological process widely applied in the feed manufacturing industry. It is also a prerequisite to obtain representative samples, necessary for laboratory procedures like for instance gas production analysis. Grinding feed samples prior to laboratory analyses is normally through a 1 mm screen. When feeds are subjected to technological processes other than grinding, like for instance expander, grinding afterwards may disturb the effect of processing, both in practice and when laboratory techniques are applied. Therefore, this study aimed to establish the possible effects of different types of the grinding and sample preparation on the degradative behaviour of expander processed barley and peas. Samples of expander processed barley and peas were subjected to 6 different types of sample preparation (intact sample, slurry sample, samples stepwise ground over a 6 and 3 mm sieve, samples stepwise ground over a 6 and 1 mm sieve, samples ground over a 3 mm sieve and samples ground over a 1 mm sieve). Pattern of gas production in these samples were studied over a period of 72 hours incubation using an automated *in vitro* gas production system. The results showed that in expander processed peas stepwise grinding leads to a faster degradation. In expander processed barley, however, the difference in the degradation pattern due to the different grinding methods was small.

Nutritional value of raw, cooked and soaked chickpeas (*Cicer arietinum* L.) with sibbald experiment

M.J. Agah[1], J. Pourreza[2] and A. Samie[2], [1]Fars Reasearch Center for Agriculture and Natural Resources, P.O.Box 71555-617, Shiraz, Iran, [2]Department of Animal Sience, Isfahan University of Technology, Iran

In this experiment, having cooked and soaked of chickpea grains, some samples of raw and processed chickpea grains were prepared to determine chemical composition and nutritional value. Crude protein, fat, crude fibre, ash, calcium and phosphorus percentages and the value of tannin antinutritional factor were measured for each samples according to standard methods. The metabolizable energy of raw, cooked and soaked chickpea grains were determined with sibbald experiment using 24 mature leghorn roosters. Chemical composition showed the effects of chickpea grains processing method on percent of ash, crude fibre and phosphorus (P<0.01) and on tannin value (P<0.05) were significant. Tannin value was the most in raw chickpea grain and the least in 10 and 20 mins cooked and 48h soaked chickpea grains respectively. Sibbald experiment showed no a significant difference between raw and processed chickpea grains in apparent protein digestibility and metabolizable energy. However 48h soaked chickpea grains had maximum value apparent metabolizable energy.

Effect of moisture heat treatment on protein degradation of canola meal

A.A. Sadeghi[1] and P. Shawrang[2], [1]Department of Animal Science, Science and Research Branch, Islamic Azad University, Tehran, Iran. [2]Department of Animal Science, Tehran University, Karaj, Iran.

This study was completed to evaluate effects of heat moisture treatment (125°C, 117 kPa steam pressure) for 15, 30 and 45 min on ruminal CP degradation parameters and *in vitro* CP digestibility of canola meal (CM). Ruminal nylon bags of untreated or autoclaved CM were suspended in the rumen of four Holstein steers for up to 48 h. Fitting data to exponential model showed that the water soluble fraction decreased ($P<0.05$) and the potentially degradable fraction of CP increased ($P<0.05$) by autoclaving. Autoclaving for 15, 30 and 45 min decreased ($P<0.05$) the effective degradability of CP at ruminal outflow rate of 0.05/h by 16, 21 and 23%, respectively. Autoclaving for 45 min decreased ($P<0.05$) *in vitro* digestibility of CP, compared to 15 and 30 min. Electrophoresis results of bag residues showed that napin subunits of untreated CM disappeared completely within 0 h of incubation period, whereas cruciferin subunits were degraded in the middle of incubation period. In autoclaved CM, napin subunits were degraded in the middle of incubation and cruciferin subunits were degraded in the longest incubation period. In conclusion, CM proteins appeared to be effectively protected from ruminal degradation by a 30 min autoclaving.

Effects of calcium salt of long chain fatty acid on performance and blood metabolites of Atabay lambs
T. Ghoorchi, A.M. Gharabash, Y. Mostafalou and N.M. Torbatinejad, Department of Animal Science, Gorgan University of Agriculture Sciences and Natural Resources, Gorgan, Iran

Calcium soaps of long chain fatty acids are relatively inert in the rumen and can increase energy density and energy consumption without altering the microbial activity in rumen. Twenty-eight male Atabay lambs with mean body weight of 24.5 ± 0.51 kg and 4-5 months of age were used in a 90-day feeding experiment. The trial was carried out using Completely Randomized Design with 4 dietary treatments containing 0, 2.5, 5 and 7.%. Ca-LCFA (DM basis). These lambs were slaughtered for the evaluation of carcass characteristics and the quality of meat after 90 days. Dressing proportions were calculated on the basis of field weight and hot carcass weight. Means for initial weight were similar (p > 0.05) among treatments. Significant differences were detected between treatments for average daily gain (ADG). These estimates were 154.7, 187.1, 149.9 and 148.7 g d^{-1} for ADG. Dry matter intake was decreased by the high fat diet. The fat level in the diet affected (P < 0.05) serum concentration of TG, glucose, but not urea and Ca. Plasma triglyceride concentration changed significantly when the high fat diet was fed. It was concluded that Ca-LCFA of 2.5% is the best level for use in diet for lambs.

Effects of gamma irradiation on cell wall degradation of wheat and barley straws
A.A. Sadeghi[1] and P. Shawrang[2], [1]Department of Animal Science, Science and Research Branch, Islamic Azad University, Tehran, Iran. [2]Department of Animal Science, Tehran University, Karaj, Iran

In the developing countries, wheat and barly straws are widely fed to ruminants. These straws have a high proportion of cell wall, which leads to a low degradation rate, principally, due to lignification. Gamma irradiation has been recognized as a method for preservation and processing of foods. The aim of this study was to evaluate effects of gamma irradiation at doses of 100, 150 and 200 kGy on cell wall degradation of wheat and barley straws. Duplicate nylon bags of untreated or irradiated straws were suspended in the rumen of four Holstein cows for up to 48 h, and resulting data were fitted to non-linear degradation model to calculate degradation parameters of DM, NDF and ADF. Gamma irradiation increased linearly ($P<0.001$) the water soluble fractions and decreased linearly ($P<0.001$) the potentially degradable fractions of DM, NDF and ADF. Effective degradability of DM, NDF and ADF of both feeds increased linearly ($P<0.001$) with increases in irradiation dose. Gamma irradiation of wheat straw at doses of 100, 150 and 200 kGy increased effective degradability of NDF at rumen outflow rate of 0.05/h by 7, 16 and 23%, and of ADF by 5, 13 and 19%, respectively. In conclusion, cell wall of straws appeared to be increased by gamma irradiation higher than 100 kGy.

Effects of gamma irradiation on protein degradation of pea and faba bean

A.A. Sadeghi[1] and P. Shawrang[2], [1]Department of Animal Science, Science and Research Branch, Islamic Azad University, Tehran, Iran. [2]Department of Animal Science, Tehran University, Karaj, Iran

This study was designed to evaluate effects of gamma irradiation on protein degradability and intestinal digestibility of pea and faba bean using *in sacco*, SDS-PAGE and mobile nylon bag techniques. Gamma irradiations of wet feeds (25% moisture content) were carried out using cobalt-60 irradiator at doses of 0, 25, 50 and 75 kGy at room temperature. Four non-lactating Holstein cows fitted with rumen fistulas and duodenal cannulae were used for *in situ* incubations. Ruminal disappearance of CP was measured for up to 48 h. Gamma irradiation at doses of 25, 50 and 75 kGy decreased ($P<0.05$) the effective CP degradability of pea by 8, 12 and 17%, and of faba bean by 13, 20 and 24%, respectively. Intestinal CP digestibility of both feeds linearly ($P<0.001$) increased with increases in irradiation dose. SDS-PAGE analyses of the bag residues of irradiated pea and faba bean showed that disappearance of proteins decreased with increases in gamma irradiation. There were cross-linked products of the degraded protein molecules that could not penetrate the running gel. In conclusion, pea and faba bean proteins appeared to be effectively protected from ruminal degradation by gamma irradiation at doses higher than 25 kGy.

A comparison between nutritive value of Iranian local feeds with NRC and MAFF

A. Abbasi, M. Zahedifar and N. Teymournezhad, Animal Science Research Institute of Iran, P.O. Box 31585-1483, Karaj, Iran

Traditionally, In Iran the NRC feed composition tables have been used to formulate rationsA national project was carried out in the past fifteen years to prepare Iranian feed composition tables. Feed samples were collected for 3 consecutive years from throughout the country according to the scientific methods and were analyzed for chemical compositions. A total number of 165 feeds were analyzed and the first feed composition tables of Iran was published. One of the examples for showing the difference between the Iranian local feeds with the foreign ones was alfalfa. The content of CP, NDF, ADF, Ca, P, Mg and K in Iranian alfalfa were: 1.46±2.66, 40.8±9.98, 33.4±6.93, 0.22±0.06, 0.32±0.11, 2.24±0.82% respectively. The figures mentioned in NRC2000 were: 19.2±3.30, 41.6±7.10, 32.8±5.10, 1.47±0.36, 0.28±0.07, 0.29±0.06, 2.37±0.42% and the ones in MAFF1990 were: 18.3±0.42, 49.3±5.91, 38.1±1.13, 1.56±0.18, 0.31±0.66, 0.17±0.03 and 2.73±0.05% respectively. There was a significant difference (p<0.05) in CP, Ca, P and Mg between Iranian and NRC data. Such difference was also observed between Iranian and MAFF data in CP, P and Mg. Results showed that in most cases there were significant differences (p<0.05) in chemical composition between Iranian local feeds and foreign feeds mentioned in NRC and MAFF. Therefore, it was necessary to prepare local feed tables and use them for balancing of rations.

Addition of freeze-dried citrus peel increases feed preservation period without disrupting in vitro ruminal fermentation
J.H. Ahn[1] and I.S. Nam[2], [1]Department of Dairy Science, College of Agriculture and Life Sciences, Hankyong National University, Ansong, 456-749, Korea, [2]National livestock research institute, Suown, Gyonggi, 441-706, Korea

The objective of this study was to investigate antimicrobial activity, during the preservation period, of animal feed and any effects on *in vitro* rumen performances by supplementing different levels (5.55, 11.11, and 22.22g/kg) of freeze dried citrus peel (FDCP) to the feed compared to the ordinary feed and the feed treated with an antifungal agent (AA) at 0.05g/kg. In a preservation test, feed supplemented with FDCP showed no deterioration over 21days. Untreated feed and AA treated feed, however, showed signs of deterioration after 16days storage. Aflatoxin was detected in untreated and AA treated feeds at 16days (8ppb and 2ppb) and 21days (8ppb and 4ppb), but aflatoxin was not detected in the feed supplemented with FDCP. In the second experiment, fermentation by rumen microorganisms of FDCP (22.22g/kg) and AA (0.05g/kg) supplemented feeds was studied *in vitro*. Feeds were incubated with buffered rumen fluid for 3, 6, 9, 12, 24, and 48h. Organic matter digestibility (OMD), ammonia-N, total, and individual volatile fatty acid (VFA) were not adversely affected by treatment. In conclusion, present results indicated that FDCP might be useful for inhibiting microbial growth of animal feed during long-term storage without disrupting rumen fermentation.

Studying the cultivation of Pleurotus fungi on the substrate of Agropyron repens
A.R. Safaei, H. Fazaeli and M. Zahedifar, Animal Science Research Institute, P.O.Box: 1483-31585. Karaj, Iran

This experiment was conducted to study the possibility of growing of Pleurotus fungi (*sajor cajo*) on dried *Agropyron repens*. To prepare culture media, *Agropyron* hay was collected from pasture around Karaj. The hay was dried and chopped (1mm) and placed in laboratory vials (5gr/vial) and sterilized with water vapor. Fungi spawn was cultured in the prepared media. Fungi growth at laboratory scale was investigated after 72 hours. Hay was sterilized in boiling water and after 24 hours the substrate was inoculated with the fungi spawn (3% of the substrate) and 10 kg was placed in each plastic bag (30×10cm) and were suspended in culture room. After 3 to 4 days, the plastic covers were cut vertically in pieces of 10×80 cm. After 2 to 3 weeks stroma was completed and the color of substrate changed into white. Finally, the fungi fruitbody was collected and weighed. Results showed that fungi composed of sufficient content of protein, fat, mineral and fiber to be used as human food. Fungi grown on *Agropyron repens* produced 40 percent more fruitbody as compared to wheat and barley straw. This increase is likely due to higher content of nutrients in *Agropyron repens* needed by fungi compared to wheat and barley straw. Chemical analysis of remained bedding material of *Agropyron repens* showed that it can be used as ruminant feed.

Body weight, feed coefficient and carcass characteristics of two strain quails and the reciprocal crosses

N. Vali[1], M.A. Edriss[2] and H.R. Rahmani[2], [1]Department of Animal Sciences, College of Agriculture Shahrekord University, [2]Isfahan University of Technology (IUT), Isfahan, Iran

A study was conducted to compare body weight (BW), feed intake, feed coefficient up to 49 days of age, and carcass characteristics of two quail strains and their reciprocal crosses for genetic groups, namely, Japanese quail (Coturnix Japanese), Range quail (Coturnix ypsilophorus), and their reciprocal in two hatches. Body weights of four groups at 1, 7, 14, 21, 28, 35, 42, 49, 56 and 63 days of ages were significantly different (P<0.01). Body weights of female at 49, 56 and 63 days of age were significantly higher than males, but there was no significant difference between male and females at the other recorded body weights (P>0.05). There was a significant effect of hatch for body weights at all ages (P<0.01). Feed intake of H2 group was also significantly larger than that other groups (P<0.01), while feed coefficient of four groups were not different (P>0.05). At 49 days of age, Carcass percent, breast percent, wing weight and giblet weight of four groups were significantly different (p<0.01), while there was no significant difference for carcass weight (eviscerated) and breast weight among them (p>0.5). Carcass weight, carcass percent, breast weight were significantly affected by sex (p<0.1), while sex was not effective for breast percent difference (p>0.5). The estimated heterosis for the body weight was maximum (18.41%) and minimum (-8.37%) at 14 and 7 days of age, respectively.

Effect of feeding the biologically treated corn stalks on performance of liver function in New Zealand White rabbits

A.A. El-Shahat*[1], A.H. Hessen[2], R.I. El-kady[1], H.A. Omar[1], M.A. Khalafallah[3] and A.A. Morad[1], Anim. Prod Dept., NRC, Dokki (12622), Cairo,[2]Agric. Botany Dept. Fac. of Agric. Al-Azhar University, Nasr City, Cairo and [3]Agric. Microbiology Dept. NRC, Dokki, Cairo, Egypt

The experimental rabbits were fed either on the commercial diet (the control group) or with ration supplemented with 10% corn stalks. At the end of the experimental period which lasted for twelve weeks, the animals were slaughtered. Blood samples were collected and placed in heparinized test tubes and centrifuged at 3000 rpm. for 15 minutes. The samples were stored at –20ºC. The various tests were performed by utilizing the proper commercial Kits. The mean values which had been obtained for the control and the biologically treated supplemented groups were 44.50 µ/L and 81.80 µ/L, respectively for Glutamic Pyruvic Transaminase (GPT). The corresponding figures for Glutamic Oxaloacetic Transeminase (GOT) were 29.73 µ/L and 61.83 µ/L respectively. Based upon the present results, however, it could be suggested that the biologically treated corn stalks increases the secretion rates of transaminases (GPT and GOT). However, The results will be discussed on the basis of animal nutrition and animal physiology sciences. The financial support of NRC, Dokki, Cairo, Egypt to perform this experiment is well acknowledged.

Using the biological treatments for improving the nutritional values of untradional animal feeds

A.A. El-Shahat[*1], A.H. Hessen[2], H.A. Omar[1], R.I. El-Kady[1], A.A. Morad[1] and M.A. Khalafallah[3],
[1]Anim. Prod. Dept. NRC, Dokki (12622) Cairo, [2]Agric. Botany Dept. Fac. of Agric. Al-Azhar University, Nasr City, Cairo, [3]Agric. Microb. Dept. NRC, Dokki, Cairo, Egypt*

The main aim of the present study was improving the nutritional values of corn stalks as an example of poor quality roughage and to improve its utilization by animals. Therefore, sixty four New Zealand white rabbits of five weeks old were randomly distributed into four nutritional groups of sixteen rabbits each. The rabbits of the first three groups were fed on rations contained corn stalks supplemented with either 10% or 20% or 30% *Trichoderma viride* (*T. Viridi*), respectively; whilst those of the fourth group were given the commercial diet. After the end of the experimental period, the digestibility trials were carried out.

The most interesting features of present investigation were that the biological treatments may affect feed consumption, may improve feed conversion values and that may increase the nutritive values (expressed as TDN and SV) of the poor quality roughage corn stalks. However, upon the basis of the present results, it could be suggested that *T. Viride* can be used successfully to enrich poor quality roughages such as corn stalks and to improve the utilization of untraditional feeds by animals. The authors thank NRC of Cairo, Egypt for providing the financial support of this research.

Influence of partial substitution of berseem hay by corn stalks treated biologically with Trichoderma viride on lipids and Bilirubin secretion rates in rabbits.

A.A. El-Shahat[*1], A.H.Hessen[2], H.A. Omar[1], R.I.El-Kady[1], A.A. Morad[1], M. Fadel[3], M.A. Kalafallah[4]. [1]Animal Production Dept. NRC, Dokki (12622), Cairo, [2]Agric. Botany Dept. Fac. of Agric. Al-Azhar University, Nasr city, Cairo, [3]Microb. Chemist. Dept. NRC, Cairo and [4]Agric. Microb. Dept. NRC Cairo, Egypt*

The experimental animals were fed *ad lib.* either on the basal diet (control group) or with rations contained corn stalks supplemented with either 20% or 30% *Trichoderma viride (T. viride)*. At the end of the experimental period, the rabbits were slaughtered. After slaughtering, blood samples were collected and placed in heparinized test tubes and centrifuged at 3000 rpm for 15 minutes. The samples were preserved in a deep freezer at (–20°C) until the time of analyzis. The various tests were calorimetrically determined by using commercial kits, following the same steps as described by manufactures. The mean values of lipids which had been obtained was 65.10 mg/dl. for the control samples, whilst those for group samples which contained corn-stalks supplemented with either 20% or 30% *T. viride* were 63.75 mg/dl. and 70.12 mg/dl., respectively. However, the corresponding figures for Bilirubin samples were 0.19 mg/dl. (control), 0.15 mg/dl. (20% *T. viride*) and 0.12 mg/dl. (30% *T. viride*), respectively. However, based upon the present data, it could be suggested that the biological treatments may increase the secretion rates of lipids and decreases those of bilirubin. NRC of Dokki, Cairo, Egypt is greatly acknowledged for its financial support of this investigation.

Determination of chemical composition and voluntary feed intake of Saffron (*Crocus sativus*) hay
A.R. Safaei, H. Fazaeli and M. Zahedifar, Animal Science Research Institute, P.O.Box: 1483-31585. Karaj, Iran

The aim of this study was to measure the chemical composition and voluntary feed inake of Saffron hay (residues). A number of 45 samples were collected from 15 different farms of South Khorasan province of Iran after last harvesting of flowers. Chemical composition was determined using the procedure mentioned in AOAC, 2000 and fermentability was assessed using gas production technique. Voluntary feed intake was measured using 8 male sheep (Zel breed) with average body weight of 52±5 Kg. The results showed that the content of organic matter, CP, Ash, CF, EE, non fibrous carbohydrates, hemicellulose, NDF, ADF, ADL, Ca and P were: 89.5, 6.9, 11.5, 4.2, 27.3, 24.8, 60.1, 35.3, 3.7, 1.32 and 0.28 percent respectively. The gross energy of sample was 4207.4 Kcal/Kg and gas production after 24 hours was 42.3 ml. The estimated digestibility of sample using *in vitro* gas production technique was 60.4 percent and the estimated ME was 3429.4 Kcal/Kg dry matter. The voluntary feed intake of Saffron was 895/2 g per day.

Effect of Saccharomyces cerevisiae on the rumen digestion, fermentation and protozoa population of bulls fed alfalfa hay or corn silage based diet
E. Ghasemi[1]*, A. Nikkhah[1] and F. Kafizadeh[2], [1]TehranUniversity Faculty of Agriculture, Karaj, Iran [2]Razi University Faculty of Agriculture,Kermanshah, Iran

To determine effect of *S. cerevisiae* on two different forages utilization, Four mature fistulated bulls were used in a 4×4 Latin square design with a 2×2 factorial arrangement of treatments (alfalfa hay or corn silage with 0 or 5 g of *S. cerevisiae*). Addition of *S. cerevisiae* caused an increment in degradability of neutral detergent fiber (NDF) of forages and organic dry matter (ODM) of total diets at 3 h of incubation. But, degradability was the same among treatments from 3 h onward. Crude protein (CP) disappearance and pH were not modified in diet supplemented with *S. cerevisiae*. NDF degradability of alfalfa hay and ODM and CP of total diet including alfalfa hay were higher than corn silage and total diet including corn silage at 3 h of incubation, respectively. Concentration of volatile fatty acids and molar percentage of propionate were higher in diets supplemented with *S. cerevisiae* at 0 and 3 h after feeding. However, there were some interactions between *S. cerevisiae* and forage sources on acetate to propionate ratio. Total numbers of protozoa were reduced with addition of *S. cerevisiae* at 3 and 12 h after feeding. The results represented transitory effect of most rumen parameters with addition of *S. cerevisiae* and initial degradability of alfalfa hay was more improved when *S. cerevisiae* was supplemented.

Effect of heat processing on nitrogen fractionations and ruminal degradability of dry matter and nitrogen in Iranian whole soybeans

M. Hassan Fathi Nasri[1]*, Mohsen Danesh Mesgaran[2], [1]Department of Animal Science, Faculty of Agriculture, University of Birjand, Birjand, Iran, [2]Department of Animal Science, Faculty of Agriculture, University of Mashad, Mashad, Iran

The effects of roasting and steep-roasting on nitrogen (N) fractionations, and ruminal degradability of DM and N of two Iranian cultivars of whole soybeans was elucidated. The seeds were roasted using a drum roaster with an exit temperature of 140 to 145°C. A fraction of the seeds were cooled immediately and the rest were held in isolated barrels for 45 minutes (steeping). Non-protein N (NPN) and buffer-soluble N (BSN) were determined using sodium tungstate solution and borate-phosphate solution, respectively. Neutral detergent insoluble N (NDIN) and acid detergent insoluble N (ADIN) was determined as N associated with neutral detergent fibre and acid detergent fibre, respectively. Ruminal degradability of feeds was determined using in situ nylon bag technique. The exponential equation of Ørskov and McDonald (1979) was used to fit the disappearance data using the NLIN procedure of SAS (SAS, 1999) to estimate the soluble (*a*) and insoluble potentially degradable (*b*) fractions, and rate of degradation (*c*) of DM and N of feeds. The N fractions of feeds were analyzed using the GLM procedure of SAS (SAS, 1999). Heat processing decreased significantly ($P < 0.001$) the NPN and BSN content of seeds. The NDIN and ADIN content of heat processed and raw soybeans were similar. The rate of degradation of DM and N was reduced severely by heat processing in both cultivars.

Effects of different levels of rapeseed meal on digestibility, growth and carcass composition of rabbits

Aziza Gasmi-Boubaker[1]*, H. Abdouli[2], M. El Hichi[2] and K. Faiza[2], [1]cité Mahrajène INAT, Tunis, Tunisia, [2]Ecole Supérieure d'Agriculture Mateur

Two experiments were conducted to study the effects of various levels of rapeseed meal in the diet of rabbits on the digestibility (Experiment 1), the growth performance and carcass composition (Experiment 2). Four levels (0, 7, 14 or 21%) of rapeseed meal were included in isonitrogenous diets. The apparent digestibility of dry matter (DMD), organic matter (OMD) and crude protein (CPD) were determined using 6 rabbits per diet. There was no effect of diets on CPD that was on average 67%.On the opposite, DMD and OMD decreased in diet containing 21 % rapeseed meal. In the experiment 2, a total of 128 weaned Hyla rabbits aged 28 days and weighing on average 1043 g were equally divided into four groups of 32 (four replicates of four male and four female rabbits each).The experimental period lasted 6 weeks. At the end of the experiment, there were no significant differences among the groups in final live weight, average daily gain and feed efficiency. However the inclusion of 21% rapeseed meal in the diets decreased the carcass yield by 2.2 units and increased the full gastrointestinal tract weight by 2.6 units compared to control group. The percentage values of skin, liver and perirenal fat were not affected by the diet. It is concluded that rapeseed meal can be given to the rabbits at the level up to 14 % of the diet without adverse effects on digestibility and growth performance.

Feeding of genetically modified (Bt) maize to laying hens over four generations
Ingrid Halle and G. Flachowsky, Federal Agricultural Research Centre (FAL), Institute of Animal Nutrition, Bundesallee 50, D-38116 Braunschweig; Germany

The objective of the study was to investigate the influence of long term feeding of genetically modified maize (Bt 176) to growing and laying hens over four generations. Such studies seem to be necessary to assess the nutritional effects of genetical modification on animal performance under special consideration of reproduction of animals. Basal diets were formulated to contain 40 (chicken and rearing hens) or 50% (layers) maize (isogenic or Bt) per kg. Feed and water were provided ad lib. to 200 growers and 32 layers of each treatment. Hatched chickens were reared over 18 weeks before the hens started the laying period. Eggs were collected at the age of laying hens of 31 weeks.

The feeding of diets containing Bt maize for four generations to growing and laying hens did not significantly influence body weight, animal health, feed intake of growers and layers (overall means of layers: 114.9 and 112.9 g/hen/day), laying intensity (83.5 and 83.3%) and hatchability (86.8 and 88.0% for control and Bt-group resp.) in comparison to the isogenic counterpart.

The results of the experiment are in agreements with data from our 10 generation experiment with growing and laying Japanese quails.

The effect of oven heating on chemical composition, degradability coefficients, ruminal and post-ruminal disappearance of crud protein of Iranian canola seeds in Holstein steers
Mohsen Danesh Mesgaran[1], M.Hassan Fathi Nasri[2] and Hassan Nassiri Moghaddam[1], [1]Department of Animal Science, Faculty of Agriculture,Ferdowsi University of Mashhad, P O Box 91775-1163, Mashhad, Iran, [2]Department of Animal Science, Faculty of Agriculture, University of Birjand, Birjand, Iran

The effect of oven heating of Iranian canola seeds (SLM) on chemical composition, degradability coefficients (in situ), and ruminal and post-ruminal disappearance (three-step enzymatic procedure) of crude protein (CP) was determined. The seeds were oven heated at three temperatures of 120, 150 and 180 °C for 2 or 3 hours. Chemical composition, and rapidly (a) and slowly (b) coefficients of CP degradability were not affected by heating. But degradation rate constant (c) of CP was reduced in heated seeds at 150 °C and 180 °C for 2 h (79 and 89 percent decrease compared with the raw seeds, respectively). The disappearance of ruminal CP was also significantly ($p < 0.05$) reduced when these temperatures were conducted (50 and 73%, respectively). Post-ruminal disappearance of non-ruminal digestible CP for all treatments was similar (except heated seeds at 150 °C for 3 h that the disappearance was increased ($p < 0.05$)). Total tract disappearance of CP was highest for raw seeds and heated seeds at 120 °C for 3 h and 150 °C for 2 h (0.91, 0.90 and 0.90, respectively) and was lowest for heated seeds at 180 °C for 2 h (0.34).

The effect of rumen pH on dry matter degradation kinetics of alfalfa hay using *in situ* technique

Mohsen Danesh Mesgaran and Alireza Vakili, Dept. of Animal Science, Faculty of Agriculture, Ferdowsi University of Mashad, P O Box 91775-1163. Mashad, Iran

Our study investigated the effect of rumen pH on ruminal dry matter degradation rate of alfalfa hay. Samples of alfalfa were incubated in the rumen of four Holstein steers (300±15 kg body weight) with ruminal fistula. Animals were fed with diets differing in concentrate:alfalfa ratios as 60:40 (T1), 70:30 (T2), 80:20 (T3) and 90:10 (T4) in a Latin Square design. Ground (2 mm) samples (5 g DM) were placed in artificial silk bags (10×20 cm, 50μm pore size) and incubated in the rumen for 0.0, 2, 4, 8, 16, 24, 48, 72 and 96 h (n=8). Data of DM degradation beyond the lag-time were further adjusted to a negative exponential model [$P=a+b(1-e^{-ct})$, where P= fraction degraded in the time t, a= rapidly degradable fraction, b= slowly degradable fraction, c= fractional degradation rate and t= incubation time]. Minimum daily ruminal pH decreased from 6.40 (T1) to 5.34 (T4) when level of concentrate was increased. There was no significant difference when ruminal dry matter degradation parameters of alfalfa were considered. Therefore, the extent of ruminal forage dry matter degradation is not a function of rumen pH.

Ensiling potato hash with hay, poultry litter, whey or molasses

B.D. Nkosi[1], H.H. Meissner[1], M.M. Ratsaka[1], T. Langa[1], D. Palic[1], R. Meeske[2] and I.B. Groenewald[3], [1]ARC -Irene, Animal Nutrition, S.Africa, [2]Outeniqua Experimental Farm, Western Cape, [3]Center for Sustainable Agriculture, UFS

The aim of the study was to determine the effect of adding combinations of hay, poultry litter (PL), molasses or whey on potato hash to produce silage. The by-product was ensiled in 1.5 L jars with either hay or PL and whey or molasses addition. Silages were produced as T1. PH (90%) + hay (10%), T2. PH (75%) + PL (25%), T3. PH (90%) + hay (10%) + whey, T4. PH (75%) + PL (25%) + whey, T5. PH (90%) + hay (10%) + molasses, T6. PH (75%) + PL (25%) + molasses. Samples were collected on day 4, 10, 20 and 40 of ensiling. DM, pH, lactic acid and WSC were determined. The PL mixed silages had higher pH values (5.349) than the hay silages (3.688). Molasses and whey silages were similar but lower significantly to the control. PL mixed silages produced significantly higher lactic acid than hay silages while treatments show no significant differences across days (P>0.05). PL and hay were not significantly different (P>0.05) in WSC while molasses has significant higher WSC than the rest. Significantly higher DM were observed for PL silages than hay silages. Molasses and whey produced higher DM values than the control (P<0.05) but were not significant across days. Further research is needed to measure other parameters and determine the digestibility and palatability of the silages.

Replacing cotton seed meal and barley by chickpea processing by-product in fattening performance of fat tailed Makui and Kizil male lambs

Sh. Mousavi, A.M. Aghazadeh, Dept. of Animal Science, Faculty of Agriculture, Urmia University, Urmia, Iran

Twenty-four intact male lambs of each of two fat tailed breed, Kizil and Makui, in equal numbers and uniform weights, four to five month of age, were randomly allocated to one of the three dietary treatments, in which three levels of chickpea processing by-products (CPB) were fed 0, 15 and 30 % on a dry matter basis. CPB was increased at the expense of decreasing cotton seed meal and barely in the diets. The lambs were individually penned and fattened for seventy days. Average daily gain (ADG) were 239.46, 233.93 and 221.61 g/day and feed conversion ratios (FCR) were 6.18, 6.33 and 6.67 kg dry matter/kg gain for three diets respectively. These results were not significant between diets (P > 0.05). In comparison of two breeds, the overall ADG for Kizil and Makui lambs were 242.26 and 221.07 g/day and FCR were 6.48 and 6.38 kg dry matter/kg gain, respectively (P > 0.05).The results suggest that increasing CPB level up to 30% of the diets did not affect ADG, dry matter intake and FCR (P > 0.05) of fat tailed Kizil and Makui male lambs.

Feeding whole wheat of different varieties with or without dietary enzyme to broiler chickens

Kh. Rahimi and M. Agazadeh, Urmia University, M. Sc & Faculty of Agriculture, Department of Animal Science, 81110 Urmia, Iran

A 2×2×2 factorial experiment with four replication of 10 chicks each was conducted to study the effect of two varieties with two wheat form(10, 25, 40 and 55% whole wheat replacing ground wheat during7-14,14-21,21-28and28-42days, respectively), and two levels of xylanase (0 and 0.1 g/kg diet) on performance, digestive tract measurement and carcass characteristics of broiler chickens. The birds were pen- weighed at 7, 14, 21, 28, 35 and 42days and feed consumption was determined for each of these periods. Feeding whole wheat in all of these periods except at 28-35 days of age did not significantly (P>0.05) influence the growth rate and feed efficiency of birds. Wheat variety and enzyme addition had not effect (P>0.05) on the bird performance. Whole wheat inclusion increased (P<0.05) the relative weight of the gizzard and liver. Xylanase supplementation had no effects (P>0.05) on the relative weight of proventriculus, gizzard, pancreas, but increased (P<0.05) the relative weight of the liver. Neither, whole wheat inclusion nor xylanase supplementation influence (P>0.05) carcass recovery and the relative weight of abdominal fat pad. The results suggest that whole wheat feeding caused significant loss (p<0.05) in total gain and feed efficiency (7-49d) as compared with the ground diets.

Assessment of nutritive value of fourteen Acacia species, in vitro

H.E.M. Kamel[1], A. Al-Soqeer[2]* and S.N. Al-Dobaib[1], [1]Department of Animal Production and Breeding, [2]Department of Plant Production and Protection, Faculty of Agriculture and Veterinary Medicine, Al-Qassim University, P.O. Box 1482- Buriedah 51431, Saudi Arabia

Yong shoots of fourteen *Acacia* species namely *A. coriacea, A. cuthbertsonii, A. ineguilatera, A. iteaphylla, A. kempeana, A. ligulata, A. microbotrya, A. nilotica, A. oswaldii, A. pruinocarpa, A. saligna, A. sclerosperma, A. seyal* and *A. victoria* were collected from Prince Sultan Research Center for Environment, Water and Desert-King Saud University (Al-Riyadh- Saudi Arabia). Chemical composition and *In vitro* gas production technique were used to assess the nutritive values of *Acacia* spp., alfalfa hay and wheat straw. The CP content among tested *Acacia* spp. ranged from 8.0 to 16.7 % and it was comparable in *A. iteaphylla* with that of alfalfa hay, the values were 16.7 and 17.1 in *A. iteaphylla* and alfalfa hay, respectively. Amongst the tested *Acacia* spp, *A. ineguilatera* had the highest CT value, the ranged from 10.4 to 77.0 mg/ g DM. No significant differences (P> 0.05) for degradable fraction were found between alfalfa hay in one hand and *A. kempeana, A. niloti*ca and *A.seyal* in the other hand. Calculated metabolizable energy (ME, MJ/ kg DM) for the tested species of *Acacia* spp. ranged from 4.35 to 6.69 MJ/ kg DM which could supply the animals with the 53-84% of the ME in alfalfa hay.

Economic possibility of collecting and processing the corn residues as ruminant feed

H. Fazaeli*[1] and M.A. Mousavi[2], [1]Animal Science Research Institute, P.O. Box 1483, 31585 Karaj, Iran, [2]Agricultural and Natural Resources Research Center of Kerman Provinces, Iran

Three farms were selected in Jiroft area of Iran, where each farm was divided into 6 equal plots. Six methods (treatments) was carried out for collecting and preparing the corn residues as ruminant feed, and the final cost per kg of prepared feed was compared among the treatments were:

T_1 = harvesting by hand and chopping by chopper machine.
T_2 = harvesting by hand and chopping by electric machine.
T_3 = harvesting by hand and chopping by tractor chaffer machine.
T_4 = collecting the residues by attachment a bag at the back of the combine and chopping by chopper.
T_5 = harvesting by combine with adjustment its head at about 18 cm up of the ground, collecting the residues by hand and chopping by chopper machine.
T_6 = same as T_5, but collecting the residues by attachment a bag at the back of combine.
The average cost per kg of prepared corn residues was 80, 192, 108, 472, 284 and 424 Rials for T_1 to T_6 respectively which were significantly (P<0.05) the lowest for T_1. It can be concluded that harvesting of corn residues by hand and chopping with chopper machine is applicable and economic in Kerman province.

Effect of ensilage and polyethylene glycol on gas production and microbial mass yield of grape pomace
D. Alipour and Y. Rouzbehan, Department of Animal Science, Faculty of Agriculture, Tarbiat Modarres University, P.O. Box 14115-336, Tehran, Iran

In this trial, the influence of ensiling process and polyethylene glycol (PEG) on the gas production and microbial mass yield of grape pomace (GP) were assessed. The water soluble carbohydrate (WSC), pH and phenolic contents of GP, which was ensiled for 5, 10 and 30 days, were determined. In vitro gas production (IVGP), estimated organic matter digestibility (OMD), ammonia concentration and microbial mass yield (MM) of each treatment with or without PEG were measured. Increasing the duration of ensilage was led to decrease the content of WSC ($P<0.01$). Total phenolics (TP), total tannin (TT), condensed tannins (CT) and protein precipitable phenolics (PPP) in day 30 were significantly lower than the other periods ($P<0.001$). Ensilage had led to decrease the amount of IVGP, OMD and MM ($P<0.01$). Adding PEG, however, resulted in increasing of IVGP, OMD, ammonia and MM content ($P<0.01$). Therefore, PEG, but not ensiling, could be used to deactivate the tannins and enhance the in vitro digestibility of GP.

Aflatoxin, heavy metal and pesticide contents of compound ruminant feeds produced in Turkey
Özlem Dağaşan and Nihat Özen, Akdeniz University, Animal Science, 07059 Antalya, Turkey

Main objective of this experiment is to determine the pollution levels and and seasonal variations of polluting agents in some ruminant compound feeds produced in Turkey and also to determine whether legal tolerance levels have been exceeded or not. In order to achieve this goal, aflatoxin B_1-B_2-G_1-G_2, lead, cadmium, mercury, arsenic and organophosphate analyses have been planned on lactating cow, fattening cattle and lamb compound feed samples collected at 3-month intervals from commercial feed compounders planted in 5 provinces where both animal and feed productions are intense. The ongoing laboratory analyses will be complemented in next June. Affinity colon and high performance liquid chromatography (HPLC) in aflatoxin analyses, gas and mass chromatographies in organophophate analyses, atomic absorbtion spectrometer in heavy metal analyses are being utilized. After the analyses, the data obtained will be evaluated through variance (ANOVA) analysis in order to compare differences among the feeds, and multiple comparison tests for seasonal variations. The overall results will be discussed in comparison to domestic and international standards.

The use of Reading Pressure Technique to estimate cumulative and rate of gas production for a range of commercially available buffalo feeds

F. Polimeno, G. Novi, V. Vitale, R.Baculo, F. Sarubbi and G. Maglione, ISPAAM-CNR, Via Argine 1085, 80154, Napoli-Italy

The objective of this study is to examine the fermentation and degradation characteristics of seven substrates (alfalfa hay, soybean flour, wheat by-product, flaked barley, ryegrass, maize silage and straw) typically used in a commercial buffalo dairy farm, using the in vitro Reading Pressure Technique [RPT] according to Mauricio *et al.* (1999). This feed evaluation system is designed to examine the rate and extent of both fermentation [gas release] and substrate degradation simultaneously. With regards to measurements, standard parameters will be evaluated i.e. dry matter (DMD) and organic matter degradation (OMD) and fermentation gas release over a 96 h incubation period.

The obtained results show that, 96 h post-inoculation, the fastest degradable feeds are soybean flour (952.28 OMD), barley flakes (894.38 OMD), ryegrass (871.04 OMD) but soybean flour produce less gas (245.54 ml v 301.07 ml (barley flakes) and 250.44 ml (ryegrass)) because of its higher content of crude protein. Also wheat by-product is highly degradable (839.44 OMD) while straw is the least degradable (667.31 OMD) because of its high content of ash and lignin that restrict degradation.

These data demonstrate that this system is useful for a rapid and efficient evaluation of different feedstuffs. Moreover is a strong research tool to predict optimal plant in animal nutrition.

The effect of early feeding on performance of broilers

N. Eila*[1], M. Eshaghian[1], M. Shivazad[2], A. Zarei[1], K. Karkoodi[3] and F. Foroudi[4], [1]Department of Animal Science, Faculty of Agriculture, Islamic Azad University, Karaj Branch, Karaj, Iran. Postcode3187644511, [2]Faculty of Agriculture,Tehran University,Karaj,Iran, [3]Islamic Azad University,Saveh Branch,Saveh,Iran, [4]Islamic Azad University, Varamin Branch,Varamin,Iran

According to an unbalanced completely randomized design experiment with a split plotted arrangement, 480 broilers were randomly distributed in 40 pens including 14 treatments (two sex × seven early feeding methods) and two or three replication. Early feeds were Soft Corn 0-2 d (SC), starter diet (ST), starter diet with drinking multivitamins and electrolytes 0-2 d as well as 0-7 d (STME2 and STME7), High-protein starter diet 0-2 d, 0-7 d (HPST2 and HPST7), and finally High-protein starter diet with drinking multivitamins and electrolytes 0-7 d (HPSTME7).In Starter period (0-10 d), weight and feed conversion ratios of chickens fed by HPST7 and HPSTME7 were significantly better than SC (P ≤ 0.05).Weight gain and feed Conversion ratios of high protein early diets and SC were not significantly different in the grower (11-28 d) and finisher (29-42 d) periods (P ≤ 0.05).The relative weight of yolk sac of chicken fed by STME7 was significantly more than those fed by HPST diets at 2 d (P ≤ 0.05). Relative weight of proventriculus (at 7 and 42 d), heart (at 7 d) and gizzard (at 42 d) were significantly affected by early feeding methods (P ≤ 0.05). Totally High protein early feeds had beneficial effects.

Chemical composition, digestibility and voluntary food intake two alfalfa variety grown in West Azerbaijan in Iran

A. Mirzaei Aghsaghali[1], A. Aghazadeh[1], N. Maheri Sis[1] and A. R.Safaei[2], [1]Department of Animal Science, Shabestar University, Shabestar, Iran,[2]Researcher of Animal Science Institute, Karaj, Iran

The aim of study was to evaluated chemical composition, digestibility and voluntary food intake of forages grown in West Azerbaijani in Iran by using *in vivo* technique. Ghareuonge and Hamedani hays were collected from two different locations in West Azerbaijan namely Uremia and Miandoab areas. To determine the chemical composition of forages, samples of Ghareuonge and Hamedani hays were analyzed for DM, OM, NDF, ADF, CP and EE. To estimate *in vivo* digestibility, 12 sheep housed in metabolic stalls during the *in vivo* study were fed in form Ad libitum feeding. The concentration of CP was higher in Hamedani hay compared with Ghareuonge hay, but OM, EE did not differ between Ghareuonge and Hamedani hays (89.66 and 1.33).
The leaf to stem ratio was higher in Hamedani hay compared with Ghareuonge hay(35.8 to 64.2 vs. 32.8 to 67.2).The rate of OM digestibility and voluntary food intake were the highest in Hamedani hay, and the lowest in Ghareuonge hay(72 vs. 50% and 1600 vs. 1200 g per kg DM per day, respectively). In conclusion, Hamedani hay seems to have the best digestibility compared with Ghareuonge.

Selenium status in sheep and goat flocks in the northeast region of Portugal

F.C. Silva[1], C. Gutierrez[2] and A. Dias-da-Silva[1]*, [1]CECAV-UTAD, Department of Animal Science, PO Box 1013, 5001-801 Vila Real, Portugal, [2]Veterinary Faculty, University of Las Palmas, 35416, Las Palmas, Spain

We have observed a number of clinical signs pointing out for selenium (Se) deficiency in sheep and goat flocks fed on native pastures in the northeast region of Portugal. White muscle disease is quite common. To test this hypothesis, a field study was made with 38 flocks of sheep and 38 flocks of goats between June and August 2000. Blood samples from 15 to 20 mature ewes or goats randomly selected in each flock were collected. To assess the Se status of the animals the activity level of the enzyme glutathione peroxidase was measured per g of haemoglobin (GSH-px/g Hb). It was found that in 55.3% of the flocks, more than 20% of the animals sampled in each flock showed values of GSH-px/g Hb lower than 60 which are considered as deficient or severely deficient. Only in 38.2% of the flocks values of GSH-px/g Hb higher than 130, considered as adequate, were observed in more than 20% of the sampled animals in each flock. This work was repeated in 2005 and extended to measure the Se content of the pastures. Although full data are not yet available, these preliminary results strongly suggest that supplementation with Se should be done in some areas to reduce the mortality of lambs and kids as well as to improve reproductive efficiency.

Comparison of different levels of perlite and zeolite on the broiler's performance, carcass, tibia and serum characteristics

Ahmad Tatar, Islamic Azad University, Gorgan, Iran

One experiment with 625 one-day-old Ross male broiler chicks was conducted to determine the effects of Perlite and Zeolite on the Broiler's performance. Five experimental diets [Control, Perlite (2.5 & 5 percent), and Zeolite (2.5 & 5 percent)] were tested in a completely randomized design with 5 replicates and 25 chicks per replicate. During the experiment feed consumption, weight gain and feed conversion ratio (FCR) were measured weekly. At 42 days of age, two chicks per replicate were slaughtered to determine carcass characteristics, tibia dry weight and serum parameters. Tibia samples (Left & Right legs) were dried for 48 h in a convention oven (60 $^\circ$C) and then weighted. Analysis of Variance and separation of means by Duncan's multiple range tests were conducted by SAS® software. The results indicated that with adding 2.5% Zeolite to diet, weight gain was increased (P<0.05), but there were not significant differences in feed consumption, FCR, carcass yield and abdominal fat percentage among treatments (P>0.05). The addition of 5% Zeolite, had significant effect to increase Phosphorus content of serum in comparison with other treatments (P<0.05).Also, 2.5% Perlite had significant effect (P<0.05) to increase the right and left tibia dry weight.

Associative effects on rumen digestibility of supplementing alfalfa hay with wheat middlings

C.M. Guedes, V. Pessôa and A. Dias-da-Silva*, CECAV-UTAD, Department of Animal Science, PO Box 1013, 5001-801 Vila Real, Portugal

To identify possible associative effects of supplementing alfalfa hay (AH; 43.6% NDF and 16.9% CP) with wheat middlings (WM; 38.7% NDF, 14.9% CP and 29.8 % starch) on NDF digestibility, a 4x4 factorial design was used in an experiment with rumen fistulated cows. The animals were fed either AH alone or AH+WM in 85:15, 70:30 and 55:45 proportions (DM basis). The nylon bag technique was used to measure the kinetics of NDF degradation in the rumen. Alfalfa samples were ground (4mmm) prior to incubation and WM were incubated without grinding. Rumen pH and N-NH3 were also measured. Either the extent (72 to 74.8%) or the rate of NDF degradation (0.05-0.06/h) of WM samples were not affected by the level of WM inclusion in the diet. However, the fraction of alfalfa NDF potentially degradable in the rumen (63 to 53.7%) decreased (P<0.001) with WM inclusion in the diet, the differences being more pronounced at higher levels of WM in the diet. The rate of degradation of alfalfa NDF (0.08 to 0.06/h) also decreased. Increasing WM inclusion in the diet decreased rumen pH (6.7 to 6.2). Rumen N-NH3 increased due to WM supplementation (P<0.001). Given that CP content of WM was lower than that of AH, this finding suggests that either rumen degradability of CP from WM was higher than that from alfalfa or that N-NH3 utilization by rumen microbes was decreased, or both.

Determination of differences in nutritive value of corn forage and silage from conventional corn hybrid and Bt hybrid

*L. Křížová[1], S. Hadrová[1], F. Kocourek[2], J. Nedělník[3] and J.Třináctý[*1], [1]Research Institute for Cattle Breeding, Ltd. Rapotín, Dept. Pohořelice, Czech Republic, [2]Research Institute of Crop Production, Praha-Ruzyně, Czech Republic, [3]Research Institute for Fodder Plants, Ltd., Troubsko u Brna, Czech Republic*

The objective of the study was to determine the differences in nutrient content of corn forage and silage from the Bt-hybrid (Bt) and its near-isogenic control conventional corn hybrid (C) grown, harvested and ensilaged under identical conditions. The experimental field of corn forage was divided into two areas of 10 m^2 – control (C) with the conventional corn hybrid and experimental containing the Bt hybrid (Bt). Entire corn plants of both hybrids were harvested at the same time at the soft dough stage of maturity, cut and immediately packed into the microsilage tubes (approximately 6.5 kg per tube, 5 tubes per treatment) and fermented at 25 °C ($\pm$1°C). A forage from the control plants (C) had significantly lower content (P<0.05) of dry matter (362.1 vs. 375.8 g/kg) and organic matter (315.2 vs. 329.7 g/kg) than the Bt hybrid forage (Bt). The dry matter, organic matter and NDF content of conventional hybrid silage (C) was significantly lower (P<0.05) than that of Bt hybrid silage (Bt) (346.7 vs. 360.8 g/kg and 297.6 vs. 315.5 g/kg, 389.8 vs. 426.0 g/kg respectively).

This study was supported by NAZV 1B53043

Determination of ruminated tablet number using soluble marker

*M. Richter[1], J. Třináctý[*1], M. Rabišková[2], T. Sýkora[2], [1]Agrovýzkum Rapotín, Ltd., Dept. Pohořelice, Czech Republic, [2]University of Veterinary and Pharmaceutical Sciences, Brno, Czech Republic*

The objective of this study was to compare the determination of the amount of ruminated tablets using the direct method (counting) and the determination of recovery of soluble marker in feaces (Yb). For this purpose tablets containing soluble indigestible marker coated by inert layer based on ethyl celulose resistent to digestive tract environment were developed. Ytterbium chloride was chosen as the marker because of its high recovery. The experiment was carried out on three ruminally cannulated lactating dairy cows in two replications. The inert tablets (2000 pcs together) were inserted into the rumen bottom, feaces were collected for 6 days and pooled in 12-hour intervals. In the feaces samples ytterbium was determined and the number of excreted tablets was determined directly after the washing of the feaces through a sieve. The irrecoverable tablet ratio (16.7 %, SEM = 2.63) was significantly higher (P<0.05, n = 6) than recovery of ytterbium (9.6 %, SEM = 1.72). Provided that ytterbium recovery represented amount truly ruminated inert tablets, in digestive tract of lactating dairy cows remained kept 7.0 % (SEM = 2.02) of tablets after a 6-day-aplication.

This study was supported by MSM2678846201 and GACR 523/02/0164.

Evaluation of rolled wheat and CCM as concentrate replacers for dairy cattle
S. De Campeneere*, D.L. De Brabander and J.M. Vanacker, ILVO, Animal Science Unit, Scheldeweg 68, 9090 Melle, Belgium

The use of home-grown concentrate replacers has gained interest to reduce the feeding costs on dairy farms. To evaluate two potential concentrate replacers, three diets were compared using 18 lactating Holstein cows in a Latin square design with 3 periods of 4 weeks. At the start of the trial, the cows were on average 117 days in milk, produced 34.7 kg milk with a fat and protein content of 4.17 and 3.04%. The control diet consisted of maize silage and prewilted grass silage (55/45 on DM base) fed ad libitum and completed with soybean meal and concentrates. At the start of the trial concentrate level was fixed to supply 105% of the net energy and digestible protein requirements, and decreased weekly to correct for the change in lactation stage. In the treatment groups, part of the concentrate was replaced (on protein and energy basis) by wheat (4.6, 4.0 and 3.4 kg for period 1, 2 and 3) or CCM (6.0, 5.3 and 4.8 kg for period 1, 2 and 3).
Preliminary results indicated that replacing concentrate with CCM or rolled wheat depressed roughage intake, milk production and the milk fat content. Protein content increased slightly for CCM and clearly for rolled wheat as compared to the control diet.

Effects of different diets on the nutrient digestibility and rumen fermentation pattern in growing goats
É. Cenkvári*, S. Fekete, H. Fébel, E. Andrásofszky and E. Berta, Szent István University, Faculty of Veterinary Sciences, Department of Animal Breeding, Animal Nutrition and Laboratory Animal Sciences, Str. 2 István., H-1078 Budapest, Research Institute of Animal Production and Animal Nutrition, Str. 1 Gesztenyés, H-2035 Herceghalom, Hungary

We performed a metabolic trial with 6 Saanen growing goat bocks to investigate the influences of 4 different diets on the rumen fermentation parameters (pH, NH_3 and volatile fatty acids) and the appearant digestibility of nutrients including the cell wall fractions (NDF, ADF and ADL). Roughage and concentrate ratio was 87:13 on dry matter basis.
The rumen fermentation pattern reflected the different composition of diets. NH_3-content of rumen juice was 2.2 times and 1.6 times higher in animals eating lucerne hay, respectively. Total volatile fatty acid production was ca. 2 times higher than in the animals fed with grass hay. Molar proportion of C2 and C3 was similar in the different feeding groups: 72.9-74.9% and 16.4-19.9%, respectively.
We found the biggest differences in the appearant digestibility of N-free extract, compared to the values determined in cattle. Based upon the N-retention, crude protein of the diets was utilized at a similar rate in the different feeding periods. Since the daily average N-excretion followed the N-intake proportionally, N-retention in the animals fed with lucerne hay was 1.8 to 2 times higher than the ones in the group eating grass hay.

The influence of urea on in vitro gas production of ensiled pomegranate peel

R. Feizi[1]*, A. Ghodratnama[1], M. Zahedifar[2], M. Danesh Mesgaran[3] and M. Raisianzadeh[1], [1]The Agricultural and Natural Resources Research Center of Khorasan Province, Mashhad P. O. Box 91735-1148, Iran, [2]Animal Science Research Institute, P. O. Box 1483-31585, Karaj, Iran, [3]Dep of Animal Science, faculty of Agriculture, University of Mashhad, P. O. Box 91775-1163

In this experiment 4 levels of urea (0, 2.5, 5 and 7.5% of dry matter) were added to pomegranate peel and ensiled for two periods of 30 and 60 days. In vitro gas production (GP) from the samples were measured during 96 h incubation with or without addition of polyvinylpolypyrrolidone (PVP) to ensiled pomegranate peel (EPP). The data from gas volume recording were fitted to the exponential equation of the form p=a+b (1-e^{-ct}). The results indicate that addition of urea and then storage decreased (P<0.05) total extractable tannins (TET) content (206, 177, 169 and 170 mg/g respectively). GP after 24 and 48 h was higher for EPP treated with 0% urea (40.37 and 45.20 ml without PVP; 43.23 and 50.29 ml with PVP respectively) and lower for EPP treated with 7.5% urea (30.21 and 34.91 ml without PVP; 31.95 and 39.24 ml with PVP respectively) (P<0.05). There was a negative correlation between the CP content of EPP and GP. Also non–fiber carbohydrate (NFC) level was positively correlated with GP potential. Tannins have negative effect on in vitro rumen fermentation and PVP could show this effect.

The effect of feeding levels on microbial protein synthesis in rumen of native mazandaran male sheep

K. Jafari Khorshidi[1] and M.A. Jafari[2], Ruminant Nutritionist, Islamic Azad University, Branch of Savadkooh, [2]Islamic Azad University, Branch of ghaemshahr, Iran

Ruminants protein requirements is provided by two ways: first, via escaped protein and the other, microbial protein synthesized in reticulo-rumen. Large amount of protein is degraded by rumen microbes and rate of degradation depends on protein sourse, processing method and retention time in rumen. In this study the effect of feeding level on microbial protein synthesis was determined. 8 native male sheep from mazandaran province (breed of *zel*) was used. Three feeding levels (600, 900 and 1200 g/day) used for estimating microbial protein synthesis. A basal diet containing 40% concentrate mixture and 60% forage (wheat straw and alfalfa hay) was fed twice daily. Estimating the amount of purine derivatives excreted via urine done using chen (1992) procedure.the results of this experiment indicated that increasing feeding level, increased rumen microbial synthesis. Average rumen microbial synthesized with 3 feeding levels of 600, 900 and 1200 gDM/day was 35, 62 and 88 g/day respectively and significant difference observed (p<0.05).

Effect of organic substrate on growth and food conversion rate of Helix aspersa during the juvenile stage

J. Perea[*1], *A. Garcia*[1], *R. Martin*[2], *J.P. Avilez*[3], *A. Mayoral*[2] and *R. Acero*[1], [1]*Department of Animal Production, University of Cordoba, Spain,* [2]*Agricultural Research Institute of Andalucia,* [3]*Catholic University of Temuco, Chile*

The effect of organic substrate on growth, mortality, food intake and food conversion rate of *Helix aspersa* during the juvenile stage was investigated under controlled laboratorial conditions. 400 snails with average initial weight of 40 ± 5 mg were used, and assigned to a randomized design with two treatments in experimental boxes (treatment I: animals reared on humus; treatment II: animals reared without humus) and five replicates of forty animals per experimental unit. Each of the ten groups of snails was fed *ad libitum* with concentrate and keep them during the 6 first weeks of life at 22 ± 4°C with a relative humidity of $70 \pm 10\%$ and illumination period of 14 h day.

The results shows significantly differences between both treatments ($p<0.001$). Snails reared on humus present higher growth rate (801 mg) and lower mortality (12.5%) than snails reared without humus (growth rate 410 mg; mortality 32.5%). Furthermore food intake was higher in snails reared on humus (359.8 mg DMI/animal per week) that snails reared without humus (158.3 mg DMI/animal per week). In this study was also observed that food conversion rate was also higher in snails reared on humus (0.49) that snails reared without humus (0.43).

Use of corn (*Zea mays*) or triticalen (*Triticum secale*) in the stage of fattening Muscovy ducks (line R51)

J.P. Avilez[1], *C. Garces*[1], *J. Perea*[2]* *A.G. Gomez*[2] and *A. Garcia*[2], [1]*Catholic University of Temuco, Chile,* [2]*Department of Animal Production, University of Cordoba, Spain*

Chile beholds auspicious conditions for the development of aviculture, nevertheless there is neither knowledge nor profitable technology. The objective of this study was to compare the corn and the triticale in the feeding of ducks. The present study was carried at Campus Dr. Luis Rivas Del Canto, belonging to Catholic University of Temuco (Chile), during November and December 2004. This is part of a project of the Fundacion de Innovacion Agraria.

20 R-51 Muscovy ducks were used in this study, separated in two poultry-yards with 10 ducks each. The study gives information about the breeding of Muscovy ducks, lineage R-51 in relation with the overage date of weight gain, food supply and conversion index. In relation with the weight gain, supported increases were observed along the time with lower overages than the original lineage. In relation with the conversion index was maintained inside the ranks of the original lineage.

According to the results of this study in relation with the weight gain, food supply and food conversion index, it is concluded that under the conditions of the present study it is possible to state that the replacement of corn by triticale in the diet makes no statistical differences ($p>0.05$).

Seasonal pattern of reproductive activity of Helix aspersa in south of Spain

J. Perea[*1], *A. Mayoral*[2], *A. Garcia*[1], *R. Martin*[2], *E. Felix*[1] and *R. Acero*[1], [1]*Department of Animal Production, University of Cordoba, Spain*, [2]*Agricultural Research Institute of Andalucia*

The aim of this study was to characterise the seasonal pattern of reproductive activity of terrestrial snail *Helix aspersa* in Andalucia, under Mediterranean semiarid climatic conditions. A sample of 400 wild adult snails was collected during the hibernation period (December 2003) from Cordoba, Spain. The animals were kept in outdoors parks at natural environmental conditions during 365 days. The experience was repeated during 3 consecutive years with different snails collected each year. During these periods, the mating rate, egg-laying rate, photoperiod, temperature and humidity, was recorded.

A clear circannual cycle in reproductive activity was observed in all experimental periods. The snails were reproductively active during autumn and spring while during summer and winter they remain inactive. The duration of autumn reproductive activity was near to 60 days. This period was the most important with 90% of matings and clutches registered. The mean date of the first mating detected was 15 September and the last egg-laying recorded was 10 November. In spring the reproductive period begins at 30 May and concludes at 3 July. The reproductive activity was concentrated at the beginning of activity periods and being slightly decreasing to the final of them.

Furthermore, temporal variations in seasonal reproductive periods were observed, in response to differences in climatic and environmental conditions.

In vitro degradation of forages by using rumen fluid from slaughtered cattle

A.S. Chaudhry, School of Agriculture, Food & Rural Development, University of Newcastle upon Tyne, NE1 7RU, UK.

This study examined the potential of slaughtered cattle to obtain rumen fluid (RF) to test the effect of adding 0, 90, 180 g/kg forage (Amount) of supplements(BS, HP) on *in vitro* degradation of grass nuts (Grass) and barley straw (Straw) over 0 to 72h of incubations. RF were collected from three freshly slaughtered cattle on three separate occasions. Each RF was mixed with a buffer (pH 7) for use as an inoculum. About 1g Straw or Grass and each BS or HP amount were weighed separately into test tubes to which 50ml inoculum per tube were added under CO_2. The tubes were incubated at 39°C for 0-72h after which their contents were washed and dried to estimate dry matter disappearance (DMD). The DMD were statistically analysed to test the main effects and their interactions excluding Animal based interactions. While, Forage, Amount and Animal effects were significant ($P<0.01$) at most hours, the Supplement effect was significant at 72h only where DMD of HP was > BS ($P<0.05$). DMD increased with time for all main effects. Grass DMD was always greater than Straw ($P<0.001$). DMD increased ($P<0.05$) at all times except 0h with increasing Supplement. Animals did not differ at 0h but differed ($P<0.05$) at other hours showing the greatest DMD at 48-72h for Animal 1 than others ($P<0.001$). It appeared that slaughtered cattle can be used to obtain rumen fluid to evaluate supplements for *in vitro* degradation of forages.

Effect of maturity stage on estimation of net energy of clover forage

P. Homolka[1], V. Koukolová[1], J. Třináctý[2], J. Forejtová[1,3], [1]Resarch Institute of Animal Production, Department of Nutrition, 104 00 Praha 114, Czech Republic, [2]Research Institute for Animal Breeding, Ltd., Department Pohořelice, Czech Republic, [3]State Veterinary Institute, Department of Chemistry, Dolní 2, 370 04 České Budějovice, Czech Republic

The aim of this study was to determine the nutrients and the energy level of different maturity stage of clover forage (divided into three miters) used in ruminant nutrition by chemical analyses, *in vitro* and *in vivo* methods. The *in vitro* digestibility of organic matter (OM) was determined using an enzymes technique. The *in vivo* digestibility of OM was determined in metabolic trials on wethers. Generally the maturation caused an increase (P < 0.05) in crude fibre, neutral-detergent fibre, acid-detergent fibre and acid-detergent lignin contents. The *in vitro* and *in vitro* calculated (*in vitro*$_{calcul}$) digestibilities of OM were in average 75.4 and 70.6 % for clover. The averaged *in vivo* OM digestibility of clover was 71.0 %. With increasing maturity of forage samples the *in vivo*, *in vitro* and *in vitro*$_{calcul}$ digestibilities of OM linearly decreased. Gross energy, digestible energy, metabolizable energy and net energy for lactation were in average 18.12, 12.41, 9.60 and 5.67 MJ/kg of absolute dry matter for clover. This study was supported by the Ministry of Agriculture of the Czech Republic (MZE0002701403).

Sugar beet tops and crown silage treated with urea and molasses and its dry matter and crude protein degradability in Iranian Balouchi sheep

M. Raisianzadeh[1], M. Danesh Mesgaran[2], H. Fazaeli[3] and M. Nowrozi[1], [1]Agricalture and Natural Resources Research Center of Khorasan, [2]Ferdosi University of mashhad, [3]Animal Science Research Institute of Iran*

Sugar beet tops and crown (SBTC) was treated with additives and ensiled for two months. Degradability characteristics were determined by in situ method. Dried samples were grounded through a 2 mm screen prior to incubation. Nylon bags were placed in the rumen for 0, 2, 4, 8, 16, 24, 48 and 72 hours. Treatments were: 1) unchopped SBTC. 2) Chopped SBTC(CSBTC), 3) CSBTC + 5 % molasses, 4) CSBTC + 5% molasses + 2% urea, 5) CSBTC + 10 % molasses, 6) CSBTC + 10 % molasses + 2% urea and 7) CSBTC + 10% molasses + wheat straw (to rich in 35% dry matter). Means were compared with L.S. means. By increasing molasses in silages, degradability of dry matter was enhanced (P<0.05). The greatest rapidly degradable coefficient (a) of dry matter was observed in treatments 3 and 5 (0.59 and 0.62 respectively). Slowly degradable coefficient (b) of dry matter in treatment 5 was decreased by increasing molasses in comparison with control silage (0.26 vs. 0.33). The increase of urea did not affect the degradability coefficients of silages. Adding wheat straw to the silage 7 decreased the rapidly degradable coefficient (a) and increased the slowly degradable coefficient (b).

The performance of meat type commercial lines under two levels of energy
A.R. Paydar[1], N. Emamjomeh[2], G. Rahimy[3] and M. Moafi[4], [1]Fars Reasearch Center for Agriculture and Natural Resources, P.O.Box 71555-617, Shiraz, Iran, [2]Department of Animal Sience, Tehran university, Iran, [3]Department of Animal Sience, Mazandaran university, Iran, [4]Jahad Agriculture Ministry Tehran, Iran

With regard to the interaction between genotype and the environment, it is always recommended to rear the pure line flocks under the same environmental conditions that their offspring would be reared commercially. Therefore an experiment was conducted to investigate the effect of two different levels of energy (high and low) in the ration on the performance of meat type pure lines. A total of 4800 male and female day old chicks of four pure lines A, B, C and D were wing_banded. At selection age all birds were individually weighed. The results showed that final body weights in four lines were higher in the high ration treatment (p<0.05). This implies that male chicks were more sensitive to the levels of nutrient concentration in the ration than females. Also the results showed that in male lines, the B line is more sensitive to the changes of nutrient concentration than line A. The interaction of sire family and ration was not significant (p>0.05). This means that, in each line, selection of sire families in two ration treatments would be the same.

Using of urea treated *Glycyrrhiza glabra* L. root pulp in the ration of fattening lambs
A.H. Karimi[1]*, M. Zahedifar[2] and M. Kamali, [1]Fars Research center for Agriculture and Natural resoueces. Shiraz, Iran, [2]Animal Sciences Research Institute P.O.Box 31585-1483 Karaj, Iran

This Experiment was carried out to determine the effects of using urea-treated *Glycyrrhiza glabra* L. root pulp (%4) in the ration of fattening lambs, The treated *Glycyrrhiza glabra* L. root pulp was included in the ration as to replace 0, 20, 40 and 60% of ration total rafage in the ration of fattening lambs. The rations (iso-nitrogenous and iso-energetic) were pelleted and each ration was fed to 12 Turky Ghashghaii ram lambs. At the end of fattening period (90 days), the lambs were slaughtered. The data were analyzed by using the SAS software, and the means were compared by using the Duncan's test. Substituting hay with urea treated *Glycyrrhiza glabra* L. root pulp, had significantly affect on the final wieght, dressing percentage and meat CP contents(P<0.05), daily weight gain, feed intake, feed conversion ratio, in the ration of 60% urea-treated *Glycyrrhiza glabra* L. root pulp with the other rations (P<0.01). The cost of one kg DM of ration was 1427, 1261, 1096 and 937 Rials for ration 1 to 4 respectively. At present costs and according to feedlot performance and carcass characteristics, inclusion of *Glycyrrhiza glabra* L. root pulp in the ration of fattenning lambs, amount %40, seemed to be cost-effective.

Improving lamb growth rate using two strategies of supplemental nutrition either ewes at early lactation or suckling lambs

N. Dabiri, Department of Animal Science, Ramin Agricultural University, Ahwaz, Iran

For improving the growth rate of suckling lambs 2 experiments were conducted in a flock of Arabi sheep of Ramin Agricultural University to determine the effects of supplemental nutritional strategy either on lactating ewes or suckling lambs. In the first experiment 50 suckling male lambs were included in the experiment from day 14 until weaning(10 lambs/group). Four groups of lambs were randomly allocated to four supplemental protein diets treatment with the fifth group fed the conventional diet in a completely randomized design. In the second experiment the effect of supplemental feed in early lactation of ewes on growth rate of suckling lambs were studied. 3 equal groups of 9 ewes allocated into 3 different levels of supplemental feed (0, 350 and 700 g/d/ewe) for a period of one month after lambing in a completely randomized design. In experiment 1, the average daily gain (ADG) of lambs fed the 4 supplemental diets were significantly higher ($p<0.05$) than control group (190 vs 148± 3.87 g/d). In experiment 2, the growth rate (g/d) of lambs reared by 2 offered supplemental feed group ewes were greater ($P<0.05$) than lambs reared by conventionally ewes. The ADG of lambs reared in above defined diets were 165±9.6, 198±9.1 and 204±9.1 g/d respectively. The results of these 2 experiments indicate that the both strategy of supplementary nutrition has a positive role for producing heavier lambs.

Evaluation of several diluents and two storage temperature on Fars native chicken sperm characteristics

M.R. Hashemi[1] and M.J. Zamiri[2],Fars Reasrarch center for Agriculture and Natural resources. Shiraz, Iran, [2]College of Agriculture. Shiraz University, Shiraz, Iran

An experiment was conducted to investigate the effectiveness of several diluents for storing native chicken sperm at either 4-5°C (Sexton; Van Wembeke; and Lake and Ravie diluents) or room temperature (19-24°C) (Chaudhuri and Lake; and Sexton diluents). Seminal characteristics (% live sperm, sperm motility and pH) were determined immediately upon dilution and at 6 and 24 h after storage. An aliquot (0.2 ml) of diluted semen (1:2 ratio) was inseminated in the afternoon, once a week for 5 weeks. Diluent type, storage time and storage temperature significantly affected seminal characteristics which were significantly higher for the semen stored at 4-5° C as compared with storage at room temperature. Alls diluents were suitable for immediate insemination after dilution (91 to 99 % fertility). Fertility of diluted semen sample stored at 4-5°C was higher than that of sample stored at room temperature. Compared with other diluents, Sexton resulted in higher fertility levels for the sperms that were stored at 4-5°C for either 6 or 24 h. Correlation coefficient of sperm motility and % live sperm with % fertile eggs and % hatchability of incubated eggs were high (r=0.80 to 0.90) and significant ($p<0.0001$).

Effect of treating whole-crop barley (WCB) with urea on degradability
B. Bazrgar, E. Rowghani and M.J. Zamiri, Research Center of Agriculture and Natural Resource of Fars Province, Shiraz, Iran

In situ dry matter and protein degradation of urea supplemented whole-crop barley in the rumen of Mehraban rams were studied. The degradable fraction CP is converted to ammonia, fatty acids and CO_2, with a portion of ammonia being used for microbial protein synthesis in the rumen. The undegradable fraction escapes digestion in the rumen and subsequently available for intestinal digestion and absorption. Degradability coefficients of DM and CP were determined by using Dacron bags in three rumen fistulated rams. Bags were made of dacron having an average mesh size of 48µm. Approximately 5g (Oven dry) of the three diets (preserved WCB with 50 and 75 g/kgDM urea and preserved WCB without urea) placed in dacron bags which were tied shut with nylon string. Then the bags suspended in the rumen of the rams for 2, 4, 8, 16, 24, 48, 72 and 96 hrr. The content of each bag was subjected to Kjeldahl N analysis. The percent disappearance of dry matter (DM) and nitrogen (N) at each incubation time was calculated from the proportion remaining after incubation. The results indicated that the addition of urea to whole crop barley at 0.75% fresh weight may be recommended in order to increase the feeding value of the preserved WCB.

The effect of dietary oils on growth performance of broilers vaccinated with La Sota Newcastle vaccine
R. Aydin, M. Karaman, H.H.C. Toprak, A.K. Ozugur and D. Aydin, Animal Science Department, Kahramanmaras Sutcu Imam University, 46060 Kahramanmaras, Turkey*

The objective of this study was to investigate the effects of oils on growth performance of broilers vaccinated with La Sota vaccine against Newcastle disease. One hundred seventy five 1-week old Ross PM3 male broiler chicks were randomly distributed into five dietary groups and fed a commercial starter, grower or finisher diets supplemented with 0.5% sunflower oil (Group A), 0.5% olive oil (Group B), 0.5 % beef tallow (Group C), 0.5% conjugated linoleic acid (CLA, Group D) or 0.5% hazelnut oil (Group E) for 5 weeks. Broiler chicks were vaccinated with La Sota Newcastle vaccine at 22 days of age. Body weights of broiler chicks were measured weekly. At the end of the study, broiler chicks were slaughtered and carcass characteristics and weight of some organ weights were determined. After vaccination, growth rates in the broilers from the Group A, Group C and Group E were negatively influenced compared to the Group D. There was no difference in the proportions of abdominal fat (%) among the groups. Also, the relative organ weights of the broilers did not differ significantly among the dietary treatments. This study showed that the broiler chicks fed a diet supplemented with CLA had better performance and carcass weights compared to other groups (P<0.05). The present study also indicated that CLA prevented weight loss against vaccination compared to other groups.

Performance of broilers fed a low protein diet supplemented with probiotic

B. Dastar*, A. Khaksefidi and Y. Mostafaloo, Department of Animal Science, College of Agriculture, Gorgan University of Agricultural Science & Natural Resources, Gorgan, Iran

Broilers are fed low protein diet may have lower performance than those are fed diets with sufficient quantity of protein. Dietary additives such as probiotic increase nitrogen retention and may improve broiler performance. Present experiment was conducted to evaluate the effects of added probiotic thepax® in a low protein corn-soy diet on Cobb-500 broilers performance. Three treatments administrated in the experiment including: 1) a control diet with dietary protein level recommended by NRC (1994), 2) a low protein diet and 3) the low protein diet supplemented with thepax® (1gr/Kg of diet). All diets were isocalloric and each of them was fed to six groups of 20 birds. The resultant data were analyzed by a completely randomized design using of SAS software. Broilers were fed low protein diet had lower weight gain (46.4 gr/bird/day) and higher feed conversion ratio (2.1 gr/gr) than those were fed sufficient protein diet (51 gr/bird/day & 1.99 gr/gr, respectively). Supplementing of probiotic to the low protein diet has significantly improved weight gain (50 gr/bird/day) and feed conversion ratio (1.97). The results of this experiment indicated that supplementing probiotic in a practical low protein corn-soy diet improve broilers performance and efficiency of protein utilization.

Pinpointing the lowest protein diet for young male broiler

B. Dastar*, Department of Animal Science, College of Agriculture, Gorgan University of Agricultural Science & Natural Resources, Gorgan, Iran

Two experiments were conducted to determine the lowest crude protein content of corn-soy diet supplemented with several essential amino acids during the starter period. The diets in the first and second experiments contained 17, 19, 21 and 18, 19, 20% CP, respectively. A positive control diet with 23% CP was used as an index in both experiments. All diets were isocalloric (3200 Kcal MEn/Kg) and supplemented with appropriate levels of synthetic essential amino acids to match the NRC (1994) recommendations. In the first experiment, reducing the crude protein up to 19% did not have a significant effect on broiler performance as compared to birds fed control diet. In the second experiment, broilers had significantly lower weight gain and higher feed conversion ratio when fed 18% CP diet as compared to 23% CP diet. The performance of broilers fed other diets was not different as compared to control ones. The results of these experiments indicating that the performance of broilers fed a corn-soy diet contain 19% CP and supplemented with appropriate amounts of essential amino acids is similar to 23% CP diet.

Effect of citric acid and microbial phytase enzyme on performance and phytate utilization in broiler chicks

Y. Ebrahim Nezhad[1], M. Shivazad[2], K. Nazeradl[1], [1]Department of Animal Science, Azad Islamic University of Shabestar, IRAN, [2]Animal Sci., College of Agric., Tehran Univ., Tehran, Iran

This study was conducted to determine the additive effect of citric acid and microbial phytase enzyme on performance and phytate phosphorus utilization in broiler chicks in corn-soybean meal diet. Under a randomized complete design in a factorial arrangement 3×2, a total of 420 (1-d-old) commercial broiler chicks (Ross-308) were randomly distributed into 28 groups consisting of 15 chicks per group. The factorial arrangement was three citric acid levels (0.0, 2.5% and 5%) by two microbial phytase levels (0 and 500 U/kg) in low available phosphorus (low-AP) corn-soybean meal diets with a positive control group in four replicates. The criteria used to assess were growth performance, alkaline phosphates enzyme activity, plasma calcium and phosphorus levels, tibia ash, and carcass yield. Citric acid addition in low-AP diets significantly increased weight gain and feed conversion ratio (FCR) ($p<0.05$). Interaction of the effect of citric acid × phytase on the body weight gain and feed conversion ratio were significant ($p<0.01$). Inclusion of citric acid to low-AP diets reduced the alkaline phosphatase enzyme activity ($p<0.01$). Adding of citric acid to low-AP diets reduced mortality of the chicks ($p<0.01$). Inclusion of microbial phytase to low-AP diets improved the growth performance. Inclusion of 500 U/kg of phytase of the low-AP diets significantly ($p<0.01$) reduced mortality.

Chemical composition and nutritive value of Gundelia tournefortii

A.H. Karimi, Fars Research center for Agriculture and Natural resoueces. Shiraz, Iran

In this study, chemical composition and digestiblility coefficient of *Gundelia tournefortii* was determined by Proximate analysis and *in vivo* method. 108 samples of thise range plant were gathered at three growth stages: preflowering, flowering and dry stage, from three different climates: cold, moderate and warm. All samples were gathered, using systematic randomized method. The percentage of moisture (MO), crude protein (CP), ether extract (EE), nitrogen free extract (NFE), crude fiber (CF), Ash, calcium (ca) and phosphorus (p) of *Gundelia tournefortii* samples at preflowering period were determined. With growing of *Gundelia tournefortii*, Cp, ash and P, had decreased but CF had increased significantly ($P<0.01$). Different growth stages had significant effects on all analysed parameters ($P<0.01$), except in EE. Different climates had no significant effects on analysis. There were no significant differences in all analysed parameters of *Gundelia tournefortii*. Digestibility coefficient of Dry matter (DM), CP, EE, NFE, CF, organic matter (OM), total digestible nutrient (TDN), digestible energy (DE) and metabolizable energy (ME)(Mcal/kg) of *Gundelia tournefortii* were determined. Different growth stages and Different climates had no significant effects on digestibility coefficient of *Gundelia tournefortii*.

Determination of nutritive value of Iran fruit and vegetable residues in summer using chemical analysis, *in vitro* and *in vivo* techniques

K. Karkoodi[*1]*,F. Foroudi*[2]*, H. Fazaeli*[3]*, N. Teymournezhad*[3] *and N. Eila*[4]*, *[1]*Department of Animal Science, Islamic Azad University, Saveh Branch, Saveh, Iran, *[2]*Agricultural Faculty, Islamic Azad University, Varamin Branch, Iran, *[3]*State Animal Science Research Institute, Karaj, Iran, *[4]*Agricultural Faculty, Islamic Azad University, Karaj Branch, Iran*

Samples of fruit and vegetable were collected during 3 months of summer every other week per month and all samples were dried. Data "except for the *in vivo* assay" were analyzed using ANOVA with a completely randomized experimental design with 3 replicates for each month. The means of DM, CP, Ash, CF, EE, NFE, NFC, NDF, ADF, ADL, Ca, P, Mg, K, Na, Fe, Mn, Cu, Zn, Pb and GE, were 11.62, 14.73, 22.55, 12.24, 1.44, 49.04, 28.55, 32.71, 20.47, 3.57, 1.41, 0.34, 0.13, 1.18, 0.28 %, 3061.22, 85.58, 13.22, 61.74, 4.16 mg/kg and 3309 kcal/kg respectively. Except for the DM, P and Mn, no significant difference was observed for the chemical composition between the months of sampling. The means of *in vitro* digestibility for DMD, OMD and DOMD were 73.68, 77.33 and 60 % respectively which were not significantly different between the months. The means of *in vivo* digestibility for DM, OM, CP, CF, EE, GE, NFE, TDN, DE and ME were 56.22, 70.24, 66.44, 55.02, 59.11, 68.38, 75.17, 50.9%, 2343 and 1922 kcal/kg respectively. It is obvious that nutritive value of dried fruit and vegetable residues may be comparable with medium quality alfalfa, except for the higher ash content in fruit and vegetable residues.

Feeding of different mulberry varieties and theirs effects on biological characters in silkworm

A.R. Seidavi[1]*, A.R.Bizhannia*[2]*, R. Sourati*[2]*, M. Mavvajpour*[2] *and M. Ghanipoor*[2]*, *[1]*Islamic Azad University, Science and Research Branch, Iran, *[2]*Iran Silkworm Research Center*

Growth, development and cocoon production of silkworm larvae dependes on nutritional elements of mulberry leaf. Different mulberry varities demonstrate various feeding nutritional effects while applied for silkworm larvae. This experiment monitored the feeding effects of 4 improved mulberry varities i.e. KM, KN, I and SI along with one local mulberry during two rearing seasons at spring and early autumn. Standard rearing conditions were used for all treatments. Each mulberry varity was considered as one treatment without any supplementary ingredients usage. The leaves of each varity were chemically analyzed by means of photometer and flamephotometer in order to determine the relationship between nutrient element (protein, nitrogen, potassium, phosphore, moisture, fiber, ash and etc) and larval performance. The data were analyzed using a complete randomized design (CRD) model with factorial arrangement by means of SAS statistical programme and DNMRT. The ontained results cleared the significant effects of different mulberry varities and rearing season on larval economic characters (P<0.05). Differernt level of nutrients were also obsereved via biochemical analysis of mulberry leaves in various varities. DNMRT indicated that Shinichinose and Kinase varieties are suitable for late autumn rearing, but Ichinose variety is the best variety in spring rearing (P<0.05). Also the most studied characters in spring season showed better performance than autumn significantly (P<0.01).

Optimization of steam pressure treatment as a method for upgrading sugarcane bagasse as feed

K. Karkoodi[*1], M. Zahedifar[2], S.A. Mirhadi[2] and M.Moradi-Shahrebabak[3], [1]Department of Animal Science, Islamic Azad University, Saveh Branch, Saveh, Iran, [2]State Animal Science Research Institute, Karaj, Iran, [3]Department of Animal Science, College of Agronomy and Animal Science, Agriculture and Natural Resources Campus, University of Tehran, Karaj-Iran*

In this study, the potential of different condition of steam pressure and different reaction time to optimize the nutritional value of sugarcane bagasse for ruminant feed was investigated. Samples were steam treated under 3 levels of pressure (14, 17 and 20 atm) and 3 levels of reaction time (120, 180 and 240s). Data were analyzed on the basis of CRD statistical design (3x3) factorial model with 3 replications. Steam treatment significantly affected all parameters except for ADF content. The least CF, NDF, ADL, hemicellulose and total phenolics contents were attributed to '20atm, 240s' samples as much as 41.67, 56.60, 9.00, 11.89 and 17.55%, respectively, and also, the most water soluble sugars, total extractable phenolics, DMD, OMD, DOMD, degradability in 48h, potential degradability, total gas production, gas production in 24h, potential gas production and gas production rate were for the mentioned treatment as much as, 29.52, 2.60, 33.60, 32.87, 31.51, 63.63, 71.10%, 61.85, 44.11, 63.31ml and 4.87 %/h, respectively. It seems that the optimal treatment results could be obtained at a steam pressure of 20atm for 240s.

The study of ensiling of Iran fruit and vegetable residues in summer

F. Foroudi[*1], K. Karkoodi[2], H. Fazaeli[3], N. Teymournezhad[3] and N. Eila[4], [1]Agricultural Faculty, Islamic Azad University, Varamin Branch, Iran, [2]Agricultural Faculty, Islamic Azad University, Saveh Branch, Iran, [3]State Animal Science Research Institute, Karaj, Iran, [4]Agricultural Faculty, Islamic Azad University, Karaj Branch, Iran*

For study ensiling residues of Iran fruit and vegetable Samples of fruit and vegetables residues were collected during 3 months of summer every other week per month. In a 3 x 4 factorial completely randomized experiment while 3 replicates, silage characteristics of fruit and vegetable residues were studied, when treatments were: 25, 30, 35 and 40 percent of DM each with 0, 2 and 4 percent of sugar beet molasses respectively. The evaluation of the appearence qualities scored a value of 3.76 to 16.96 (based on a scale of 0-20), which were significantly different.The pH values were from 4.81 to 5.47 and DM content from 25.5 to 42.93% that significantly differed among the treatments. There were also significant differences between among the treatments for the content of OM, Ash, Total Nitrogen(TN), NH_3-N, NH_3-N/ TN, Total VFA and *in vitro* digestibility. In general,the silage contained 35 % of DM and 4 % of molasses showed to be superior treatment for ensiling of fruit and vegetables residues.In this treatment the means of DM, OM, Ash, TN, NH_3-N, NH_3-N/ TN, DMD, OMD, DOMD, pH, score and Total VFA were 38.52, 79.13, 20.87,1.23, 0.07, 6.03, 60.44, 62.57, 49.48% (DM basis), 4.98, 16.58 and 52.00 mMol/100g (DM basis) respectively.

Feed quality of wheat *(Triticum aestivum)* from ecological and conventional farming
Z. Mudrik[1]*, B. Hucko[1], A. Kodes[1], J. Havlik[1] and V. Christodoulou[2], [1]Czech University of Agriculture Prague, 165 21 Prague, Czech Republic, [2]NAGREF, 58100 Giannitsa, Greece

We performed a study in which wheat cv. Estica, cultivated under similar conditions in two farming systems: ecological and conventional, was evaluated by the means of feed value and nutrient content. The experiment was conducted on rats, which were fed by both conventional (EstC) and ecologically (EstE) grown wheat. Feed quality was evaluated by growth- and balance experiments.

Content of essential nutrients of tested grains were on similar levels, except for the fibre (44 g/kg for EstC and 33 g/kg for EstE), which was considered as significant nutritive factor, participating on its feeding value. The crude protein was 128 g/kg for EstC vs. 103 g/kg for EstE and the brutto energy values were more or less similar for both samples (16.9 KJ/kg for EstC and 16.4 kJ/kg for EstE).

It has been determined that the biosynthesis of significant nutritive proteins in the grain depends on factors that are related to the way of cultivation. All parameters measured except the biological value (54.3 for EstC snd 61.4 for EstE) were higher in the conventionally grown wheat cv. Estica. It might be concluded that the biological value of ecologically grown wheat was higher in our tests than that of the wheat grown conventionally.

This work was supported by grants: VZ MSM 6046070901 and MZP 1C/4/8/004.

Intraction between fattening periods and levels of energy on growth and carcass composition in Iranian Chalishtory lambs
M. Karami and F. Zamani, Scientific members of agriculture and natural resource research center of shahrekord, Iran

This study was investigating intraction between different levels of metabolizable energy (DME) (2.3 and 2.5) and fattening period (FP) (90 and 120 days) on carcass composition of Chalishtori male lambs. Rations were isonitrogenous (14% CP/DM) and used of completely random design with factorial method (2*2). Finally 32 lambs randomly slaughtered and data analyzed by SAS. Interaction final weight, metabolic weight and carcass weight between (FP 120 days) in (DME) had significant ($p<0.05$). The mean of daily weight gain was 162.37g/d, interaction between (FP) and (DME) did not influence on daily weight gain and feed intake. The mean of dressing percentage was 53.67% and interaction between (FP) (120 days) in (DME) for dressing percentage was significant ($p<0.05$). Interaction between (FP) in (DME) for surface of loin area and back fat thickness were significant ($p<0.05$). The mean of total carcass meat, total bone, and total subcutaneous fat and fat tail percent were 46.94, 11.71, 15.96 and 22.87 percent respectively and interaction between (FP) and (DME) on carcass compositions. However, interaction was between (FP) and (DME) influence on some traits and result of this experiment shown that the best fattening period and metabolizable energy are 90 days with 2.3 ME that recommended.

Effect of a lactic probiotic on kids growth
M.A. Galina, M. Delgado and M. Ortíz, FES-Cuautitlan National Autonomous University of Mexico, Ciencias Pecuarias, Km 2.5 Carretera Cuautitlan Teoloyucan San Sebastian Xhala, Cuautitlan Estado de Mexico, 54714, Mexico

Fiftysix Alpine kids 22.6 (±.450) kg. plus four cannulated goats were placed in two diets 129 d. All animals were fed 55% alfalfa and 45% concentrate. T1 (n=29 plus two cannulated animals) 19.2 (±.750) kg/BW fed basal diet. T2 n=27 plus two cannulated goats diet was speread with a lactic probiotic (*Lactococcus lactis; Lactobacilus brevis; helveticus and delbrueckii; Leuconostoc lactis* and *Bifidus essences, molasses and poultry litter.* Supplemented with 400 g/d 15% CP concentrate. Kid growth was 123 g/d (±18) T1 and 162 g/d (±26) T2 (P<0.05). Total dry matter intake (DMI) were $1,127 \pm 183$ g/d for T1 and $1.272 \pm$ for T2 (P<0.05). Ammonia concentration were augmented in T2 (P<0.05). *In vivo* nitrogen digestibility was higher (P<0.05) in T2 diet (79.12%), T1 (56.14%). Fiber digestibility was higher (P<0.05) for T2. Digestion rate of NDF constant (k_d/h) favored T2 diet (P<0.05). Passage rate (k_p/h) for NDF was 0.059/hr for T1 to 0.080/hr for T2 (P<0.05). True digestibility was higher in T2, 48.33% from T1 34.11% (P<0.05). Time of disappearance of cellulose in T1 (17.54 hr) was less (P<0.05) than in T2 (30.34 hr). Digestion rate was higher (P<0.05) in T1. Passage rate higher in T2 (0.080/hr) from T1 (0.059/hr). True digestibility in T2, (48.33%) was higher than that of T1 (34.11%) (P<0.05). Half-time (t ½) disappearance for hemi cellulose was higher for T2 31.14 hr (P<0.05). It was concluded that probiotic supplementation growth and ruminal physiology.

Investigation on nutritional management in dairy farms
M. Nafisi[1], A. Abbasi[2] and S. Sohraby[1], [1]Agricultural and Natural Resources Research Center of Tehran, Iran, [2]Animal Science Research Institute of Iran

This project was conducted in order to investigate feed management in dairy farms (comparison between amounts of nutrient consumed with standard nutrient requirements). Information about dairy farms, number of animals and milk production and milk parameters in Tehran province were obtained and by this information, eighty dairy farms were chosen. Amount and kind of consumed feedstuffs and milk production of each farm by questioner forms were recorded and the percentages of milk fat were measured in lab. Sample of feedstuffs were analyzed with AOAC method (1990) for Crude Protein(CP),Dry Matter(DM), Calcium(Ca) and Phosphorous(P) and amount of Net Energy Lactation (NEL) were estimated by feedstuffs proximate analysis equations (NRC 1989).Dry Matter and nutrient requirements of dairy cows were estimated by NRC(1989).The resulted indicated that amount of consumed DM, NEL, CP, Ca and P in dairy farms, were 24.75 Kg, 1.59Mcal/Kg, 3.55 Kg, 188 g and 111 g respectively and amount of nutrient requirements for above parameters were 20.97 Kg, 1.48 Mcal/Kg, 2.98Kg, 117g and 75 g, respectively. There were significant differences between all nutrient requirements (DM, NEL, CP, Ca, P) and consumed by dairy cows. The results indicated dry matter and nutrients intake were more than requirement in dairy farms.

Separation and determination of sulfur containing amino acid enantiomers by high performance liquid chromatography

Zs. Csapó-Kiss[1], Cs. Albert[2], K. Lóki[1], É. Varga-Visi[1], P. Sára[1], J. Csapó[1,2], [1]University of Kaposvár, Faculty of Animal Science, Department of Chemistry-Biochemistry, Guba S. u. 40., H-7400 Kaposvár, [2]Sapientia, Hungarian University of Transylvania, Csíkszereda Campus, Department of Food Science, Szabadság tér 1., Csíkszereda, Romania

Performic acid oxidation of cysteine and methionine resulting in the formation of cysteic acid and methionine sulphon has been applied in order to avoid the loss of these two sulfur containing amino acids during the acidic hydrolysis of proteins that is necessary prior to amino acid analysis. The aim of the research was assigned by the increasing demand for the determination of the amount amino acid enantiomers: the applicability of performic acid oxidation was evaluated in this point of view. Racemization of L-cysteine and L-methionine was found not significant during oxidation with performic acid; therefore this process can be applied before hydrolysis during quantification of cysteine and methionine enantiomers. Additionally, the quantification of cysteic acid and methionine sulphon enantiomers was accomplished in the form of their diastereoisomer derivatives via the development of a reversed phase high performance liquid chromatography method.

New possibilities for the determination of the tryptophan enantiomers

K. Lóki[1], Cs. Albert[2], É. Varga-Visi[1], P. Sára[1], Zs. Csapó-Kiss[1], J. Csapó[1,2], [1]University of Kaposvár, Faculty of Animal Science, Department of Chemistry-Biochemistry, Guba S. u. 40., H-7400 Kaposvár, [2]Sapientia – Hungarian University of Transylvania, Csíkszereda Campus, Department of Food Science, Szabadság tér 1., Csíkszereda, Romania

Diastereoisomers of L- and D-tryptophan were formed with a chiral reagent 1-thio-β-D-glucose tetraacetate (TATG) and o-phthaldialdehyde (OPA) and they were separated from the derivatives of the other amino acids that occur in food proteins on an achiral column by high performance liquid chromatography. Mercaptoethanesulfonic acid that is an adequate agent for hydrolyzing proteins made the OPA-TATG derivatization impossible, contrary the reaction completed in the presence of p-toluenesulfonic acid. During boiling, the racemization of tryptophan can be detected after 12 hours above pH=9, but the rate of conversion was lower than expected (<1%). The concentration decrease of L-tryptophan after 24 h was 2-5% depending on pH. Beside racemization other reactions e. g. oxidative deterioration may played a role in the loss of L-Trp.

Producer profiles, production characteristics and disease control applications at dairy herds in Konya, Burdur and Kırklareli Provinces, Turkey

C. Yalcin[1], S. Sariozkan[2]*, A.S.Yildiz[1], A.Gunlu[3], [1]Department of Animal Health Economics and Management, Veterinary Faculty, Ankara University.Ankara, Turkey,[2]Department of Livestock Economics, Veterinary Faculty, Erciyes University, Kayseri, Turkey, [3]Department of Livestock Economics, Veterinary Faculty, Selcuk University, Konya, Turkey

The producer & production characteristics and disease control applications of randomly selected 91 dairy herds in Burdur, Kırklareli and Konya provinces of Turkey were investigated in July 2004. Average figures for 3 provinces with respect to the formal education level and the proportion of the producers having job training were 7.2 years and 21% respectively. The majority of the producers were not aware of the EU regulations related to dairy farming, and had no idea about the likely impact of Turkey's integration to the EU on their business. Amongst the producers interviewed, 13% used antiseptic when cleaning udder and 37% dried udder after washing. The proportion of producers using post milking teat dip, dry cow therapy, vaccination against mastitis, California Mastitis Test, and regularly keeping records for clinical cases were 18%, 62%, 29%, 15% and 20% respectively. The study concluded that the producers in general, applied well-known methods for treatment and control of mastitis, but had lack in knowledge related to basic rules of hygiene applications and disease preventions. These problems are likely to have strong links with the level of formal education and job training of the producers.

Influence of management and genetic value for milk yield on the oxidative status of dairy cow plasma

S. De Smet[1], N. Wullepit[1], M. Ntawubizi[1], B. Beerda[2], R.F. Veerkamp[2] and K. Raes[1], [1]Laboratory for Animal Nutrition and Animal Product Quality, UGent, Proefhoevestraat 10, 9090 Melle, Belgium, [2]Animal Sciences Group, Wageningen University and Research Centre, Division Animal Production, 8200 AB Lelystad, The Netherlands

This study was part of a larger study that addresses whether milk production levels affect health risks in dairy cows taking into account the effects of genotype, environment and interactions between these. Plasma samples were collected from 80 Holstein Friesian heifers at 2 weeks pre-partum and at 4 and 8 weeks post-partum in a balanced 2x2x2 factorial design with the factors breeding value for milk production (high or low), milk frequency (2 or 3 times a day) and feed energy level (high or low). The following parameters indicative of the oxidative status were measured by spectrophotometric methods: ferric reducing antioxidant power (FRAP), glutathion peroxidase activity (GSH-Px) and two measures of lipid oxidation, namely malondialdehyde (MDA) concentration and paraoxonase activity. Significant effects occurred only for FRAP and GSH-Px. FRAP, i.e. a measure for the total antioxidant capacity of plasma, was lower before calving than 4 and 8 weeks after calving (P<0.001), and was lower in the high vs. the low feed energy level group (P<0.05). The plasma GSH-Px activity was higher 4 weeks after calving compared with 2 weeks before calving (P<0.05). These results indicate changes in the plasma oxidative status around parturition but minimal influences of genetic merit for milk yield, feed quality, milking frequency and, consequently, milk production level per se.

Session M19   Theatre 3

Postweaning Multisystemic Wasting Syndrome (PMWS) in pigs: An attempt to bring together the pieces of the puzzle

F. Madec and N. Rose, AFSSA, Zoopole, BP 53 Ploufragan, France

The disease was first reported in North America and then in western Europe in 1996. Since then, most of the main pig producing countries around the world got concerned. The piglet when 8-13 weeks of age is the typical target and mortality can be up to 30%. Contrary to other acute infectious diseases, PMWS does not typically impact on sow productivity despite experimental trials tend to show potential consequences. The lesions are especially severe in the lymphoid tissues. A small DNA single stranded virus (Porcine Circovirus type 2) is playing a pivotal role in disease expression. However the virus (without any obvious genomic difference) can also be found in healthy pigs from healthy farms with no history of PMWS. Archived tissues stored in the freezers far before 1996 were also found PCV2 positive. We experimentally reproduced the disease in a mild form in our facilities with the PCV2 alone. When the immune system is stimulated in a certain way more severe clinical signs and a heavier PCV2 load is detected in the tissues. At the farm level, analytic epidemiology showed a combination of risk factors for PMWS. They roughly relate to hygiene and herd management at large including vaccination. Backed to the research of the authors and that of other teams, the presentation will make a sort of state-of-the-art of the knowledge available regarding this devastating disease. It will also deal with control perspectives. Finally the reasons for disease emergence will be discussed in the light of the current scientific knowledge.

Session M19   Theatre 4

Floors and crates induced injuries in sows and piglets in farrowing pens

J. Troxler[1], K. Putz[1] and M. Schuh[2], [1]University of Veterinary Medicine, Institute of Animal Husbandry and Animal Welfare, 1210 Vienna, Austria, [2]University of Veterinary Medicine, Clinic of Swine, 1210 Vienna, Austria

Modern farrowing pens developed in the last years show a high incidence of injury problems of unknown causes. The goal of the project was investigating incidences and causes of injuries on 39 sow breeding farms. Skin damage in piglets and teat damage in sows were scored in 652 sows and 5977 sucklers on ten different floor types without straw bedding. Crusted blood on the carpal junction was present from the day of birth and increased over the first two weeks (max 80%) with significant differences between floor types. Injuries on the sole were present in nearly all piglets in the first days of life. All these alterations started to heal in the third week.

45% - 70% of the sows had teat damages related to the floor type at least in one teat. The main incidence of teat damage involved the fifth, sixth and seventh teat pairs and appeared related to sow parity.

The factors causing injuries are: slippery floors, solid concrete surface, sharp edges and inaccurately laid perforated floor elements. This investigation shows the necessity of testing housing systems in view of animal welfare before using in practice.

The effect of weaning age on health and lifetime performance of growing pigs

H.L. Edge[1], K. Breuer[2], K. Hillman[3], C.A. Morgan[3], A. Stewart[4], W.D. Strachan[3], L. Taylor[5], C.M. Theobald[6] and S.A. Edwards[1]*, [1]University of Newcastle, NE1 7RU, UK, [2]ADAS Terrington, PE34 4PW, UK, [3]SAC, Edinburgh EH9 3JG, UK, [4]Harper Adams University College, Newport TF10 8NB, UK, [5]MLC, Milton Keynes MK6 1AX, UK, [6]BIOSS, Edinburgh EH9 3JZ, UK

Increasing weaning age may be one strategy to promote better pig health following the removal of in-feed antimicrobial growth promoters (AGPs) within the EU. The AGEWEAN co-ordinated study is investigating the effects of weaning age (4, 6 or 8 weeks), in both indoor and outdoor lactation environments, on the biological and economic efficiency of pig production systems where diets contain neither AGPs or supra-nutritional levels of copper and zinc. Six experimental sites, representing a range of geographical locations and production systems within the UK, are monitoring a total of 190 contemporary sows on each weaning age treatment over four parities. The health and performance of their progeny through to slaughter are also being monitored. Results show that, although later weaned pigs grow faster in the immediate post-weaning period, lifetime growth rate is similar for all treatments. Mortality, prevalence of health problem requiring veterinary treatment and microbiological indicators of gut health show few treatment effects. It is concluded that, provided four-week weaned pigs are subject to good nutritional and environmental management, later weaning confers no significant health benefit.

Effects of n-3 polyunsaturated fatty acids supplementation associated with stressing breeding conditions on lipoperoxidation in ruminants

C. Gladine*, D. Bauchart and D. Durand, INRA-URH, 63122 Saint Genès Champanelle, France

Dietary supplementation of ruminants with n-3 polyusaturated fatty acids (n-3 PUFA) improves the nutritional quality of meat lipids. However, this can favour plasma and tissue lipoperoxidation leading to deterioration of animal health. The question is whether such deleterious effects would be more marked in stressing situations. Adult Texel sheep (n=6) were given for 7 wks a concentrate diet (1.5 kg/d) supplemented with extruded linseed (29g of 18:3n-3/d) (Step 1). This period was followed by a stress test (step 2) consisting in 20 min of transport in cattle truck followed by a 30 min forced run in pasture. Plasmas were collected before and after step 1 and after step 2. Linseed supplementation reduced by 11% the total antioxidant status (TAS, $P<0.05$) and slightly increased the production of final lipoperoxidation metabolites (MDA, +10%, NS). The resistance capacity of plasma lipids to lipoperoxidation (lag phase) evaluated *in vitro* decreased by 11% (NS) favouring the production of intermediates lipoperoxidation metabolites (conjugated diene) (+22%, $P<0.05$). The stress test (step 2) made worse the fall of TAS (-4%, $P<0.05$) and dramatically increased the production of MDA (+41%, $P<0.05$). We concluded that PUFA supplementation make animals very susceptible to lipoperoxidation especially during stressing situations and should be associated with an intake of antioxidants to preserve animal health and products quality.

Effect of a 9-h journey by road, a 12-h rest period, a 9-h road journey, followed by a 2-h rest and a 9-h road journey (9-12-9-2-9-h) on physiology and liveweight of bulls

D.J. Prendiville, B. Earley, M. Murray and E.G.O'Riordan, Teagasc, Grange Beef Research Centre, Dunsany, Co. Meath, Ireland*

The objective was to examine the effect of transporting bulls (spatial allowance $1.3m^2$/head) for 9-h by road, followed by a 12-h rest period (unloaded), a 9-h road journey, 2-h rest (on truck), and a 9-h road journey on physiology and performance. Continental x beef bulls (n=30; mean BW=472 ± s.d. 56.7kg) were allocated to one of two treatments; Transport (T for 9-12-9-2-9h; n = 15) and Control (C on slats; n = 15). Liveweights and blood samples (jugular venipuncture) were collected from T and C bulls, before, immediately after the first 9-h journey (J1), after the 12-h rest period, after the second (J2) and third 9-h journeys (J3) and at 4, 12 and 24-h post-transport. Bulls travelling for the first 9-h had lower ($P\leq0.05$) liveweight compared with baseline. Lymphocyte numbers were lower ($P\leq0.001$) and neutrophil numbers were higher ($P\leq0.001$) in T versus C bulls. Blood protein and creatine kinase concentrations were higher ($P\leq0.001$) in T bulls and returned to baseline within 24-h. In conclusion, liveweight and physiological responses of bulls returned to pre-transport levels within 24-h. Transport of bulls for the sequence 9-h (J1), 12-h (rest), 9-h (J2), 2-h (rest) and 9-h (J3) did not impact negatively on animal welfare.

Effect of banding or burdizzo castration on plasma testosterone and growth of bulls

W. Y. Pang[1,2], B. Earley[1], D.J. Prendiville[1], M. Murray[1], V. Gath[2] and M.A. Crowe[2], Teagasc, Grange Beef Research Centre, Co. Meath, Ireland; [2]School of Agriculture, Food Science & Veterinary Medicine, University College Dublin, Ireland.*

The objective was to assess the effect of banding or burdizzo castration of bulls, performed on farms, on plasma testosterone and growth. 195 Continental × Friesian bulls (12 months; 397.8±5.83 kg) from three farms were allocated randomly to one of three treatments (n = 65): banding castration (Band), burdizzo castration (Burd), or controls (Con). Band and Burd had lower (P<0.001) plasma testosterone concentration than Con 28 d post-castration. From 1 to 2 and 5 to 8 weeks post-castration, Band and Burd castrates had lower ADG (P<0.05) than Con. From 3 to 4 weeks, Band castrates had lower (P<0.05) ADG than Burd castrates and Con, while there was no difference (P=0.76) between Burd castrates and Con. From week 9 to 12, Burd castrates had lower (P=0.02) ADG than Con, while the ADG of Band castrates was not different (P>0.12) from Burd and Con. The ADG of Band and Burd castrates over week 1 to 16 was lower (P<0.05) than Con. In conclusion, Band or Burd castration reduced plasma testosterone concentration; retarded ADG mainly during the first two weeks, which was not compensated during the subsequent 16 weeks; Burd showed an advantage over Band in ADG during 3 to 4 weeks following castration.

A model to investigate the interaction between host nutrition and gastro-intestinal parasitism in lambs

D. Vagenas[1]*, S.C. Bishop[2] and I. Kyriazakis[1,3], [1]Animal Nutrition and Health, SAC, West Mains Road, Edinburgh, EH9 3JG, UK, [2]Division of Genetics and Genomics, Roslin Institute, Roslin, EH25 9PS, U.K, [3]Veterinary Faculty, University of Thessaly, Karditsa, Greece

Gastrointestinal parasitism in sheep is a significant source of income loss for farmers. Although usually it is manifested by subclinical infections, it causes significant loss in production in terms of growth. It is usually treated with anthelminthic drugs, however, this strategy is threatened by the development of resistance by the parasites and therefore alternative control strategies need to be considered. These include dietary supplementation of hosts. A computer simulation model has been developed to account for the interaction of parasitism with host nutrition, and their combined impact on host growth. The host is described in terms of its capacity to grow and its resistance to parasites, i.e. its ability to control parasite establishment, fecundity and mortality. Other inputs include the description of the amount and quality of the feed as well as daily larval intake. The model describes the utilisation of nutrients and their partitioning to growth or immunity, and the impact of parasitism on these processes. Outputs include the host food intake, growth rate, worm burden and faecal egg counts. Thus, this model gives us the opportunity to explore the impact of nutrition and genotype on the performance of parasitised lambs kept in different environments.

Clinical and biochemicalstudies of macro and micro-elements profile in the rumen liquor and serum of camels

T.A. Baraka and T.A. Abdou, Department of Medicine and Infectious Diseases, Faculty of Veterinary Medicine,Cairo University, Giza, Egypt

This investigation was carried out to study the level of rumen and serum macro and micro-elements in camels in order to build a base for camel data in the field of internal camel medicine; with special reference to the effect of gastrointestinal parasite infection. Thirty seven camels was used (12 healthy camels and 25 gastrointestinal parasites infected camels). The normal rumen sodium, potassium, calcium, magnesium, copper and zinc was 92.727 ± 5.414 mmol/L, 11.143 ± 1.509 mmol/L, 0.421 ± 0.169 mmol/L, 0.326 ± 0.199 mmol/L, 0.321 ± 0.032 μmol/L and 0.752 ± 0.154 μmol/L respectively; while their levels in serum were 140.180 ± 1.525 mmol/L, 4.881 ± 0.208 mmol/L, 1.758 ± 0.208 mmol/L, 0.539 ± 0.073 mmol/L, 8.848 ± 1.033 μmol/L and 10.235 ± 1.940 μmol/L respectively. The major infections in the camels were identified as *Trichostrongylus, Nematodirus, Heamonchus, Trichuris, Cooperia* and *Monezia*. The recorded data confirm and explain the tight interaction between the blood and rumen constituents under the control of rumen pH, type of feed stuffs gained by the animal, stress factors accompany the infections and their own pathological alterations in the different systems of the camels. The obtained data were statistically analyzed using the SXW statistical computer Software. Copy writes 1996 version 1.0.

Effect of rearing ducks on fish farming pond as a polyculture of Chinese carps

*F. Foroudi[*1], M.A. Canyurt[2], K. Karkoodi[3] and N. Eila[4], [1]Agricultural Faculty, Islamic Azad University, Varamin Branch, Iran, [2]Fisheries Faculty, Ege University, 35100, Bornova, Izmir, Turkey, [3]Agricultural Faculty, Islamic Azad University, Saveh Branch, Iran, [4]Agricultural Faculty, Islamic Azad University, Karaj Branch, Iran*

Beijing ducks were reared on one-hectare fish farming pond containing a polyculture of Chinese carps. Growth of 4 different carp species in this system was compared to the fish yield of an adjacent control pond containing the same species of carps. The average daily gain (ADG) for each speices of carp for duration of 5 months from April to August in ponds of with and without duck for Silver carp, Bighead carp, Common carp and Grass carp were 8.32 vs 5.90, 11.76 vs 8.20, 12.93 vs 12.29 and 12.32 vs 12.46g respectively. The ADG of Silver carp and Bighead carp indicated the growth rate of these fish in the integrated pond was higher than of the without duck pond($p<0.05$). The higher growth rate of the Silver carp and Bighead carp in the duck-fish pond indicated remarkable positive effects of duck excreta on growth and propagation of phytoplankton and consequently on zooplankton, since phytoplankton is the main feed of Silver carp and Bighead carp consumes the zooplankton. The ADG of Common carp and Grass carp in with and without duck ponds, showed no significantly difference ($p>0.05$). Because the Common carp is bottom feeder and Grass carp only consume aquatic plants.

Performances of six Iranian silkworm hybrids under four different environmental conditions

M. Mavvajpour[1], S.Z. Mirhoseini[1], M. Ghanipoor[1] and A.R. Seidavi[2], [1]Iran Silkworm Research Center, Rasht, Iran, [2]Islamic Azad University, Science and Research Branch, Iran*

The rearing performances of six different Iranian silkworm hybrids including 151×152, 151×110-32, 151×154, 103×104, 31×32 and 107×110 were investigated under standard as well as warm/humid, warm/dry and temperature fluctuating rearing conditions. Four climate treatments (each in three replications) were employed for 3rd molted larvae during two rearing seasons of spring and early autumn. The obtained results revealed significant effects of environmental conditions on inspecting traits. The hybrids of 151×152 showed the highest value of good cocoon percentage (76.26) significantly. Middle cocoon percentage was also lower in this hybrids (18.52) compare to others. Therefore 151×152 can be declared for high quality cocoon production. Larval mortality were higher in 103×104 (6.2%) and 31×32 (5.09%) and lower in 151×110-32 and 107×110 (2.85). 103×104 and 31×32 also showed the highest pupal (13.13 and 12.96% respectively) and total mortalities (17.76 and 16.86% respectively). The fluctuation of temperature left no effect on cocoon weight reduction in 151×110-32, 31×32 and 107×110, while caused great raise in cocoon weight of 103×104 hybrid. The weight of cocoon shell in 151×110-32 and 103×104 did not fall because of temperature fluctuating rearing condition. In case of humidity enhancement, only the hybrid of 151×154 showed decrease of cocoon shell weight while this character showed no significant difference in humid and dry rearing condition for other hybrids.

Behavioural effects of different breeding systems in three calf breeds

P. Pregel[1], E. Bollo[1]*, G. Brizio[2], E. Maggi[1], S. Origlia[2], M. Francesconi[2] and P.G. Biolatti[3]
[1]University of Turin, Department of Animal Pathology, Via L. da Vinci, 44 – 10095 Grugliasco, Italy, [2]Public Veterinary Service 17, Via Trento, 5 - 12037 Saluzzo, Italy, [3]Institute for Zooprophylaxis, Via S. Pertini, 11 – 12100 Cuneo, Italy

The aim of the study was to investigate the influence of breed and breeding systems on behaviour of calves. Three breeds (Friesian, Piedmont and crossbred) characterized by different breeding conditions (type, time and way of feeding) were examined. Five animals per breed, living in the same box, were observed during 3 observation cycles, with 3 observations per cycle every two days.

The behaviours were video recorded every 2 minutes during the following moments: 1 hour before, during and after every meal, and in between the meals. The recorded behaviours were the following: lie and stand (combined with the following), inactive, move, eat, drink urine, ruminate, non-nutritive chewing, lick self, social, suckle, suck calf, other (tongue playing/rolling). The time-length of each behaviour was measured. Statistical analysis (ANOVA or Kruskal-Wallis test with post-test) showed significant differences between the groups. Friesian calves spent more time eating than crossbred, and more time ruminating compared to Piedmont calves. The latter lied longer than crossbred and Friesian calves, and ate for a longer time compared to crossbred calves, which were the group more involved in movement. The Piedmont group exhibited a minor number of stereotyped behaviours.

Diprosopus with some other defects in a lamb

A. Asadi[1], B. Shojaei[1],H. Moradi Shahrbabak[2]*, [1]department of veterinary medicine, shahrbabak Islamic Azad University, Shahrbabak, Iran, [2]Department of Animal Science, Faculty of agriculture, University of Tehran, P.O. Box 31587-11167,Karaj, Iran

A pregnant ewe was referred to the veterinary clinic of shahid Bahonar University of Kerman, for dystocia. The dicephalic lamb was conducted to a C.T. Examination and transverse images were prepared from the whole body by 8-mm intervals. Also, 3-dimensional images of the skull were prepared for detailed studies of the bones. The lamb was carefully dissected after macroscopic and C.T. examination and a photographic record was made of all recognizable anomalies. pinkish membranous part was observed. The vertebral column was involved in extreme kyphoscoliosis at the thoracic region. The heart had no defect but left subclavian artery was branched separately from the aortic arch. The palate was grossly complete in the C.T. images; the bony roof of the skull was not formed between the common orbit of medial eyes to the foramen magnum; which was larger than normal. Microscopic examination should haemoragic spelenitis, degeneration of somniferous tubules of the testes, hepatocellular degeneration and severe fatty change of the liver and acute tubular necrosis with intratubular hemorrhage and granular casts in the kidneys. Lungs, forestomaches, abomasum, lymph nodes and epididymis were normal.Malformation of the head and face are common among sheep (3). However, among the ovine conjoined twin anomalies duplication of the caudal parts of the body is most frequent (2).

Haematological and immunological parameters in calves of different breeds, kept in different breeding systems

E. Bollo[1]*, G. Brizio[2], P. Pregel[1], A. Rampazzo[1], E. Maggi[1], S. Origlia[2], P.G. Biolatti[3], R. Guglielmino[1] and A. Cagnasso[1], [1]University of Turin, Department of Animal Pathology, Via L. da Vinci, 44 – 10095 Grugliasco, Italy, [2]Public Veterinary Service 17, Via Trento, 5 - 12037 Saluzzo, Italy, [3]Institute for Zooprophylaxis, Via S. Pertini, 11 – 12100 Cuneo, Italy

The aim of the investigation was to evaluate the influence of breed and breeding systems on haematological parameters and humoral immunity of 192 calves (Friesian n=138; crossbred n=54). A complete blood count was performed and serum IgG concentrations were evaluated.

A statistical analysis by using the unpaired t-Student or Mann-Whitney tests was performed, as well as the Fisher's exact test to evaluate the relative risk.

RBC resulted lower in Friesian calves compared to crossbred (8.50 x $10^6/\mu l$ vs 8.95 x $10^6/\mu l$, $P<0.05$), while MCV was higher in Friesian calves compared to crossbred (32.35fl vs 31.39fl, $P<0.05$). Neutrophils, monocytes and basophils resulted lower in Friesian calves compared to crossbred (neutrophils: 2.61 x $10^3/\mu l$ vs 3.25 x $10^3/\mu l$, $P<0.01$; monocytes: 0.35 x $10^3/\mu l$ vs 0.46 x $10^3/\mu l$, $P<0.001$; basophils: 0.06 x $10^3/\mu l$ vs 0.07 x $10^3/\mu l$, $P<0.005$). Serum IgG concentration was lower than 10 mg/ml in 31.9% of Friesian calves compared to 9.3% of crossbred. Friesian calves showed a higher risk of anemia and Failure of Passive Transfer of colostral IgG ($P<0.001$).

Content of inhibitory substances in cow milk after antibiotic treatment of mastitis

R. Toušová, L.Stádník* and J. Vodička, Czech University of Agriculture in Prague, Department of Animal Husbandry, Kamýcká 129, 165 21, Prague 6 – Suchdol, Czech Republic

Quality milk, acquired from health dairy cows, belongs to basic foodstuffs. Milk quality is evaluated from different points of view with many traits. Content of residuum of inhibitory substances (RIS) is one of this indicators. RIS are detected in cows milk more frequently after antibiotics treatment of mastitis. RIS have a negative impact to bacteria of milk fermentation and negative allergic reaction is possible in consumers. RIS occurence in milk is dependent on many factors, for example meeting criteria of protection period, changes in animals metabolism in time of disorder, way of medicament aplication, numbers of aplications. The goal of this work was to determine occurence of RIS in dairy cows milk after end of protection period. Length of protection period wasn´t sufficient in all cases. Deal of 15,4% was positive in content of RIS, it means threehold overfullfilment of indicating limit of producer (Biopharm), which is to 5% of all samples. Results was confirmed with SAS GLM using.

Efficiency of pig raising outdoors in concrete yards
V. Juskiene and R. Juska, Institute of Animal Science of LVA, R. Zebenkos 12, LT-82317, Baisogala, Radviliskis distr., Lithuania

The purpose of the study was to determine the efficiency of pig raising outdoors in concrete yards in comparison with pig raising in common pig-house with controlled microclimate. The study was conducted in the period of summer and autumn. Two analogous groups – control and experimental – were formed. Control pigs were raised in a pig-house in the pens of 18 m² area (1 m² per pig). While the experimental pigs were kept outdoors in concrete yards of 23 m² area (1.27 m² per pig) and equipped with 13.7 m² area shelters.
The study indicated that in the summer time pigs raised outdoors gained weight worse, and at the end of the experiment the average weight of a pig was 7.18 kg (P=0.014) lower that of pigs raised indoors. However, pig raising outdoors resulted in 5.9% higher survival. On the contrary in autumn time pigs raised outdoors gained weight slightly better. At the end of the experiment, the average pig weight was 4.8 kg (P=0.086) higher than that of pigs raised indoors. Moreover, the survival of pigs raised outdoors was even 14.3% higher. No noticeable differences were found after behaviour analysis between pigs raised outdoors and indoors.

Evaluation of udder health in relation to lactation stage by means of ultrasonography
O. Amerlingova, A. Jezkova* and M. Parilova, Czech University of Agriculture in Prague, 165 21, Czech Republic

The project was carried out on 108 Czech Pied cattle cows. There were observed changes on teat and teat tissue, which were caused by milking, and their return to condition prior to milking. Using cutimeter teat length (TL) and diameter (TD) were measured. Changes in teat canal length (TCL) and teat end width (TEW) were assessed by ultrasound. Each parameter was measured before milking, just after milking and two hours after milking. Tie stall milking system and tandem parlour milking system were compared as well as lactation stage (7 days post partum, 50-60, 100-110, 200-210 and >300 days). In the tie stall system the greatest changes of all parameters occurred at the beginning of lactation. In the parlour greater changes in the periods between 100-110 and 200-210 days were detected. In the tie stall both front and rear teats became about 0,21 cm longer. Furthermore, front teats as well as rear ones became narrower (0,18 and 0,07 cm, respectively). TCL became about 0,02 cm longer and TEW about 0,02 cm wider in both front and rear teats. In the parlour the elongation of teat was about 0,60 cm both in front and rear teats and constriction was about 0,17 cm in the both. TCL became about 0,10 in front and 0,05 cm longer in rear teats. TWT became about 0,06 cm wider.

Effect of 12 day sea transport from Ireland to the Lebanon on physiological, liveweight and temperature of bulls

B. McDonnell[1]*, B. Earley[1], M. Murray[1], D.J. Prendiville[1] and E.G.O'Riordan[1] and M.A. Crowe[2], [1]Teagasc, Grange Beef Research Centre, Dunsany, Co. Meath, Ireland [2]Faculty of Veterinary Medicine, University College Dublin, Belfield, Ireland

The objective was to examine the stress response in bulls induced by the stages of transport during a sea journey of approximately 12 days from Ireland to the Lebanon. Holstein x Friesian bulls (n = 121; mean BW = 429 ± s.d. 59.2kg) were allocated to one of three treatments; T (transport by sea) (n=57) assigned to 6 pens on five decks of the shipping vessel (spatial allowance of 1.7m²/animal. Control bulls were housed in 3 pens at 1.7m² (n=9/pen) and 3 pens at 3.4m² (n=9/pen) at Grange Research Centre. Animals had ad libitum access to hay and water and were fed 2kg of concentrates/head/day. Liveweights, rectal temperatya and blood samples (jugular venipuncture) were collected from T and C animals before the journey (day-1) and on days 3, 6, 9 and 11. Bulls travelling by sea for 12 days had greater (P≤0.05) liveweight gain compared with control animals remaining in Ireland. Blood protein and creatine kinase concentrations were higher (P≤0.001) following transport and returned to baseline by day 3. In conclusion, physiological and haematological responses of transported bulls returned to pre-transport levels by day 11. Transport of bulls for 12 days by sea did not impact negatively on animal welfare.

Emergency vaccination and pre-emptive culling in classical swine fever epidemics

I. Witte[1], S. Karsten[1], J. Teuffert[2,] G. Rave[3], J. Krieter* [1], [1]Institute of Animal Breeding and Husbandry, Christian-Albrechts-University, D-24118 Kiel, [2]Friedrich-Loeffler Institute, D-16868 Wusterhausen, [3]Institute of Variation Statistics, Christian-Albrechts-University, D-24098 Kiel

A spatial and temporal Monte-Carlo simulation model was developed to evaluate the emergency vaccination for classical swine fever epidemics in contrast to pre-emptive culling. The present study includes 18.323 pig farms in Lower Saxony, Germany. The primary outbreak was set on a farrow-to-finishing farm within a region with a high farm density (1.66 farms/km²). Pre-emptive culling was compared to emergency vaccination scenarios varying time until vaccination starts (1, 18, 25, 32 days), type of vaccine (live/ marker vaccine) and vaccination radius around the primary outbreak (0-3, 0-10, 0-20 km).
Pre-emptive culling resulted in 10 infected, 21 culled and 742 banned farms. Vaccination reduced the number of culled herds (21 to 8) but additionally 25 farms were vaccinated. Using marker vaccines did not reduce the effectiveness of vaccination although time until immunity was three times higher. Time until vaccination started had no influence on the number of infected, culled and banned herds, but less farms were vaccinated when vaccination starts later (24, 14, 10, 7). Increasing vaccination radius hardly reduced the number of infected and banned farms but increased enormously the number of vaccinated farms (24, 308, 1138). Assuming a high farm density and a farrow-to-finishing farm as primary outbreak emergency vaccination is hardly more efficient than pre-emptive culling.

Campylobacter spp. and Yersinia spp. in pork production

T. Wehebrink[1], N. Kemper[1], E. grosse Beilage[2], J. Krieter[1], [1]Institute of Animal Breeding and Husbandry, Christian-Albrechts-University, 24118 Kiel, Germany, [2]Fieldstation for Epidemiology, 49456 Bakum, Germany*

The objective of this study was to get more information about the prevalences of *Campylobacter* spp. and *Yersinia* spp. at different stages in the pig production-chain. Faeces were collected from 68 sows and from 256 suckling piglets of 4 farrowing herds. Further samples were taken from 362 growing and 354 finishing pigs of 12 fattening herds. Additionally, 56 feed and environmental samples were collected. During slaughtering 122 pigs were sampled three times. Finally, 86 raw meat samples were taken from 34 retail stores.

Campylobacter spp. were isolated in sows (33.8%), in piglets (80.9%) and in growing (89.2%) and finishing (64.7%) pigs. *Yersinia* spp. were detected only in growing (15.2%) and finishing (13.3%) pigs. During lairage, *Campylobacter* spp. were identified from faeces of pigs from all farms whereas *Yersinia* spp. were detected in pigs from only two herds. After 12 h chilling neither *Campylobacter* spp. nor *Yersinia* spp. were ascertained. In raw meat samples *Campylobacter coli* was isolated from only one liver sample and *Yersinia enterocolitica* from two meat samples (ground pork and cutlet). Common slaughter technique and hygiene procedures may be effective tools to reduce the risk for contamination or recontamination of meat products since *Campylobacter* spp. and *Yersinia* spp. pathogens were found only sporadically in raw meat samples.

Influence of Ascaridia galli infections and anthelmintic treatments on the behaviour and social ranks of laying hens (*Gallus gallus domesticus*)

M. Gauly[2], C. Duss[1] and G. Erhardt[1], [1]Institute of Animal Breeding and Genetics, University of Giessen, Ludwigstrasse 21B, 35390 Giessen, [2]University of Goettingen, Albrecht Thaer Weg 3, 37075 Goettingen, Germany

The effects of an experimental *A. galli* infection on the behaviour and social status of layers (Lohmann LSL; Lohmann Brown; LB) were studied. The helminth-naive hens were artificially infected with *A. galli* at an age of 27 weeks. Some were kept as uninfected controls. 11 weeks *post infectionem* they were slaughtered, worm burdens counted and faecal *Ascaridia* egg counts (FEC) performed. Throughout the experiment behavioural parameters were recorded by focal animal observation (n = 10 per group), according to the time-sampling method. All agonistic interactions were recorded simultaneously. An individual social rank index was estimated. The following results were obtained: Infections with *A. galli* resulted in significant behavioural changes as the infected birds showed a higher food intake and lower locomotion activity during the prepatent and patent periods. Infected LSL hens showed also changes in ground pecking and nesting activity not only during the prepatent and patent periods. Social rank did not significantly change as a consequence of *A. galli* infection, but infected animals had a tendency to display more agonistic activity than the non-infected controls. This study showed that even sub-clinical *A. galli* infections can have an impact on animal behaviour.

Study of genome instability in Ancient Autochthonous Genetic Type (AAGT) 'Casertana' pig by using Micronucleus Test

D. Matassino*[1,2], D. Falasca[1], N. Castellano[1], G.Varricchio[1] and D.Fornataro[1], [1]ConSDABI. NFP.-FAO- Center of omics Science for Nutritional Quality and Excelence - 82100 Benevento, Italy, [2]Sannio University, Dep. of Biological and Environmental Science – 82100 Benevento, Italy

Micronucleus Test (MN) is quick and easy method to reveal DNA oxidative damage from chemical and/or physical agents. Our study was carried out on single samples (20 males and 13 females) of 'Casertana', pig AAGT mostly present in Campania (Italy), reared at ConSDABI experimental farm. The age of examined subjects ranged from 20 months to 96 months. MN were detected according to Matassino et al. (1994) method, opportunely modified, using lymphocytes from peripheral venous blood. The main results could be so summarized: *(i)* presence of polynucleated cells, with one, two, three, four, five, six, seven, eight nucleus and one cell with thirteen nucleus and one cell with fifteen nucleus; *(ii)* the number of MN changes from a minimum value of 1 MN to a maximum value of 8 MN; *(iii)* the percentage of cells with MN showed a positive correlation with the age by functions changing from linear to harmonic (P<0,07 ÷ P<0,40); *(iv)* the percentage of cells with MN showed a linear positive correlation with number of nucleus per cell (P<0,06). The value of mean percentage of cells with spontaneous MN, being in the normality range, could be a valid index of animal welfare status in its microenvironment.

Preliminary results on the genome instability evaluated by Micronucleus Test in four italian Ancient Autochthonous Genetic Types (AAGT) pigs

D. Matassino*[1,2], D. Falasca[1], G.Gigante[1], G.Varricchio[1] and D.Fornataro[1], [1]ConSDABI. NFP.-FAO- Center of omics Science for Nutritional Quality and Excelence - 82100 Benevento, Italy, [2]Sannio University, Dep. of Biological and Environmental Science – 82100 Benevento, Italy

The genome instability (natural or induced) may determine the lose of some portions of genome and therefore it could favourite mutations, rearrangements, delections, subsequent gene activation and inactivation and hence changes in the gene expression regulation mechanisms (Franceschi, 1994). The present study concerned the evaluation of micronucleus (MN) frequency on four pigs AAGT reared at ConSDABI experimental farm: *Casertana* (*CT*), normally present in Campania (Italy); *Calabrese* (*CL*), normally present in Calabria (Italy); *Nero Siciliano* (*SC*), normally present in Sicilia (Italy); *Cinta Senese* (*CS*), normally present in Toscana (Italy). MN were detected according to Matassino, et al. (1994) method, opportunely modified, using lymphocytes from peripheral venous blood. Within the limits of the observation field, *CL* shows a mean frequency of binucleated cells with MN inferior to *SC* (P < 0.01; 2.06 *vs* 3.26) and to *CT* (P<0.005; 2.06 *vs* 3.43). Furtheremore, considering two age ranges: 1. (30 ÷ 40); 2. (> 40), it was possibile to evidence a trand to the increase of binucleated cells with MN passing from the firsth to the second range of age.

Effect of post-milking teat disinfection on new infection rate of dairy cows over a full lactation

D.E. Gleeson*, W.J. Meaney, E.J. O'Callaghan and B. O'Brien, Teagasc, Moorepark Dairy Production Research Centre, Fermoy, Co. Cork, Ireland

Previous studies have demonstrated the benefits of post-milking teat disinfection in reducing mastitis incidence when an artificial bacterial challenge was used. The objective of this study was to measure the effect of omitting post-milking teat disinfectant over a full lactation using natural bacterial challenge. Fifty-six Holstein-Friesian cows were milked, with right-sided teats disinfected post-milking (TD) by submerging teats in a solution containing 4250 ppm chlorohexidine gluconate. Left-sided teats were not dipped (NTD). Quarter milk samples were taken fortnightly for somatic cell count (SCC) analysis and pathogen typing. The SCC was lower (P<0.01) over the full lactation for TD (153,000/ml) compared to NTD (261,000/ml). Clinical mastitis (CM) incidence was higher (P<0.001) for NTD teats compared to TD. NTD tended to have more *Staphylococcus aureus* present at day 100 and a higher (P<0.05) number of *non-haemolytic staphylococci* present on day 230 compared to TD. When data from all sampling dates were pooled NTD had a higher (P<0.01) number of *Staphylococcus aureus* and *non-haemolytic staphylococcus* and less quarters (P<0.001) with no pathogens present than TD. There were no differences between treatments for sub-clinical infection, non-specific (NS) sub-clinical infection, transient sub-clinical infection, transient NS sub-clinical or in the number of teats with *Streptococcus dysgalactiae* or *Streptococcus uberis*. Omitting post-milking teat disinfectant resulted in higher SCC, CM and numbers of milk pathogens.

Expression of the mitochondrial tricarboxylate carrier is associated with high level of intramuscular fat in cattle

G.V. Gnoni[1], L. Siculella[1], D. Bauchart[2], D.W. Pethick[3] and J.F. Hocquette[2], [1]University of Lecce, Italy, [2]INRA, Herbivore Research Unit, Theix, France, [3]CRC, Murdoch University, Perth, 6150, Western Australia

Intramuscular fat (IMF) deposition influences many quality attributes of beef meat. Among the key metabolic pathways involved in IMF accumulation, the mitochondrial tricarboxylate carrier (TCC) may play a significant role since it transports, in the form of citrate, glucose-derived mitochondrial acetyl-CoA into the cytosol, where lipogenesis occurs. To test this hypothesis, TCC expression was assessed in bovine muscles by ribonuclease protection assay as well as 28S RNA for normalisation purpose. Samples of three muscles, namely *rectus abdominis* (RA), *longissimus thoracis* (LT) and *semitendinosus* (ST) were taken from 12 Limousin steers and samples of LT from eight Angus steers. Animals were slaughtered after a long finishing period with a cereal-rich diet to increase IMF deposition. IMF content was in the decreasing order: LT from Angus, then RA, LT and ST from Limousin (100.3, 15.0, 10.7 and 5.5 mg/g fresh tissue respectively). TCC expression was 3.22-fold higher in LT of Angus than of Limousin. It was also higher in RA than in LT and ST of Limousin (1234, 774 and 681 units respectively). In conclusion, variability in IMF is positively associated with variability in TCC expression in bovine muscles supporting the idea that glucose may be a significant precursor for *de novo* lipogenesis in bovine muscles.

Session Ph20

Theatre 2

Histamine and neurohypophyseal hormone control of HPA axis in sheep

H.R. Rahmani*, Dept. of Animal Sciences, Isfahan University of Technology (IUT), Isfahan 84156, Iran

The roles of histamine and neurohypophyseal hormones (OT and AVP) in central nervous system functions have been well established. Recent studies consider these effects in relation to each other and in this regard the role of histamine in neurohypophyseal hormone secretion and HPA axis has been evaluated in laboratory animals. In three separate experiments four intra-cerebroventricularly cannulated Nainee (fat tailed, Iranian breed) rams were used as a model which in turn received; 1) 100 μl of PBS containing 0 (control), 100, 200 and 400 nM histamine chloride, 2) 300 nM of OT, AVP or AVT analogue, or 3) 0 (control), and 200 nM histamine followed by 300 nM OT, AVP or AVT each dissolved in 50 μl of PBS, for at least four times. Blood collected via a jugular vein was analyzed for cortisol concentration by RIA method. The data were analyzed by GLM and comparing the means by Duncan's range test revealed that; 1) histamine significantly ($P<0.05$) increases ram cortisol level dose and time dependently, 2) AVP and AVT, but not OT, significantly ($P<0.05$) increases cortisol concentration, 3) histamine pre-treatment potentiates cortisol secretion in response to OT, AVP and AVT significantly ($P<0.05$), but histamine do not affect the cortisol concentration in pre-infused rams with either OT, AVP or AVT. It seems that the priority of secretion or infusion of amine or peptide is important in this relation for synaptic signalling.

Session Ph20

Theatre 3

Effects of feeding gossypol in cottonseed meal on some hematological parameters in Dallagh rams

F. Ghanbari[1]*, Y. J. Ahangari[2], T. Ghoorchi[2] and S. Hassani[2], [1]Islamic Azad University of Baft, I R Iran, [2]Faculty of Animal Science, Gorgan University of Agricultural Science and Natural Resources, I R Iran

An experiment was conducted to investigate the effects of gossypol in cottonseed meal (CSM) on some hematological parameters in Dallagh rams. Eight Dallagh rams of 2 years old with an average body weight of 58±6.09 kg were used. Rams were divided in two control and treatment groups (N=4(. The isocaloric and isonitrogenous diets calculated for the control and treatment groups contained 10% soybean meal (SBM) and 15% CSM (containing 850 ppm free gossyopl), respectively. Blood samples were collected for 12 consecutive weeks to measure erythrocytes osmotic fragility and plasma level of potassium. Blood was also collected 3 times within 12 weeks of experiment to measure plasma level of testosterone, red and white blood cells counts, hemoglobin concentration and hematocrit. The data were analyzed with the nested design using SAS software. The results showed that differences between the control and treatment groups for blood parameters were not significant ($P< 0.05$) except erythrocytes osmotic fragility ($P< 0.05$). Erythrocytes osmotic fragility was higher in the treatment group than in the control group (6.01 g/L vs. 4.97 g/L). Because the increase in erythrocytes osmotic fragility precedes other known gossypol-induced physiological changes, it can be concluded that long time feeding of Dallagh rams with CSM containing relatively high levels gossypol should be limited.

Application of enzymeimmunoassays for understanding growth, reproduction, behavior and reproductive health monitoring in mithun (*Bos frontalis*)

M. Mondal*, NRC on Mithun, Jharnapani, Medziphema, Nagaland-797 106, India

Mithun (*Bos frontalis*), a descendent from wild gaur, is used mainly for beef in India and other mithun inhabited areas of the world. Its socio-economic, cultural and religious importance demands its scientific rearing in a big way. Endocrine information on growth, reproduction and behavior is of paramount importance for scientific rearing of this species. The remote location of the institute makes it difficult to get access to radiochemicals etc. required for conventional RIA, and due to multi-advantageous nature of enzyme immunoassay (EIA), EIAs for GH, LH, estradiol-17β, total estrogen, prolactin, oxytocin and PGFM were developed for mithun, and were applied to understand the growth, reproduction, behavior and reproductive health monitoring in mithun. Some of the interesting results were: a) higher GH level, b) control of aggression by blood GH, c) less pronounced behavioral estrus and unique trend of endocrine changes during the estrous cycle and peri-estrous period, d) pulsatile nature of LH release during preovulatory LH surge, e) possibility of automatic detection of estrus by using standing heat as an ovulation predictor, f) need of higher optimum LH pulses ($\geq$9 pulses/24h with an amplitude $\geq$1.56 ng/ml) to attain puberty, g) higher blood oxytocin and its possible role to control temperamental behavior and h) use of blood PGFM as an indicator of reproductive health. In conclusion, all these results suggest that mithun is a unique species of Indian origin.

Effect of CIDR with various doses of eCG hormone on fertility of Zel ewes in breeding and non-breeding seasons

Y.J. Ahangari, University of Agricultural Sciences & Natural Resources, Gorgan, Iran

66 ewes of 2-3 years old of Zel breed were selected for estrus synchronization using CIDR. After 12 days, CIDRs were withdrawn and various doses of eCG hormone of 0, 200, 300 and 400 IU injected intramuscularly. Teaser rams were introduced to the flock and 24 to 36 hours after eCG injection ewes showed signs of estrus. Rams were introduced to ewes for mating with a ratio of 1 to 10. Records of pregnancy rates, lambing rates and prolificacy of treated ewes were collected. Fertility results showed that differences between those ewes treated with eCG hormone and the control ewes in autumn and spring seasons were significant (P<0.01). Fertility differences among treated ewes with various doses of 200, 300 and 400 IU eCG were not significant (P>0.05). Mean of lambing rates for control treatment and for 300 IU eCG treatment were 70 and 78% in breeding season and 55 and 70% in non-breeding season. Mean of prolificacy rates for control ewes and for 300 IU eCG treated ewes were 0.87 and 1.21 in breeding season and 0.82 and 1.20 in non-breeding season, respectively. In conclusion, an injection of eCG hormone immediately after withdrawal of CIDR caused an increase in fertility efficiency of Zel ewes.

Effect of honeybee royal jelly on fertilization of in vitro matured bovine oocytes

A.G. Onal[1]*, Y.Z. Guzey[1], S. Kariptaş[1] and Z. Gocmez[2], [1]Mustafa Kemal Universities, Faculty of Agriculture, Department of Animal Science, Hatay, [2]Institute of Cukurova Agrucultural Research, Adana, Turkey

Royal jelly supplementation has previously been used as an alternative source to serum for ovine and bovine oocyte maturation. Considering the disturbing effects of serum during embryonic development, the aim of the present study was to investigate the effects of royal jelly on fertilization rate of *in vitro* matured bovine oocytes. Aspirated bovine oocytes (2 to 8 mm diameter) were matured either in 10% (v/v) fetal calf serum (n=247 oocytes), 0.62 % (w/v) honeybee royal jelly (RJ1; n=241 oocytes) or 1.25% (w/v) honeybee royal jelly (RJ2; n=232 oocytes) supplemented with TCM-199 in the presence of FSH (10µg/ml) and LH (10µg/ml) under a humidified atmosphere of 5% CO_2 at 38.6 °C for 20 h. All oocyte were fertilized following maturation. The ratio of oocytes reaching Metaphase-II stage of nuclear maturation did not differ between treatment groups (P>0.05) and it was 77.9% in serum-supplemented group, 74.7% in RJ1 and 64.8% in RJ2 supplemented group. Furthermore, fertilization rates did not differ between treatment groups (P>0.05); it was 41.8% in serum-supplemented group, 38.9% in RJ1 and 33.3% in RJ2 supplemented groups. In conclusion, bovine oocytes matured in 1.25% and 0.62% royal jelly supplemented conditions can successfully be fertilized.

The effect of KSOM and R1ECM-BSA culture medium in SSV vitrification in Spraque Dawley rat pronuclear-stage embryos

T. Akkoc*[1], H. Bagis[1] and I. Soysal[2], [1]TÜBİTAK- Research Institute for Genetic Engineering and Biotechnology (RIGEB), Transgenic Core Facility, Gebze/Kocaeli, [2]Trakya University Agriculture Faculty of Tekirdag, Department of Animal Science, Tekirdag, Turkey

The cryopreservation of pronuclear-stage (PN) embryos is one of the important supporting technologies for advanced animal husbandry, transgenic technology, human embryology and gene banking. In this study a novel vitrification technique (solid suface vitrification, SSV) was used in order to get optimal culture medium after cryopreservation of PN rat embryos and control groups as well. Rat embryos in PN were transfered to equilibration medium containing 4% (v:v) Ethylene glycol (EG) in base medium for 15 minute. Then these PN rat embryos were exposed to vitrification solutions containing 35% EG, 5% PVP 0.4 M Trehalose in base medium for 20 seconds. Then they were dropped in 2 µl drops onto pre-cooled (-150 to -180°C) metal surface. Following the vitrification of PN rat embryos melted in thawing medium. In this study two culture media (KSOM and R1ECM-BSA) were compared. In four cell division stage there was no statistical differences between R1ECM-BSA vitrification and R1ECM-BSA control groups (11%). Both two and four cell stage percentage was higher in KSOM-vitrification (68% and 26%, respectively) and only medium with R1ECM-BSA reached the eight cell stage. Our results demonstrate that R1ECM-BSA vitrification is a more promising method than KSOM-vitrification in embryo culture.

Prediction of fertilizing capacity of rabbit sperm using Annexin V assay

A.V. Makarevich[*], V. Parkanyi,, L. Ondruska,, J. Pivko, L. Riha, E. Kubovicova and J. Rafay, Slovak Agricultural Research Centre, Nitra, Slovak Republic

Success of methods for sperm testing is consisted in the correlation between its viability and fertilizing capacity. The aim of this study was to analyze, using annexin V method for the detection of apoptosis, fresh or differently influenced rabbit sperm in relation to conception rate (CR) of females fertilized with semen from groups: 1. Fresh semen, 2. Incubated in sperm diluent, 3. Incubated in cryoprotective medium, 4. Frozen in floating freeze rack (Minitub), or 5. Frozen in manual regimen. A higher rate of annexin V-positive (AnV+) cells was observed in both cryopreserved groups (4, 5), compared with groups 1, 2 and 3 ($p < 0.05$, u-test). Most of frozen-thawed sperm showed membrane damages in acrosomal region, equatorial segment, posterior ring or tail. Sperm incubated in cryomedium did not reveal these damages. Electron microscopy of frozen-thawed sperm confirmed the above mentioned membrane injuries. The highest CR of females was obtained with semen from groups 3 (65.4%), 2 (56%) or 1 (44.4%), but it was minimal using sperm frozen in floating freeze rack (16.7%), whilst no conception was seen in 13 females when using manually frozen sperm. In conclusion, Annexin V method is able to detect early membrane changes on sperm affected by different influences. A negative linear dependence between An-V+ cell index and female CR was found. Annexin V might be considered as a method of choice to predict fertilizing capacity of rabbit sperm.

Relationship of Plasma IGF-I to growth performance and carcass quality of Chaal, Zandi and their crosses with Zel breed fattening lambs

S. Dashti[1] and S.D. Sharifi[2], [1]Management and planning organization, Iran, [2]Abureyhan College of Agriculture, Tehran University, Iran

The relationships of plasma IGF-I to growth performance and carcass traits of *Chaal* and *Zandi* breeds and their crosses with *Zel* were studied. Twelve lambs were used in this experiment. The lambs were weaned at 90 days of age and were fattened for 114 days. Growth performance of lambs was recorded weekly. Blood samples were collected at 4, 6, 8, 10, 13, 17, 22 and 26 weeks of age and the plasma IGF-I concentrations were determined. After slaughter, the carcass chemical composition and the weight of visceral organs were determined. Results showed that the plasma IGF-I concentration was significantly different between breeds ($P < 0.01$). Male lambs had higher plasma IGF-I concentrations than females ($p < 0.01$). There was significant correlation between weaning weight and plasma IGF-I at 4 week of age ($r = 0.75$, $P < 0.05$). Weight gain during suckling and fattening periods was significantly correlated with plasma IGF-I at 26 weeks of age ($r = 0.81$, $P < 0.01$). There were significant correlations between percentage of fat or protein of carcass and plasma IGF-I at four weeks of age ($r = -0.61$ and 0.58, $P < 0.05$, respectively). From the results obtained, it is suggested that plasma IGF-I concentrations could be used as a marker in breeding programs for selecting animals with high protein and low fat at age of slaughter.

The influence of the season on the fatty acid composition and conjugated linoleic acid content of the milk

R. Salamon[1], É. Varga-Visi[2], P. Sára[2], Zs. Csapó-Kiss[2] and J. Csapó[1,2], [1]Sapientia – Hungarian University of Transylvania, Csíkszereda Campus, Department of Food Science, Szabadság tér 1., Csíkszereda, Romania, [2]University of Kaposvár, Faculty of Animal Science, Department of Chemistry-Biochemistry, Guba S. u. 40., H-7400 Kaposvár, Hungary

The purpose of the research was to determine the fatty acid composition of milk of general breeds in Hungary that is Hungarian Simmental, Red Holstein Friesian and Black Holstein Friesian and the changes in the fatty acid composition of their milk fat throughout the year with special respect to the conjugated linoleic acid content. The amount of the saturated fatty acids dropped to a minimum level during the summer months (butyric acid: 2.8–2.9 relative %; myristic acid: 10.9–11.0%; palmitic acid: 28.1–28.2%; stearic acid: 10.4–10.5%), and reached a maximum during winter and in early spring (butiric acid: 3.6–3.7 relative %; myristic acid: 11.5–11.7%; palmitic acid: 28.7–28.8%; stearic acid: 10.7–10.8%). The amount of oleic (summer: 26.7%; winter: 25.0%), linoleic (summer: 3.0%; winter: 1.7%), and linolenic acid (summer: 1.4%; winter: 0.9%) and conjugated linoleic acid (summer: 1.4%; winter: 0.8%) was highest in summer. The biological value of the fat in summer milk, according to a higher essential fatty acid content, is higher than that of winter milk.

The determination of luteinizing hormone response to different doses of lecirelin in Tuj ewe-lambs

M. Kaya, M. Cenesiz and S. Yildiz[*], Kafkas University Faculty of Veterinary Medicine, Department of Physiology, 36280 Kars, Turkey

The aim of the current study was to determine LH response to different levels of lecirelin acetate (GnRH agonist, Dalmarelin, Fatro S.p.A., Italy) in the Tuj ewe-lambs. Twenty ewe-lambs, 5-month-old, were divided into 4 groups (n=5) according to BW and BCS. Lecirelin was injected into jugular vein of lambs in group 1, 2, 3, and 4 at 0.1, 0.2, 0.5 and 1.0 µg/kg BW respectively (Application I). Blood samples were collected from the jugular vein at 30 min intervals for the determination of plasma LH concentration for 300 minutes. Described application was repeated 3 days later, to determine the replenishment of pituitary stores of LH in the same lambs (Application II). In both applications, LH response was dose-dependent and the dose of 0.5 and 1.0 µg/kg gave a significantly greater response than the dose of 0.1 and 0.2 µg/kg (P<0.05). On the other hand, LH release characteristics were similar between first and second applications (P>0.05). The results suggest that (1) LH stores were replenished within 3 days, and that (2) higher LH responses were obtained at 0.5 and 1.0 µg/kg dose levels.

Effect of PEG administration to growing lambs fed sulla

M.S. Spagnuolo[1], R. Baculo[1], P. Abrescia[2], F. Polimeno[1], A.Carlucci[2], M. Sitzia[3], N. Fois[3] and L. Ferrara[1], [1]ISPAAM-CNR, 80147-Napoli, [2]Dipartimento Scienze Biologiche-Università Federico II, 80134-Napoli, [3]IZCS, 07040 Omedo, Italy

Increased frequency or amount of metabolic processes, during growth, is associated with enhanced production of reactive oxygen species (ROS). The imbalance between ROS production and antioxidant defence mechanisms can lead to "oxidative stress". Retinol and α-tocopherol scavenge ROS, and prevent lipid oxidation in growing tissues. Condensed tannins, as reducing protein degradation in rumen and increasing the flow of essential amino-acids toward the intestine, improve protein utilization. Polyethylene glycol (PEG) prevents the formation of tannin-protein complexes. We analysed the effect of a diet, containing condensed tannins, on redox and energy status in growing lambs. Twelve lambs were fed, for two months, by grazing on *sulla* (*Hedysarum coronarium* L.) (5 hours/day) and on Italian ryegrass sward. Twelve lambs received the same diet, supplemented with PEG. Plasma samples, collected weekly, were analysed for retinol and α-tocopherol levels. Plasma nitro-tyrosine concentration was measured and used as a marker of protein oxidation. The levels of retinol, tocopherol, and nitrotyrosine were not different between the two groups. Similarly, plasma levels of cholesterol, triglycerides, and glucose, titrated to evaluate the energetic status, did not differ between the groups. Our data demonstrate that plasma levels of the analysed parameters were not affected by PEG administration and suggest that condensed tannins, as not influencing energy and redox status, might regulate other physiological functions of growing lambs.

Effects of growth hormone (GH)-releasing factor, thyrotropin-releasing hormone and GH-releasing hexapeptide on prolactin and GH in gilts

D. Outor-Monteiro[1], R.B. Mestre[1], A.A. Colaço[1], L.F. De-La-Cruz[2] and A.L.G. Lourenço[1]*, [1]CECAV-UTAD Department of Animal Science, PO Box 1013, 5001–801 Vila Real, Portugal, [2]USC Facultad de Veterinária, 27002 Lugo, Spain

Six Large White x Landrace gilts (body weight=130 ± 12 kg) were used to analyse the influence of a single dose of growth hormone (GH)-releasing factor (GRF(1-29)NH$_2$) thyrotropin-releasing hormone (TRH) (0.2 µg kg^{-1}, each), growth hormone-releasing hexapeptide (GHRP$_6$) (0.8 µg kg^{-1}), the first two (II) and all three (III) in combination on serum prolactin (PRL) and GH concentrations. A saline solution was used as control. Samples were taken from –15 until 120 minutes after administration. Treatments were arranged in a 6 x 6 Latin square design, with treatment, day and animal as sources of variation. Area under curve (AUC) analyses showed that PRL was not significantly affected (P>0.05) by GRF(1-29)NH$_2$, GHRP$_6$, or TRH, although peaks obtained with the last two hormones were higher (P<0.05) than basal concentrations. Treatment II improved (P<0.05) PRL AUC and treatment III amplified the former effect. Peak of GH was higher (P<0.05) than control with GHRP$_6$ and III, but GH AUC only increased (P<0.05) with treatment III. The results suggest that there is a complementary effect between TRH and GRF(1-29)NH$_2$ with GHRP$_6$ on GH secretion and between TRH and GRF(1-29)NH$_2$, improved with GHRP$_6$, on PRL secretion.

Circadian rhythmicity of leptin in Holstein bulls

A. Towhidi, Department of Animal Science, University of Tehran, P.O.Box 4111, Karaj, Iran

Leptin is a hormone which is primarily produced and secreted by adipocytes. Little is known about circadian rhythm of leptin in ruminants. The objective of this study was to determine variation of plasma leptin over 24 hours. Six Holstein bulls, 12 months of age, were selected from the herd of animal science department, university of Tehran in Karaj. The animals were fed with diet formulated based on NRC (1996) for beef cattle. Blood samples were collected through a jugular vein with heparinized venipuncture at 2 hours interval for 24 hours. The samples were centrifuged at 3,000 x g for 60min and plasma was separated and frozen. Plasma leptin concentrations were measured by RIA Kit (Tabeshyarnoor Ltd, Iran). Data were analyzed by ANOVA for repeated measures using the PROC MIXED of SAS. Mean plasma concentration of leptin increased gradually from 1 a.m. and reached a peak at 5 a.m. Results indicated that leptin has a circadian rhythm with a peak during hours with lower temperature similar to others mammals. The higher levels of leptin during early morning probably result in an increase in thermogenesis.

Heat stress during follicular growth may affect subsequent pregnancy rate in the first insemination after calving in lactating Jersey dairy cows

E. Sirin[1], E. Soydan[2] and M. Kuran[1], [1]Universities of Gaziosmanpasa, Tokat and [2]Ondokuz Mayis, Samsun, Faculty of Agriculture, Department of Animal Sceince, Turkey

It is well accepted that heat stress has an adverse effect on reproductive performance of dairy cows in addition to high milk yield level. In this study, we investigated whether high temperature humidity index (>78 THI) during any time of the 10 days before Artifical insemination (AI) has any effect on the subsequent pregnancy rate in the first insemination after calving in lactating Jersey dairy cows differing in their lactation milk yield level. A total of 1068 lactation and insemination record of high (4253 kg), avarage (3153 kg) and low (2157 kg) milk yielding Jersey cows were analysed and THI >78 was considered as high. Pregnancy rate in the first insemination after calving subjected to khi-square analysis for high (>78) and low (≤78) THI groups. High THI during any time of the 10 days before AI which coincides with the development phase of the ovulatory dominant follicle reduced (χ^2=4.43; P<0.05) the subsequent pregnancy rate (67 vs 57 % for high vs low THI group) in the first insemination after calving. There was a similar trend in the pregnancy rate of all high, avarage and low milk yielding cows. In conclusion, heat stress during the follicular development may reduce the subsequent pregnancy rate in the first insemination after calving in lactating Jersey dairy cows irrespective of their milk yield level.

Session L21

Theatre 1

Sheep for meat farms in plain: Diversity and evolution over 16 years

M. Benoit* and G.Laignel, Laboratoire Économie Élevage INRA Theix 63122 France

This work describes, by a multifactorial analysis, the diversity of a group of sheep for meat farms in plain, in the centre of France, and their evolution over a long term period. Profitability, work, and demand of the market are the main factors concerned with farms adaptation. A chart based on 14 synthetic variables presents the individual trajectory of the farms. These variables allowed us to identify 5 groups of farms (year 2003) that are analysed according to their interest for the community: market, territory occupation, and environment. The Sheep-and-Crop group has been strongly encouraged by the CAP since 1992; the net income is rather low, as the environmental benefits, but the work is lower than in specialised sheep farms. The Mixt-Self-Sufficient farms, with large flocks and lambings from November to May are very interesting for the processors; their environmental benefits are good. Meanwhile, the levels of work and capital are very high. Traditional-Grazier farms, with lambings only in spring, have a very good profitability over the 16 years and they are very sustainable from environmental and work point of view. However, they don't produce lambs in winter, when the demand is the highest. The extensive farms have good income and environmental interest but they use very large areas with a low added value per ha and so they have a very small contribution to the local economic activity.

Session L21

Theatre 2

Typology of extensive goat farming systems in Spain

A. Garcia*[1], J. Perea[1], R. Acero[1], F. Peña[1], M. Herrera[1] and J.P. Avilez[2], [1]Department of Animal Production, University of Cordoba, Spain, [2]Catholic University of Temuco, Chile

The typology of extensive goat farms from native local breeds (Negra Serrana, Blanca Celtiberica, Retinta Extremeña, Moncaina, Blanca Andaluza, Azpi Gorri and Pirenaica) located in Spain have been characterised from a survey including 111 farms (58% of population). Multivariate analysis (ACP and cluster analysis) was used to understand the relationships between variables and to establish farm typologies.

Three groups of goat systems have been identified. The first group (26%) is constituted by farms of non-commercial objectives and includes the smallest farms (24 goats) with breeds of high mountain zones (Azpigorri, Moncaina and Pirenaica). The second and third groups are integrated by farms of commercial objectives and include breeds of low mountain zones (Negra Serrana, Blanca Celtiberica, Retinta Extremeña y Blanca Andaluza). The second group (42%) is integrated by farms of intermediate size (159 goats) and shows high technological level and low farming costs. The strategy followed by this group is the intensification of the production system and the complementary of rents with the inclusion of other livestock species in the farm like sheep and beef cattle. The third group (32%) is constituted by farms of large size (400 goats) and exclusively dedicated to goats. This group has followed a strategy of specialization in meat production.

The relationship between benefits obtained from livestock and preferred selection traits by smallholder resource-poor farmers in the Limpopo Province of South Africa

F.J.C. Swanepoel[1]*, N.D. Nthakeni[2], A. Stroebel[1] and A.E. Nesamvuni[2], [1]Centre for Sustainable Agrculture, University of the Free State, P O Box 339, Bloiemfontein, 9300, South Africa, [2]Limpopo Department of Agriculture, Polokwane, South Africa

A farming systems study was conducted in six rural villages located in the Limpopo Province of South Africa. Data was collected by means of a general questionnaire and participatory rural appraisal (PRA) methods. The objective of the study was to develop an understanding of the functions, benefits, and important and preferred selection traits of livestock by smallholder resource-poor farmers. The results showed that adaptability, fertility (regular calving), disposition, traction utility, and colour are the preferred selection traits. Furthermore, the results confirmed that livestock owners link benefits and functions, i.e. selling and meat consumption; wealth, status and savings; cultural activities and draft power obtained from livestock, to these preferred selection traits. It is concluded that benefits and functions obtained from livestock in smallholder systems form the basis for decisions with respect to selection and culling. Therefore, recommended breeding programmes for smallholder, resource-poor livestock producers should include their preferred selection traits.

Structure of organic beef cattle farms in Spain

J. Perea*, A. Garcia, R. Acero, E. Felix, T. Rucabado and C. Mata, Department of Animal Production, University of Cordoba, Spain

The structure of organic beef cattle farms located in Spain have been characterised from a survey including 52 farms. The farms present an intermediate size, both in terms of surface (389 ha) and number of animals (80 cows), with an average stocking rate of 0.50 animals ha-1. Likewise, the mean productivity of these farms is scarce (with an index of commercial calves of 80% and a mortality rate of 6.4%) and the total workforce amounts to 2.3 AWU per farm.

Furthermore, taking account the strategy of production it is possible to discriminate two groups of farms: farms specialized in raising steers; and farms in which raising and finishing systems. Both farming systems are extensive and feeding depends largely on grazing, with seasonal supplementation.

Raising and finishing farms are located in northwest of Spain and they present small size (282 ha). The productivity period lasts around 15 months. The unitary costs (€ kg-1) of organic meat production are higher than conventional meat production in a 31%.

Raising farms are concentrated in southwest (Andalucia, Castilla La Mancha and Extremadura). The average farming surface of this system is 461 ha. In this farms sell all calves after weaning (average BW of 200 kg of 6 month of age) to livestock traders, who take them to other regions for conventional fattening until slaughter.

Spring calving versus spring and autumn calving in suckler cow production
F. Szabó, Gy. Buzás, University of Veszprém, Georgikon Faculty of Agriculture, Hungary, H-8360 Keszthely Deák F. str. 16.

Most beef producers manage their cows to calve late winter and early spring. Others may have both spring and autumn calving program to extend the use of their bulls and use their labour and forage more efficiently furthermore to increase calf crop. As each calving system is in practice in Hungary a study was carried out to compare one season, spring calving (programme A) and two season, spring and autumn calving (programme B) for economic point of view. The comparison was based on calf crop, retail product, price, income, feed consumption, feeding costs data. 50-50 cows as a family farm in each programme kept at extensive conditions were considered as a unit of the evaluation. Only spring calving happened in A, while 50% spring-, 50 % autumn calving took place in programme B. According to the results the calf crop of A was lover (83%), than that of B (93%). Similarly the retail income of A (507 €/cow) was lower than that of B (570 €/cow). The difference is 63 €/cow. At the same time the nutrition and veterinary costs of A (268 €/cow) was by 68 €/cow lower than that of programme B (336 €/cow). Consequently no considerable difference (5 €/cow) of gross margins was realised between the one season, spring and two season spring and autumn calving programme.

Evolutions, scatters and determinants of the farm income in suckler cattle charolais farms
P. Veysset, M. Lherm and D. Bébin, INRA Clermont-Theix, Economie de l'Elevage, 63122 Saint Genès Champanelle, France*

The monitoring of a constant group of 69 suckler cattle charolais farms from the north Massif Central over 15 years (of 1989 to 2003) shows a high increase of the farm area and herd size with a constant workforce. This increase in the labour productivity, response to the constant drop in the meat sale price only partly compensated by subsidies, was accompanied by an evolution of the farming systems (decrease in the rate of fattened males and shortening of the production cycles). The variability of our sample is expressed by the size (area and herd), also by the specialisation rate (livestock or mixed crop-livestock farming) and by the choice of the production system (animals fattened or not). Overall, the average farm income per worker remained stable, with an increased dependence of this one facing the subsidies. The average of the incomes is almost the same whatever the production system, but the scatter of the incomes intra-system is relatively significant. The income per worker remains always correlated with size criteria but with the following distinction: before the first CAP reform, the size of the livestock was important because of the outputs that it generated, after 1992 the size has effect on the farm income by the subsidies that it makes it possible to perceive.

Structural characterisation of organic dairy farms in northwest of Spain

A. Garcia*, E. Felix, J. Perea, R. Acero, V. Rodriguez and C. Mata, Department of Animal Production, University of Cordoba, Spain

The aim of this study was to characterise the structure of organic dairy farms located in northwest of Spain. The area of study includes three Spanish regions (Asturias, Cantabria and Galicia) that concentrate the 60% of the organic dairy farms registered in Spain. The sample was obtained using a stratified sampling by regions and comprised 42% of the official census in these regions.

The farms are small (an average of 46 cows and 45 ha), with an average stocking rate of 1.12 animals ha -1. Feeding depends largely on supplementation, with 0.38 kg of concentrate and 0.17 kg of forage kg -1 of milk. This farming system are characterised by high mortality rates (9%) and scarce productivity (less than 5600 Kg of cow/Milk and 0.78 annual calves/cow). The mean capital investment by farm was 216000 €; where buildings, machinery and cows represent 83%. The total workforce amounts to 2.3 annual work unit per farm on average, and is usually of family origin. This group of farms can be assigned to the category of extensive or semi-extensive systems. The production system contributes to the maintenance of certain ecosystems, reduces the risk of forest fires, soil erosion and loss of biodiversity. Organic production is often the only significant activity in some rural areas: it therefore contributes to sustainability of rural population.

The some approaches for the future of the ecological livestock farming

T. Aygün, C. Zer and A. Yılmaz, Yüzüncü Yıl Universityi, Faculty of Agriculture, Department of Animal Science, 65080 Van, Turkey

In this review, we evaluated studies regarding ecological livestock farming. In addition, we discussed the problems which may arise in the future by observing details during the practice of ecological farming.

The increase of production in livestock has required artificially changes in production factors. These reasons make the use of technology inevitable in animal production. On the other hand, using the technology in agriculture has resulted in considerable ecological problems. This has necessitated development and interrogating of the different production systems. The ecological livestock farming as an alternative production system is a production type which requires sustainability in using the production factors. This situation points out that using modern production techniques intensively, may not provides sustainability of production factors.

The aim of the ecological livestock farming is not making profit in the short-term but modeling the variables like human, animal and environment in the long-term and also providing the sustainability in agricultural production by achieving suitable and healthy food production for human and next generations. However, owing to the animal rights in many countries, animal welfare has increasingly become important in societies. In recent years, increasing ecological awareness has required replacement of current production systems with natural formations. However, it appears that the ecological farming has critical dilemmas when considered in many different aspects.

Technical competitiveness for cattle fattening on Argentina Pampas
*R. Acero[1], A. Garcia[*1], J. Perea[1], J.P. Avilez[2], F. Peña[1] and N. Ceular[1], [1]Department of Animal Production, University of Cordoba, Spain, [2]Catholic University of Temuco, Chile*

The winter cattle-fattening farms of the north-eastern Pampas are classified according to their level of competitiveness. A group of isoquants of efficiency and another of isocost lines are established, on which the farms are situated. A crossed analysis of the two groups (isoquants versus isocosts) determines zones of competitively that classify the farms in different pastoral subsystems of winter cattle fattening:

- Competitive subsystem (36%): The farms of this group present mean daily gains exceeding 550 g, with quick winter cattle fattening lasting around 15 months. They use strategic supplementation and rotational grazing, and implement available technology (permanent technical advice, electric fencing, division into small plots, nocturnal corralling, etc.). This group aspires to long-term permanence in the market.
- Pre-competitive subsystem (45%). The model is normal winter cattle fattening characterized by daily gains of 430 g, and lasting 21 months. Feeding is based on grazing and seasonal forage, supplementing (20 $/ha) in periods of shortage and at the end of fattening.
- Non-competitive subsystem (29%). Is a model of slow winter cattle fattening, lasting longer than 31 months. It is based on a traditional pastoral system, with continuous grazing, and without regard to time for new shoots. There is scant or no use of reserves or advice.

Future of agriculture in Iran;case study: Animal products
F. Mirzaei, Department of Animal Production and Management, Animal Sciences Research Institute, P.O.Box:31585-1483, Karaj, Iran

Animal production situation is changed in past decade positively,so its production is increased from 5390000 tons in 1991 to 8107000 tons in 2001 that is indicated to 3.8% annual growth in the mentioned term. Production capacities of animal products sub-sector is 120 millions animal unit,820 millions commercial poultry,2440000 honeybee colonies,598 dairy factories,134 livestock slaughter-houses,112 meat processing manufactories,551 Pelt,Pickle and Leather processing firms,214 feedstuffs manufactories. Also,its export contribution from Agriculture sector is 11.7%.
Longterm quantitative targets in livestock sub-sector(from 2005-5009):

- increasing of animal protein supply per person to 5.7 gram until future five years based on 3.14gr from milk,0.32gr from red meat,1.84gr from poultry meat,0.4gr from eggs.
- enhancement of animal production from 8834000 tons in 2003 to 12911000 tons in 2009.
- enhancement of milk production from 6316000 tons to 9556000 tons.
- increasing of red meat from 752000 tons to 921000 tons.
- increasing of poultry meat from 1100000 tons to 1605000 tons.
- enhancement of egg production from 628000 tons to 789000 tons.
- enhancement of honey production from 29000 tons to 40000 tons.
- appropriate enhancement of other products same as wool,hair and leather

Competitiveness of the Iranian egg industry in the Middle East region
F. Mirzaei[1]*, S. Yazdani[2] and A. Gharahdaghi[1], Animal Science Research Institute, P.O.Box:31535-579, Karaj, Iran, Tehran University, Agriculturae college, Agricultural economics department, Karaj, Iran

Trade policies are believed to have influenced a country's comparative advantage.Iran is the first producer and exporter in the Middle East and main producer of Egg in the World(FAO2004).This paper looks at the performance of egg export and examines the revealled comparative advantages indices in the Middle East region economics over the period of1990-2004. The changes in Iran's export structure were compared with Middle East ones.The results of the research showed,the trade and production policies and economic behaviour of producers and exporters have been in such a way that they could not show a appropriate and timely response of region,also the revealed instabilities in the trends of RCA and RSCA(revealed symmetric comparative advantage)indices during the study period led to the conclusion that there is no well defined strategy and plan for utilization of low price production factors,effectiveness human power,knowledge of bargaining power and target markets

---

Floristic composition and farming management of hay meadows in central Pyrenees
M. Santa-María[1], C. Chocarro[2], J. Aguirre[1] and F. Fillat[1]*, [1]Pyrenean Institute of Ecology, CSIC. Av. Rto. Galicia s/n. E-22700. Jaca (Huesca), Spain, [2]ETSEA- Lleida University. Av. Rovira Roure, 177. E-25198. Lleida, Spain

The main factor which determine the floristic composition of Pyrenean hay meadows is the livestock management according to seasonal movements. The aim of this paper is to establish a correlation between floristic composition and different management groups following techniques related with fertilisation, irrigation, grazing and cutting. An amount of 255 hay meadows in Broto Valley of Aragonees Pyrenees were sampled during June – July 2001. The species abundance data were used in order to obtain six groups of meadows by TWINSPAN classification. Each group was characterized by one or more indicator species. The grassland production yield and vegetation diversity (number of species, Shannon index and Evenness) appear as opposite parameters from the results of a DECORANA analysis. Their relation appears to be directly and indirectly influenced by the management techniques.

Effect of progesterone supplement prior to pessary removal on reproductive performance of ewes out-of-season

M.Q. Husein[1],*, M.M. Ababneh[2] and H.A. Ghozlan[1], [1]Department of Animal Production, [2]Department of Veterinary Clinical Sciences, P. O. Box 3030, Jordan University of Science and Technology, Irbid 22110, Jordan

An experiment was conducted to examine the effects of progesterone (P_4) supplement on enhancing reproductive performance of seasonal anestrous ewes. Twenty-eight Awassi ewes were randomly allocated into treatment (n=14) and control (n=14) groups and were administered with intravaginal P_4 pessaries (CIDR-G) for 12 days. Ewes in the treatment group were given 25 mg P_4 injections 24 h prior to CIDR-G removal (day 0, 0 h) and those in the control group were administered with saline solution. At 0 h, three rams were introduced and ewes were checked for breeding marks at 6-h intervals for 3 days. Blood samples were collected for analysis of P_4 and LH. Occurrence of estrus (100 vs 64.3%) and LH surges (92.9 vs 57.1%) were greater (P<0.05) in P_4-supplemented than control ewes, respectively. The intervals from 0 h to estrus (45.4±2.4 vs 35.3±1.9 h) and to the LH surge (46.6±2.6 vs 36.8±2.4 h) were longer (P<0.01) in P_4-supplemented than control ewes, respectively. The magnitudes of LH surges were higher (P=0.01) in P_4-supplemented (51.1±7.1 ng/ml) than in control (24.6±3.2 ng/ml) ewes. P_4 supplement resulted in elevation of P_4 concentrations between days −1 and 0 from 1.8±0.1to 4.2 ±0.3 ng/ml (P<0.001). Pregnancy (92.7 vs 50%) and lambing (92.7 vs 42.7%) rates were greater (P<0.01) in P_4-supplemented than control ewes. In conclusion, P_4 supplement prior to CIDR-G removal produced higher reproductive performance in seasonal anestrous ewes.

Reproductive characteristics of yearling and adult Damascus female goats

C. Papachristoforou, G. Hadjipavlou and A.P. Mavrogenis, Agricultural Research Institute, P.O.Box 22016, Lefkosia, Cyprus

Data collected over the period 2000 to 2006 on 717 adult and 313 yearling Damascus goats kept at the Athalassa experimental station of the Agricultural Research Institute, were analysed in terms of reproductive performance. Using the male effect to induce oestrus and ovulation, adult goats were mated during anoestrous Period1 (June-July); adults not conceiving during Period1 together with yearlings (7 to 10 months old) were mated during the first two months (September-October) of the reproductive season of adult goats (Period2). Of those put to the buck, the proportion of adult goats mated in Period1 (73,2%) and Period2 (69,2%), was lower than that of yearlings (94,9%). However, the overall pregnancy rate of adults (cumulative of Periods 1 and 2) was 81,6% and similar to that of yearlings (78,0%). Of adult goats present at mating, 9,6% showed no oestrus in either period. Conception rate (CR) at first service was 76,5% and it was higher than CR at second service (59,3%). Adult goats mated in Period2, had higher litter size at birth (lsb:2,24) and litter born live (lsbl:2,07) than those mated in Period1 (lsb:2,06, lsbl:1,95). Yearlings had lower lsb (1,74) and lsbl (1,58) than adults. Regardless of age and mating period, goats conceiving during first service, had higher lsb and lsbl than those conceiving during second service. In adults and yearlings respectively, abortions accounted for 5,5 and 11,1%.

The effect of breed, stage of lactation and parity on sheep milk fatty acid profile and milk fat CLA content under the same management and particularly feeding practices

E. Tsiplakou*, K.C. Mountzouris and G. Zervas, Department of Animal Nutrition, Agricultural University of Athens, Iera Odos 75, GR-11855, Athens, Greece

The fatty acid profile and the CLA content in milk fat of four pure sheep breeds (Awassi, Lacaune, Friesland and Chios), were examined. All sheep were kept indoors all year round under the same feeding practices, without any grazing at all. A total of 237 individual milk samples were collected at three sampling times (December, January and March) from sheep of different parity and different days in milk, for fatty acids profile determination. The results showed that : a. there was a large variation in milk fat CLA content among individuals consuming the same diet, b. the CLA content of milk fat was significantly (P=0.000) lower in Friesland breed, compared to the other three breeds that did not differ, c. the CLA content of milk fat was not affected by ewes parity d. the CLA declined as a rule during the lactation period, but only in Chios breed the results were significant, e. there was a negative but not significant correlation between milk fat and CLA concentration and f. the CLA content of milk fat was not correlated with milk yield.

Impact of butchers and market trends on the integrity of small ruminant genetic resources

M. Djemali[1]*, S. Bedhiaf-Romdhani[2], M. Wurzinger[3] and L. Iniguez[4], [1]Laboratoire des Ressources Animales et Alimentaires, INAT, 1082 Cité Mahrajène, Tunisia, [2]Laboratoire des Productions Animales et Fourragères, INRAT, Rue Hédi Karray, 2049 Ariana, Tunisia, [3]BOKU- University Gregor-Mendel- Str.33, A-1180 Vienna, Austria, [4]ICARDA. P.O. Box 5466, Aleppo, Syria

In response to market trends, farmers are crossing the Barbarine (a fat tailed breed) with thin tailed breeds (Algerien Ouled Djellel and Black Thibar). The objectives of this study were to identify reasons for these changes and assess the degree of crossing between the Barbarine breed and thin tailed breeds. A total of 394 surveys was conducted in six major sheep regions including 259 sheep owners, 64 butchers and 71 consumers. Main results showed that 53% of livestock owners were breeder-fatteners, 35% were breeders and 12% were fatteners. Breeders' preferences were in 70% for the fat tail breed. Fatteners were in 55% in favour of thin tail breeds. Butchers preferred mainly thin tailed breeds and crosses over the Barbarine breed. Butchers preference is affecting 26% of Barbarine owners to shift to thin tail breeds and crosses. Surprising, consumers' answers showed that the majority and mainly in rural areas is still in favour of the Barbarine breed due to their diet habits. These results showed that the market changes for thin tailed breeds are dictated by butchers and not consumers.

Session S23

Theatre 5

Electronic miniboluses vs. visual tags for tracing lambs from suckling to fattening: Performances and digestibility effects

J.J. Ghirardi[1], G. Caja[1], C. Flores[1], D. Garín[2], M. Hernández-Jover[1] and F. Bocquier[3], [1]Grup de Recerca en Remugants, Universitat Autònoma de Barcelona, Bellaterra, Spain, [2]Universidad de la República, Uruguay; [3]UMR Élevage des Ruminants en Régions Chaudes, Campus ENSAM-INRA-PHASE, Montpellier, France

Three types of miniboluses (B1, 13.8 g; B2, 16.2 g; B3, 20.1 g), containing 32 mm half-duplex passive transponders, were administrated to 545 suckling lambs as early as possible after birth. Each lamb also wore 2 types of plastic visual ear tags (E1, 1.5 g; E2, 4.1 g). Growth, bolus retention and health were recorded weekly. Minibolus location in the forestomachs and retrieval was evaluated at slaughter (24 kg BW). Moreover, 8 male lambs were used to study the digestibility effects of B2 under intensive feeding conditions (concentrate and straw ad libitum). On average, B1 and B3 did not differ in (27 d) and weight (9.2 kg) at administration, but B2 required older and heavier lambs. Miniboluses did not affect lamb performances, but final readability differed by minibolus (B1, 97.7%; B2, 95.2%) and B3 (100%), and by ear tag (E1, 98.1%; E2, 100%). Although 100% miniboluses were recovered at slaughter, location in reticulum (B1, 90.0%; B2, 83.3%; B3, 93.8%) varied by bolus type, and 1.6% of B1 were recovered from abomasum. No differences in intake, average daily gain and nutrient digestibility were reported. In conclusion, B3 proved to be efficient devices for lamb identification before weaning (> 4 wk; and >10 kg BW), allowing traceability of lambs from suckling to harvesting.

Session S23

Theatre 6

An accelerated lamb production system in New Zealand: Results from two years

G. deNicolo[*], S.T. Morris, P.R. Kenyon and P.C.H. Morel, College of Sciences, Massey University, Private Bag 11 222, Palmerston North, New Zealand

Lamb production in New Zealand is driven by the seasonal pattern of pasture growth. The resulting compact pattern of lamb production has several negative consequences including the inconsistent supply of lamb for slaughter and poor utilisation of meat processing plants throughout the year. A trial was set up in New Zealand that compared the conventional once-a-year lamb (CL) production system with an accelerated lamb (AL) production system in which each ewe had the opportunity to lamb five times in three years. The AL production flock mating dates each year were 14 January, 28 March, 9 June, 21 August and 2 November. Lambs in the AL system were weaned at 73 days from the planned start of lambing. Results from the first two years indicate that low out-of-season pregnancy rates are the limiting factor in the success of the AL production system. In the AL system, pregnancy rates ranged from 36% out-of-season to 95% in season, compared to 98% in the CL system (P<0.05). Lamb live weights at weaning were lower in the AL system (19 kg) compared to the CL system (27 kg; P<0.05). The AL production system produced more weaned lambs and more total kg of lamb per year compared with the CL system, and has potential for further improvement if the low out-of-season reproductive performance can be improved or overcome.

Session S23   Theatre 7

Relationships between morphological and functional udder characteristics in improved Valachian, Tsigai and Lacaune dairy sheep breeds

M. Milerski[1], M.Margetin[2], A. Čapisträk[2], D. Apolén[2], J. Špánik[2], M. Oravcová[2], [1]Research Institute of Animal Production, Prague-Uhřiněves, Přátelství 815, 10400 Praha, Czech Republic. [2]Slovak Institute for Agricultural Research, Hlohovská 2, 949 92 Nitra, Slovakia.

Udder morphology traits were measured and subjectively assessed by the use of linear scores, cistern cross-section areas were scanned by ultrasound machine and milking fractions were examined in 266 dairy ewes of Tsigai (TS), Improved Valachian (IV) and Lacaune (LC) dairy ewes. Animals were recorded repeatedly within and between lactations, therefore 772 sets of measurements were collected in total. Analysis of variance was conducted with the mixed procedure of SAS statistical package. The model included effects of experimental day, parity, day in milk, random effect of animal and residual error. Subsequently correlations between random animal effects for udder measurements, linear scores and milking fractions were computed. Subjectively assessed linear scores for udder depth, cistern depth, teat angle and teat size showed high correlations with actual measurements of appropriate traits on udder in all examined breeds (r_p=0.65-0.80). Linear scores for cistern depth and teat position were highly correlated (r_p=0.84; 0.77 and 0.90 for TS; IV and LC ewes), suggesting that they are nearly identical traits. Machine milking yield and total milk yield were significantly correlated with udder external size characteristics and with cistern cross-section areas measured by ultrasound technique.

Session S23   Theatre 9

A comparative study between desert sheep and goats in biological efficiency of meat production under semi intensive system

M. Ibrahim[1], Mona M.M. Mokhtar[1] and M.A.I. Salem[2], [1]Desert Research Center, Animal and Poultry Production Division,Cairo,P.O..Box 11753, [2]Faculty of Agriculture, Cairo University*

This study was carried out over one full year utilizing two flocks one of sheep (346 ewes plus 12 rams) and the other of goats (103 does and 4 adult bucks) to compare Barki sheep and goat in their biological efficiency of meat production. Information on reproductive and productive traits for both sheep and goats were recorded. Average live body weight at birth and at weaning (at 90 days) for lambs and kids were found to be 3.53, 2.4, 18.65 and 7.65 kg, respectively. Average daily gain from birth to weaning for lambs and kids were found to be 140 and 63 g / day, respectively. Total livebody weight of weaned lambs and kids were found to be 4923.6 and 795.6 kg, respectively. Average DM intake over the experimental period (12 months) were 125218.0 and 28533.0 kg for sheep and goat, respectively. Biological efficiency in terms of kg DM needed to produce 1kg weaned lambs (kids) was 25.4 and 35.9 kg DM for sheep and goat respectively. Different proposals to improve such low efficiency, especially for goats, were forward and discussed. Gross margins per head for both sheep and goats were obtained.

A comparison of small ruminant farms in GAP region of Turkey and EU

G. Keskin[1], İ. Dellal[1]* and G. Dellal[2], [1]Agricultural Economics Research Institute, Ankara, Turkey, [2]Ankara University, Faculty of Agriculture, 06110 Ankara, Turkey

Livestock farming in Turkey is generally small-scale family enterprise, carried out in the vast majority of cases, a sideline by arable farmers. Small ruminant farms are genarally located in Central and Soutestarn part of Turkey. The Southeastern Anatolia Project (GAP) of Turkey, covers 9 administrative provinces, is a multi-sector and integrated regional development effort approached in the context of sustainable development. The GAP had originally been planned in the 70s consisting of projects for irrigation and hydraulic energy production on the Euphrates and Tigris, but transformed into a multi-sector social and economic development program for the region in the 80s. Small ruminant farms are generally located in GAP Project region as 15% of small ruminant population in Turkey.

In this paper, It was determined that the economic size of small ruminant farms in GAP region and it was compared with EU smalll ruminant farms with the respect of structural indicators, economic size, Gross Margin, incomes from sheep and goats and total production value. The data which was used in this paper obtained from Farm Data Network System (FADN) in EU and survey result which was conducted by researches in GAP Region.

Genetic parameters for body weight and ultrasonic measured traits for Suffolk sheep in the Czech Republic

J. Maxa[1,2]*, E. Norberg[1], P. Berg[1] and M. Milerski[3], [1]Department of Genetics and Biotechnology, Danish Institute of Agricultural Sciences, Research Centre Foulum, P.O. Box 50, 8830 Tjele, Denmark, [2]Department of Special Animal Breeding, Czech University of Agriculture, Kamycka 129, 165 21 Prague, Czech Republic, [3]Research Institute of Animal Production, Prátelství 815, 104 01 Prague-Uhrineves, Czech Republic

Suffolk is the most common sheep breed in the Czech Republic, and in this study, heritabilities and (co)variance components for body weight at 100 days (BW), muscle depth (MD) and fat depth (FD) were estimated. Data from 1996 to 2004 were extracted from the sheep recording database of the Czech Sheep and Goat Breeding Association. Average values for BW, MD and FD were 27.91 kg, 25.5 mm and 3.3 mm, respectively. Direct and maternal heritability for BW were 0.17 and 0.08, respectively, direct heritabilities were 0.16 for MD and 0.08 for FD. Maternal heritability estimates for ultrasonic measurements were generally low. Direct genetic correlations between BW and MD and maternal genetic correlations between BW and MD were positive and favourable. Both direct genetic correlations between BW and FD and maternal genetic correlations between BW and FD were negative, but not significantly different from zero. Favourable genetic correlations between BW and MD make ultrasound measurements a valuable tool in breeding programs focusing on growth and carcass characteristics.

Evaluation of some economical traits in south of Khorasan province cashmere goat

D.A. Saghi* and S.A. Shiri, Iran-Mashhad, Agricultural & natural resources research center of khorassan, P.O.BOX. 91735-1148, Iran

Goat mainly rear in arid, semi-arid and mountainous countries. There are 700 million head of goat in the world and 22 million in Iran. Khorasan province located in north-east of Iran with arid and semi-arid condition has about 3 million head of goat. In this research phnotypic characteristics and performance of goat under grazing condition of arid rangeland in south of khorasan inculading Nehbandan, Birjand, Ghayen and Gonabad were studied. In order to data 1373 collected and studied traits were herd size, birth weight, weaning weight, sixth month weight, daily milk production, fleece weight, cashmere, length and diameter fiber have been recorded and analysed as statistical data were 235 ± 35 heads, 2.2 ± 0.28, 11.48 ± 2.28 and 13.23 ± 1.01 g, 434.43 ± 115 (g/day), 319.47 ± 125 and 148.49 ± 58.3 g, 47.5 ± 8.9 mm and 16.6 ± 1.2 micron respectively. The goats are milked by hands four months. Dahiring is done in spring with special sisors. Results indicated this goat has small body size with short legs and feet, small triangle face. The eye-socket is some embossend. This animal has simple or ensiform horns. Ear are long/or short tubular shape. Goats obsereved as black, light or full brown colours. Overall the goat has a good potential for producing of milk and cashmere in Iran condition.

Synchronization of estrus in Sanjabi sheep using CIDR or FGA sponges during out breeding season

H. Hajarian[1], M. Moeini[1], A. Moghadam[2] and H. Bahrampour[3], [1]Razi University,College of Acriculture, [2]Veterinary College,Kermanshah, Iran, [3]Agricultural Ejihad Division, Lorestan, Iran

Genetic improvement has a tremendous progress in milk, meat and fiber production. The use of controlled internal drug releasing device (CIDR) in estrus Synchronization protocols in sheep is well established. Treatment with intravaginal pessaries impregnated with 300 mg progesterone (CIDR) or 45 mg fluorogestone acetate (FGA) for a period of $13 - 14$ days has been used in Sanjabi ewes to assess their effect on ovulation and lambing rate. One hundred eighty ewes (3-4 years old) randomly divided into two groups at April 2004 and CIDR or FGA intravaginale sponges were introduced for 13 days. At FGA or CIDR Withdrawal, each animal received 400 IU. PMSG, then 48 to 72 hours after sponge's removal; estrus was detected using vasectomiezed rams. Animal in estrus were inseminated using fresh semen of selected valuable rams. The overall proportion of ewes exhibiting estrus was %72 in CIDR group, which was significantly higher than FGA group with % 58 estrus detection. The lambing rate was % 61 with % 17 twinges in CIDR group which significantly was higher than FGA group (% 47 with % 9 twinges).
In conclusion, using CIDR appears to be more effective than FGA for the induction of synchronized estrus, better fertilization and lambing rate in Sanjabi ewes.

A comparison between a 4-day and a 12-day FGA treatments with or without eCG on reproductive performance of ewes bred out-of-season

M.Q. Husein*, S.G. Haddad, H.A. Ghozlan and D. Abu-Ruman, Department of Animal Production,Faculty of Agriculture, Jordan University of Science and Technology, P.O. Box 3030, Irbid 22110, Jordan

Reproductive responses of ewes were evaluated out-of-season by comparing intravaginal FGA sponges inserted for 4 or 12 days with or without eCG. Forty-eight seasonally anestrous Awassi ewes were divided in equal numbers in a 2x2 factorial design. Treatments were timed to end together (day 0) and were 12-day-FGA-eCG, 12-day-FGA, 4-day-FGA-eCG and 4-day-FGA. Five rams were introduced for breeding and estrus detection. Blood samples were collected to compare progesterone (P_4) concentrations and for pregnancy diagnosis which was determined on day 30 using ultrasonography. Overall estrus expression was similar among groups and occurred in 43/48 ewes. Intervals to estrus were similar between 12-day-FGA-eCG and 4-day-FGA-eCG and were shorter (P<0.05) in eCG-treated than non-eCG-treated, with no differences between 12-day-FGA- and 4-day-FGA-treated ewes. P_4 levels were basal from day -12 through day 4 (P>0.1) and increased gradually thereafter until day 15. P_4 remained elevated through day 19 in 11/12, 8/12, 9/12 and 6/12 ewes, respectively, which were confirmed pregnant on day 30 by ultrasonography. Of the 34 ewes that became pregnant, 31 lambed 150 days following day 0. Pregnancy and lambing rates and the number of lambs born were similar between either 12-day-FGA groups and the 4-day-FGA-eCG. In conclusion, a 4-day FGA treatment with eCG produced reproductive responses similar to those produced using FGA for 12 days and eCG is essential for application of such a 4-day-FGA protocol out-of-season.

Induction of synchronized oestrous in Sanjabi ewes outside the breeding season

Aliasghar Moghaddam, Large Animal Clinical Scienct, Veterinary Faculty, Razy University, Kermanshah, Iran

Synchronized oestrous was induced in Sanjabi ewes, outside the breeding season, using CIDR (300mg, Progesterone, Inter Ag, Canada), plus GnRH (100mg, Fertagyl®, Intervet, Holland). A total of 80 Sanjabi ewes of mixed ages (2-5 years) were divided rondomly into 3 treatment (T1,T2,T3) and a control (C) groups. The females in groups T1 (n=20), T2 (n=20) and T3 (n=20) were treated with CIDR for 10, 12 and 14 days, respectively, also half of them received an i.m., injections of GnRH at CIDR withdrawal. Animals in group C (n=20) received no treatment.The results indicate that, oestrus was induced between 30-58 h (mostly 30 h) post-CIDR withdrawal in 75%, 95% and 85% of animals in groups T1,T2 and T3 (P>%5), respectively, whereas, only 10% of animals in group C showed oestrus (p<%5). Conception rates of animals with (n=26) and without (n=25) receiving GnRH at CIDR withdrawal, were 15.38% and 0%, respectively, (p<%5). The conception rate in group C (10%) was more than animals without receiving GnRH (P>%5) and less than (p<%5) animals with receiving GnRH at CIDR withdrawal.The results saggest that it is possible to induce synchronized oestrus in Sanjabi ewes outside the breeding season, using CIDR and GnRH. Although, because of obtaining low conception rate in treament groups, probably due to using GnRH, further experiments are warranted.

Seasonal changes in semen quality, scrotal circumference and blood testosterone of Persian Karakul rams

M. Safdarian[1]*, M. Hashemi[1] and M. Kafi, [1]Fars Research Center for Agricultural and Natural Resources, 71555-617, Shiraz, Iran, [2]Department of Clinical Sciences, Faculty of Veterinary Medicine, Shiraz University, Shiraz 71342, Iran

Six rams (3-4 years old) were trained to serve the artificial vagina. Semen collection was performed every 2 weeks, commencing in January (at onset of winter) to December (end of autumn) 1999. Semen ejaculates were evaluated for volume, total sperm output, mass motility and the percentage live sperm. Moreover, changes in the scrotal circumference, body weights, and plasma testosterone concentrations of the rams were also recorded at 2-week intervals. The semen was characterized by a mean ($\pm$S.D.) ejaculate volume of 1.2 ± 0.3 ml, a mean ($\pm$S.D.) mass motility of 3.8 ± 0.8, and a total number of 4442 ± 1247 ($\times 10^{6)}$ sperm per ejaculate for the observation period. The overall mean ($\pm$S.D.) scrotal circumference recorder was 32.0 ± 1.2 cm. Seasonal variations (p< 0.05) were recorded for semen volume, mass motility, scrotal circumferences and plasma testosterone concentrations, but not for the total sperm output. In conclusion, albeit there were seasonal changes in semen characteristics of Karakul rams, the semen has the capability and quality to be used for artificial insemination all year round. Semen of superior quality and quantity was especially collected in late summer and throughout autumn.

Out of season breeding in Moghani ewes using CIDR

F. Farivar, Gorgan Univercity of Agricultural sciences and natural resources, Dept of Animal Science, Iran

This experiment was conducted to evaluate effect of three PMSG doses (300, 400, and 500 IU) and two insemination method (artificial or natural) on out of season lambing rate of ewes. Estrous was inducted in 450 moghani ewes of an industrial farm (khorasan province-north east Iran) by intra vaginal insertion of CIDR (12 days) during anestrous season. After CIDRs removal experimental animals were divided randomly to three groups of 150 ewes and each group received an injection of 300, 400 or 500 IU PMSG. Then, each group randomly divided to two groups. In each group, ewes in one sub-group were artificial inseminated using fresh semen and in another sub-group were naturally mated. Lambing data were collected for each group and analyzed by chi-square test. Lambing percent of natural mated ewes was greater than artificially inseminated ewes (35% vs. 25%), but this difference was not statistically significant (p> 0.50). Also, higher doses of PMSG in this experiment had no significant effect on lambing percent of both groups. Results of this experiment showed that artificial insemination can be considered as an alternative for natural mating and using lower doses of PMSG can reduce out of season breeding costs without significant effect on performance.

Effect of progesterone supplement post-mating on pregnancy and lambing rates of ewes out-of-season

M.Q. Husein*, J.F. Hijazi, S.G. Haddad and H.A.Ghozlan, Department of Animal Production,Faculty of Agriculture, P.O. Box 3030, Jordan University of Science and Technology, Irbid 22110, Jordan

The experiment was conducted to examine the effects of progesterone (P_4) supplement post-mating on pregnancy and lambing rates ewes out-of-season. Thirty-nine ewes were induced to estrus using CIDR-G for 12 days. At the time of CIDR-G removal (day 0 and 0 h), four rams were turned-in with ewes for breeding and heat checking for 3 days. After mating ewes were randomly allocated to four groups to be treated with i.m. injections of 20 mg P_4 supplement once daily between days 5 and 10 (Tr1, n=10), days 10 and 15 (Tr2, n=10), days 5 and 15 (Tr3, n=10) or did not receive P_4 injection (control, n=9). Blood samples were collected to verify P_4 levels and to determine pregnancy, which was confirmed on day 30 using ultrasonography. Estrus was observed in 94.9% of the ewes at 35.7±2.8 h. Occurrence of estrus and P_4 levels until day 4 were similar among groups. P_4 between days 5 and 15 differed significantly ($P<0.001$) among groups. Maximum P_4 concentrations were reached between days 11 and 15 among ewes of Tr2, Tr3, and control groups. Group Tr1 ewes had their maximum concentrations between days 7 and 9. Pregnancy and lambing rates were similar among groups and were 53.8 and 41%, respectively. In conclusion, post-mating P_4 supplement was not effective in improving pregnancy and lambing rates of ewes out-of-season. The overall acceptable pregnancy and lambing rates obtained maybe attributed to the use of CIDR-G.

Reproductive failures in female goats submitted to the male effect

J.A. Delgadillo* and R. Rivas-Muñoz, Centro de Investigación en Reproducción Caprina, Universidad Antonio Narro, México, Periférico y Carretera a Sana Fe, Torreón, Coahuila, México

Fertility of does submitted to the male effect varies between 60 % and 75 %. To investigate the causes of this fertility, we studied the sexual activity and the evolution of pregnancy in 29 anoestrous does exposed to three bucks. Males were rendered sexually active during the non-breeding season by exposure to 2.5 months of long days. Estrous behavior of does was recorded for 15 days after male introduction. Pregnancy was determined by echography 26 and 60 days after last estrus detection. All does displayed an estrus behavior, but only 16 gave birth. Of the 13 females that failed to give birth, six goats returned to anestrus less than 26 days after the last displayed estrus, four lost their embryos between 26 and 60 days after the last estrus, and two more did so at a later stage of pregnancy. One doe was successfully stimulated and detected pseudopregnant on day 60 after the last estrous behavior. These results indicate that fertility in goats exposed to the male effect is associated with three types of reproductive failures: 1) goats that were successfully stimulated but returned to anestrous 26 days after the last estrous behaviour detected (46% of failures), 2) post-implantation loss females in first and last third of pregnancy (46% of failures) and 3) pseudopregnancy (8%).

Cryoprotection of ram semen at 5 °C
Mustafa Numan Bucak and Necmettin Tekin, Ankara University, Veterinary Faculty, Department of Reproduction and Artificial Insemination, 06610 Diskapi-Ankara, Turkey

Frozen-thawed semen of ram shows extensive cryodamage as seen by low motility, membrane damage, and, as expected, low in vivo fertility.Alternative studies to improve preservation of ram semen have, as a result, been focussing on liquid storage of ram sperm cooled to 5 °C.Oxidative damage of spermatozoa resulting from reactive oxygen species (ROS) during liquid storage is the one possible cause of the decline in motility and fertility in storage; the other cause is low temperature destabilisation of membrane structure. The aim of this study was to determine the effects of the addition of the antioxidants taurine and GSH, and the membrane structure stabiliser, trehalose, on sperm viability survival during low temperature liquid storage.Semen was collected using an artificial vagina from two Sakız rams and was diluted with Tris-based extender containing additives and without additives as controls.Percent motility, percent abnormal forms, percent dead spermatozoa, and hypoosmotic swelling test (HOST) percentages were determined after 0, 6, 24 and 30 h. liquid storage at 5 °C.Trehalose at 50 mM provided the best maintenance of motility compared to the other additives during 30 h. of storage, and gave the lowest percent dead sperm at 24 and 30 h. Trehalose at 50 mM also gave the highest percent membrane-intact sperm as assessed by HOST.The antioxidants as additives had no significant improvement on sperm survival during 30 h. liquid storage at 5 °C.

Improvement of reproduction performances on sheep production systems associated with agriculture in Algerian semi-arid areas: First lessons for a sustainable development
K. Abbas[1]*, T. Madani[2] and Ah. Djennane[3], [1]INRA Algeria, 19000 Sétif. Algeria, [2]Université Sétif, Alegria, [3]CEVA Algeria

In order to accompany the intensification tendency of the breeding systems, and for better understanding the rules of its operation, this work tries to highlight the first lessons concerning the potential use of the "*mélovine* " in real production farm. This led us to put in reproduction sukling ewes witch lambed in November-December 2003. The test was carried out on 75 adult ewes divided into 3 groups and 24 rams. The treatments took place at 04/02: M = "*mélovine*", (20.1mg) and ms = "*Mélovine + selécure*" (Selenium + cobalt + zinc). A control group was also studied. All the ewes were recovred at 11/03. The body score scales were estimated at 16/3. The data was analysed under a multivaried variance model. The fertility rates are very low because of body state on average very weak (general average = 1.5) this is probably the result of a winter food deficit and sukling. Dates of lambings, although significantly different, didn't inform on a significant advance of the sexual season (difference in one week). The treatment M gave a significantly higher TP. Our results showed also that the effect of "*mélovine*" is a function of the body score at the beginning of the recovring. This factor is an additional element of success of the "*mélovine*" treatment.

The fleece characteristics of pure Karayaka, Sakız x Karayaka (F1) and Sonmez x Karayaka (F1) crossbred young sheep

M. Olfaz and M. A Çam, Ondokuz Mayis University Department of Animal Science, 55139, Kurupelit Samsun

A study was conducted to determine fleece characteristics of pure Karayaka (K), Sakiz x Karayaka crossbred (SaK, F1) and Sonmez x Karayaka crosbred (SK, F1) young male and female sheep. Young lambs were shorn at 6 months of age and fleece samples were taken at the left mid-side of animal. In this study, liveweight at shearing and fleece characteristics such as greasy fleece weight, fibre diameters, fibre length, camp fibre ratio and medulation ratio were measured. The liveweights of K young sheep (24.14±40 kg) were lower than those of SaK (25.92±1.09 kg) and SK 28.97±0.91 kg) young sheep. The fibre lenghts of K young sheep (13.09±0.31 cm) were higher (P<0.05) than those of SaK (9.84±0.65 cm) and SK 9.85±0.58 μ) young sheep. The fibre diameters of K young sheep (32.22±0.50 cm) were higher (P<0.05) than those of SaK (25.58±0.70 μ) and SK (26.87±0.90 μ) young sheep. The greasy fleece weights of K (963.59±25.03 g) and SK young sheep (920.69±52.02 g) were higher (P<0.05) than those of SaK (768.52±64.34 g) young sheep. Consequently, these results suggested that pure Karayaka young sheep have coarser fibre than those of its crossbreds with SaK and SK.

Meat characteristics from three Ethiopian indigenous goat breeds

Ameha Sebsibe[1], N.H. Casey[1]*, W.A.van Niekerk[1] and Azage Tegegne[2], [1]Department of Animal and Wildlife Sciences, University of Pretoria, Pretoria 0002, South Africa, [2]International Livestock Research Institute, Addis Abeba, P.O.Box, 5689, Ethiopia

The study evaluated seventy-two intact males of three Ethiopian indigenous goat breeds, the Afar, Central Highland (CHG) and Long-eared Somali (LES), using three grainless diets under stall-fed conditions. The diets varied in concentrate: roughage ratios, as Diet 1: 50: 50 (8.5), Diet 2: 65:35 (9.2), Diet 3: 80:20 (10 MJ ME/kg DM), respectively. The feed resources were native grass hay, wheat bran and noug cake (*Guizotia abyssinica*). Genotype influenced the carcass fat and crude protein (CP) and the values ranged from 10.3-14.0 and 19.3-21.1 %, respectively. The effect of diet was significant on CP %, but was similar on fat content. Cooking and drip losses differed between genotypes. Genotype and diet influenced the composition of most muscle fatty acids. An interaction effect of genotype and diet was also exhibited on certain fatty acids. Compared to CHG, the Afar and LES breeds had higher unsaturated fatty acids. Diet 1 was recommended due to a higher proportion of roughage, better feed efficiency and higher meat protein. The relatively higher carcass fat, the absence of C12:0, lower concentration of C14:0 and higher C18:1 and lower cooking losses than some other goat breeds are some of the distinguishing characteristics observed in these Ethiopian indigenous goats.

Improving value addition: Simple technological changes in the processing of milk in Northern Syria

M. Hilali*, L. Iñiguez and M. Zaklouta, International Center for Agricultural Research in Dry Areas (ICARDA), P.O. Box 5466, Aleppo, Syria

Set type sheep milk yogurt is a highly demanded product in Syria and its production increased from 59 to 91 thousand ton during 1999-2003. Furthermore, it contributes progressively more to the economy of dairy sheep producers. This is the case of El-Bab region in northern Syria, where 50-60% of the farmer's income derives from yogurt and cheese production. In general yogurt prices are affected by taste and texture and the main marketing constraints identified by farmers are sourness and weak texture caused by processing failures. To solve firmness problems that affect the transportability of the products, three commercial starters were tested with farmers: S1 (Very mild flavour, high viscosity), S2 (mild flavour, high viscosity) and S3 (strong flavour, medium to low viscosity). Yogurts made with these cultures using Awassi sheep milk were compared with the farmers' traditional yogurt as control. The different yogurts were evaluated at the market for price differences. Yogurt made using S1 was sold 2 SL/kg less compared with control, however yogurt elaborated with S2 and S3 were paid 2 SL and 5 SL more that the control. Viscosities of S3 yogurt and S2 yogurt were 60.2% and 71.6% respectively higher than the control. Furthermore regarding to firmness S3 yogurt and S2 yogurt were 29.9% and 19.7 stronger than the control.

Milk yield and composition of Chios and Farafra sheep under subtropical Egyptian conditions

H. Hamdon[1]*, M. N. Abd El Ati [2], M. Zenhom[3] and F. Allam[2], [1]Environmental Studies & Research Institute, Minufiya University, Sadat, Egypt, [2]Animal & Poultry Production Dept, Faculty of Agriculture, Assiut University, Egypt, [3]Sheep & Goats Research Dept., Animal Production Research Institute, Giza, Egypt

Chios ewes produced 0.703, 49.90, 16.77, 66.67 kg in 96.15 days, while, Farafra ewes produced 0.675, 45.96, 13.43, 59.40 kg in 87.93 days, for daily milk yield (DMY), milk yield pre-weaning (MBW), milk yield post-weaning (MAW), total milk yield (TMY) and lactation length (LL), respectively. Breed differences were significant ($P< 0.01$) with TMY and LL. Ewes reared twin lambs were produced significant ($P< 0.01$) much milk than those ewes reared single lambs. Ewes lambed at the spring season had greater milk yield compared to ewes lambed at the summer season and autumn season. Ewes of (> 4 - ≤5) years old showed relatively higher milk yield than younger or older ones. Fat% averaged 6.17 for Chios and was significant ($P< 0.01$) higher than 5.59% for Farafra milk. Also, Chios had slightly higher (TS%) than Farafra (16.04 *vs* 15.59%), respectively. Protein content for Chios and Farafra milk differed significant ($P< 0.01$) and averaged 5.62 and 5.31%, respectively. Chios ewes had higher milk energy (4.59 MJ/kg) than Farafra once milk energy (4.34 MJ/kg). The autumn lambing ewes had a significant and higher fat%, TS% and milk energy (MJ/kg) than both summer lambing and spring season.

The influence of sex and carcass weight on carcass composition of Serrana kids

S. Rodrigues[1]*, V. Cadavez[1], R. Delfa[2] and A. Teixeira[1], [1]Escola Superior Agrária, Instituto Politécnico de Bragança, Apt. 1172, 5301-855, Bragança, Portugal, [2]Unidad de Tecnología en Producción Animal, Servicio de Investigación Agrária, Diputación General de Aragón, Zaragoza, Spain

This work aims to evaluate sex and carcass weight effects on carcass composition of kids. Sixty Serrana kids (31 male and 29 female), a Portuguese breed, were used. Kids were slaughtered after 24 h fasting. Carcasses were cooled at 4°C for 24 h, halved and left side was dissected into muscle, subcutaneous fat, intermuscular fat, bone and remainder (major blob vessels, ligaments, tendons, thick connective tissue sheets). Kidney, knob and channel fat (KKCF) was considered as a carcass component since in Portugal kid's carcasses are commercialised with KKCF included. Female kids presented higher (P < 0.05) intermuscular fat proportion, muscle/bone ratio and KKCF than males, however, male kids had higher (P < 0.05) bone proportion and muscle/fat ratio. All fat depots increased and bone proportion decreased (P < 0.05) with carcass weight increasing. Carcass weight increasing induced muscle/bone ratio increasing (P < 0.05) and muscle/fat ratio decreasing (P < 0.05). Female kids should be slaughtered at lower live weight in order to minimize the carcass fatness development. In spite of the differences in tissue proportions induced by carcass weight, heavier kids (8 kg) weren't excessively fattened, indicating that they can be slaughtered at higher live weights without compromise carcass quality.

Consumer preferences on sheep and goat meat in the world

F. Karakuş, Yüzüncü Yıl University, Faculty of Agriculture, 65080 Van, Turkey

The aim of this paper is to define differences concerning sheep and goat meat consumption between cultures and traditions in various geographic areas. Sheep and goat meat consumption per capita has a tendency to decline over the last 40 years. However, ethnic changes in the populations may lead to increase demand for sheep and goat meat in some of the countries especially in those with large numbers of Muslim immigrants.

Specific consumption patterns and preferences for sheep and goat meat are dictated by cultural and traditional backgrounds and the socio-economic status of the community. Also quality parameters (flavor, odor etch.) affect the use and customs of the populations.

In Central and South-Eastern Asia, sheep meat is generally disliked because of its strong flavor and odor, whereas in the Middle East and Northern Asia, people consume great quantity of sheep meat. In North America the strong flavor of the meat limits its consumption. In Africa, the consumption of adult animals is popular, while in Oceania lamb and mutton are a significant part of the diet. In Europe, although there is a traditional sheep meat cuisine, products with strong flavor are not preferred especially by young people. Goat meat is little consumed in the Northern European countries except by emigrant populations. On the other hand, in the Mediterranean countries, seasonal consumption of unweaned kids seems to be a deep-rooted custom.

Relationship between somatic cell counts and some technological characteristics of goat milk

A. Gursel*, F. Cedden, E. Sanli and O. Koskan, Ankara University Faculty of Agriculture, 06110 Ankara, Turkey

Goat milk is used alone or as a mixture with cow milk in Turkey for yoghurt and cheese production. In general, the overall quality of yoghurt and cheese are affected by the chemical composition of milk and the processing conditions employed. Milk usually contains somatic cells. In many countries, somatic cell counts (SCC) are widely used as monitors of udder health and milk quality. Maximum levels of SCC in milk are held as standards by health officials to ensure quality milk. The limits of SCC for goat is established as 750,000-1,000,000 in some country, whereas SCC value is much higher in Turkey. In this study, the relationship between milk SCC and such technological characteristics of goat milk as coagulation properties with rennet and starter culture and titratable acidity were investigated. Average SCC, titratable asidity (TA), rennet coagulation time (RCT) and starter culture activity (SCA) were found as 3,481,837 ± 420,499 cells/ml, 6.78± 0.128 °sH, 57.23 ± 1.638 mn and 17.62 ± 0.575 % lactic acid, respectively. TA was found negatively correlated with SCA (-0.379) and RCT, whereas no correlation was observed with SCC. In conclusion, SCC value along does not evaluate goat milk availability for yoghurt and cheese production.

Physical and sensorial traits of meat from different ovine categories

Américo Garcia da Silva Sobrinho, Rafael Silvio Bonilha Pinheiro, Hirasilva Borba Alves de Souza and Sandra Mari Yamamoto, UNESP – São Paulo State University, Zootecnia, Via de Acesso Prof. Paulo Donato Castellane s/n, Jaboticabal, SP 14884-900, Brasil

For evaluating qualitative traits of meat from different ovine categories and muscles from different carcass cuts, ½ Ile de France ½ Polwarth animals (uncastrated lambs, discarded ewes and wether mutton) were used. Animals grazed *Cynodon dactylon* and were daily concentrate supplemented with 1% of their live weight. Lambs were slaughtered at 32 kg and ewes and wethers with 55 kg of live weight. There were no interactions between animal categories and carcass muscle cuts for pH_{45min} and pH_{24h}, with values of 6.49 and 5.58, respectively. In lambs, *Longissimus lumborum* and *Semimembranosus* luminosity was higher (40.42) than that *Triceps brachii* (36.17). In relation to water holding capacity, there was no interaction between animal categories or muscles (56.40%). Cooking losses were not affected by ovine categories, with exception of the losses observed in *Longissimus lumborum*, which were higher (46.44%) in lambs than in adult sheep (38.82%). The shearing force was higher (2.77 kgf/cm^2) in adults meat, but it was equal in *Longissimus lumborum* from different categories (1.65 kgf/cm^2). The sensorial analysis did not differ in relation to flavor, color and preference, but lamb meats received higher scores (8.13) for tenderness than meat from adults (6.90). Meat of adult animals was tougher and darker than lambs meat, but other traits as pH, water holding capacity, flavor, color and preference were similar among lambs, ewes and wether.

Nutritive value of meat from Pomeranian lambs and crossbreeds by Suffolk and Texel rams

H. Brzostowski*, Z. Tański, Z. Antoszkiewicz and K. Kosińska, University of Warmia and Mazury in Olsztyn, Faculty of Animal Bioengineering, Poland

Parameters of the nutritive value of meat from 100-day-old Pomeranian lambs (P) and crossbreeds by Suffolk (PS) and Texel (PT) rams were determined in the study. Samples were taken from *m. longissimus dorsi* to determine the chemical composition and physicochemical properties of meat, as well as muscle fiber thickness, energy value, cholesterol content, collagen content, protein amino acids and the fatty acid composition of intramuscular fat. It was found that meat from PS and PT lambs contained significantly more protein and had a more favorable ratio between essential and non-essential amino acids, as compared with meat from P lambs. It also contained more collagen and less cholesterol, was darker in color and had a greater diameter of muscle fibers. Meat from PT lambs, in comparison with meat from P and PS lambs, contained less intramuscular fat and more monounsaturated fatty acids (MUFA), and was characterized by a better water-holding capacity. Meat from PS lambs had a higher calorific value and higher concentrations of polyunsaturated fatty acids (PUFA). Due to a high content and biological value of protein, as well as a small diameter of muscle fibers, a low calorific value and low levels of cholesterol and collagen, meat from the lambs tested in the study, especially from PS and PT crossbreeds, can be recommended as a component of a healthy, low-fat diet.

The effect of different weaning ages on fatty acid composition of Akkeçi (white goat) male kids

S. Karaca[1], Ü. Üzümcüoğlu[2], A. Kor[1] and A. Soyer[2], [1]Yuzuncu Yıl University Faculty of Agriculture, 65080 Van, Turkey, [2]Ankara University Faculty of Engineering 06110 Ankara, Turkey

The study aimed to determine the effect of different weaning ages on fatty acid composition and nutrition value of chop (6-12 ribs) area of Akkeçi (White Goat) male kids carcass. The study consisted of total 15 heads of kids weaning 45, 60 and 75 days of age before fattening period. All kids were slaughtered at 125 days of age from each group. There were no significant differences for the nutritional value and ligament tissue quantities between groups. However, there were significant differences on fatty acid composition between the groups. Kids weaned at 45, 60 and 75 days; C14:0 were found 3.15, 3.25 and 4.73 %, respectively. C14:0 was found higher at kids weaned at 75 days than the others ($P<0.05$). Odd-chain fatty acids like C17:0 % was 1.75, 1.52 and 1.54 %, respectively. The group weaned at 45 days had higher C17:0 % than the others ($P< 0.05$). There were no significant differences between the groups for the total saturated and unsaturated fatty acids of the fat tissue of chop area. However, the groups weaned at 45 and 60 days of age had higher mono-unsaturated fatty acids than the kids weaned at 75 days ($P< 0.05$). C18:1 % greatly affected the differences between the groups for the mono-unsaturated fatty acids ($P < 0.05$).

Effect of slaughter weight on carcass characteristic in Argentine Criollo kids
A. Bonvillani[1], F. Peña[2], J. Perea[2], A.G. Gomez[2], R. Acero[2] and A. Garcia[2]*, [1]National University of Rio Cuarto, Argentine, [2]Department of Animal Production, University of Cordoba, Spain

The effect of slaughter weight was evaluated using 60 Argentine criollo kids. The management of all kids was similar in extensive farming system. After weaning, the kids were fed by grazing until slaughter. The kids were slaughtered at 10, 11.5 and 13.5 kg live weight, according to the methodology of Colomer-Rocher et al. (1988).
The hot carcass weight increased significantly according to slaughter weight (4.7 kg, 5.6 kg and 6.8 kg respectively). Although the weight of non-carcass component increased according to slaughter weight, the proportions of non-carcass component diminished significantly (44.7%, 42.9 % and 41.8% respectively). Furthermore carcass yield (46%, 47.5% and 49.4% respectively), carcass conformation and carcass measurements increased significantly with slaughter weight. As the slaughter weight increased, the percentage of muscle also increased and the percentage of fat diminished. Nevertheless the percentage of bone was not affected by slaughter weight. The commercial cuts showed values of 32% for the leg, 21.2% for the shoulder, 15.1% for the ribs and 9.3% for the neck. Slaughter weight showed no significant effect on the proportion of commercial cuts.

Effect of different production systems on lamb meat quality
B. Panea, M. Joy, G. Ripoll and R. Delfa*, CITA, P.O. Box 727, 50080, Zaragoza, Spain

The influence of four different production systems on lamb meat sensory characteristics was analysed. The treatments were: INT, lambs were kept indoor with free access to concentrate while dams were only indoor from 1500 to 0800 h. EST, Lambs and ewes were kept indoor with free access to concentrate. ALF, lambs and ewes grazed alfalfa without concentrate; ALF+SUP, animals grazed alfalfa and lambs had cereal supplement. Lambs from INT and EST were weaned at 45 days old. Six non-castrated male lambs from each treatment were slaughtered at 20-22 kg live weight and refrigerated at 4°C. At 24 h post-morten, loins from left half carcasses were excised, vacuum packed and aged for a total period of three days. Samples were tested by a trained 8 member panel using a structured scale of 10 points, from 1 (lowest) to 10 (highest), for the following attributes: lamb odour, wool lamb, tenderness, juiciness, lamb flavour, liver flavour, fat flavour, metallic, flavour and overall appraisal. A comparative multisample test in a balanced incomplete block design was carried out and results were analysed by GLM Procedure (SAS, 1988). No differences were found among treatments for any considered attribute. Nevertheless, a tendency in tenderness was observed, being higher in ALF. Results would imply that grazing alfalfa allows the prodution of light lambs without modifiy sensory meat quality in comparation to those produced under indoor systems.

Freezing time on meat quality of lamb fed with diets containing fish residue silage

Sandra Mari Yamamoto, Américo Garcia da Silva Sobrinho, Hirasilva Borba Alves de Souza, Rafael Silvio Bonilha Pinheiro, UNESP – São Paulo State University, Zootecnia, Via de Acesso Prof.Paulo Donato Castellane s/n, Jaboticabal, SP, 14884-900, Brasil

Eighteen 7/8 Ile de France 1/8 Polwarth lambs averaging 17 kg, confined in individual cages and were distributed among the following treatments: control diet; 8% freshwater fish (*Oreochromis niloticus*) residue silage diet; 8% sea fish *(Lophius gastrophisus)* residue silage diet. The fish residue silages have parcially substituted the soybean meal and forty percent of corn silage was used as roughage. Lambs were slaughtered at 32 kg of body weight and carcass was cut after *rigor mortis* (24 hours after slaughtering). The muscle *Tríceps brachii* was removed and analyzed *in natura* and after frozen for 30 and 90 days in a –18°C freezer. There was no interaction between pH, coordinate a* (red intensity), water holding capacity and exudation losses *in natura* and frozen meat. However, the meat of lambs fed with the control diet showed higher a* mean (14.60). Also, the red intensity of the meat *in natura* was higher (14.38) in comparison to when it was frozen for 90 days (13.11). Unfrozen meats showed higher water holding capacity (63.26%). For the number of reactive substances to 2-thiobarbituric acid (TBARS), there was an interaction between diet and freezing time. The muscle frozen for 30 and 90 days showed higher TBARS than unfrozen muscle in all diets studied here, making evident the oxidative aging process in meat freezing.

Effect of sheep breed on the quality of lamb meat stored under modified atmosphere conditions

Z. Tański, H. Brzostowski, K. Kosińska and Z. Antoszkiewicz, University of Warmia and Mazury in Olsztyn, Faculty of Animal Bioengineering, Poland*

The quality of meat from 100-day-old single ram lambs of the Pomeranian (P) and Ile de France (IF) breed was analyzed in the study. The quality of fresh meat (evaluated 48 h postmortem) was compared with the quality of meat stored in a modified atmosphere (80% N_2 and 20% CO_2) for 10, 20 and 30 days at + 2 °C. Samples of the quadriceps muscle of the tight (*quadriceps femoris*) were taken to determine the chemical composition, physicochemical and sensory properties of meat, as well as the fatty acid profile of intramuscular fat. It was found that meat from IF ram lambs P had higher concentrations of dry matter, protein, crude fat and saturated fatty acids, as well as a higher calorific value, and lower concentrations of unsaturated fatty acids, a lighter color and a lower pH, as compared with meat from P ram lambs. Modified atmosphere packaging of fresh lamb meat, aimed at extending its shelf-life, caused a decrease in water-holding capacity and pH. In addition, meat became somewhat lighter in color and received lower grades for some sensory parameters, especially for taste desirability, juiciness and tenderness. Storing lamb meat under modified atmosphere conditions (80% N_2 and 20% CO_2) enables to improve the ratio between unsaturated fatty acids (UFA) and saturated fatty acids (SFA).

Carcass yield and loin tissue composition of lambs feedlot fattened with diets containing fish residue silage

Américo Garcia da Silva Sobrinho, Sandra Mari Yamamoto, Rose Meire Vidotti, Antônio Carlos Homem Junior and Rafael Silvio Bonilha Pinheiro, UNESP – São Paulo State University, Zootecnia, Via de Acesso Prof.Paulo Donato Castellane s/n, Jaboticabal, SP 14884-900, Brasil

Eighteen 7/8 Ile de France 1/8 Polwarth lambs averaging 17 kg, confined in individual cages, were used in this study. Animals were distributed among the following treatments: T1- control diet; T2- 8% freshwater fish (*Oreochromis niloticus*) residue silage diet; and T3- 8% sea fish (*Lophius gastrophisus*) residue silage diet. Fish residue silage partially replaced soybean meal, and each diet included 40% corn silage. Lambs were slaughtered at 32 kg of body weigh, and analyses showed that fish residue diets did not affect quantitative carcass traits, with cool and warm carcass weights of 14.95 and 14.52 kg, respectively, and biological and commercial yields of 55.31% and 47.47%, respectively, and weight lost by chilling of 2.92%. The mean loin eye area was 12.05 cm^2 and muscle and bone proportions 57.61 and 16.67%, respectively. The percentage of fat (19.77%), was lower in loins of lambs that received the control diet, in comparison to lambs fed the fish residue silage diets (T2 = 24.68% and T3 = 24.01%). The composition of the muscle *Longissimus lumborum* was not affected by the diets, with crude protein, ether extract and ash levels of 25.66%, 3.48% and 1.03%, respectively. The replacement of soybean meal with fish residue silage resulted in increased loin fat, with no effect on quantitative carcass traits.

Women roles in the small ruminant sector of Turkey, special reference to Mediterranean area

N.Darcan and D.Budak, Çukurova Univ. Agriculture Fac. Adana/Turkey

The aim of this study was to determine the gender role in small ruminant sector of Turkey.
Women had very heavy burden in small ruminant activities like milking by hand, feeding, mating, producing cheese and cleaning. In other words, women performed labor-intensive work in division labor while men or conducted selling, vaccination and health care activities related out of home or market. In the other words, women were seen as generally unpaid family labor from the point of view division of labor. They were seen helper status most of the time. Actually, they performed productive, reproductive and community roles without interruption women did not aware of their primary status in rural life and they defined their job as a housewife while they were main component in productive activities.

Production characteristics of the Mutton sheep breeds Charollais and Suffolk under field conditions

Mohamed Momani Shaker[1], Martina Malinová: Czech University of Agriculture in Prague, Institute of Tropics and Subtropics, Czech Republic.

The objective of this study was to evaluate the effect of genotype on selected production characteristics. In addition the effect of litter size, sex of lambs, sires and year of rearing on birth lamb weight and on growth ability of lambs from birth to 100 days lambs' age was analysed. During the years 2002 to 2004 following traits in ewes of Charollais and Suffolk were evaluated: percentage of conception, fertility percentage per ewe and per ewe lambed. Reproduction characteristics were evaluated by using analysis of variance according to the Stat-graphic programme. The live weight of lambs at birth and live weight of lambs at the age of 100 days (± 10 days) was recorded by weighing on digital scales to the nearest 0.1kg. The obtained data were processed by SAS using a least squares model equation with fixed effect. A conception rate of 91.6% was achieved in the whole flock, the fertility rate per ewe of the foundation stock 167.3%. Conception and fertility rates were influenced by genotype and year of rearing ($P < 0.05$). Average daily gain (ADG) of lambs from birth to the age of 100 days was 315 ± 0.07g. Live weight of lambs at birth, live weight of lambs at the age of 100 days and ADG were influenced by effect of sires, litter size, sex and rearing year ($p < 0.05$; $p < 0.001$). These results indicate that both mutton sheep breeds Charollais and Suffolk have excellent similar performance.

Relationship between body condition score and muscle and fat depths measured by ultrasound in meat and dairy ewes

S. Silva, V. Santos, C. Guedes, A.L.G. Lourenço*, A.A. Dias-da-Silva and J. Azevedo. CECAV-UTAD Department of Animal Science, PO Box 1013, 5001–801 Vila Real, Portugal

The suitability of body condition score (BCS) to predict fat and muscle reserves was assessed after regressing in vivo real-time ultrasound muscle and fat depth measurements and BCS. This study was conducted with 17 Ile-de-France (IF) mature meat type ewes and 47 Churra da Terra Quente (CTQ) mature dairy type ewes. The animals were scanned with an Aloka 500V using a linear probe of 7.5 MHz. The probe was placed perpendicular to the backbone over the 13th thoracic vertebra and between the 3rd and the 4th lumbar vertebrae. The subcutaneous fat (SC13 and SC34) and the Longissimus thoracis et lumborum muscle (MD13 and MD34) depths were measured over these points. BCS was assessed using the methodology proposed by Russel et al. (1969) with a half point interval. BCS ranged from 1 to 5 and 2 to 4.5 for CTQ and IF ewes, respectively. The subcutaneous fat depth was (SC13 and SC34) accurately predicted by BCS (r^2 varied between 0.58 and 0.80; RSD < 1.0 mm). Muscle depth was positively but inconsistently related to BCS ($r^2 = 0.013$ and 0.03 for IF and $r^2 = 0.42$ and 0.23 and CTQ for MD13 and MD3, respectively). BCS reflects the subcutaneous fat depth variations, although the assessment of muscle variation by BCS is inadequate.

Relationship between body condition score and body composition in Churra da Terra Quente dairy ewes

S. Silva, A.L.G Lourenço*, C. Guedes, V. Santos and J. Azevedo, CECAV-UTAD Department of Animal Science, PO Box 1013, 5001–801 Vila Real, Portugal

The relationship between body condition score (BCS) and physically dissected body composition was assessed in 47 non-pregnant and non-lactating Churra da Terra Quente (CTQ) dairy ewes. BCS was assessed using the methodology proposed by Russel et al. (1969) with a half point interval. After slaughter, the internal fat depots (kidney knob and channel, omental, and mesenteric fat) were weighed and weights were expressed in grams and in percentage of carcass weight. Carcass composition was assessed after dissection on bone, muscle and fat depots (subcutaneous and intermuscular fat), which were weighed and weights were expressed in grams and in percentage of carcass weight. Over the range of 1.5 to 5.0, BCS was positively correlated to the internal fat depots weights (r varied between 0.528 and 0.788, P<0.01) and to its percentage in carcass (r varied between 0.207, P<0.05 and 0.759, P<0.01). The highest correlation coefficients were observed between carcass fat depots, expressed both in grams and in percentage of carcass weight, and the BCS (r varied between 0.813 and 0.890, P<0.01). The BCS was positively correlated with the muscle expressed in grams (r= 0.561, P<0.01) but it was negatively correlated with the muscle expressed in percentage of carcass weight (r=-0.852, P<0.01). On the range of body composition studied the BCS well mirror the body fat reserves of CTQ dairy ewes.

Analysis of lamb mortality in Sumava, Suffolk, Charollais, Romney, Merinolandschaf and Romanov sheep in the Czech Republic

M. Milerski, Research Institute of Animal Production, Prague, Uhříněves, Přátelství 815, 104 00, Praha 10, Czech Republic

The objective of the study was to estimate effects of type of birth and dam age on lamb mortality to weaning in sheep breeds Šumava (S), Suffolk (SF), Charollais (CH), Romney (RM), Merinolandschaf (ML) and Romanov (R) in the Czech Republic. Data were collected on 35 036 offsprings during the years 1995-2006. Data were analysed with the mixed model with fixed effects of birth type and age of dam and random effects of dam and HYS for individual breeds separetely. LSMs for lamb mortality rates ranged from 8.2 % (R) to 15.2 % (CH) for single-born lambs, from 11.8 % (R) to 20.5 % (ML) for twins and from 18.1 % (R) to 35.8 % (RM) for triplets. Romanov lamb excelled in survival rate to weaning in all types of birth. The highest mortality of single born lambs was detected in meat sheep breeds CH and SF. This fact is probably in relationship with higher birth weights of lambs of these breeds. The mortality rate increased in lambs born as triplets or greather. This fact was significant especially in RM, where animals are kept mostly outdoor during the whole year with minimum of individual care. LSMs for lamb mortality rates for singletons, twins and triplets were 10.7 %, 18.1 % and 35.8 % in RM.

Feedlot performance and carcass characteristics of fat-tailed Karakaş ewe-lambs fed intensively for 70 days

T. Aygün[1], M. Bingöl[1] and S. Arslan[2], [1]Yüzüncü Yıl University, Faculty of Agriculture, Department of Animal Science, 65080 Van, Turkey, [2]Ondokuz Mayıs University, Faculty of Veterinary Medicine, Samsun, Turkey

The aim of this study was to evaluate the fattening performance and carcass characteristics of fat-tailed Karakaş ewe-lambs. A total of 15 Karakaş ewe-lambs (10 single, 5 twins) were used as material of this study at barns of Agricultural Faculty of Yüzüncü Yıl University. Fattening of the ewe-lambs started when they were approximately 7-8 months old. The single and twin Karakaş ewe-lambs intensively fed for 70 days had the mean initial weight of 24.3 and 26.3 kg; the mean weight at the end of fattening period of 36 and 38.4 kg; the mean daily weight gains of 166 and 172 g; the mean cold carcass weight of 19.5 and 18.7 kg; the mean dressing percentage of 51.1 and 51.7%, respectively. The results of this study indicated that differences between single and twin were not significant for live weight, slaughter and carcass characteristics of Karakaş ewe-lambs.

The internal and external effects on lamb's growth ability of mutton Charollais sheep breed

Mohamed Momani Shaker, Renáta Mudruňková, Martina Malinová and Karel Srnec: Czech University of Agriculture in Prague, Institute of Tropics and Subtropics, Czech Republic

The goal of the study was to observe and evaluate the internal and external effects on lamb's growth ability of mutton Charollais sheep breed. In the years 2004 and 2005 the live weight was determined in 328 lambs at birth and subsequently every fortnight until 100 days of age. Average live weight of lambs at birth was 3.66 kg and at the age of 100 was 26.45 kg. The genotype of lambs affected average daily gain and live weight of lambs at 100 days of age highly significantly ($P \leq 0.001$). The live weight of lambs at birth was not affected by genotype. Investigation of the effect of sex on BW and live weight of lambs at 100 days of age showed that the differences between males and females were statistically highly significantly ($P \leq 0.001$). The litter size influences weights gains and live weight at birth and at the age of 100 days highly significantly ($P \leq 0.001$). Live birth weight of lambs was 11.17 % higher in singles compared with twins. Live weight of single lambs at 100 days was 29.074 ± 0.53 kg compared with 25.265 ± 0.36 kg in twins. Likewise, ADG until 100 days of age and live weight at 100 days of age were affected by the year of lambing ($P \leq 0.001$).

Factors affecting growth of the Balkan goat kids from birth to 90 days
V. Bogdanovic[1]*, N. Memisi[2] and F. Bauman[3], [1]Institute of Animal Sciences, Faculty of Agriculture, University of Belgrade, Nemanjina 6, 11080 Zemun – Belgrade, Serbia, [2]Dairy Plant, Subotica, Serbia, [3]Assocciation of Sheep and Goat Breeders, Belgrade, Serbia

Data of 96 Balkan goat kids (48 male kids and 48 female kids) originated from 4 herds was studied to estimate herd, sex and order of dam lactation or dam milk production effects on birth weight, monthly weights measured at 30, 60 and 90 days as well as ADG from birth to 90 days of age. Two statistical models were used: model with 3 categorical independent variables (herd, sex and order of dam lactation), and model with 2 categorical independent variables (herd, sex) and one continuous predictor variable (dam milk production). Average birth weight and average weight at 30, 60 and 90 days were 2.7 kg, 6.6 kg, 10.25 kg and 13.63 kg, respectively. Average daily gain from birth to 90 days of age was around 0.12 kg. Herd of origin was the most important (from $P<0.02$ to $P<0.001$) nongenetic sources of variation for almost all growth traits. Order of dam lactation had significant effect only on birth weight and daily gain up to 60 days of age while sex has no effect on variability of analysed traits. On the other hand, dam milk production was significant source of variation for all traits except for ADG during the first month of kid's life.

Effect of genotype and slaughter weight of lamb males on sensory properties of meat
I. Dobeš*[1], A. Jarošová[2] and J. Kuchtík[1], [1]Mendel University of Agriculture and Forestry Brno, [1]Department of Animal Breeding, MSM 4321 00001,[2]Department of Food Technology, Zemědělská 1, Brno 613 00, Czech Republic

The aim of our study was to evaluated the effect of different genotypes (F_1 and F_{11} crossbreeds of Charollais (CH), Suffolk (SF) and Improved Walachian (IW): CH x SF (n = 8), (CH x SF) x SF (n = 7) and IW x SF (n = 7) and different live slaughter weights (below 30 kg (n = 7), 31 – 39 kg (n = 8) and higher than 40 kg (n = 7), of lamb males on sensory properties (colour, flavour intensity, odour intensity, stringiness, tenderness and juiciness) of meat. Before slaughter all lambs were kept in identical environmental and nutritional conditions. Musculus biceps femoris was used for sensory analysis by a trained taste panel of 10 members. Genotype had a significant effect only on colour and stringiness while F_1 crossbreeds of CH x SF were evaluated best regarding colour, stringiness, odour and taste. However, the best taste was found out in IW x SF. Live slaughter weight (LSW) had a significant effect on colour, tenderness and juiciness, while lambs with LSW up to 30 kg were evaluated best regarding colour, stringiness, odour and taste. However, the most tender and juicy meat was found out in lambs with LSW from 31 to 39 kg.

Some breeding characteristics of Norduz sheep

M. Bingol[1] *F. Cedden*[2] *and I. Daskiran*[1], [1]*Yuzuncuyil Univ.Faculty of Agriculture, Dep.of Animal Scie, Van, Turkey,* [2]*Ankara Univ., Faculty of Agriculture, Dep.of Animal Scie., Ankara, Turkey*

There are 25,173,706 of sheep which produce 771,716 ton milk and 216,730 ton meat yearly in Turkey. The largest part of sheep population is composed of native breeds which are well adapted to restricted feeding conditions of Anatolia but they sustain their productivity.Among principal sheep breeds, Akkaraman is well known. One of its varieties, called Norduz sheep appear in Norduz area situated in Van province in the eastern part of Turkey. Fleece is usually white, sometimes ash and grey-white scarcely. Also, black flecks can appears in head, chest and legs which are relatively longer and covered entirely by fleece, beside the neck. Males have usually horns, whereas half of females have it. They have fat tail composed of three parts.The birth weight for female, male, single and twin are 3.99-4.06, 4.31-4.34, 4.57-4.65 and 3.73-3.75 kg respectively. Average body weight, height at withers, body length, chest width, chest depth,chest girth, hind leg circumference for adult females are 57.9-63.5kg, 70.8-71.2cm, 67.6-67.7cm, 18.3-18.4 cm, 33.20cm, 97.8-98.0cm and 66.10-66.9cm.They have coarse mixed wool. greasy fleece weight, trees and fibre length, fibre diameter and wool efficiency are 567 g, 4.92cm, 10,9 cm, 36.6μ and 86.8% respectively. Average length and milk yield of lactation are 183 days and 133 kg. fertility 89-99%, twin birth 11%.The viability rate of lambs until 180th day is 91%.

Some breeding characteristics of Norduz goat

I. Daskiran[1], *F. Cedden*[2] *and M. Bingol*[1]*, [1]*Yuzuncuyil Univ., Faculty of Agriculture, Dep. of Animal Scie., Van, Turkey,* [2]*Ankara Univ., Faculty of Agriculture, Dep. of Animal Scie., Ankara, Turkey*

There are more than 6.5 millions of hair goat which produce 274.350 tons milk yearly in Turkey. Goats raised in South-Eastern part of Turkey, involves about to 26% of total goat population in Turkey. There is a goat breed raised in Norduz area of Van province which is described differently from hair goat. This breed is called Norduz goat. The main color of Norduz goat is black. But, white, cream, black-white, grey, ash, roan and Brown may appear. Males have robust, long and upward horns, whereas females have gently curved ones. Average birth weight in female, male, single and twin kids are 2.7 ± 0.08, 3.0 ± 0.06, 3.0 ± 0.08, 2.8 ± 0.06 kg; weight at weaning 18.4 ± 0.84, 21.1 ± 0.56, 20.0 ± 0.84 and 19.2 ± 0.56 kg respectively. Body weight, body lenght, height at withers, chest width and hind leg circumference of adult female goats are 43.7-55.7 kg, 66.1-69.6 cm, 63.9-66.8 cm, 79.8-88.6 cm and 52.9-55.4 cm respectively. Some physical characteristics of down and coarse fibre are found as follows: Fibre diameter 18.65 and 65.95; fibre length 27.88 and 73.6; breaking strength 4.69 and 18.86; elongation rate 33.79 and 20.79.

Genetic parameters estimation of milk production and prolificacy from Serra da Estrela sheep breed using bayesian analysis and Gibbs sampling

J.B. Oliveira[1], L. Varona[2]*, R. Dinis[3] and J.A. Colaço[4], [1]Agrarian School of Viseu, 3500-606 Viseu, Portugal, [2]UdL-IRTA, Lleida, Spain, [3]National Association of Serra da Estrela Sheep Breeeders, 3400 Oliveira do Hospital, Portugal, [4]University of Trás-os-Montes e Alto Douro, 5000 Vila Real, Portugal

The Serra da Estrela sheep breed is considered as the best Portuguese milk sheep breed, with both the White and Black strains.

The main products resulting for the exploitation of these animals are milk (transformed in the Serra da Estrela Cheese) and meat (lambs with about 30-40 days old). Both of them are Protected Designation of Origin (PDO).

The main goal is the joint estimation of genetic parameters for prolificacy (PROL) and standard 120 days milk yield (PL120) applying a threshold-linear mixed animal model, with a latent continuous variable through data augmentation associated to the first trait, by a Bayesian approach and Gibbs sampling.

The estimated posterior means of the genetic parameters for the White/Black strains were, respectively, to the PROL and PL120 variables: 0.06/0.07 and 0.12/0.14 for heritabilities; 0.09/0.10 and 0.14/0.18 for repeatabilities; 0.44/0.28 for the additive genetic correlations; 0.83/0.85 for the permanent environmental correlations and -0.01/-0.02 for the residual correlations.

From the results of this study, we can conclude that it is possible to achieve genetic progress for milk yield with positive genetic correlation in prolificacy.

Results of crossbreeding between Belgian Texel and Moroccan locale breeds of sheep: Carcass characteristics and meat composition

M. El Fadili[1] and P.L. Leroy[2]. [1]Institut National de la Recherche Agronomique, Maroc, [2]Faculté de Médecine Vétérinaire, ULg,, Belgique

An experiment was carried in order to evaluate the performances of Belgian Texel (BT) rams and their progeny when mated to Moroccan local ewes. 50 lambs sired by BT, Timahdite (T) and D'man (D) rams were slaughtered and compared for carcass characteristics using the GLM-SAS procedure. Meat composition was obtained for each genotype using a half-carcass complete dissection of a male lamb.

Results show that, crossbred lambs sired by BT rams were youngest (-7 days) and heaver (+1.7 kg) at slaughtering and have a better dressing-out percentage (+1.2%), when compared to T and D purebreds lambs. They also deposit less mesenteric fat (-197 g) and their carcasses were more compact (+4%) and well conformed (+1.5 points). The BT crossbred carcasses were also shorter (-6 cm) and larger (+2 cm). Furthermore, the half-carcass complete dissection for each genotype indicated that the meat composition by genotype were respectively for mussel, fat and bone: 62.4, 7.62, and 20.10% for TB x DT, 64.96, 7.88 and 19.03% for TB x T, 60.71, 8.51 and 22.35% for T and 57.67, 11.16 and 20.70% for D. These results indicate that BT breed and their progeny have well performed under Moroccan management conditions. Thus BT breed could be considered to improve sheep meat quality in Morocco especially by its utilization in crossbreeding.

Effects of sheep housing on productive traits in village managementt system
M. Karami and F. Zamani, Scientific members of agriculture and natural resource research center of shahrekord, Iran

This study was comparing the industrial (ISH), reconstructed (RSH) and traditional sheep housing (TSH) for different characteristics constant: building (construction Index), healthy, ventilation and etc on sheep productive traits in village management system, we randomly selected 13 flocks with 3578 sheep for recorded: 3 flocks on TSH, 6 flocks RSH and 4 flocks ISH respectively. The data were used about construction and equipment of sheep house and productive traits and for statistical analysis used of SAS package. Total mean of surface close area was 2.05 m^2/head. The mean of closed house in TSH was 0.94 m^2/head (smallest) and TSH and ISH were 3.51 and 3.71 m^2/head respectively (largest), between TSH and RSH, ISH were significant (P<0.05). The least of yard area was TSH (0.73 m^2/head) and two other groups RSH (3.36 m^2/head), ISH (5.13 m^2/head) were maximum. The mean of length, high and depth of feeding trough and feeds storage ISH were higher than other housing (P<0.05). Reproductive of ISH were higher than TSH (P<0.05). Number of feedstuff (forage and grain) used of ration ISH and RSH were higher than TSH. In conclusion in this experiment obtained that the better and the first housing was (ISH) and then the second was (RSH) and in total two houses were better than (TSH).

Comparison of ranging rams after experimental and natural infections with *Haemonchus contortus*
J. Lüdicke[1], G. Erhardt[1] and M. Gauly*[2], [1]University of Giessen, Institute for Animal Breeding and Genetics, 35390 Giessen, [2]University of Göttingen, Institute for Animal Breeding and Genetics, 37075 Göttingen, Germany

Estimations of resistance to endoparasites in sheep have been made either based on experimental or natural infections. The results may be different and their interpretation for breeding programs may be difficult. Therefore Merinoland sheep (n = 49), Rhönsheep (n = 62) and their reciprocal crosses (n = 180) descending from 7 Merinoland and 7 Rhönsheep rams were included in an experimental resp. natural infection. One group was experimentally infected with 5000 *Haemonchus contortus*-l_3-larva and kept indoors while a second group was turned out on pasture after weaning at an age of 12 weeks. After 4 and 8 weeks individual blood and feacal samples were taken for feacal egg count (FEC) and haematocrit. The influence of the group was highly significant on haematocrit and log FEC after 4 weeks as well as on weight, daily gain and log FEC after 8 weeks. Interaction between group and week was significant for log FEC for the second sampling. Differences between breeds and sexes were equal in both groups. On the basis of their offspring the rams were not ranked significantly different in both groups concerning log FEC. Therefore it can be concluded that the estimation of genetic parameters for *Haemonchus contortus* infection can be done on experimental or natural infection.

Phenotypic and genetic parameters of milk traits of German fawn goats in Serbia
M. Ćinkulov[1]*, S. Trivunović[1], M. Krajinović[1], A. Popović-Vranješ[1] I. Pihler[1], K. Porcu[2], [1]University in Novi Sad, Faculty of Agriculture, 21000 Novi Sad, Serbia and Montenegro, [2]University „Ss. Cyril and Methodius" Faculty of Agriculture, 1000 Skopje, R. Macedonia

Records of milk traits (length of lactation, milk, fat and protein yield, fat and protein contents) of German Fawn goats (100) which were bred in Serbia were analyzed with mixed linear models in order to estimate the effects of year and season of kiding, number of born kids and order of lactation on milk traits. All factors considered in the models affected milk traits significantly. Significant influence of year and season of kiding was just to length of lactation. Yield of milk, fat and protein was lower in first parity goats than in latter parities. Fat and protein contents showed the same trend. Goats with more than one kids at parturition had higher milk, fat and protein yield than goats with one kid.

The flock with 50% of goats in first lactation and with 622 kg average milk yield present good base for further selection. The same suggest obtained high values of heritability coefficients for milk, fat and protein yield (0.727, 0.704 and 0.888).

The effects of North American and South African genotypes on Turkish Angora goats
H. Gunes, Istanbul University, Faculty of Veterinary Medicine, Department of Animal Breeding, 34320 Avcilar, Istanbul, Turkey

Angora goats have developed their characteristics around Ankara which was used to be known as "Angora". Angora goats are being bred for the production of mohair. Although its origin is Anatolia (Ankara) the production characteristics of Angora goats in other countries, particularly in North American (Texan) and South African are higher than Turkish Angora goats. Studies are being carried out since 1982 in Anadolu (Cifteler) State Farm to improve the production characteristics of Turkish Angora goats. The aim of the study here is to investigate the effects of firstly North American and then South African genotypes separately on Turkish Angora goats. In the statistical analyses of data general linear models (GLM) procedure was used.

The use of North American genotype provided an evident improvement in the production and quality of mohair but not in the body improvement and live-weight of the animals in this population. Similarly, South African genotype had an improvement effect on mohair production but not on live-weight in the crossbred generation

As a result, it was determined that to improve the production characteristics of Turkish Angora goats, besides using the advantages of North American and South African genotypes, more emphasis should be given on the selection and culling procedures, two shearings per year should be applied and husbandry and feeding practices should be improved.

Investigation of the sheep production systems in Greece using multivariate techniques
Ch. Ligda[1], *G. Kliambas*[2] *and A. Georgoudis*[2], [1]*National Agricultural Research Foundation, 57 001, Thessaloniki,* [2]*Aristotle University of Thessaloniki, Faculty of Agriculture, Dept. of Animal Production*

In this study the results from the analysis of technical and economical data of certain sheep farms of Greece are presented. In total data from 108 sheep farms of 10 local breeds were analysed, mainly located in mountain and marginal areas. The variables used were referring to the total area of the farm, the altitude of the farms, the number of sheep and/or goats, the percentage of income coming from agricultural activities, months of grazing, information on feeding, on the veterinary expenses. Principal components analysis was used in order to restructure the observed variance to linear combinations of the selected variables. Four principal components were extracted explaining the 72.7% of the total variance. In the next step, the cluster analysis of the new uncorrelated variables has resulted in five groups of farms, where in the two of them the great majority of the farms was placed.

The reproductive and productive performance of Zaraibi goats under two production system in Egypt
W.M.S El-Desokey[1], *M.R. Anous*[1], *M. Mourad*[1] *and E. B. Abdalla*[1], [1]*Ain Shams University, Faculty of Agriculture, P.O. Box 68 Hadayek Shoubra, 11241 Cairo, Egypt*

A total of 20 does of different ages and 66 male and female kids of the Egyptian Zaraibi breed, from an experimental flock belonging to the Faculty of Agriculture, Ain Shams University, were used for a period of 2 years to evaluate their reproductive and productive performance. They were raised under two production systems; intensive and semi-intensive. The criteria used in the evaluation were litter size, total milk yield and lactation period for does and average daily gains and body conformation indices for kids. GLM procedure and the mixed model procedure with repeated statement were performed to analyze the data using different models according to the type of the studied traits. Production system had the greatest effect on all the studied traits (P<0.01) compared to the effect of the other considered factors. Does raised under the intensive production system had greater litter size, higher total milk yield and longer lactation period than those raised under the semi-intensive production system. They also produced kids with higher growth rate and body fleshness and better body compactness.

Maternal behaviour in sheep and the effects on lambs productivity

C. Schichowski, E. Moors and M. Gauly, Institute of Animal Breeding and Genetics, University of Goettingen, Albrecht Thaer Weg 3, 37075 Goettingen, Germany

After lambing ewes (n = 602) of different breeds were placed with their offspring in a lambing pen for 48 hours. At the second day of life lambs (n = 1003) were taken out of pens, placed in front of them and were then returned. Based on the observations three scores for maternal behaviour were given: a score during separation (BdS), a vocalisation score dS (VdS) and a score after separation (BaS). 71.8 % of the ewes showed agitation after lambs were separated. 87.2 % of the ewes gave sound during separation and 96.2 % were sniffing and licking their offspring after they were returned. Breed and age of ewes had a significant ($p < 0.001$) effect on BdS and VdS. Ewes being between one and two years old showed in average the lowest interest in offspring during and after separation. Highest interest were recorded for ewes being five years and older. Littersize only significantly effected the BaS ($p < 0.01$). Sex of the lambs didn't significantly ($p > 0.05$) effect any of the scores. The lowest birth weights (4.39 kg) were recorded for lambs whos mothers showed no agitation in the BdS, whereas the highest birth weights were associated with ewes showing agitation and fence climbing during the test.

Activity and gene expression of the main lipogenic enzymes in growing lambs

A. Arana*, B. Soret., J.A. Mendizábal, J. Egozcue and A. Purroy, Departamento de Producción Agraria. Universidad Pública de Navarra. Pamplona. Spain

The aim of this work was to study the changes in the adipose tissue metabolism in the perirenal (PR), omental (OM) and subcutaneous (SC) through the development of growing lambs. The amount of fat, the size and number of adipocytes, the activity of the lipogenic enzymes G3-PDH, FAS and G6-PDH and the gene expression of the enzymes G3-PDH, G6-PDH, ACC, LPL and SCoA in Navarra breed lambs slaughtered at 12 kg LW (n=6) and at 25 kg LW (n=6).

Results show that the development of PR depot was mainly due to the hypertrophy of the adipocytes, as there was a strong correlation between the amount of fat and the diameter of the adipose cells (r=0.88; P<0.001) but there were no changes in the number of adipocytes. In the OM and SC depots the increase in the amount of fat was achieved by both an increase in cell size (r=0.86 and r=0.85, respectively; P<0.001) and in cell numbers (r=0.90 and r=0.64; P<0.001 and P<0.05; respectively). It was observed that in the three depots, nevertheless it occurred hypertrophy and/or hyperplasia, there was an increase in the lipogenic enzymes activity (P<0.05) which was not mirrored in all cases by an increase in the gene expressions.

Estimating body weight of Creole and Creole x Nubian goats from morphometric measurements

S. Vargas[1], A. Larbi*[2], M. Sánchez[1], [1]Department of Animal Production, University of Cordoba, Avda Medina Azahara, 14005, Cordoba, Spain, [2]International Center for Agricultural Research in the Dry Areas (ICARDA), P. O. Box 5466, Aleppo, Syria

The aim of the study was to quantify the morphological characteristics and to determine the relationship between morphological characteristics and liveweight of native Creole goat breeds and the first (F_1) and second (F_2) generations of Creole does and introduced Nubian bucks. Live-weight (LW), head length (HL), head width (HW), body length (BL), trunk length (TL), chest girth (CG), abdominal perimeter (AP), rump length (RL) and height at withers (HW) of 435 does and 34 bucks of Creole and crosses between Creole does and Nubian bucks aged 18-60 months were recorded from 24 goat herds in six small-scale agro-pastoral communities in Puebla State of Mexico from 1997-2000. Bucks had higher (p<0.05) body measurements than does. The F_2 does had higher (p<0.05) body measurements and LW (28.8±1.5 vs. 27.7±0.5 kg) than the native Creole does, with the exception of TL. Body traits of the F_1 does were similar (p>0.05) to Creoloe does with the exception of HL, BL, and HR. The live-weight of the Creole and Creole x Nubian crosses was positively correlated with TL, HW, CG, and AP partly suggesting that LW could be estimated from body traits.

Effect of body measurements on lactation and reproduction traits in Hungarian goats

T. Németh[1]*, A. Molnár[1], G. Baranyai[2], S. Kukovics[1], [1]Research Institute for Animal Breeding and Nutrition, Gesztenyés u. 1. Herceghalom 2053, Hungary, [2]Hungarian Goat Breeders'Association, Gesztenyés u. 1. Herceghalom 2053 Hungary

The authors collected two years' (2003 and 2004) lactation yield and reproduction data of 552 does belonging to 5 milking goat breeds (Hungarian Milking White, -Milking Brown, -Milking Multicolour, Alpine, Saanen). At the same time the following measurements were taken: body weight, wither height, body length, thorax depth, thorax width, pelvic width, hip width, head length, ear length and width, and distance between eyes. Correlations among lactation yields/ reproduction traits and body measurements, the effects of breed and age on lactation yield, on litter size and on weaning weight, as well as the repeatability of these traits were studied.
The data collected was processed by using SPSS for Windows 10.0 software (correlation, univariatre and multivariate analysis of variance, paired-samples t-test).
Repeatability of lactation yield was low, significant differences (P≤0.05) were observed between the results of the two studied years. Correlations among body measurements (except wither height) and lactation yields were found in the case of does above 3 years of age. According to our results, the body weight, body length and the thorax measures (depth and width) are elemental factors in the selection for higher milk production in all studied breeds.

The effect of rapeseed meal treated with heat and formaldehyde on Sanjabi sheep performance
M. Moeini, M. Souri and A. Heidari, Razi university, Iran

Sixteen male Sanjabi sheep were used in 110 days study in December 2004. A randomized complete block design with live weight as block was performed. Treatments were soybean meal, untreated, heat and formaldehyde treated rapeseed meal. Rapeseed meal (RSM)) were treated by either dry heat (125 °C for 10 min. in oven) or by formaldehyde (0.4 g / 100 g crude protein).
The diets contained : basal diet + (% 19 soybean meal), (% 25 untreated RSM), (% 25 heat treated RSM), and (% 25 formaldehyde treated RSM). All treatments had 159g kg^{-1} CP, and 2.53 Mcal/kg ME. Dry matter intake, digestibility, nitrogen retention and live weight gain were measured. Wool growth measurements were carried out by taking patch samples (10×10 cm) from the mid-side area of the sheep every 28 days.
The results indicated that the coefficient digestibility of CP, Dm, CF, EE,OM, dry matter intake, daily live weight gain were not significantly different between treatment groups. Fiber diameter in treated group increased (from 32 to 38.5 $^{\mu}$) but was not significantly different. Wool quantity and quality were not significantly affected by using heat or formaldehyde treated rapeseed meal.

Relative feed value of dual-purpose cowpea varieties fodder fed to West African Dwaf sheep
I. Etela[1], A. Larbi*[2] D.D. Dung[3], U.J. Ikhatua[4], U.I. Oji[1] and M.A. Bamikoe[4], [1]Rivers State University of Science and Technolgoy, Port Harcourt, Nigeria, [2]International Center for Agricultural Research in the Dry Areas, P. O. Box 5466, Aleppo, Syria, [3]National Animal Production Research Institute, Zaria, Nigeria, [4]University of Benin, Benin City, Nigeria

In the semi-arid zone of West Africa, cowpea (*Vigna unguiculata*) grains and hulms are main protein source for human and livestock respectively, and source of household income in crop-livestock systems. Improved varieties with potential for both grain and fodder production (dual-purpose) have been developed, but little information on their feed value is scanty. Hulms of four improved varieties were fed as sole diets to West African dwarf (WAD) rams over a 60-day period using a randomized complete block design with 6 animals per variety to determine differences in feed value among varieties. Significant (P<0.05) differences were recorded among the varieties in organic matter intake (50.5-78.9 g kg W$^{-0.75}$ d^{-1}) and digestibility (377-591 g kg^{-1} DM), crude intake (6.8-14.7 g kg W$^{-0.75}$ d^{-1}) and digestibility (400-613 g kg^{-1} DM), neutral detergent fibre intake (30-39 g kg W$^{-0.75}$ d^{-1}) and digestibility (469-646 g kg^{-1} DM), nitrogen balance (0.03-10.5 g d^{-1}), and daily weight gain (3-55 g d^{-1}). The cowpea varieties could be ranked as: IT81D-994 > IT86D-719 > IT89KD-391 > IT86D-716 in decreasing order of feed quality based on daily gain of WAD rams.

Effecting of treated or supplemented wheat besats with urea and molasses on efficiency mixture feeding in Baluchi male lambs

S.A. Shiri* and D.A. Saghi, Iran-Mashhad, Agricultural & natural resources research center of khorasan, P.O.BOX. 91735-1148, Iran

A two steps expriment was conducted to evaluate the effect of wheat besats(W.B) on digestibility and fattening in Baluchi lambs.W.B was mixted with solutions (1 litter/kg,DM) containing 5% urea or 10% molasses. the mixture used aftrer 24 houres (supplemented method) or ensiled for 30 days (treated method). Digestibility patterns of treatments were studied at the first step of the expriment when 20 Baluchi rams were fed with only the treatments(W.B.T with urea and molasses, W.B.T with urea, W.B.S with urea, W.B.S with urea and molasses and W.B). The results showed that the digestibility of DM ranged from 577 to 615, OM from 539 to 614, CP from 592 to 705, CF from 401 to 577, EE from 517 to 649g/kg and GE from 43.6 to 56.9% was influenced by the supplemented or treated method. In the second step the treatments were mixed with a basal diet containing Alfalfa 27%, barly 56.7%, cottenseed meal 6%, suger beet pulp 10%, lime stone %0.2 and mineral %0.1 at the rate of 30:70 and fed in 35 lambs 21.6±3.8 kg for 12 weeks. Daily gains ranged from 111.31 to 151.36g (treatments W.B.S with urea and W.B.T with urea and molasses respectively). The results indicated that the using of supplemented or teated W.B increased apperient digestibility, daily gain and feed intake significantly ($P<0.05$).

Effects of cottonseed meal replacement by Canola meal in the diets of finishing lambs

V. Rezaeipour[1], T. Ghoorchi[1], S. Hasani[1], G. Ghorbani[2] and M. Norozi[3], [1]Agricultural sciences and natural resources of gorgan university, department of animal science,Gorgan, Iran, [2]Isfahan technology of university, department of animal science, Iran, [3] Zanjan university, department of animal science, Iran

An experiment was conducted to investigate the effects of use of canola meal in Atabai finishing lamb diet and some nutritional characteristics of the canoal meal compared with cottonseed meal. A completely randomize design was used with 4 treatments and 5 replications (lamb) in each treatment. The experimental groups had different levels of 0, 33, 66 and 100 percent of canola meal instead of cottonseed meal. Different groups had the same nutrients in the diet. The finishing period lasted 84 days and finally carcass analysis was done for every treatment. The results showed that the effects of different levels of canola meal on daily gain during the whole experimental period, feed efficiency and other carcass characteristics, but liver weight, was not statistically significant ($p>0.05$). However the effect of canola meal on amount of daily feed intake was statistically significant ($p<0.05$). The canola meal used in this experiment contained 14.75 micromole per gram DM aromatic glucosinolates. According to the results in this experiment, as canola meal not only had a diverse effect on the finishing ability of lambs, but also it had good effects in some casas, it is suggested to replace the cottonseed meal by canola meal in the diets of finishing lamb.

Monitoring nutritional status of the goats in organic production
Z. Antunović[1]*, M. Šperanda[1], Đ. Senčić[1], M. Domaćinović[1], Z. Steiner[1], B. Liker[2] and V. Šerić[3], [1]Faculty of Agriculture, Trg sv. Trojstva 3, 31000 Osijek, Croatia, [2]Faculty of Agriculture, Svetošimunska 25, 10000 Zagreb, [3]Clinical Hospital, 31000 Osijek, Croatia

Analyses of body condition scores (BSC) and metabolic profile (MP) were made on 30 French alpine goats on the organic production during early lactation period (in the first 30 days). Goats were fed on the cereals mixture (wheat, rye, corn and wheat bran) and meadow hay. Milk yield was measured at evening and morning milking before blood collection. Blood samples were collected from the jugular vein. BSC was assessed according to Santucci and Maestrini (1985). Average daily milk yield of goat was 2.5 l/day. BSC values were 2.7 (from 2.2 to 3.5). The average blood glucose level (3.17 mmol.l^{-1}) and milk lactose level (4.47%) in the goats confirmed sufficient energy supply, but plasma urea level (2.76 mmol.l^{-1}) and milk urea level (10.97 mg.dl^{-1}) showed lower protein delivery. Mineral concentrations (Na, K, Cl, Ca) and anion gap in the goats blood showed adequate mineral supply. Acid-base status values (pH, pO_2, pCO_2, HCO_3^-) ranged within physiological limits for dairy goats. Results refered on low protein ration composition during early lactation period in organic goats production. BSC, MP and milk composition can be useful parameters for evaluation nutritive and health status of dairy goats.

Cholesterol and fat acids profile in the meat of lambs fed with diets containing fish residue silage
Sandra Mari Yamamoto, Américo Garcia da Silva Sobrinho, Hirasilva Borba Alves de Souza and Carolina Buzzulini, UNESP – São Paulo State University, Zootecnia, Via de Acesso Prof.Paulo Donato Castellane s/n, Jaboticabal, SP, 14884-900, Brasil

Eighteen 7/8 Ile de France 1/8 Polwarth lambs averaging 17 kg, confined in individual cages, were distributed among the following treatments: control diet; 8% freshwater fish (*Oreochromis niloticus*) residue silage diet; and 8% sea fish (*Lophius gastrophisus*) residue silage diet. The fish residue silages have parcially substituted the soybean meal and 40% of corn silage was used as roughage. In the production of fermented fish silage, *in natura* fish industry remains were ground and placed in plastic containers, with 7.50% sugarcane molasses, 2.50% natural yoghurt and 0.125% sorbic acid. Lambs were slaughtered at 32 kg of body weigh and the cholesterol in the muscle *Longissimus lumborum* it was of 53.97 mg/100g. The fat acids found in larger amount in the fat intramuscular of same muscle there were oleic (41.46%), palmitic (25.93%), stearic (19.75%), linoleic (2.96%) and miristic (2.81%). The total of fat acids saturated was larger in the lambs meat fed with fish residues silages, with average of 50.75%, and the meat of the ones what received the control diet was 47.61%. The relationships polyunsaturated:saturated and ω6:ω3 fat acids were not affected by diets, with values of 0.115 and 5.06, respectively. The use of fish residues silages as alternative proteic source in the lambs feeding did not influence the cholesterol and the fat acids composition in the lambs meat.

Digestibility and net energy value of chicory pulp and corn gluten feed: Pregnant sows versus fattening pigs

M.J. Van Oeckel*, N. Warnants, J. Vanacker, M. De Paepe and D.L. De Brabander, ILVO, Animal Science Unit, Scheldeweg 68, 9090 Belgium

The nutrient digestibility, digestible energy (DE) and net energy (NE) of chicory pulp (CP) and corn gluten feed (CGF) for pregnant sows and fattening pigs was studied by means of in vivo trials. Within each animal category, two cross-over trials were carried out with a control diet and a diet composed of 75% control diet + 25% CP or 25% CGF. In each trial 6 pregnant hybrid sows (on average 5.8[th] parity and 211 kg for CP and 3.0[th] parity and 228 kg for CGF) and 6 Piétrain x hybrid pigs (on average respectively 51 and 44 kg for CP and CGF) were involved. Sows were fed at requirement and pigs at 3 times maintenance level. The nutrient digestibility coefficients were not significantly different between sows and pigs, except for the higher digestibility of gross energy, organic matter and crude fibre in CGF by sows vs. pigs. The DE and NE of CGF was 13 and 10% higher for sows vs. pigs, with respectively a DE-value of 12.4 and 11.0 MJ/kg and a NE-value of 7.8 and 7.1 MJ/kg. A similar DE and NE of CP for sows and pigs was found, with on average a DE-value of 7.7 MJ/kg and a NE-value of 6.8 MJ/kg.

Chicory pulp (CP), corn gluten feed (CGF) and rape seed meal (RSM) as energy intake restrictors in gestation diets of sows

M.J. Van Oeckel*, N. Warnants, J. Vanacker, M. De Paepe and D.L. De Brabander, ILVO, Animal Science Unit, Scheldeweg 68, 9090 Belgium

The effect of CP, CGF, RSM and the combination of CP and CGF on the voluntary feed intake of pregnant hybrid sows was studied in three Latin square design trials. The first trial (15 sows) consisted of a control diet, one with 25% CP and one with 50% CP; the second (15 sows) of a control diet, one with 55% CGF and one with 25% RSM; the third (18 sows) of a diet with 44% CP, one with 47% CGF and one with 22% CP and 23.5% CGF. While the net energy (NE) requirement for pregnant sows is about 22 MJ/day (CVB, 2004), the NE intake on respectively the 25%, 44% and 50% CP diets, 47% and 55% CGF diets, 25% RSM diet and 22% CP + 23.5% CGF diet was 30 (s.d. 8), 26 (s.d. 5), 19 (s.d. 9), 35 (s.d. 14), 21 (s.d. 12), 37 (s.d. 10) and 32 (s.d. 9) MJ/day. Only a minority of sows on the 44% and 50% CP diets ate less than 50% or more than 150% of the NE requirements, while this was worse for the CGF, RSM and CP/CGF ration. In conclusion, CP is most suitable as energy intake restrictor for voluntary fed pregnant sows.

Pen fouling and ammonia emission in organic fattening pigs

S.G. Ivanova-Peneva[1]*, A.J.A. Aarnink[2] and H.Vermeer[2], [1]Agricultural Institute, 9700 Shumen, Bulgaria, [2]Wageningen UR, Animal Sciences group, P.O. Box 17, 6700 AA Wageningen, The Netherlands

The aim of this project was to test different designs of outside yards, which influence pen fouling and ammonia emissions in organic pig production. 238 fattening pigs (GYxNL) were studied in each of two growing periods, divided in 8 groups with 15 animals. The following treatments were studded: outside yard without anything in it (NY); outside yard with a rooting trough (RY); outside yard with a rooting trough and a drinking bowl (R+DY); outside yard with a dinking bowl (DY). The fouled floor areas with urine and faeces inside and at the outside yards were scored from 0 to 4. Ammonia emissions were calculated on the base of pen fouling and according to a previous study on practical farms. Inside pen fouled floor area was on average 19.0%, which was much lower than area outside - 69.3%. Clear differences in calculated ammonia emissions from the floor of outside yards between treatments were found ($P<0.001$). Outside air temperature and air speed had significant effects on calculated ammonia emissions from the floor of the outside yards ($P<0.001$). It is concluded that an outside yard with a partly slatted floor, drinker in one of the corners and a rooting trough in the other could be recommended in practical organic pig farms, because of the least pen fouling and ammonia volatilization.

The environmental impact of increasing weaning age

K. Breuer[1]*, C. Docking[1], F. Agostini[2] and K. Smith[2], [1]ADAS UK Ltd, Terrington St Clements, King's Lynn, Norfolk, PE34 4PW, UK, [2]ADAS UK Ltd, Woodthorne, Wergs Rd, Wolverhampton, WV6 8TQ, UK

Increasing weaning age may be a means for dealing with the ban on antibiotic growth promoters, enabling piglets to develop a more mature gut to cope with the weaning process. The implications for the environment in increasing weaning age are unknown. It was the objective of this study to assess the effect of weaning age on slurry volume and composition, and ammonia emissions.
A total of 36 first parity sows and their litters (Large White × Landrace × White Duroc) were studied over 12 weeks with three weaning age treatments: four (4ww), six (6ww) or eight (8ww) weeks of age. The volume and composition of slurry, ammonia emissions, feed input and growth of animals was recorded. The effect of treatment was analysed with analysis of variance.
Total feed consumed did not differ between treatments. Weaning age did not affect daily ammonia emissions, but less NH_3-N was emitted by the 8ww system for at least part of the production cycle. However, there were fewer animals in this system in our study due the removal of one sow and coincidentally smaller mean litters. An 8ww system may require more breeding animals (to produce the same number of weaned pigs as the earlier weaning systems, hypothesis to be tested) which may increase excretal N output and, therefore, increased potential for environmental emissions. The results suggest that increasing weaning age will not increase the nitrogen load to the environment.

Does an A→G transition in the promoter of Myostatin have an influence on the expression of the gene?

A. Stinckens[1]*, J. Bijttebier[1], T. Luyten[1], M. Georges[2] and N. Buys[1], [1]Laboratory of Livestock Physiology, Immunology and Genetics, K.U.Leuven, Kasteelpark Arenberg 30, 3001 Leuven, Belgium, [2] Unit of Animal genomics, Faculty of veterinary medicine, University of Liège, Boulevard de Colonster 20, 4000 Liège, Belgium

Myostatin (MSTN), a member of the TGF-β family, is an essential factor for growth and development of muscle mass. The protein functions as a negative regulator of muscle growth in mice and is related with the so-called double muscling phenotype in cattle. One particular breed of pigs, the Piétrain pig, also shows a heavily muscled phenotype. The similarity of muscular phenotypes suggested that myostatin could also be a candidate gene for muscular hypertrophy in pigs.

In this study we used Taqman Real-Time assay with Sybr Green to investigate whether or not a previously reported A→G transition in the promoter of the porcine MSTN gene had an influence on the expression of the gene. The real-time study was executed on 24 pigs at 110 kg of different genotypes.

In our population we found that there is an effect of the transition on the expression of the MSTN gene. Whether this difference is consistent in different populations is currently under investigation.

Osteochondrosis lesions: Heritabilities and genetic correlations to production and exterior traits in Swiss station-tested pigs

H. Luther, D. Schwörer and A. Hofer, SUISAG, CH-6204 Sempach, Switzerland

Since 2002, SUISAG records seven osteochondrosis lesions (OC) at front and hind legs within a random sample of all station-tested pigs in Switzerland. OC examinations were performed on 2622 of at total 9511 pigs by trained persons. OC were scored from 1 – 4 at five locations and 1 – 6 at two joints where 1 means "no lesions" and 4 respectively 6 denotes "severely affected". Appreciable phenotypic variation was observed at four locations. At the other three locations less then 2% of animals were gently affected (score 2), respectively. For all locations, there were only small and inconsistent differences between the breeds (Large White and Swiss Landrace).

Linear mixed models were used for variance component estimation. Only three of the four locations with appreciable phenotypic variation showed also considerable genetic variance. Heritability ranged from.16 to.18 for OC at condylus medialis humeri, distal epiphyseal cartilage of ulna and condylus medialis femoris. Genetic correlations of OC at these locations to daily gain (ADG) and percentage of premium cuts (PPC) were low (≤ ±0.21). Genetic correlations to linear description of exterior traits ranged from low to moderate values and were nearly all favourable.

Therefore, it seems to be possible to breed a more productive pig (ADG, PPC) without an increase of OC, if the selection criteria accounts for OC scores or appropriate exterior traits.

Adaptation of the English Large White pig breed in Lithuania
R. Klimas and A. Klimienė, Šiauliai University, P. Višinskio 19, 77156 Šiauliai, Lithuania

The goal of this work was to investigate influence of new environmental conditions on the process of adaptation of English Large White (ELW) pig breed.
The work has been carried out in the years 2000-2005. Reproductive traits of primaparous sows of imported and born and raised in Lithuania ELW of the first (F_1) and second (F_2) generation (n=72; 24 primaparous sows in each group) have been analyzed, as well as fattening performance and meatiness traits of delivered F_1 (n=30), F_2 (n=55) and F_3 (n=54) progeny. It was indicated that litter size of primaparous sows and number of piglets at 21 days of age statistically was not reliably different after environmental conditions changed for imported and born and raised in the country ELW. However depending on adaptation to new conditions, rate of piglets losses was declining. Preserving of offsprings of second generation (F_2) ELW, comparing to imported and F_1 sows, improved by 9.3 % (P<0.01) and 5.8% (P<0.05) correspondingly. Speeder grow of progeny of this breed was indicated, starting from the second generation (F_2). Average daily gain for F_2 and F_3 progeny was correspondingly 15.9 % (P<0.001) and 9.7 % (P<0.01) higher than of F_1 progeny. Though difference is not reliable, however starting from F_2 generation decreasing of muscularity by 0.8 % can be noticed. Thus, new environmental conditions didn't have negative influence on adaptation of pigs of ELW.

Usage of the pig breeds of German origin in commercial pig breeding of Ukraine
A. Getya[1]*, L. Flegantov[2], H. Willeke[3], [1]Institute of pig breeding named after O.V.Kvasnytskyy UAAS, 36006 Shvedska Mogyla 1, Poltava, 36006, Ukraine, [2]Poltava State Agrarian Academy, Skovorody str.,1/3, Poltava, 36003, Ukraine, [3]University of Applied Sciences Weihenstephan, Department Triesdorf, Germany

The pig breeds of German origin are not presented in the breeding programs of Ukraine. The aim of this study was to analyse the effect of usage of the most common German breeds in Ukrainian commercial pig breeding.
In experiment the sows of Ukrainian Large White (ULW) breed were artificially inseminated with sperm of boars of 3 German breeds: German Large White (GLW), German Landrace (DL) and Pietrain (Pi). Backfat thickness over 6/7 thoracic vertebra (BF, mm), average daily gain (ADG, grams), lean cuts (LC, %) and fat cuts (FC, %) were measured.
In the presented study was proved, that the usage of German breeds led to decrease of BF in all cases. The average BF of pure ULW was 35.45±0.40 mm meanwhile the BF of ULW x GLW and ULW x DL were 26.38±0.70 mm and 30.09±0.97 mm respectively. The thinnest BF was obtained by crossing ULW with Pi – 25.39±0.46 mm. No significant differences in ADG between the genotypes could be proved. However, the increasing of LC and decreasing of FC by hybrid pigs were noted. The analysis of variance has shown, that factor of genotype statistically influenced only trait BF (by F=7.883341, p-level=0.000328 and P=0.05) and no other traits.

The effect of growth rates on carcass performance in pigs

J. Citek, R.. Stupka, M. Sprysl and E. Kluzakova, Czech University of Agriculture Prague Department of Animal Husbandry, 165 21 Prague 6-Suchdol, Czech Republic

One hundred and nineteen crossbred pigs were housed 2 per pen and fed ad libitum. The animals were fattened from 30 to approximately 107 kg body weight. Pigs were sorted into four groups according to their ADG (G1: under 849; G2: 850-949; G3: 950-1049 and G4: over1050 g/day). The pigs were slaughtered and carcass data were collected. The analysis suggests differences in carcass performance among the 4 groups of growth rates. G1 had the highest level of meatiness expressed by the proportion of lean meat share (56.81%) compared to G4 (54.44%) (P<0.01). With increasing growth rate from G1 to G4 the proportion of main meat parts in the carcass was decreased (52.61-49.87%) (P<0.05) as well as the proportion of ham (22.23 – 20.75%) (P<0.05) and shoulder (10.37 – 9.82%) (P<0.01). The proportion of fatty parts was increased (18.78 - 20.36 %) from G1 to G4. Proportions of bone parts were equal in all groups (11.48-11.65%). The highest proportions of fat from main meat parts were in G4 (16.90%) while in other groups were equal (15.43-15.46%). On the base of obtained results one could say that growth rate of pig has significant effect on the carcass performance.

Effect of the IGF-II and stress genotype on carcass quality in pigs

K. Van den Maagdenberg[1]*, E. Claeys[1], A. Stinckens[2], N. Buys[2] and S. De Smet[1], [1]Laboratory of Animal Nutrition and Animal Product Quality, Department of Animal Production, Ghent University, Proefhoevestraat 10, 9090 Melle, Belgium, [2]Division of Gene Technology, Department of Biosystems, K.U. Leuven, Kasteelpark 30, 3001 Heverlee, Belgium

A new QTN in the paternally imprinted IGF-II gene with effects on carcass lean content was recently discovered in the pig. The aim of this study was to investigate the effect of this IGF-II allele (Qpat vs. qpat) on carcass parameters in relation with the stress genotype (Nn vs. NN). The animals (n=117) were the progeny of two sires and consisted of castrates and females (n=7-19 per group of IGF-II*stress*sex). Pigs were slaughtered according to live weight (109±7kg). Data were analysed with a univariate general linear model with IGF-II and halothane genotype, sex and sire as fixed factors, their two-way interactions and carcass weight as covariate. Carcass lean content measured by the CGM apparatus, cross-sectional area and weight% of the longissimus and ham weight% were significantly higher for Qpat vs. qpat animals (Δ resp. 4.4%, 2.87cm^2, 0.26%, 0.89%), whereas fat thickness measurements and belly weight% were significantly lower (Δ 3.7mm, 0.48%). Carcass length was not affected by the IGF-II genotype. Significant differences were found between Nn and NN animals as expected, but differences were smaller than the IGF-II genotype effect. No significant interaction effects between stress and IGF-II genotype were found.

Traceability of swine meat: An esperience by EID + DNA in "Suino Tipico Sardo"

W. Pinna[1], M.G. Cappai[1], P. Sedda[1], A. Fraghi[2], S. Miari[2] and I.L. Solinas[3], [1]Sezione di Produzioni Animali del Dipartimento di Biologia Animale, Sassari; [2] Istituto Zootecnico Caseario della Sardegna, Bonassai (SS); [3]Joint Research Center(EU) – Institute for the Protection and the Security of the Citizen- Traceability and Vulnerability Assessment Unit, Via Enrico Fermi 1, 21020 Ispra (VA), Italy

20 suckling piglets, "Suino Tipico Sardo" branded, at 1-7 days old and at an average weight of 2514±895 g, have been electronically identified (EID) by means of intraperitoneal transponders (HDX 32.5×3.8 mm, 134.2 kHz) and from each of them an auricle sample has been kept for DNA analysis. The readability of the transponder of each piglet has been detected by static and dynamic readings, *in vivo* and *post mortem* and was 100%. In the slaughtering chain, from each piglet, slaughtered at 29-35 days of age and at an average weight of 8902±1944 g, a second auricle sample has been kept for DNA analysis. The matching between the two samples of each piglet has been investigated by a genetic test, deploying n. 6 microsatellites ascribed in the FAO-ISAG panel for swine species. The DNA analysis of each electronically identified piglet confirmed the correspondence between the *in vivo* and *post mortem* samples. The pairing of EID+DNA can improve the traceability in the commercial net of the "Suino Tipico Sardo".

A new selection strategy for nucleus gilts to optimize genetic gain and the increase of relationship in the Swiss pig breeding program

H. Luther and A. Hofer, SUISAG, CH-6204 Sempach, Switzerland

Especially in small nucleus populations, it is important to control the increase of relationship effectively to preserve genetic diversity. Using software based on the optimal contribution theory (e.g. GENCONT) would be the best way to optimize genetic gain and rate of inbreeding. But if one has to select new tested gilts within several nucleus farms daily, GENCONT needs too much computing time, especially if all young animals are included.

Therefore SUISAG develops an adapted selection index to select gilts for female replacements in the Swiss nucleus farms. The index will combine the total breeding value and the average relationship of the animal to the complete nucleus population. This index favours gilts with high breeding values and/or few relatives in the population. The index weights will be defined in such a way, that the gilts will be ranked approximately the same way as GENCONT proposes. Calculating huge relationship matrixes is a heavy task. However, it will be possible to calculate the required values on weekends and to store them in the database, because the composition of the population doesn't change a lot within a week.

The new selection strategy and recommendations for breeders will allow us to optimize genetic gain in the Swiss pig breeding program while controlling the increase of relationship within the nucleus herds more accurately and effectively.

Comparison of genetic improvement for litter size at birth by multiple trait BLUP selection in swine herd

M. Satoh, National Institute of Agrobiological Sciences, Tsukuba-shi 305-8602, Japan

Responses to selection for number of piglets born alive (NBA) by total number of piglets born (TNB), NBA, and NBA plus number of piglets born dead (NBD) were compared using accuracy of selection and expected genetic gain calculated from selection index with family information and real response to selection using data generated by Monte Carlo computer simulation. Accuracy for NBA selected by the combination of TNB and NBA was highest in the three selection methods in each family structure. Selection by TNB resulted in the greatest expected genetic gain for TNB among the selection methods. In best linear unbiased prediction (BLUP) selection, genetic gain for NBA accumulated by NBA tended to be similar to that accumulated by the combination of NBA and NBD, and both genetic gains at generation 10 were significantly larger than that by TNB (P<0.001). Accumulated responses selected by two-trait animal model BLUP estimated from genetic parameters with errors were similar to those estimated from true parameters, and there was no significant difference between them. These results indicate that selection by NBA or by NBA and NBD gives more genetic improvement in NBA than that by TNB.

Influence of MYOD family on meat performance of Czech Large White and Czech Landrace pigs

P. Humpolíček*, J. Verner and T. Urban, Department of Animal Morphology, Physiology and Genetics, Mendel University of Agriculture and Forestry Brno, Czech Republic

Since the meat performance has considerable effect on the economic efficiency of pigs companies the genes which can influence that are intensively studied. The objective of our study was to analyse the associations of *MYOD* family polymorphic variants with meat production in Czech breeds of pigs. To verify associations between these polymorphisms and selected meat traits, Czech Large White and Czech Landrace breeds of pigs were tested. There were studied these meat characteristics: weights of neck, loin, shoulder and ham, lean meat content, backfat thickness, intramuscular fat, remission, dry matter content and average daily gain. A mixed linear model (REML) in SAS for Windows 9.1.3 was used to find the association between genes of *MYOD* family and chosen traits. Significant differences were observed between *MYOD1*, *MYF4* and *MYF5* genes and meat production traits. No significance between *MYF6* and performance traits was found out.

This work was supported by Czech Science Foundation (Project No. 523/03/H076) and the Ministry of Education, Youth and Physical Training of the Czech Republic (Project FRVS No. 239/2005).

Studying of some multiple traits' influence on plasma ascorbic acid concentration
E.V. Kamaldinov and L.A. Bykova, Novosibirsk State Agrarian University, The Research Institute of Veterinary Genetics and Selection, 160 Dobrolubova, 630039 Novosibirsk, Russia

It is completely described in literature that it is impossible to take into account the all indices that can influence on trait by using of factor analysis. We supposed that vitamin C level in pig's plasma can depend on multiple factors. Factor analysis (Kim D.O., 1989) was applied to extract the group of indices (hidden factor) and then to classify them. It was also presumed that the hidden extracted factors are affected on other labile organic and inorganic compounds.
The studies were conducted in Tulinskoye farm and Kudryashovskoe pig-raising complex of Novosibirsk region on piglets at two weeks after weaning and on adult animals. Ascorbic acid (AA) plasma level was established by method described by I.P. Kondrakhin (1985) using 2,2'-dipyridyl.
Thus, the purpose of our research was to define factors (by level of influence) that possibly can influence to mean AA value in pigs'. The results of studies are presented by four factor in order that explain the importance of multiple traits (correlating with each other) influence on vitamin. The extracted factors were characterized by the following factors in order of importance: hereditary factor; physiological factor; factor of ascorbic acid oxidation in presence of the catalysts and behavioral factor. Thus, we conclude that hereditary factor most of all influence on vitamin C plasma content in pigs.

Influences of inbreeding depression on the initial growth and blood elements* in Duroc pigs
*K. Ishii[*1], O. Sasak[1]i, O. Kudoh[2], K. Fukawa[2], E. Kobayashi[3], T. Furukawa[1] and Y. Nagamine[1], [1]National Institute of Livestock and Grassland Science, Tsukuba 305-0901, [2]Pig Breeding Center, Central Research Institute for Feed and Livestock, Zen-noh, Kamishihoro 080-1406, [3]National Livestock Breeding Center, Nishigoh 961-8511, Japan*

We investigated the influences of inbreeding depression on the initial growth and blood elements using the piglets, which had different inbreeding coefficient from the same litter. Piglets with high (HIP: inbreeding coefficient=25%) and low (LIP: inbreeding coefficient =0%) inbreeding coefficient were produced in the same litter from SPF breed Duroc sow using mixed semen that was adjusted with the same number of sperm from full sib and non-relative boars to dams. The piglets, HIP and LIP, were identified by DNA markers. The traits, body weight at 1 moth, daily gain(0-30days), hematocrit at birth, total serum protein, serum albumin, serum amylase, and serum urea nitrogen, of LIP were significantly higher than these of HIP. On the other hand, total cholesterol of HIP was significantly higher than that of LIP.

Carcass quality assessment of pigs raised in Lithuania with the aim of carcass classification system improvement

D. Ribikauskienė, Institute of Animal Science of LVA, R. Zebenkos 12, LT-82317, Baisogala, Radviliškis distr., Lithuania

The study was carried out in 2004. Altogether were evaluated 58 000 data of pigs from 61 farm. Studies should be carried out again to evaluate the pig population of the country for the carcass quality, as the average lean meat amount in the carcass has considerably improved; the other reason is to make the new regression equation according to which the output of lean meat in the carcass will be calculated. The purpose of this work is to evaluate pig carcasses, which could precisely reflect the population of pigs breeding in Lithuania and according to this improve pigs' carcasses classification system. In 2004 average pig carcass muscularity was 55.65%. When measured by Fat-o-Meat'er the average back fat thickness at 3rd-4th last ribs was 16.73 mm and the average carcass weight was 75.21 kg. The biggest percentage (40.1%) was of pig carcasses weighting 65-80 kg and having 13-20.9 mm of fat depth.

Effect of the share of meat and of the sex on selected quantitative and qualitative indices of the carcass value

M. Okrouhlá, R. Stupka, J. Čítek, M. Šprysl and E. Kluzáková, Czech University of Agriculture, Faculty of Agrobiology, Food and Natural Resources, Deparment of Animal Husbandry, Kamýcká 129, Prague, Czech Republic

The objective of the work was to determine the influence of the attained share of meat and of sex on selected quantitative and qualitative indices of the carcass value.
116 pcs of abattoir swine of the following genotypes were included into the experiment: (CLW$_s$xPN) x (CLW$_m$xCL), PIC x (CLW$_m$xCL), (HxPN) x (CLW$_m$xCL) and line 38 x (CLW$_m$xCL). The pigs were divided according to the grading criteria – „share of lean meat" into 3 groups, it is 60.0 % and more, 55.0 – 59.9 % and 50.0 – 54.9 % of lean muscle. The right abattoir half was cut into individual „meat parts"- leg, roast meat, shoulder, neck and side. From the main meaty parts (MMP) the samples were withdrawn, homogenized and further subjected to the chemical analysis for of the determination of the content of intramuscular fat (IMF).

Reproducibility of the classification method in the Czech Republic
M. Sprysl, R. Stupka, J. Citek and L. Stolc,Czech University of Agriculture, Department of Animal Husbandry, 165 21 Prague, Kamycká 129, Czech Republic

For lean meat measuring there are used probes based on indirect measuring back fat as well as loin eye area depth (reflectance, optical or ultrasound). Unless the indirect measure is related perfectly, the predicted values deviate from the true value because there are various measurement techniques and skills. To manage these deviations, repetitions and statistical analysis can be used. From this point of the view the aim of the trial was to determine the reproducibility of the classification methods FOM/ZP used in abbatoirs in the Czech Republic. Equation for lean meat prediction (where M = depth-muscle; S = depth-fat) were for
FOM $y = 81.8909 + 0.2006 * M + 14.1911 * \ln S$,
ZP $y = 76.6722 - 1.0485 * M + 0.00794 * M2 - 0.002884 * S2 + 9.0151 * \ln (M/S)$.
The trial has been done without respect to equipment (differences between FOM/ZP) as well as with respect to equipment in one abbatoir where 2 groups of 60 pigs were measured by use of ZP, 2 FOMs' apparatus and by 2 operators. On the base of obtained results one could say, that:
In the range of measuring there were documented a large differecnce between equipments FOM1-2 (2.55%) caused mainly of depth-fat-difference (2.14mm) and depth-muscle (5.17mm). The equation for CR prefer mainly depth-fat as it is obvious.
The size of differences are not affected of sequence of measuring apparatus.
The classification of pig carcass by abbatoir-equipment was always worse.

The effect of growth rate to carcass formation in hybrid pigs
R. Stupka, M. Sprysl, J. Citek and M. Okrouhla, Czech University of Agriculture, Department of Animal Husbandry, 165 21 Prague, Kamycká 129, Czech Republic

The effect of various growth rate on achieved carcass lean meat share characteristics as well as a loin eye area formation (MLLT) during station test was the objective of the trial.
For that purpose 216 pigs of common genotypes used in the Czech Republic were tested. The housing of the animals followed the metodology for testing of purebred and hybrid pigs complying with the principle of 2 pigs/pen (1barrow/1gilt). For the duration of test 1625 measuring were realized. All pigs were fed ad-libitum by common feeding mixtures and were sorted according their growth rate (ADG) into 5 groups (less 750g, 750-850g, 850-950g, 950-1050g, above 1050g), whereas obtained values were listed in the 7 live weight (LW) categories (less 60kg, 60-70kg, 70-80kg, 80-90kg, 90-100kg, 100-110kg, above 110kg). From the obtained results is evident that lean meat share in the carcasses significantly ($P \geq 0.05$resp. $P \geq 0.001$) dropped down with respect to increasing growth rate by all LW-categories as well as depth of loin eye area of MLLT in the spot between 2-3 last rib ($P \geq 0.05$resp.0.001). On the other hand the bacfatthickness in all LW-groups increased significantly ($P \geq 0.05$resp. 0.001).

Carcass value and intramuscular fat content of pigs in dependence on sex and breed type
I. Bahelka, P. Demo, E. Hanusová, L. Hetényi*, Slovak Agricultural Research Authority, Research Institute of Animal Production, Hlohovská 2, 949 92 Nitra, Slovak Republic

Carcass traits and intramuscular fat content of three different pig genotypes [WM - White Meaty x (Hampshire x Pietrain), n = 98; WM x Landrace, n = 38 and WM x (Yorkshire x Pietrain, n = 22] were evaluated. Number of barrows and/or gilts was equal (n = 79). Pigs were slaughtered at average live weight of 109 kg. Day after the slaughter, detailed dissection of four main cuts (shoulder, loin, ham, belly) according to Walstra and Mekus method was done. Carcass weight (CW), weight of half carcass side (WHS), average backfat thickness (BF), lean meat content (LMC), percentage of fatty cuts (PFC) were determined. Intramuscular fat content (IMF) was analysed from samples of musculus longissimus dorsi taken at the last rib.
Significant differences between barrows and gilts in BF (28.95 and/or 25.66 mm, resp.), LMC (52.95 and/or 57.67 %, resp.), PFC (21.55 and/or 17.47 %, resp.) and IMF (2.48 and/or 1.96 %, resp.) were found. Differences between breed types were significant in BF (29.45, 23.62 and 24.09 mm) but not in LMC (55.52, 55.63 and 53.79 %) and IMF (2.22, 2.11 and 2.40 %). The correlation coefficients between traits evaluated were calculated. BF correlated positively with CW (0.488**) and IMF (0.285**) but negatively with LMC (-0.472**). PFC correlated positively with IMF (0.534**) whereas correlation between LMC and IMF was negative (-0.517**).

Effect of rosemary oil administration on the physical properties of the meat of Nero Siciliano pigs
L. Liotta[1], B. Chiofalo[1], E. D'Alessandro[1] and V. Chiofalo[1,2], [1]Dept. MOBIFIPA, University of Messina, 98168 Messina, Italy, [2]Consortium of Meat Research, Sicilian Region, Italy

The influence of dietary rosemary extract (*Rosmarinus officinalis L.*) was evaluated on the physical properties of the meat of 30 Nero Siciliano pigs. During the finishing period (90-days), the animals, 16 castrated males and 14 females, were divided into two homogeneous groups of 15 each one (8 males and 7 females, LW 42±2 kg), which received the basal diet (3% of B.W.) supplemented with (ROX group) or not (CTR group) $1 g \cdot kg^{-1}$ a rosemary oil extract (ROX P® - Sevecom S.p.A.). After the slaughter, pH_1 (45' *post mortem*), pH_u (24 hours *post mortem*), Colour (L*,a*,b*, Chroma and Hue), oxidative stability (TBARs), Cooking losses and Warner-Braztler shear force (WBS) of the *Longissimus dorsi* muscle (T8-L1) were determined. Dietary treatment has positively influenced a* index (CTR 14.51 vs. ROX 13.31; P<0.05), Hue (CTR 0.81 vs. ROX 0.88; P<0.05) and TBARs values (CTR 0.055 vs. ROX 0.044; P=0.25); whereas, pH_1 (CTR 6.19 vs. ROX 6.27), pH_u (CTR 5.54 vs. ROX 5.53), cooking losses (CTR 25.65% vs. ROX 23.08%), WBS (CTR 2.91 kgf/cm^2 vs. ROX 3.67 kgf/cm^2) and Chroma values (CTR 21.22 vs. ROX 21.08) as well as L* (CTR 61.58 vs. ROX 63.11) and b* indices (CTR 15.41 vs. ROX 16.23) showed no significant differences (P>0.05). Results suggest that the rosemary extract could represent a valid alternative to synthetic antioxidants for pig feeding.

Use of ultrasound measurements of the longissimus lumborum muscle to estimate meat percentage in pigs

M. Tyra, B. Orzechowska and M. Różycki, National Research Institute of Animal Production, 32-083 Balice n Krakow, Poland

The study involved a total of 412 animals (Polish Large White, Polish Landrace, Pietrain and line 990 gilts) tested in 2002-2005 at the Pig Slaughter Testing Station.

Ultrasound measurements of the *longissimus lumborum* muscle and backfat thickness were performed at 100 kg body weight. The measurements were taken using an ALOKA SDD 500 device with a 3.5 MHz linear array transducer. Ultrasound images were archived on a computer disk for further dimensioning using Multiscan software.

Based on the measurements obtained, meat percentage was estimated using the station (based on dissection), live (Piglog 105) and ultrasound methods. In the ultrasound method, cross-section measurements of the *longissimus lumborum* muscle (backfat thickness, area of fat around the muscle and loin eye height) were taken into account. The results of meat percentage measured on live animals (Piglog and Aloka) were similar at 57.1 and 57.7%, respectively, while meatiness based on dissection was slightly higher (60.6%).

The correlations estimated between live and dissection measurements for meat percentage showed a slight advantage of the ultrasound technique over Piglog 105 measurements. The respective correlations were .788 and .756, which is evidence that these devices are useful for measuring meatiness in pigs.

Effect of daily gains on the histochemical profile of *M. lonissinus lumborum* in Polish Landrace pigs

B. Orzechowska[1], D. Wojtysiak[2], W. Migdał[2] and M. Tyra[1], [1]National Research Institute of Animal Production, 32-083 Balice n Krakow, Poland, [2]University of Agriculture. Al. Mickiewicza 24/28, 30-059 Krakow, Poland

Polish Landrace pigs, tested in pig slaughter testing stations were investigated. Animals were fed and housed according to the station methodology. The study involved three groups of animals differing in daily weight gains during fattening (25 to 100 kg). Average daily gains were above 880 g, 771-879 g and below 770 g in groups I, II and III, respectively. The purpose of this study was to determine histochemical characteristics of *m. longissimus lumborum* in relation to average daily gain of Polish Landrace pigs. Samples of muscle were sectioned in a cryostat and stained by histochemical reaction for the mATP-ase activity to distinguish the types of muscle fibre: I – red fibres, IIA – intermediate fibres and IIB – white. Muscle fibre percentage and diameter were estimated. The results of these investigations showed no significant differences in muscle fibre percentages between the examined groups of fatteners. On the other hand, daily gains significantly affected fibre type and size. Muscles from slow-growing fatteners (group III) had significantly ($p < 0.05$) smaller diameter of type IIB fibres (67.52 µm) compared to the muscles from fast-growing pigs (group II – 81.21 µm and group I – 84.54 µm), between which no significant differences were observed. Moreover, the increasing average daily gain of fatteners significantly ($p < 0.05$) increased the diameter of type IIA fibres to 52.01, 56.38 and 64.74 µm in groups III, II and I, respectively.

Relationships between growth rate and intramuscular fat content of the *M. lonissinus lumborum* in Polish Landrace and Puławska pigs
M. Tyra, B. Orzechowska and M. Różycki, National Research Institute of Animal Production, 32-083 Balice n Krakow, Poland

The study involved a total of 516 animals (Polish Landrace and Puławska gilts) tested in 2003-2005 at the Pig Slaughter Testing Station in Chorzelów. Animals were kept in individual pens and fed *ad libitum* with finely pelleted mixture provided from automatic feeders.
Animals were divided into groups according to weight gains achieved during the test (25 to 100 kg of body weight). Groups I, II and III contained gilts with daily gains of >900, 800-900 and <800 g, respectively. Animals were slaughtered in accordance with the station methodology after reaching 100 kg of body weight. The intramuscular fat content of the *longissimus lumborum* muscle was determined with the method of Soxhlet using samples taken 24 h postmortem.
The average content of intramuscular fat for the Polish Landrace breed was 1.83%. Animals with the highest gains (group I) had the intramuscular fat content of 1.65%, compared to 1.89 and 1.92% in those from groups II and III, respectively. The observed differences between the animals from group I and the other groups of animals were highly significant.
The content of intramuscular fat was higher for Puławska than for Polish Landrace animals and averaged 2.46%. In this breed, no animals with weight gains above 900 g were found. This parameter was 2.36% in group II and 2.50% in group III, although differences between the groups were not significant.

Relationship between meatiness traits of live pigs
A. Klimienė and R. Klimas, Šiauliai University, P. Višinskio 19, 77156 Šiauliai, Lithuania

In selection of pigs it's important to consider such selectional and genetic parameters as correlation coefficient. According to the State Pig Breeding Station data, analysis of meatiness traits (backfat and loin lean thickness, lean meat percentage) of purebred pigs, grown in the Lithuanian breeding centres and evaluated by apparatus *Piglog 105,* has been carried out. Correlation (r) of meatiness traits and its dependence on the age and live weight of breeding progeny was determined and analysed for Lithuanian Whites (LW), Swedish Yorkshires (SY), Danish Large Whites (DLW), German Large Whites (GLW), English Large Whites (ELW), German Landraces (GL), Finish Landraces (FL), Norwegian Landraces (NL), Danish Landraces (DL), Pietrains (P) and Durocs (D). 7620 pigs of mentioned breeds were used for that purpose. At the breeding centres measurements were taken for breeding pigs at 85-110 kg live weight.
Analysis of correlation between meatiness indicators of progeny of different breeds, evaluated in breeding centres with Piglog *105,* showed that lean meat percentage of pigs is more related to their backfat thickness (r= from −0.78 to −0.95, P<0.001), than to loin lean thickness (r= from 0.11 to 0.47). Gaining of backfat thickness in one point of the back was followed by analogous process in the other point of the back (r= from 0.62 to 0.84, P<0.001). It was also indicated, that live weight of pigs has more influence on mentioned meatiness traits than age.

Effect of stalling type on pork meat quality
S. Moro[1]*, G.L. Restelli[1] A. Montironi[2], A. Stella[2], G. Pagnacco[1], [1]V.S.A. Faculty of Veterinary Science, 20133 Milan, Italy, [2]Parco Tecnologico Padano, 26900 Lodi, Italy

The type of housing, in particular the flooring, may be among the management factors that have a major effect on the quality of pork meat. To test this hypothesis, we collaborated with a single farm that had facilities with both slatted and non-slatted concrete floors. A sample of 234 animals was split into two groups and raised on the two flooring types. The animals were from 47 sows and 7 sires and paired by sex and initial health status prior to group assignment. Animals were assigned to treatments after weaning. At slaughtering, 4 carcass measurements (carcass weight, mean ham weight, and thickness of the backfat and the subcutaneous fat covering the hams) and 8 subjective scores (describing colour, marbling and tenderness) were recorded for each individual. Statistical analysis was performed using a mixed model that included the fixed effects of age at slaughtering and flooring type and the random effects of sire and dam. Highly significant ($p < 0.0001$) effects were found for the type of flooring on carcass weight, mean of leg weight and ham-fat covering. Significant values ($p < 0.05$) were also found for leg-fat covering and marbling. Slatted floors were associated with poorer meat quality.

The interrelation of the piglets' response to the low level laser (LLL) and their behavioural reactions
A.I. Serzhantova, Novosibirsk State Agrarian University, The Research Institute of Veterinary Genetics and Selection, 160 Dobrolubova, 630039 Novosibirsk, Russia

The paper regards the study on infrared LLL influence (890 nm) upon Large White suckling pigs of different behavioural status. We aimed to study a dependence of the piglets' physiological and biochemical status changes (under the LLL effect) on their behavioural reaction to the researcher. At the time of the experiment we took into account the animals with the extreme reactions – calm piglets and strongly excitable ones. Thus, there were experimental (irradiated) and control (intact) groups with 2 subgroups inside (calm and excitable, respectively). The experimental group piglets were irradiated thrice in the biologically active lung meridian area.

After the course of irradiation and observation biochemical analyses of blood samples were conducted, which showed the distinction in cholesterol level of calm and excitable piglets in experimental group (2.2 and 3.2 mmol/l, respectively, $p<0.05$).). As the high cholesterol value could be the sign of the felt stress, we can see that the laser therapy procedure serves as a stress factor to the excitable animals while areactive piglets stand it easy enough.

Moreover, phenotypic variability of other biochemical indices of the calm piglets was visibly lower than that of the excitable animals.

It is possible that the calm animals respond to the laser energy with more quick adaptation of metabolic processes than reactive ones.

Effects of housing wild boar during farrowing and lactation on sow and litter behaviour, piglet survival and growth rates

G.E. Maglaras[1,2]*, G. Vatzias[1], A.P.Beard[2], E.Asmini[3] and S.A. Edwards[2], [1]Laboratory for Reproduction of Domestic Animals, T.E.I. of Epirus, Greece, [2]School of Agriculture, Food and Rural Development, University of Newcastle upon Tyne, U.K, [3]Laboratory of Statistics, T.E.I. of Larisa, Greece

Piglet mortality is an important issue for both welfare and production. The aim of the present study was to characterise wild boar sow and piglet behaviour with respect to piglet survival. Data from 30 multiparous wild boar sows and their 164 piglets, in two different trials, were used. Sows were allocated between different farrowing environments (unimproved outdoor group paddocks – OUT, or fenced individual pens with housing -IN). Detailed behavioural observations were made for both sows and piglets until weaning. IN sows were more often observed lying, sitting and drinking ($P<0.005$) in both trials. OUT sows performed more walking (20.7% vs 5.1% and 24.3% vs 5.3% of observations) in both trials, standing (13.0% vs 8.9%) in trial one and nosing (4.1% vs 2.4%) in trial two. Piglet mortality was higher for OUT piglets (12.1% vs 8.6%) in trial one but lower ($P<0.05$) for trial two (8.4% vs 14.1%). Growth rates of IN and OUT piglets were similar in both trials ($P>0.05$). These results indicate no simple relationship between wild boar sow and piglet welfare and their initial farrowing environment.

The damage and management of wild boar (*Sus scrofa* L.) in agricultural areas of Turkey

A. Yilmaz, Ankara Plant Protection Research Central Institue, Entomology, Bağdat Cd. No 250 Yenimahalle, 06172 Ankara, Turkey

Wild Boar (*Sus scrofa* L.) cause the important damage in cultural plants in Turkey. Any chemical treatment has not been applied against the pest because of the negative effects to wild life. In this article, the cultural plants and the area damaged by wild boar, the number of wild boar killed by driving and the bullet number used were presented for last three years across Turkey.

Genetic associations among traits from foals and mares evaluations with regard to optimizing the breeding program in the Trakehner Population

R. Teegen*, A. Jedruch, C. Edel, G. Thaller and E. Kalm, Institute of Animal Breeding and Husbandry, Christian-Albrechts-University, 24098 Kiel, Germany

The objective of this study was to estimate genetic parameters of traits from the conformation tests of foals and mares in the Trakehner breed. The analysed data consisted of 12,540 records of foals and 5,980 records from mare conformation tests. Heritabilities for traits in the foals' evaluations ranged from 0.19 (walk) to 0.48 (type), in the mares' evaluations from 0.22 (correctness of legs) and 0.49 (trot). Genetic correlations between traits of foal evaluations were high (0.43-0.93). The same was true for traits of mare evaluations (0.42-0.88). Highest genetic correlations in both test forms were found between the trait 'overall impression' and the conformation of the body (foals: 0.93, mares: 0.88). Genetic correlations between corresponding traits from foals' and mares' tests were high (0.70-0.91), but indicate a different genetic background, whereas phenotypic correlations were low (0.20-0.30). The results indicate that the information from the foals' might be of some value in a general breeding framework, if some emphasis on conformation traits. Not only because of high genetic correlations but also because that this information is available early in life and even on animals not entering the breeding scheme later. However the true value of this information can only be appreciated in the context of optimizing the breeding strategy in this breed.

Early criteria for selection of jumping ability

B. Langlois, C. Blouin and E. Barrey, INRA-CRJ-SGQA 78352 Jouy-en-Josas France

The aim of this study was to evaluate whether early measurements of morphology and gaits can be used to predict jumping performance. One hundred seventy-two 3 year-old French saddle horses participating in breeding events from 1998 to 2000 were tested for morphology and gaits. Their performance records in jumping were collected from 1999 to 2003. The Equimetrix™ gait analysis system provided 74 variables (10 for walking, 10 for trotting, 18 for free jumping and 36 for morphology). We report here a discriminant analysis of these horses. Three groups were made according to the jumping competition results: (a) those horses having no earnings (b) those belonging to the 50% lower earnings and (c) those having 50% higher earnings. The three groups could be discriminated with less than 3% of mistakes when using all the information on the 74 variables. The first axis allowed isolating the (c) horses and the second axis allowed separating (a) and (b) horses. However this result must not be over-weighted. The same analysis, on a stepwise mode, when introducing only significant variables led to a mean canonical R^2 of 0.36 only, instead of 0.94 when all 74 variables were analysed. For the 14 significant variables then retained, only 5 concerned conformation, the other variables concerned gaits (6) and free jumping (3). In conclusion, for saddle horses, predicting jumping ability in competition just by a morphological exam is illusory. The accuracy increases by measuring kinetic variables of gaits and free jumping. However, even if the morphology and gait test could be improved, the predictive value is limited.

Genetic parameters of Hungarian Sporthorse mares

J. Posta*, I. Komlósi and S. Mihók[1], University of Debrecen, H-4032 Debrecen Böszörményi út 138, Hungary

Mare performance tests for the Hungarian Sporthorse population were evaluated. Data from 1993-2004 were used, covering scores of 435 3-year-old and 240 4-year-old mares. Eighteen subjectively scored traits were considered, that were scored on a 0–10 scale. The animal model for the evaluation of the test results included the fixed test year, age, owner and the random animal and error effects. Variance and covariance components were estimated with VCE-5 software package. Heritabilities ranged from 0.28 (neck) to 0.53 (saddle region) for conformation traits, from 0.29 (jumping style) to 0.52 (jumping ability) for free jumping traits and from 0.22 (walk) to 0.51 (canter and test rider's score) for movement analysis traits. Phenotypic correlations ranged from 0.10 to 0.63 for conformation traits, from 0.20 to 0.82 for free jumping traits and from 0.26 to 0.66 for movement analysis traits. Genetic correlations for conformation traits varied from low to high. For free jumping traits genetic correlations were high. Genetic correlations between movement analysis traits were moderate to high. Positive genetic correlations were found between movement analysis traits and jumping style and jumping ability (0.42–0.87), thus breeding for both characteristics is facilitated.

Determination of horse jump parameters over upright and spread fences by use computer's scanning technique

S. Pietrzak, K. Strzelec and K. Bocian, Agricultural University, Horse breeding and use, Akademicka str. 13, 20-605 Lublin, Poland

The aim of this paper was to define the jump parameters in horses jumping over upright and spread fences. The research was conducted by means of computer scanning of digital images and it focused on 371 horses and 995 starts in competitions with obstacle height ranging from 100 to 140 centimeters organized in Poland between 2004 and 2005. Digital record of the whole performance helped us establish the general number of strides in the course and note horses' faults. Images with 5 stages of jump (over oxer) and with 3 stages (over vertical) were saved to computer by means of MultiScanBase. It was concluded, that older horses made fewer strides and the percentage of rhythm loss was lower than with younger horses. We managed to calculate a statistically crucial difference between the number of fore legs faults and hind legs faults. Moreover, it was proved that the landing distance of horses jumping over oxer was indeed statistically shorter than over vertical. We also noted essential differences in horse's head position in relation to the ground in vertical and oxer jumps. The average degree of angle between the position of horse's hind part and the ground was smaller in oxer jumps than in verticals. Computer image scanning can be effectively used in the selection process of jumping horses.

Session H25

Theatre 5

Development of a linear type trait system for Spanish Purebred horses (preliminary analysis)

M.D. Gómez[1], M. Valera[2], I. Cervantes[1], M. Vinuesa[1], F. Peña[3] and A. Molina[1], [1]Department Genetics, University Cordoba. [2]Department Agro-forestal Science, University Sevilla, [3]Department Animal Production, University Cordoba, Spain

In this study 160 Spanish Purebred horses (5.6±0.21 year-olds) were evaluated on a linear scoring system (from 1 to 9) and were measured objectively for related physical components of conformation. The analysis included 49 zoometric measures, 5 general conformation variables and 78 linear type traits.

Genetic parameters were estimated by a multiple trait animal model using VCE software v.5.0.. Evaluator, year of birth and season of evaluation were included as fixed effects. The pedigree file was produced using four generations (including 1,204 animals).

Average scores for the 78 linear traits ranged from 2.60 to 6.80. Heritabilities were from 0.02 to 0.52, with standard errors between 0.019 and 0.060. The average heritability of zoometric measures is higher than the heritability of linear traits.

The biological relationships among conformation traits were analysed by principal component analysis and phenotypic and genetic correlations (the genetic correlations generally being of the same sign and larger than corresponding phenotypic ones).

Our results allowed the reduction of the number of analysed linear traits selected for the genetic evaluation. We choose 38 traits (8 general and 30 linear) by their genetic parameters, their relationship with dressage and their ease of measurement and repeatability.

Session H25

Theatre 6

Estimation of genetic parameters for biokinematic variables at trot in Spanish Purebred horses under experimental conditions

M. Valera[1], M.D. Gómez[2], A. Galisteo[3], F. Miró[3], M.R. Cano[3], E. Agüera[3], A. Molina[2] and A. Rodero[2], [1]Department Agro-forestal Science, University Seville, [2]Department Genetics, University Cordoba, [3]Department of Anatomy and Pathologic Anatomy, University Cordoba, Spain

It is necessary to identify efficient tools for indirect selection in dressage performance, because of its low heritability values. Gait analysis could provide an early criterion for selection in horses. The aim of this study is to estimate the genetic parameters (heritability and genetic correlations) of 20 biokinematic variables (5 linear, 5 temporal and 10 angular) in Spanish Purebred (SPB) horses at trot on treadmill, in order to select those to be included in the breeding scheme of this breed.

The kinematic variables of 130 SPB horses (4.6±1.5 years-old) proceeding from 24 studs were recorded at trot (4m/s). Stride parameters were obtained from digitised videos in a semiautomatic analysis system (SMVD.2.0). Genetic parameters were estimated by VCE program, using a bivariate mixed animal model including age and stud as fixed effects and animal additive genetic and residual error as random effects.

In general, the heritabilities obtained were high (0.33-0.88). Temporal variables have shown the highest heritabilities, whereas angular presented the lowest ones. The genetic correlations were also very high and this allows us to reduce the number of variables included. According to our results: stride duration, hindlimb swing phase duration, range of stifle and elbow angles, minimal angle of carpus and minimal retraction-protraction angle of hindlimb were selected.

Genetic parameters for growth traits of Holstein warmblood foals

S. Walker[1]*, J. Bennewitz[1], E. Stamer[1], G. Thaller[1], K. Blobel[2] and E. Kalm[1], [1]Institute of Animal Breeding and Husbandry,Christian-Albrechts-University, 24098 Kiel, Germany, [2]HIPPO-Blobel, 23812 Ahrensburg, Germany

Longitudinal records of four growth traits (bodyweight, wither height, foreleg length and cannon bone length) were taken for 291 breeding warmblood weanling colts between October 2004 and May 2005. For the genetic analysis average daily gain, respectively average monthly growth rates for the other selected traits were calculated for each colt. Genetic parameters of all traits were estimated multivariate with an animal model that included the fixed effects breeding farm, season of birth (early/late) of the colts and the random additive genetic effect of the animal. The heritability estimates were 0.23, 0.32, 0.05, and 0.22, respectively, for daily gain, monthly growth of wither, monthly growth of foreleg length and monthly growth of cannon bone length. High genetic correlations were estimated between monthly growth of foreleg and monthly growth of cannon bone ($r_g = 0.91$), respectively monthly growth of wither ($r_g = 0.86$). Medium correlations were found between monthly growth of wither and monthly growth of cannon bone ($r_g = 0.57$), accordingly for daily gain ($r_g = 0.52$). Negative or no genetic correlation was estimated between DG and MGCL ($r_g = -0.41$), respectively for MGFL ($r_g = 0.01$).

Demographic and reproductive parameters in the Spanish Arab horse

I. Cervantes[1]*, M.D. Gómez[1], A. Molina[1], C. Medina[1] and M. Valera[2], [1]Department of Genetics, University of Cordoba, [2]Department of Agro-forestal Sciences, University of Seville, Spain

The evolution of demographic and reproductive parameters of the Spanish Arab Horse breed are reported in this paper. The first registration in the Spanish Arab Studbook was in 1847. Since then a total of 18,880 animals have been registered.
Parameters were computed using ENDOG v 3.0 program.
The values for parents' age at the first and at last foaling, productive life and average of number of offspring were: for stallions 8.12±0.09, 11.43±0.14, 3.27±0.10 years and 6.5±0.3 descendents and for mares 6.78±0.05, 10.93±0.08, 4.31±0.07 years and 3.15±0.05 descendents respectively. The average of generation length was 8.58±0.12 years and the value for foaling periods was 2.14±0.03 years.
The average inbreeding coefficient and the average relatedness were 7.00% and 9.08 % for the whole population and 9.81% and 11.63% for animals born in the last decade respectively. F-statistics were computed considering studs as subpopulations ($F_{is} = -0.0721$, $F_{st} = 0.0909$ and $F_{it} = 0.0254$). F_{st} value indicates that around 9% of total genetic variation was explained by studs differences.
The level of heterozygosity in the population is due to the use of foreign stallions and mares, but the increase in the inbreeding coefficient proves that to avoid the loss of genetic variability a breeding program could be useful.

The use of computer scanning techniques to define the parameters of rider's position in show jumping events
S. Pietrzak, K. Bocian, K. Strzelec, S. Pietrzak, K. Strzelec and K. Bocian, Agricultural University, Horse breeding and use, Akademicka str. 13, 20-605 Lublin, Poland

The aim of this paper was an attempt to use computer scanning techniques in order to define some of the parameters of rider's position in jumping over upright and spread fences (from 100 to 140 cm of height). The research focused on 360 riders mounting 371 horses during national competition between 2004 and 2005. All the events were recorded with digital cameras. From all the digital data obtained (saved with MultiScanBase software) we chose images presenting 3 stages of a jump over vertical and 5 stages over oxer. In each of the tested stages of the jump we analysed the position of the rider's particular body parts by measuring two distances (1-from bit to rider's hand, 2-from hip to cantle) and five angles (1-opening of ankle joint, 2-opening of knee joint, 3-between torso and thigh, 4-opening of elbow joint, 5-degree of torso's slope). It was concluded that the length of distance from a bit to rider's hand depends, to a large extend, on the class of the competition, its speed and the size of the arena. The riders taking part in the less advanced events differed tremendously from the competitors in higher classes in terms of the gape between particular joints. This theory was confirmed by the number of faults, caused mainly by the lack of balance in the jump, which resulted in forestalling of the horse's movement so typical of lower class competitors.

The pedigree analysis of Arab horses in Poland and France in relation to their endurance predispositions
G.M. Polak, J. Dybowska, M. Wieczorek, Warsaw Agricultural University, Department of Genetics and Animal Breeding, ul Ciszewskiego 8, Wraszawa 02-786, Poland

The increase of endurance races is followed by a demand for the improvement of suitable horses. French breeders, who widely used Arab purebred sires originating from Poland in their breeding programme, achieved particularly good results of the improvement towards endurance races. The study contains the analysis of Polish and French endurance horses pedigrees taking into consideration those with Polish-Arab ancestors. The research involved 286 French horses (group A) which took part in endurance competitions from 1999 to 2003 and were classified from the 1st to the 5th place and 90 Polish horses (group B) which took part in the endurance competitions from 1990 to 2005, both groups on the distance => 100 km. The analysis of the French group appeared 111 horses with one or more Polish ancestors in the II, III or IV generation. Out of total of 2288 ancestors, in IV generation 221 constituted the Polish horses - 9,66%, 61 in III – 5,33% and only 8 in II – 1,4%. The gens share of Polish ancestors estimated with Gen Drop methodology in examined French group was 13,65%. The most frequently found Polish ancestors were the stallions: Comet – 89 times, Aquinor – 73, Arax – 70 and the mares Eleonora (Exelsjor mare) – 79, Bajdara (Baj mare) – 72. They were the same founders of family in Poland and in France: Comet, Negatiw, Aswan, Aquinor, Ofir.

Session S26

Theatre 1

Principles of adaptation in domesticated ruminants to life in harsh environments

N. Silanikove, Inst. Anim. Sci., Agricultural Research Organization (ARO), PO Box 6, Bet Dagan 50250, Israel

Considerable part of ruminants' evolution occurred during periods of prolonged drought on earth. This is reflects in the anatomical adaptations, particularly the developments of a specious forestomach (the reticulorumen or simply rumen), which serves as a fermentation vat that enable ruminants to digest efficiently lignocellosic plant material. Compared with most monogastric mammals, in which water loss over 15% of body weight is fatal, all ruminants, including modern dairy cows, can tolerate a water loss in the range of 18 to 40% of body weight, depending on species, breed and adaptation to desert life. The rumen plays an important role in the above evolved adaptation by serving as a huge water reservoir, which is used during dehydration, and as a container, which accommodates the ingested water and regulates water distribution upon rehydration. Domestic ruminants are selected and bred to attain maximal productive trait/s, such as milk and meat yields, in a given ecosystem. As metabolism and productivity run parallel, selection imposed an extra burden on the ability of domesticated ruminants in comparison to wild ones to attain energy, thermoregulatory and water balances in harsh environments. This review is concerns with over viewing basic principle of adaptation in domestic ruminants to life in harsh environments in the level of genera, species and breeds. It is hoped that this review will contribute to devise an effective breeding-selecting and feeding schemes in harsh environments that are consistent with sustainability, ethic and welfare considerations.

Session S26

Theatre 2

Effects of feed scarcity and of alternative feeding strategies on ruminants meat quality

A. Priolo and V. Vasta, University of Catania, DACPA Animal Production Department, Via Valdisavoia 5, 95123 Catania, Italy*

This manuscript reviews the meat quality of ruminants exposed to feed scarcity or fed alternative feed resources in harsh climate regions. Under-nutrition reduces meat fat content due to higher mobilization of body fat stores compared to meat of well-nourished ruminants. Moreover, the negative metabolic energy balance in underfed animals causes high meat ultimate pH values, producing detrimental effects on meat quality attributes. The exploitation of bushes and browses in natural rangelands and of agro-industrial by-products is an effective solution to overcome feed scarcity. Some of these feedstuffs contain secondary compounds, as condensed tannins, resulting in meat lighter in colour compared to meat of animals offered tannin-free resources. Special attention is given to shrubs (e.g. cactus, *Opuntia ficus indica*), trees (e.g. Argan fruit pulp and leaves) or forbs (e.g, *Salicornia bigelovii*) that seem to increase intramuascular fat content of conjugated linoleic acid (CLA) and of poly-unsaturated fatty acids (PUFA), which are beneficial to human health. Meat flavour is variously affected by animal nutritional condition and diet, depending on the accumulation of odour-active compounds which are transferred from feed into animal tissues, or which are originated by animal metabolism.

Lamb meat production in alpine regions
F. Ringdorfer, Agricultural Research & Education Centre, Raumberg – Gumpenstein, Austria

In Austria about 330.000 sheep are kept. More than 50 % are kept in alpine regions. In this regions, the sheep stay on pasture during the vegetation period, this is about 6-7 months, from April to October, and about 6-5 months they have to stay in stable. The alpine pastures are in a sea level up to 2000-3000 m. For the wintertime the farmers have to produce feed conserves like hey or silage. The main aim of sheep breeding is to produce lambs with high quality. High quality is defined as young lambs (3-5 months), high proportion of meat and less fat. The problem is not feed scarcity, but the harsh environment during summer. This includes cold temperature, sometimes snow, predators like golden eagle, fox or lynx. Most endangered are the lambs, therefore no lambs are born in this time. Due to the steep grassland areas, the maintenance requirement is much higher than on lowland pastures. The daily gains are only 150 to 200 gram. If spring born lambs go to the alpine pastures, they are 6 to 7 months old when they come back. A better production system is to wean the lambs before the ewes go to the alpine pasture and feed the lambs indoor with concentrate and hay until their final body weight.

Alternative feed resources to improve ruminant performances in harsh environments
H. Ben Salem, INRA-Tunisia, Laboratory of Animal and Forage Productions, Rue Hédi Karray, Ariana 2049, Tunisia

Feed scarcity and poor quality of feeds are the main constraints affecting productive and reproductive performances and sanitary conditions of the farm animals. Several cost-effective technologies (feed blocks and pellets, ensiling agro-industrial by-products, etc.) and fodder trees and shrubs based approaches were developed and were shown to be highly effective in improving livestock performance. A wide range of fodder trees and shrubs (FST) could be used to replace forage (cultivated green forages and hays) or common protein (soybean, rapeseed, and cotton seed meals) and energy (barley, maize and wheat grains) sources. Tanniniferous FST could be used advantageously in livestock feeding, for protecting good quality dietary proteins from rumen degradation, for reducing methane production and for controlling nematodes. Cactus (*Opuntia* spp.) and *Moringa oleifera* are considered as two miracle plants which can survive in dry areas and improve farmers' income through different ways. Cactus is a major cash crop, which could be used as energy and water source for ruminants. It is also used for fruit production and has many medicinal uses. The leaves and defatted kernels of *Moringa oleifera* are high in proteins and could provide cost-effective solution to improve animal production and to overcome human malnutrition problems. Promising feeding strategies based on alternative feed resources should be transferred to target dry areas.

Research developments on cactus pear (*Opuntia* spp.) as forage for ruminant feeding systems in dry areas: Review

Firew Tegegne, C. Kijora and K.J. Peters*, Animal Sciences Institute, Humboldt University of Berlin

Under environmental conditions unfit for conventional crops, cactus pear (*Opuntia ficus-indica*) has shown its potential. It is an extremely drought tolerant, highly productive, multipurpose and succulent plant with incomparably high water and land use efficiencies. Consequently, it is found widely distributed predominantly in arid and semiarid regions of the world, including the Mediterranean regions of Turkey, Greece, Italy, Spain and Portugal. Though there was encouraging expansion in research and development on cactus pear in the 1990s, emphasis was given to the food aspect while its forage uses continued to receive little attention. Though not supported with research findings, the use of cactus pear cladodes in livestock feeding is a very old practice. Related to its excessive feeding during periods of feed scarcity, diarrhoea and bloat have been reported. These nutritional disorders have to be identified if farmers are to be advised on the optimum supplementation level. Compositionally, most researchers agree, cactus pear is high in water, readily digestible organic matter and Ca contents but low in crude protein, Mg, Na, K and especially P contents in relation to ruminant requirements or to common forages. Its high water-use efficiency and readily digestible organic matter content links nitrogen-fixing forages (legumes) and (urea-treated) roughages. The paper reviews the recent developments related to research on the forage aspect of the plant.

Fodder trees utilization and constraints to adoption in the Ethiopian highlands

A. Mekoya[1]*, S.J. Oosting[1] and S. Fernandez-Rivera[2], [1]Wageningen University, Animal Production Systems Group, P.O.Box 338, 6700 AH Wageningen, The Netherlands, [2]International Livestock Research Institute, P.O.Box 5689, Addis Ababa, Ethiopia

Fodder trees have been promoted in Ethiopia to alleviate feed shortage and improve crop residues utilization. A single visit survey of farm-households was undertaken between August and December 2004 in three districts and two production systems (cereal and horticulture-based livestock production systems) of the Ethiopian highlands to identify farmer's perception on management practices, nutritional value, and constraints to adoption of fodder trees. The dominant fodder tree developed was *S.sesban* (88.4%) followed by *C.calothyrsus* (8.6%). On average 587 trees were found household [-1] and 68.3% of respondents' posses below 500 trees. Livestock, cross-breeds number and land size were significantly ($P<0.05$) correlated with number of fodder trees owned. The attributes of fodder trees were improving intake of straw diets, body condition and milk production. Average age of cutting was 10.4 months after plantation; frequency of cutting was 2.64 times in a year mainly from September to May targeting dry season supplementation. Initial age and frequency of cutting were affected by number of cattle and fodder trees owned, rainfall and feeding practice. About 89.8% and 10.2% of respondents were willing and against continuing fodder development. 44.5% of willing farmers had preference for local fodder trees. The major constraints of adoption were agronomic problems, low multipurpose value, and land shortage to continue/begin fodder plantation.

Adaptation strategies of small ruminants production systems to environmental constraints of Semi-Arid areas of Lebanon

C.I. Dick, A.M.Ghanem and S.K.Hamadeh, American University of Beirut. Animal Science Department, Riad ElSolh 1107-2020, Beirut, Lebanon

In Lebanon, sheep and goat production has traditionally been an integral part of farming systems; however, this sector has been undergoing drastic changes in response to serious constraints, mainly feed availability and seasonality of rainfall.

This study focuses on the major small ruminant production systems in marginal semi-arid lands of Lebanon, and the interaction between these systems and the environment. A formal survey covering 16 farmers, representative of 4 small ruminant production systems (semi-nomadic, transhumant, sedentary and semi-sedentary), was conducted in 1997 and 2002 and generated information regarding size, composition and movement of herds, feed calendars including grazing sites and patterns and livelihood strategies. All systems studied revealed major adaptation strategies to environmental constraints. Comparative feed charts indicated that herders decreased their overall herd size, reduced the movement of flocks and cut down on the use of concentrates. Producers also modified their livelihood strategies and diversified their sources of income. Finally, a conceptual driving force-state-response model was built to identify action and research needs and revealed the importance of secure access to pastures coupled with range rehabilitation measures.

Intensification of integrated crop-livestock systems in the dry savannas of West Africa

H.H. Hansen[1,2] and S.Tarawali[2], [1]The Royal Veterinary and Agricultural University, 1870 Frederiksberg, Denmark, [2]International Livestock Research Institute (ILRI), Addis Ababa, Ethiopia*

On-going population changes and fluctuating climatic conditions in sub-Saharan Africa are predicted to result in intensification and expansion of agricultural production. Traditional production systems and the management of natural resources within them are breaking down. This appears to be occurring at a faster rate in the dry savannas in part because of the fragile, sandy soils, erratic rainfall and shortening growing periods. Unsustainable farming practices are emerging with potentially disastrous results for poor people, food security and the environment.

This paper reports results of a Danish funded, ILRI led multi-institutional, inter-disciplinary project to promote intensification of sustainable agriculture and livestock production in Nigeria, Mali, Niger, Burkina Faso and Ghana. The project was based upon research that began in 1998 with funds from ILRI, International Institute of Tropical Agriculture (IITA) and International Crops Research Institute for Semi-Arid Tropics (ICRISAT), and the System-wide Livestock Programme (SLP). The use of improved and/or dual-purpose cowpea, millet, sorghum and groundnut varieties gave more fodder and more or equal amounts of grain in all countries. This increased plant production was only translated into increased animal gains in Nigeria where adjusting the level of cowpea and millet bran supplements increased ram daily gains compared to farmers' strategy. There is a need to find innovative ways of translating increased fodder production into livestock production.

Researches to elaborate a cheap and easy method for silage inoculation applicable in small farms

T. Vintila, University of Agricultural Science and Veterinary Medicine, Calea Aradului 119, Timisoara 300645, Romania

This work consists of three experiments. In a first experiment was tested the capacity of lactic bacteria from the Bank of Industrial Microorganisms of FZB Timisoara – Romania to grow in different culture media and temperatures. The CFU, pH and OD evolution of cultures were followed. Parameters of biosynthesis were established. Liquid whey resulted from enzymatic coagulation of milk was choused as culture medium for lactic bacteria. It is a valuable culture medium because of content in lactose, minerals and growing promoters for lactic bacteria. In the second experiment, the selected bacterial strains were used to inoculate two types of forages (one easy to ensilage and other difficult to ensilage, respectively sorghum and alfalfa). After inoculation, the ensiling was carried out in laboratory, in anaerobic conditions. Control probes were made for each type of forage. The evolution of lactic bacteria (in microaerophilic conditions), pH value and acidity was determined. In the third experiment the inoculant consisting of whey and lactic bacteria was applied to control fermentation in silages in farm conditions. Obtained results recommend whey as culture medium for silage inoculant production, used to initiate lactic fermentation in silage. This method can be applied especially in small farms, where milk curdling is made in place, although whey can be purchased at low price from dairy processors.

Effect of protein supplementation on growth performance and carcass characteristics of sheep under grazing conditions

A. Sepehri*, M. Aghazadeh and A.R. Ahmadzadeh, Islamec Azad University Shabestar branch. Shabestar, Iran

Extensive data were generated on growth performance of finisher lambs maintained under intensive feeding. While such information for extensive range management is lacking in literature. Present study was conducted on Mogani male lambs maintained on varying levels of Protein supplementation in addition to free grazing to assess their performance and carcass characteristics. Lambs (20) 7 to 8 months old (36.2 kg), were divided into 4 groups of 5 each and were grazed for 75 days. The first group (G_1) of 5 lambs was solely maintained on grazing (conventional method of fattening in the region), whereas the group 2, group 3 and group 4 was supplemented with 12, 14 and 16% protein diet respectively. The concentrate mixture was supplemented in the evening after grazing for 8 hr daily and confined overnight individually. Average daily gain (gr), slaughter weight (kg), warm dressing percentage, loin eye area (cm^2), kidney fat (gr), pelvic fat (gr), abdominal fat (gr), fat tail weight (kg), and all wholesale cuts weight (kg) were significantly ($P<0.05$) increased by increasing levels of crude protein from 12% to 16% in supplementary diets. The protein-supplemented groups were significantly ($P<0.05$) superior to the control group (G_1) in all of these traits. These results indicate that supplementary feeding of high protein diet significantly increased growth performance under grazing conditions.

Session S26

Diversification of lamb production in Spanish dry mountain areas: Carcass characterization

A. Sanz*, R. Delfa, L. Cascarosa, S. Carrasco, R. Revilla and M. Joy, CITA-Aragón, Apdo. 727,50080 Zaragoza, Spain

To prospect the diversification of lamb production, forty-two single male lambs of Churra-Tensina breed were slaughtered at three different live-weights: 9-12 (LW1), 20-24 (LW2) or 28-35 kg (LW3; castrated at 35 days old). Lambs were reared with their dams in natural meadows and mountain pasture (3 last month of LW3) until slaughtering. LW2 and LW3 lambs had *ad libitum* access to concentrate (except in mountain pasture). Animals presented 34, 62 and 183 days old at slaughter; 236, 309 and 158 g/d average gain; and 6.2, 11.7 and 16.2 kg carcass weight; for LW1, LW2 and LW3, respectively. Carcass classification, conformation and fatness degree were carried out at 24h post-mortem.
Carcasses of heavier lambs showed better conformation (O-, O+ and R; $p<0.001$), higher fatness degree (2, 2+ and 3; $p<0.001$) and redder meat ($p<0.001$). Carcase shrink was similar (2.63, 2.30 and 2.33%), but carcass dressing percentage decreased as slaughter LW increased (55.2, 52.9 and 50.7%; $p<0.001$). The lightest animals reached the highest pelvic-renal ($p<0.001$) and the lowest mesenteric-omental fat percentage ($p<0.001$). Joints percentage of first commercial category increased with slaughter LW (59.9, 60.1, 62.6%; $p<0.001$).
The range of carcass characteristics found in the current study would confirm the commercial viability of the three slaughter LW, suggesting the possibility of opening new markets and recovering certain traditional products whose production had been abandoned.

Session P27

Bodily manipulations in poultry and pigs: Balancing between convenience and necessity in managing animal welfare

H. Hopster* and T.G.C.M..Fiks-van Niekerk, Animal Sciences Group of Wageningen UR, Animal Production Division, PO-Box 65, 8200 AB, Lelystad, The Netherlands.

Various manipulations like beak trimming, beak treatment, despurring, toe clipping, and dubbing in poultry and tail docking, teeth clipping and castration in pigs are widely inflicted on animals in intensive animal production systems. Societal criticism, due to animal welfare deterioration and impairment of the animals' integrity, motivated the Dutch government in 1996 to make explicit which manipulations were allowed and which manipulations should be banned in the future. Beak treatments, toe clipping, dubbing, and despurring will no longer be allowed from this autumn onwards. To evaluate the possible welfare consequences of a non-manipulation practice, the Dutch Organisation of Poultry keepers (NOP) initiated a review of the literature and of practical experiences. The results of this study will be discussed in view of the welfare consequences for the birds of both manipulating and non-manipulating management strategies. In addition, the Dutch regulations on manipulations in pigs will be discussed. We will address how the pig industry complies with these regulations, and ongoing research that aims at finding management alternatives for manipulations. A general overview of the differences in regulations between the European nations regarding manipulations will be presented. The paper will further prompt some discussion and questions on the science and ethics of various procedures on animals, and particularly the scientific justification for such practices.

Anaesthetic solutions for castration of newborn piglets
S.M. Axiak*, N. Jaeggin and U. Schatzmann, Section of Anaesthesia, Vetsuisse-Faculty University of Bern, Postfach 8466, 3001 Bern, Switzerland

Castration is a documented painful surgical intervention in newborn piglets and should ethically and lawfully (in some countries) be performed under anaesthesia. Castration is performed to increase meat quality by preventing boar taint, avoid indiscriminate breeding and maintain general control of stock. The anaesthetic technique used should result in significant reduction or elimination of pain and stress for the piglets and should also be short acting, economical, ecologic, residue free, easy to use and have a large therapeutic range. Currently reported injectable anaesthetic techniques are effective but prolonged induction and recovery periods remain a problem. Local anaesthetic techniques (currently in practice in Norway) are also effective but do not completely block the nociceptive response. Inhaled carbon dioxide provides fast induction, recovery and complete analgesia for a short period but unwanted side effects including hyperventilation, agitation and gasping make this technique unacceptable. Isoflurane, via a special double mask to reduce pollution, has been shown to be a fast, safe and practical method. Handling, equipment costs and exposure concerns have to date limited its acceptance. Investigations are underway at our institution concerning intranasal administration techniques. A combination of ketamine, azaperone and climazolam has been found to provide short acting anaesthesia when administered intranasally or intramuscularly, but inconsistent effectiveness remains with the intranasal technique likely due to problems with administration.

Effect of supplementation with natural tranquillizers in pig diets on the animal behaviour and welfare
N. Panella*, A. Dalmau, E. Fàbrega, M. Font i Furnols, M.A. Oliver, J. Tibau, J. Soler, A. Velarde and M. Gil, IRTA-17121 Monells, Spain

Sixty-one animals with different Halothane gene (homozygous halothane positive-HP, n=34; and homozygous halothane negative-HN, n=27) were fed with three diets (control group, with no supplement; Mg group with 1.28g $MgCO_3$/Kg and Tryptophan group with 5g L-Trp/Kg) during the last 5 days before slaughter. Animals were submitted to minimal stress *antemortem* conditions. Pig behaviour was recorded in the raceway to the CO_2 stunning system and during the stunning period; corneal reflexes were recorded after stunning. Statistical analysis was performed using mixed and glm procedures. There were no differences in feed intake among diets (p>0.05). HP group had lower intake than HN (p<0.01). The behaviour of the pigs in the raceway did not differ (p>0.05) among treatments or halothane genotype. A significant (p<0.001) interaction diet*Halothane was found in the first retreat attempt during the exposure to the CO_2 system. In HP group, the time to perform the first retreat attempt was higher in the Mg (p<0.05) than the Control group. Moreover, in Mg group, HP had a later (p<0.05) first retreat attempt than HN. Thus, Mg supplementation could have a positive effect on animal welfare of HP pigs. HP had a lower proportion of animals that showed corneal reflexes after stunning than HN, indicating a higher effectiveness of the stunning method in HP pigs.

Immunological behaviour and laying performance of naked neck and normally feathered chickens

S.A. El-Safty*, U.M. Ali and M.M. Fathi, Ain Shams University Faculty of Agriculture, Cairo, Egypt

The study was undertaken to evaluate and measure some immunological traits and laying performance for two genotypes (Nana and nana) of chicken under low ambient temperature. The main results can be summarized as follows: The Nana hens had heavier body weight and slightly higher body temperature as compared to nana ones. According to H/L ratio and leucocyte percentage, it could be seen that the nana birds were more stressed than Nana counterparts. The results of PHA-P assay showed that the Nana hens had a significantly greater dermal swelling compared to normally feathered ones. Additionally, the normal plumage hens had a higher mortality and culling rate than heterozygous naked neck hens. Concerning egg production and eggshell quality measurements, the nana hens had a better performance than Nana ones. In naked neck hens, there was a positive relationship among egg mass, egg number and cell-mediated response occurs at 48 and 72hrs post-injection. While, in normally feathered genotype, there was a highly positive correlation between egg weight and cell-mediated response at 72h post-injection.

Water-cooled floor perch effect on broiler performance in high temperatures

A.G. Karaca* and T.N. Chamblee, T.K.B. İl Kontrol Laboratuvarı Antalya, Turkey, Mississippi State University Mississippi, USA

High ambient temperature coupled with high humidity has a deleterious effect on broiler production. The effect of water-cooled floor perches on productive performance of broiler chickens was studied in a high temperature environment. Three trials were conducted in the MSU Agriculture and Forestry Experiment Station Poultry House during August 1996, July and August 1997. Chickens received four treatments: single line PVC perch, four line PVC perches, single line steel perch and four line steel perches. The statistical procedures in this study were Completely Randomized Design and Mixed Model Correlation Analysis.
The effect of high environmental temperature with high humidity was recorded on weight gain in the second trial. The highest weight gain was in the SLSP group and the lowest weight gain was seen in the control group. There were no significant differences among treatment groups in 42 day body weight, feed consumption, feed gain or mortality rate. Perched birds number increased as dry, wet bulb, and outlet temperatures increased, and decreased as inlet temperature increased. The FLSP were preferred as ambient temperature increased. As dry bulb, wet bulb, inlet/outlet temperatures increased the number of birds panted. These results show that better productive performance at broiler chickens can be obtained in water-cooled steel perches. Water-cooled steel perches could assist in dissipating excess body temperature by conductive heat loss.

Influence of feed withdrawal and pre-slaughter conditions on carcass and meat quality traits

M. Gispert, J. Soler, J. Cros, E. Fàbrega, T. Velarde and J. Tibau, IRTA 17121 Monells, Spain

Seventy-five male pigs were used in the study (50% Pietrain, 25% Duroc and 25% Landrace). The objective was to determine the effect of the fastening period in farm (DJ) and the lairage time (TE) in the slaughter plant on the yield, carcass and meat quality parameters. Two DJ (2 and 12 h) and three TE (0, 5 and 10 h) were studied. Variables of yield (loss of weight depending on DJ and TE, gut weight and carcass yield), carcass quality and meat quality (pH, caecum pH, colour and drip losses) were studied. The results in the variables of yield were in accordance with the literature: loss of weight with the DJ and TE depended significantly on the time. The gut weight (without cleaning) depended significantly on DJ (2=6.01, 12=4.90) and TE (0=5.97, 5=5.42, 10=4.96). Carcass yield was affected by TE (0=79,0%, 5=79,9%, 10=80,7%). Carcass quality variables were not affected significantly by DJ and TE. Satisfactory results were found for the meat quality variables for all the treatments DJ*TE. Although in some cases pH24h and CE24hSM were affected by TE, in practice the differences were not important. The pH at the level of caecum showed significant differences depending on DJ (2=6.13, 12=6,55) and TE (0=5.98, 5=6.40, 10=6.64). When the TE increased the pH increased and this would be in favour of *Salmonella* proliferation.

Labour and economic aspects of dairy farming

B. O'Brien[1], L. Shalloo[1], D. Gleeson[1], S. O'Donnell[1,2] and K. O'Donovan[1,2], [1]Teagasc, Moorepark Dairy Production Research Centre, Fermoy, Co. Cork, Ireland; [2]School of Agriculture, Food and Veterinary Medicine, University College, Dublin, Ireland*

This study investigated interrelationships among labour-use, scale of dairy enterprise and net farm profitability on a sample of Irish dairy farms. Farm labour input data were collected from 171 full-time dairy farmers, over a 2-year period. The farms were grouped into three categories; < 50 cows (small), 50-80 cows (medium) and > 80 cows (large). Labour input data were analysed using a repeated measures analysis within the SAS Statistical Package. Financial analysis of the farms was carried out using the Moorepark Dairy Systems Model, which is a stochastic budgetary simulation model of a dairy farm. An imputed cost for operator and family labour was included at the current agricultural wage rate of 12.50 Euro/h. Small, medium and large farms of average herd size 44, 62 and 147 cows, respectively, had an average dairy labour input of 49.7, 42.2 and 29.3 h/cow/yr. The number of cows managed per labour unit (1848 h/yr) on these farms was 36, 43 and 62, respectively, with an associated cost of labour of €621, €528 and €366 per cow/yr, while production levels of 204, 217 and 330 tonnes of milk per operator were observed. Increased labour efficiency at increased herd size was due to labour saving technology, dairy system specialisation and economies of scale.

Session C28

Theatre 2

Profitable dairy cow traits for future production circumstances

A. De Vries, Department of Animal Sciences, University of Florida, Gainesville 32611, USA

Dairy farmers worldwide are constantly searching for dairy cows that are profitable under their specific production circumstances and provide few problems. In dairy cattle, traits influencing the profitability of production are broadly categorized as production traits (milk, beef) and functional traits (health, reproduction, efficiency, milkabilility). Economic weights of different traits depend on milk markets, farming systems, feed supply and cost, availability of data and industry goals. The economic values of many functional traits vary with the quality of herd management. Production in the United States is not constrained by production quotas or availability of land, unlike in many European countries. The net effect for the United States has been more emphasis on milk volume and fat, and less on protein and functional traits. Selection of profitable dairy cows is shifting from production traits towards functional traits. New functional traits will include stillbirth and metabolic disease resistance. The trend towards large farms and continuous milking will place more emphasis on uniform cows that improve parlor throughput. There is a significant interest in crossbreeding to overcome some of the weaknesses in functional traits. Heat stress tolerance is improved, for example by incorporating the slick-hair gene. Increased environmental pressures may lead to more focus on feed efficiency and smaller body sizes. Dairy farmers may have individualized selection indexes that allow them to attain unique dairy cows for their specific circumstances.

Session C28

Theatre 3

High-tech and low cost farming: What is the future?

M.H.A. de Haan, C.J. Hollander and S. Bokma, Animal Sciences Group of Wageningen University and Research Centre, P.O. Box 65, 8200 AB Lelystad, The Netherlands

The Low cost farm and High-tech farm, both one person enterprises, are experimental farms in The Netherlands established to study the perspectives of these farming systems. Main objective for both farms is to realise a low cost price for milk: € 0,34 or less per kg. The High-tech farm with cows all year round inside has a milk quota of 800.000 kg, which is relatively large for Dutch circumstances. To lower the cost price many laborious jobs as possible have been automated, eg milking and feeding. The milk yield is 10.000 kg.

The low cost farm has a milk quota and herdsize equal to the Dutch average (about 450.000 kg quota). The strategy is to use cheap alternatives to keep the cost price low, in which grazing is essential. Montbeliarde and "low producing" HF-cows are kept on this farm. The milk yield is 7800 kg, in spite of low concentrate feeding (50% lower than Dutch average).

The Low cost farm has reached its economic goal already a few years, and realized a cost price about 15% better than the Dutch average. The High-tech farm is managed in less than 50 working hours per week, which is 40% less than comparable farms.

Both systems give solutions for farmers to cut back costs and produce efficient on labour.

Environment friendly and sustainable dairy systems in the Atlantic area

A. Pflimlin et al., Institut de l'Elevage, France

Green Dairy is an European Interreg III B project started in 2003 for a 3 years period. It brings together 10 R&D partners from 5 countries and 11 regions of the Atlantic Area which range from Scotland to Portugal. It aims to provide a better understanding of the impact of intensive dairy systems on the quality of the environment in order to develop ways of improving practices in the different regions. This is expected to encourage more rapid responses to the problems of deteriorating water and a more appropriate response within or between regional contexts that could be used as proposals for implementation of regulations or advice. The project brings together three key actions:

- monitoring of mineral flows in research stations to identify the critical points for water pollution,
- developing discussion groups with motivated farmers, optimising their environmental practices,
- modelling the contribution of the dairy farming industry to the mineral surplus at the region scale and the link with the water pollution.

The systems studied cover a wide range of intensification and are based on maximum use of grazing in the northern regions or on maize silage only with high quantity of concentrate in the South.

These results will be discussed in relation with the economical data from the different dairy regions (FADN). On basis of this, perspectives for dairy farming in these regions will be pictured.

Fertility monitor: Management tool to improve fertility and farm economics

R.M.G. Roelofs and F. Hamoen, NRS b.v, P.O. Box 454, 6800 AL Arnhem, The Netherlands*

Reduced fertility is one of the major concerns in todays dairy farming. In order to give the farmer some insight in the status of the cow at the moment of insemination, technicians of CR Delta in The Netherlands are collecting three extra characteristics. Uterine tone (quality of heat), uterine discharge (hygienic circumstances) and body condition score (energy balance). All three have a proven effect on the fertility where bad uterine tone and uterine discharge both show a 6% lower non-return rate at 56 days.

Fertility monitor is a 1-page sheet which on herd-level combines the three new characteristics with the well-known fertility indicators like calving interval, interval calving to 1st insemination, non-return rate at 56 days and number of inseminations per inseminated cow. An economic indicator, presenting loss due to fertility is calculated by evaluating calving interval, number of inseminations and culling due to fertility. A knowledge based system (KBS) determines a wordly summary based on the indicators in order to help the farmer to comprehend the message. The KBS first determines if the main problem is heat detection or pregnancy rate. Secondly the two main causes for the problem are presented.

Based on a survey, 86% of the farmers was actively going to change their fertility management, varying from increasing time spent on heat detection, changing feeding or increasing hygiene at calving.

Mastitis detection in dairy cows

C. Henze*, D. Cavero-Pintado and J. Krieter, Christian-Albrechts-Universität Kiel, Faculty of Agricultural and Nutritional Science, Institute of Animal Breeding and Husbandry, Olshausenstraße 40, 24098 Kiel, Germany

The aim of the present research was to investigate the usefulness of neural networks and fuzzy logic in the early detection and control of mastitis in cows as a decision support for farmers.

A data set from the automatic milking system of the experimental farm "Karkendamm", containing 403,537 milkings of 478 different cows was used. Mastitis was determined according to three definitions: udder treatment (1) or somatic cell counts (SCC) over 100,000/ml (2) or SCC over 400,000/ml (3). Definitions 2 and 3 also in combination with udder treatment. Electrical conductivity, milk production rate, milk flow and days in milk were used as information for the study. With the fuzzy logic tool sensitivities from 83.2 % to 92.9 % and specifities from 75.8 % to 93.9 % (depending of the definition of mastitis) have been reached, but also high error rates from 41.9 % to 95.5 %. The results of the neural network were sensitivities from 78.6 % to 84.2 %, specifities from 51.1 % to 74.5 % and error rates from 51.3 to 80.5 %. Further investigations with several day-datasets of approximately 15,000 milkings from a milking parlour at the experimental farm "Karkendamm" with different methods of cluster analysis are planned, combining time series data with improved sensor informations.

Dairy farmers' plans and communication under new EU policies

M. Klopcic[1], S. Kavcic[1], J. Osterc[1], D. Kompan[1] and A. Kuipers[2], [1]University of Ljubljana, Biotechnical Faculty, Zootechnical Department, Domzale, Slovenia, [2]Expertisecentre for Farm Management and Knowledge Transfer, WageningenUniversity and Research Centre, Wageningen, The Netherlands

After accession to EU, farmers in the new-member states have to adjust to the EU agricultural policies and market. In Slovenia an analysis is made of the plans and communication aspects of farm development under quota and CAP. As tool a questionnaire was distributed to dairy farmers. 1000 questionnaires, about 20 % of the distributed ones have been returned anonymously, implying that 10 % of the dairy farmers' population is part of the analysis. Strategies of the farmers, their interest in information and routes to collect this info are analysed in relation to base parameters, like age of farmer, size of farm, milk control or not, breed(s), milk production level, less favoured area or not and importance of disciplinary knowledge fields to farmer. Results of this research will be presented and the importance of underlying parameters indicated. Preliminary results show a very significant demand for info about EU premium programs, especially CAP direct payments, a considerable activity in farm planning and perhaps less interest in diversification as expected. Communication channels are divers, but frequency of direct contact with advisors may be less than predicted. However this picture may very well change with completion of data set and study. Moreover this picture will be detailed to various targets groups of farmers. Also, a comparison of results can be made with an identical study for suckler cow farmers.

Session C28

Theatre 8

Entrepreneurship and technical competencies
Abele Kuipers[1], Ron Bergevoet[2] and Wim Zaalmink[2], [1]Expertise Centre for Farm Management and Knowledge Transfer, [2]Agricultural Economics Institute, Wageningen University and Research Centre, The Netherlands

Differences in income between farms do not so much depend on size of farm, etc., but on the competencies of the farmer. Studies are performed to characterise the farmer in relation to successful management and entrepreneurship. This is done by studying farmers who are active in discovery and demonstration projects, called the front runners, as well as farmers who belong to the large group, called the followers. Also "learning styles" of farmers and their interest in the various technical knowledge fields are analysed. Technical knowledge and management competencies are the base for a successful farmer. The demand for technical knowledge has become more specific and directed to the local environment However, entrepreneurship becomes a more important characteristic when farm size grows and complexity and risk of environment increases. A successful strategy is based on the right match of competencies, current situation (farm structure and results) and external factors. In pilot projects a method is developed to facilitate farmers in strategy making. Too, the ability to utilise communication channels determines the vitality of the farm, accepting that the traditional extension service system is gradually disappearing. Work contacts of farmers are studied indicating that networks within and between sectors differ significantly. New initiatives of communication will be described: one is a network approach aiming at links between practice and knowledge partners stimulating farmers to learn from each other; another counts on innovation through confrontation by inviting outsiders.

Session C28

Theatre 9

Quantification of disease production losses in a dairy farm: A software to support on-farm estimation
C. Fourichon*, N. Bareille, F. Beaudeau and H. Seegers, Veterinary School of Nantes – INRA, BP 40706, 44307 Nantes Cedex 3, France

Disease costs impair the revenue of a dairy farm to a large extent. Quantifying economic consequences of diseases is pivotal to support decision making for herd health. Expenditures for disease control are traced in invoices, but there are no data to measure losses consecutive to disease cases. This paper presents a software to support cattle consultants for providing information to their clients on economics of production diseases in their farm.
A partial budgeting model was developed to calculate economic losses for a given farm. Technical consequences of diseases are modelled based on effects of diseases at the cow level: loss in milk yield, milk withdrawal consecutive to treatments, reduction on milk quality, increased calving interval, increased culling rate and mortality. Economic consequences are calculated on a yearly basis at the farm level as the sum of variation in incomes and in variable costs. Prices and economic values are defined by the user for the farm and period of concern. The model is available on the internet. On-farm data are needed on cases of production diseases (22 questions), technical characteristics and results of the dairy herd (18 questions) and economic values and prices applicable to the farm (16 questions).
Results provides a basis for discussion on possible investment by a farmer to improve performance.

Financial losses from clinical mastitis in Turkish dairy herds

C. Yalcin[1]*, S. Sariozkan[2], A.S.Yildiz[1] and A.Gunlu[3], [1]Department of Animal Health Economics and Management, Veterinary Faculty, Ankara University, Ankara, Turkey,[2]Department of Livestock Economics, Veterinary Faculty, Erciyes University, Kayseri, Turkey, [3]Department of Livestock Economics, Veterinary Faculty, Selcuk University, Konya, Turkey

Financial losses from clinical mastitis in Turkish dairy herds were assessed using the field data obtained from 78 dairy herds in 3 provinces (Burdur, Konya and Kirklareli) between June 2003 and September 2004. Firstly, averages financial loss per infected cow were estimated in different prognosis of the infection (mild, medium and severe infections) taking account of incidence rates, probabilities of culling due to mastitis, re-infection, and full and partial recovery after the infection. Then, the total average avoidable and unavoidable losses in different level of aggregates (cow, herd, provinces) were calculated.

The average losses under mild, medium, severe cases and their weighted average were estimated to be US$67/case, US$142/case, US$491/case and US$233/case respectively. Average total losses of all herds were calculated to be US$700/herd (US$70/cow), and US$552 of this (% 82) was regarded as avoibable losses considering the average incidence of the lowest 10% of the herds as the target incidence rate.

Precision farming and models for improving cattle performance and economics

I. Halachmi, ARO, Israel

Daily management and farm design in new farms in Israel are supported by the Precision Dairy Farming (PDF) approach. Each milking parlor or robot is equipped with antenna that read the data from sensors attached to the animal. Information on the individual animal includes: (1) walking activity, (2) laying behavior, (3) body weight, (4) body conditions, (5) location, (6) milk yield and (7) milk composition (fat, protein, lactose, somatic cells) measured on-line during a milking session, (8) ruminate disorders, (9) hard rate and (10) feed intake. Sensors 1-8 are already commercially available or in the latest phase of development in the ARO research farm. Sensors 9-10 were developed for research purposes.

The challenge in the PDF research program is how to automatically apply the huge amount of data in order to take the right decisions without investing too much time looking at the data.

The new model, to be presented, gives an alarm if a cow's behavior exceeds a predefined dynamically updated norm defined in the model. The model supports decisions regarding an individual animal such as concentrate feeding, insemination period, culling a cow, as well as farm design levels such as where and how to install cooling system. The presentation will include both, the sensors and the models.

Opportunities for cattle husbandry in East Europe
A. Svitojus[1], I. Kiissa[2] and G. Saghirashvili[3], [1]Baltic Foundation HI, Konarskio 49, LT-03123 Vilnius, Lithuania, [2]Minsk, Belarus, [3]GNAAP, Tbilisi, Georgia

The disintegration of the Soviet Union at the start of last decade significantly affected East Europe countries rural economy. The production chain was severely disrupted. Basic production inputs were difficult to obtain, services such as breeding supporting schemes disappeared and prices stagnated, resulting in no incentives for farmers to sell.
Presently the situation starts showing indications of improvement due to the increase in the production volumes of domestic animal husbandry products. The quality of domestic products has also improved somewhat.
Currently, there are three main types of livestock farming systems on which domestic animals can be raised: Cooperatives replacing former collective farms and; Family farms; Small holders.
The emphasis should be put on the family farms. Very important is to improve the land base: quality and amount of harvest need to be increased. However the care of the land is a topic of concern. Complicated ownership relations (like common pastures) lead to poor fertilization of the land. Simple economic calculations may lead to more understanding of how to increase profits. A well-trained advisory service can help in these aspects.
Registration of cattle and breeding programs are required. This perhaps can be done the best by regional cooperation or by utilizing existing computer centers in Europe. The improvement of quality of product remains permanently of importance.

Investigation the effects of dietary rumen protected choline on performance of lactating dairy cows
A.H. Toghdory, Islamic Azad university of Gorgan, Iran

This study was designed to Investigation the effects of dietary rumen protected choline (RPC) on intake, milk production and composition and blood metabolites of dairy cows. Sex Holsteins randomly assigned with in a change over design with three periods of 21 days (14 d for adaptation and 7 d for data collection). The treatments were: 1) control 2) rumen protected choline 25 g/d and 3) rumen protected choline 50 g/d. Cows were individually fed diets as total mixed rations and choline top dressed in the first meal. Cows were milked thrice daily and milk yields were measured electronically. During two day of each period milk sampled for determine milk composition. Blood samples were collected at the end of the period for determine plasma glucose, triglyceride, cholesterol, and blood urea nitrogen. The result showed that RPC (50g/d) increased dry matter intake, milk yield, fat corrected milk yield and fat percent and production. Production and percent of milk lactose, protein, total solid and solid no fat were not affected by RPC (50 g/d). choline (50 g/d) had no a significant effect on blood metabolites. Rumen protected choline (25 g/d) had no a significant effect on performance of dairy cows. In conclusion our result indicated that RPC (50 g/d) had a significant effect on performance of dairy cows in early lactation.

Evaluation of Spartan dairy ration balancer and net carbohydrate and protein system for predicting dry matter intake of Holstein cows during the midlactation

M. Tghinejad Rudbane[1], *A. Nikkhah*[2], *H. Amanloo*[2] *and M. Kazemi*[2], [1]*Tabriz Islamic Azad University, Tabriz, Iran,* [2]*Department of Animal Science, University of Tehran, Karaj, Iran*

Because nutrient intake is a function of dry matter intake (DMI), predicting DMI precisely and accurately is important to prepare balanced rations. Lack of accuracy in prediction of feed intake can affect on animal performance, animal health or environmental impacts of dairy industry. DMI prediction in Spartan is based on empirical equations (NRC 1989) but CNCPS predicts DMI using the mechanistic model. Sixteen lactating Holstein cows averaging BW=622± 23kg, DIM=55± 13d and MY=32± 0.8kg were randomly assigned in a CRB design. Two rations using the same feedstuff and similar inputs were formulated by Spartan and CNCPS for each group of cows. The formulated rations fed individually for 90 days three times a day to allow 10% feed refusal. Daily DMI and other trait were recorded. Data were analyzed using GLM procedure of SAS (1999). The variables that used to evaluate DMI prediction accuracy are: MSPE = mean square prediction error (kg2/d), were calculated to quantify the inaccuracy associated with the prediction of DMI; MPE = mean prediction error (kg/d); RPE = relative prediction error (%). By increasing the RPE rate, accuracy of the DMI prediction decreases. The results of the present study showed that Spartan (RPE = 8.1%) predicts DMI more accurately than CNCPS (RPE = 10.8%).So it has more conformity to actual DMI monitored in farm.

Comparison of lactating Holstein cows performance fed rations formulated by National Research Council 2001 and Cornell Pennsylvania Miner softwares

A. Nikkhah[1], *Kazemi*[1], *H. Mehrabani*[1], *M. Taghinejad*[2], [1]*Department of Animal Science, University of Tehran, Karaj, Iran,* [2]*Tabriz Islamic Azad University, Tabriz, Iran*

A variety of software programs are available that designed to evaluate and formulate rations for dairy cattle. These programs differ greatly in their components and the types of objective functions available for ration formulation and completeness and versatility of report writing. Sixteen lactating Holstein dairy cows averaging BW=642± 23kg, DIM=55± 11d and MY=32± 0.9kg were randomly assigned in a CRB design. Cows allocated in individual pens. The same inputs used for formulating the rations by NRC 2001 and CPM-dairy softwares with the same feedstuffs. The formulated rations fed to cows for 105 days. Milk yield and composition data were collected. Individual daily DMI measured. Data were analyzed using GLM procedure of SAS (v8.2).The cows fed ration formulated by NRC had higher crude protein intake than those fed ration formulated by CPM_Dairy (*P< 0.05*). However, differences for dry matter intake (DMI) and net energy intake were not significant. As well, differences for MY, milk composition and milk efficiency were not significant between two groups. The cost of feed intake per cow/day and per kg of MY, was significantly affected by the different treatments (rations). Results of this study demonstrate that CPM_Dairy could decrease the need for protein supplement in lactating dairy cow's diet and feed cost per Kg of milk yield.

Prediction of the production of milk from the dairy control of the first third of the lactation
J.P. Avilez[1], M.V.C. del Valle[1], H. Miranda[1], A. Garcia[2], J. Perea[2] and R. Acero[2], [1]Catholic University of Temuco, Chile, [2]Department of Animal Production, University of Cordoba, Spain

With the aim of predicting the total production of complete individual lactation from dairy cattle, to prove its discriminatory capacity to identify homogeneous groups of lactations and select one method based in identification on not supervised classification of groups of lactations, or one classification method based in a discriminating model that requires first a supervised classification based in an expert judgement. The study used a selective sample of 153 representative complete individual lactation conformed by 45.900 daily data of production. To obtain de tendency of milk, we interpolated the daily data with the numeric method of flexible cubic Spline. From the estimations for each cow, the productive indexes where elaborated, this where evaluated for discriminatory capacity by the factorial analysis of the principal components, discarding the regression coefficients of Wood model, and selecting as more appropriated Ts35_105, P_105, Mxsp_f, Dmx_p105, Dmx_s_w, Min_spf, Mx_min, Dt_s_w. The results of the classification lactation curves from the step wise discriminating analysis with the evaluating method of data exclusion, presented an internal quality classification of 82% for both types of lactation. Finally is possible conclude that investigation results provide a effective and valid systematization prediction of the individual total production for homogeneous groups.

Genetic parameters for milk production for Tunisian Sicilo-Sarde dairy ewes estimated in an animal model using AG-REML
M.H. Othmane[1] and L. Trabelsi[2], [1]Institut National de la Recherche Agronomique de Tunisie, INRAT, Tunisie, [2]Institut National Agronomique de Tunisie, INAT, Tunisie

Genetic parameters were estimated for total milk yield from 881 lactations of Sicilo-Sarde dairy ewes belonging to the Tunisian National Institute of Agronomic Research (INRAT), Experimental Unity of Lafareg. Milk yields were calculated and adjusted to 180 days, from 8220 bimonthly test-day records, using the Fleischmann method. Animals included in the relationship matrix were 363 ewes and 83 rams. Estimates were from the use of a univariate, analytical gradients, restricted maximum likelihood procedure (VCE package). Least squares analyses showed significant effects year-season, age of ewe, and lambing type on milk yield, with the year-season being the most important factor. Genetic estimates agreed closely with those obtained previously in this population and were lower than those obtained in other dairy sheep populations. Heritability and proportion of permanent environmental variance were 0.14 ± 0.06 and 0.19 ± 0.06, respectively. Genetic and phenotypic variation coefficients reached 7.6% and 23.4%, respectively. Our results indicated that efforts should focus on improving the level of management and starting an adequate selection programme, in order to improve milk yield and to repair decreases in milk production level observed in the actual conditions of Sicilo-Sarde breed.

Survey on repeatability and genetic trend of some economical traits in north of khorassan kordian sheep

S.A. Shiri, D.A. Saghi and M. Mohammadzadeh, Iran-Mashhad, Agricultural & natural resources research center of khorassan, P.O.BOX. 91735-1148.*

In order to survey on genetic progress changes of some economical *traits* in Khorassan north kordian sheep, 6741 collected records of animal breeding station of Shirvan during 14 years were used. studied *traits* were birth weight, weaning weight, 6 month weight, 9 month weight and yearling weight and average daily gain(before and after weanig, from birth to yearling and 6 to 12 month) that the related mean and standard deviation were 4.15 ± 0.24, 20.57 ± 4.31, 30.27 ± 4.30, 31.85 ± 3.70, 40.45 ± 3.62, 192.65 ± 57.24, 75.87 ± 25.20, 98.35 ± 11.57 and 58.50 ± 18.72 gram respectively. Effects of sex, age of dam, birth type, locate and birth year were studied to design statistical model, and they were significant($p<0.05$) for growth traits. and so repeatability of weaning weight, Pre-Weaning gain, lamb fleece weight and adult fleece weight were 0.40 ± 0.11, 0.25 ± 0.1, 0.12 ± 0.06, 0.28 ± 0.11 respectively. Variance component in animal model estimated by DFREMEL software. genetic trends of studied traits detemined on the basis of mean breeding values per year. Birth weight genetic trend had fluctuation during these years and genetic trend for other growth traits was positive and wool traits was negative.

Genetic parameters for milk yield in dairy goats across lactations in Germany

B. Zumbach[1], S. Tsuruta[1], I. Misztal[1], K.J.Peters[2], [1]University of Georgia, Athens, GA; [2]Humboldt University Berlin, Germany

BVE for dairy goats in Germany is still based on herd mate comparison within breeding society. This study presents genetic parameters for milk yield based on a test day model as basis for a new national evaluation.

35.463, 29.871, and 23.103 test day records from lactations 1, 2 and 3 from 5.217, 4.125, and 3.133 animals, respectively, were obtained between 1987 and 2003 from 6 Goat breeding societies.

The multiple trait (lactation 1-3) repeatability model included the fixed effects of herd-year (1239 levels), litter size (1, $2 \geq 3$), lambing season (1, 2,.., 11), and days in milk of 3rd order (1-270) nested within herd-year, the random effects of animal additive and permanent environment. The 3-trait random regression model included also the random regressions based on 2nd order Legendre polynomials for animal additive and permanent environmental effects.

Heritability estimates in the repeatability model were 0.27 ± 0.02, 0.20 ± 0.02, and 0.37 ± 0.02 for the 1st, 2nd and 3rd lactation, respectively. Genetic correlations between lactations were 0.69 (1-2), 0.77 (2-3), and 0.43 (1-3), respectively.

Heritability estimates from the random regression model decreased continuously during the course of 1st and 2nd lactation, with average values of 0.28 and 0.26. Estimates in the 3rd lactation showed a maximum in the middle of lactation, averaging 0.36. Genetic correlations between lactations averaged 0.60 (1-2), 0.65 (2-3), and 0.44 (1-3), respectively.

Session G29

Theatre 4

Statistical modelling of main and epistatic gene effects on milk production traits in dairy cattle

J. Szyda*[1], J. Komisarek[2], [1]Department of Animal Genetics, University of Natural Sciences, Kożuchowska 7, 51-631 Wrocław, Poland [2]Department of Cattle Breeding and Milk Production, August Cieszkowski Agricultural University of Poznań, Wojska Polskiego 71 A, 60-625 Poznań, Poland

Recent studies show that epistasis, a phenomenon commonly neglected in earlier studies, is an important factor in the genetic background of quantitative trait variation.

Based on a data set of 252 Black-and-White AI bulls, born between 1990 and 1998, main and epistatic effects of selected genes on milk production traits were analysed. The following functional polymorphisms were considered: F16Y, P35Q and K468R in the bytyrophilin (*BTN1A1*) gene located on bovine chromosome 23, K232A in the diacylglycerol acyltransferase 1 (*DGAT1*) gene on chromosome 14, T945M in the leptin receptor (*LEPR*) gene on BTA3, and Y7F, R25C, A80V as well as promoter C/T substitution at position -963 in the leptin (*LEP*) gene on chromosome 4. Two measures of milk, fat and protein yields were considered: (i) breeding values estimated on a national basis using a random regression test day model, which are expected to contain only polygenic additive effects and (ii) daughter yield deviations derived from the above model, which are composed of additive polygenic and residual effects. A mixed model accounting for an additive polygenic effect is used to estimate effects of the genes and their interactions.

Session G29

Theatre 5

Analysis of genetic polymorphisms in the Egyptian goat CSN1S2 using polymerase chain reaction

Othman E. Othman* and Sahar Ahmed, Cell Biology Department - National Research Center - Dokki – Egypt

The dairy industry continually strives to improve the quality of its products. It has long been known that the manufacturing properties of milk are influenced by the relative composition of its proteins. The genes that encode the major milk proteins are thought of as candidate genes for the observed variation in protein composition. In goat milk, four caseins (αs_1, αs_2, β and k-casein) coded by four autosomal genes (CSN1S1, CSN1S2, CSN2 and CSN3, respectively) are present. The four Ca-sensitive caseins of goat exhibit both quantitative and qualitative variations arising from genetic polymorphism in the encoded genes.

Forty-five animals belonging to four Egyptian goat breeds were undertaken in this study to analyze the different alleles of CSN1AS2 gene. The PCR products of five different primers were digested by restriction enzymes *Mse*I, *Pst*I, *Nco*I, *Nla*III and *Alw26*I for detecting the presence of seven alleles A & B, C, D & 0, E and F, respectively.

Our results showed that the homozygous genotypes AA and BB were observed at frequencies 28.9% and 26.7%, respectively and the heterozygous genotype AB was displayed in 18 animals (40%) whereas the genotype AF was present in two animals (4.4%). The alleles C, D, E and 0 were not displayed in tested animals.

Restriction fragment length polymorphism and gene mapping of two genes associated with milk composition in Egyptian river buffalo

Othman E. Othman, Cell Biology Department, National Research Center, Dokki, Egypt

Gene maps and genetic polymorphisms of genes related to the economic quantitative traits as tools for developing more efficient breeding strategies have as targeted animal improvement by the genomic approach. This study aimed to identify the genetic polymorphism and gene mapping of two genes, BTN and ODC, are associated with milk composition in river buffalo. By means of PCR, the amplified fragment of BTN obtained from all tested buffalo DNA was at 501-bp. After digestion of the PCR products with *Hae*III, we can distinguish between A and B alleles. The allele A is the most common allele in the Egyptian river buffalo, where the estimated frequencies were 0.89 and 0.11 for alleles A and B, respectively. The PCR products of ODC with a 796-bp fragment were digested with *Msp*I. Fifty-four of 70 animals (77.14%) have MspI(+/+) genotype and 13 animals (18.57%) have MspI(+/−) whereas three animals (4.29%) displayed the homologous MspI(−/−) genotype. The assignment of BTN and ODC to river buffalo chromosomes was done by calculating the correlation coefficient (φ) between these two genes and markers representing the river buffalo chromosomes. The results showed that BTN is assigned to the bi-armed buffalo chromosome 2, whereas ODC is assigned to the acrocentric river buffalo chromosome 12.

Fine-mapping of QTL affecting mastitis resistance in Nordic dairy cattle

J. Vilkki[1]*, G. Sahana[2], L. Andersson-Eklund[3], H. Viinalass[4], N. Hastings[5], A. Fernandez[3], T. Iso-Touru[1], B. Thomsen[2], S. Värv[4], J.L. Williams[5] and M.S. Lund[2], [1]MTT Agrifood Research Finland, [2]Danish Institute of Agricultural Sciences, [3]Swedish University of Agricultural Sciences, [4]Estonian University of Life Sciences, [5]Roslin Institute, UK

Genetic improvement of mastitis resistance based on phenotypic selection is inefficient because of the low heritability of the trait and the difficulties in obtaining phenotypic information. The extensive records on clinical mastitis in the Nordic countries were used for fine-mapping QTL affecting mastitis and developing of tools for marker-assisted selection. A granddaughter design of 34 families with 1098 bulls from three related red breeds (Finnish Ayrshire, Swedish Red and White, Danish Red) was used to confirm previous mapping results and then to fine-map the QTL. Initially, six candidate chromosomes were examined using a common marker map, and the two chromosomes with the best evidence for a QTL with major effect were selected for fine-mapping. Dense marker maps were developed for the QTL regions. Information on linkage disequilibrium within and between breeds and multiple traits was exploited in LA/LD fine-mapping. Haplotypes that are predictive for mastitis in the general population were identified. A QTL on BTA9 was mapped within a 2-cM region. The marker haplotypes associated with differences in mastitis incidence can be used in marker-assisted selection in the Nordic red breeds. This work was funded in part by EC FP5 project MASTITIS RESISTANCE (QLK5-CT-2002-01186).

Detection of quantitative trait loci for response to an insulin challenge in Meishan x Large White F2 pigs

J.P. Bidanel[1], N. Iannuccelli[2], J. Gruand[3], Y. Amigues[4], D. Milan[2] and M. Bonneau[5], INRA, [1]SGQA 78352 Jouy-en-Josas, [2]LGC 31326 Castanet Tolosan, [3]GEPA, 86480 Rouillé, [4]Labogéna 78352 Jouy-en-Josas, [5]UMR SENAH 35590 Saint Gilles, France*

A quantitative trait locus (QTL) analysis of Growth Hormone response to insulin-induced hypoglycaemia was performed on 431 F2 Meishan x Large White male pigs issued from 6 F1 boars and 23 F1 sows. A whole genome scan was performed using 137 markers covering the entire porcine genome. Traits analysed included pre- and post-challenge plasma glucose (GLa and GLp) and growth hormone (GHa and GHp) levels, as well as their variation (ΔGL and ΔGH, respectively). Analyses were performed using both line-cross regression and half-/full-sib maximum likelihood methods. Genome wide significant additive QTL were detected for GLa and ΔGL on SSC3 (SW72-SW102 and SW102-S0372 intervals, respectively), for GLa on SSC7 (SLA-S0102 interval) and for GLp on SSC9 (close to SW174). Meishan QTL alleles had positive effects for ΔGL on SSC3, for GLa on SSC7, and, to a lesser extent, for GLp on SSC9. Conversely, Meishan QTL alleles on SSC3 had negative effects on GLa. Five additional chromosome-wide significant QTL were detected for plasma glucose levels on SSC 1, 7, 14, 15, 18. Six chromosome-wide significant QTL affecting plasma growth hormone levels were also mapped on SSC 2, 4, 5, 12, 15 and 18.

Functional analysis of the pig CYP2E1 promoter and identification of the transcription factors required for CYP2E1 expression

O. Doran, J.D. McGivan, R.-A.Cue and J.D.Wood, Department of Clinical Veterinary Science, University of Bristol, Langford, Bristol, BS405DU, U.K.*

Cytochrome P4502E1 (CYP2E1) is an enzyme involved in metabolism of skatole in pig liver. Low CYP2E1 expression results in repression of hepatic skatole metabolism with consequent accumulation of skatole in adipose tissue and development of boar taint. The aim of the present study was to investigate the molecular mechanism regulating CYP2E1 expression in pig liver. Functional analysis of the pig CYP2E1 promoter identified two activating elements. One of the elements contained the consensus sequence for the transcription factor HNF-1. The other element contained the sequence which binds a transcription factor, identified as COUP-TF1 by gel supershift assay. Binding of HNF-1 and COUP-TF1 to the CYP2E1 promoter was demonstrated by electrophoretic mobility gel shift assay. Cloning and sequencing of HNF-1 cDNA and estimation of HNF-1 mRNA levels did not reveal any differences between animals with high and low skatole deposition. The CYP2E1 promoter activity could be inhibited by the other compound of boar taint, androstenone. The effect of androstenone was mediated via inhibition of binding of the transcription factor COUP-TF1 but not HNF-1. This study has identified for the first time the transcription factors involved in the regulation of pig CYP2E1 expression and reported androstenone as a factor affecting CYP2E1 promoter activity *in vitro*.

Pig peripheral blood mononuclear leucocyte (PBML) sub-sets are heritable, and genetically correlated with performance

M. Clapperton, S.C. Bishop and E.J. Glass. Roslin Institute (Edinburgh), Midlothian, UK

Markers used to select pigs for increased resistance to infection must be heritable and, preferably, be associated with improved performance. We measured the heritability of a range of immune traits and their genetic and phenotypic correlations with performance. We measured immune traits in 495 pigs and performance in 1568 pigs from eight farms. All pigs were apparently healthy. Immune traits were total white blood cells, and peripheral blood mononuclear leucocyte (PBML) sub-sets positive for CD4, CD8, $\gamma\delta$ TCR$^+$ cells, NK cells (CD11R1$^+$ cells), B cell and monocyte markers at start- and end-of-test. At both timepoints, all immune traits were moderately to highly heritable except for CD8$^+$ cells. At end-of-test, heritability estimates (h^2±se) were (0.25±0.11) for total white blood cells, and for PBML sub-set proportions, (0.52±0.14) for $\gamma\delta$ TCR$^+$ cells, (0.62±0.14) for CD4$^+$ cells, (0.44±0.14) for CD11R1$^+$ cells, (0.58±0.14) for B cells and (0.59±0.14) for monocytes. There were also negative genetic correlations between numbers and proportions of CD11R1$^+$ cells, B cells and monocytes and end-of-test performance (-0.30 to -0.72) and similar effects at start-of-test for CD11R1$^+$ cells and B cells. There were also negative, although weaker, phenotypic correlations between these cell types and performance. Genetic correlations were strongest between CD11R1$^+$ cells and performance. These results suggest that selecting boars with lower levels of some immune traits, especially CD11R1$^+$ cells, could lead to higher performing progeny.

High rate of null allele occurrence in OLA-DRBps microsatellite locus confirms hypervariability of the motif

S. Qanbari[1], H. Mehri[1], R. Osfoori[2] & M. Pashmi[3], [1]Research Institute of Agricultural Physiology & Biotechnology, University of Zanjan, P.O. Box, 313, Zanjan, Iran, [2]Research Center of Agricultural-Jihad, Zanjan Province, [4]Department of Animal science, Abhar Azad University, Iran

Immunological importance of the MHC genes and their possible role in disease resistance has been a major impetus for research on the MHC system in sheep. In this study, the allelic variation at a microsatellite locus in intron 5 of the ovine Pseudo-Gene *DRB2* was investigated in 144 unrelated Afshari sheep sampled from 22 different flocks in Zanjan province. Microsatellite was amplified by polymerase chain reaction (PCR), followed by 7% polyacrylamide gel electrophoresis. Interestingly, a high rate of null alleles as 42% of genotyped animals was observed. The observation could be explained by accumulation of mutations in the *OLA-DRBps*. MHC-DRB2 microsatellite showed a high level of polymorphism; as such 11 alleles were detected and the most frequent alleles in Afshari sheep were *5 (42.3%), *2 (21.2%) and *1 (13.3%), respectively. The values for heterozygosity coefficient and polymorphic information content were 0.87 and 0.83 respectively. Corresponding to high rate of null alleles indicating that the locus *OLA-DRBps* could not be used as an informative genetic marker in Afshari sheep. Study of MHC-DRB2 allelism to estimate host resistance to parasites in sheep, is a possible application of this technique, although these experiments were not designed to test this.

Six novel alleles in exon 2 of BoLA-DRB3 gene in Azerbaijan Sarabi cattle

M. Pashmi[1]*, M. Moini[1], S. Qanbari[2], A. Javanmard[3], [1]Department of Animal science, Abhar Azad University, Iran, P.O. Box 22, [2]Research Institute of Agricultural Physiology & Biotechnology, University of Zanjan, Iran, [3]West and North-West Agriculture Biotechnology, Department of Genomics, Tabriz, Iran

Major histocompatibility complex (MHC) class II locus DRB3 was investigated by PCR based restriction fragment length polymorphism (PCR-RFLP) assay. A total of 136 Sarabi cows were sampled from Sarabi Breeding Station. A two-step polymerase chain reaction was carried out in order to amplify a 284 base-pair fragment of exon 2 of the target gene. Second PCR products were treated with three restriction endonuclease enzymes *RsaI*, *BstYI* and *HaeIII*. Digested fragments were analyzed by 8% polyacrylamid gel electrophoresis. Twenty-six BoLA-DRB3.2 alleles were identified. Six new allele types (BoLA-DRB3.2*34, *35, *42, *46, *51, *vaa) observed in this study have not been reported previously. The six most frequent alleles (DRB3.2 *6, *16, *23, *46, *kba and *vaa) accounted for 64.7% of the alleles in the population. With respect to the high number of the observed alleles in this survey and the novelty of some alleles with no previous record of reporting, it is plausible to conclude that the BoLA-DRB3.2 locus is highly polymorphic in Azerbaijan Sarabi cows.

Survey of diversity genetic growth hormone and growth hormone receptor genes in Iranian indigenouse sheep breed (kordy sheep) using a non-radioactive SSCP

S.A.K. Shiri, D.A. Saghi, M.R. Nasiri, H. Omrani, A.k. Masodi, F. Montazer Torbati and M. Mohammadzade, Iran-Mashhad, Agricultural & natural resources research center of khorassan, P.O.BOX.91735-1148

Evaluation of the genetic diversity for two genes in 100 animals of Iranian indigenouse sheep breed (kordy sheep) was done. A non-radioactive method to allow single-strand conformation polymorphism (SSCP) detection was optimized, starting from genomic DNA and PCR amplification of two fragments: exon 4 of the growth hormone gene and exon 6 of the growth hormone receptor gene. In our study, the fragment exon 4 of growth hormone showed two conformational patterns. Exon 6 of growth hormone receptor showed no polymorphism. The absence of diversity does not imply that genes are not polymorphic. It only means that the primers used do not delimitate a polymorphic region. These Data provide evidence that Iranian indigenouse sheep breed (kordy sheep) has genetic Variability, which opens interesting prospects for future selection program and also preservation strategies. Also our data show that PCR-SSCP is an appropriate tool for evaluating genetic variability.

Direct DNA test to detect *Callipyge* mutation in some Iranian native sheep

S. Qanbari[1], R. Osfoori[2] and M. Pashmi[3], [1]Research Institute of Agricultural Physiology & Biotechnology, University of Zanjan, P.O. Box, 313, Zanjan, Iran, [2]Research Center of Agricultural-Jihad, Zanjan Province, [3]Department of Animal science, Abhar Azad University, Iran*

A trait with financial impact on sheep production is the growth efficiency and *Callipyge* gene is a well known major gene concerned with the muscle hypertrophy as well as growth traits in sheep. A direct test to detect *Callipyge* mutation originated in the Dorset sheep was carried out using the PCR based restriction fragment length polymorphism assay in Iranian Afshari, Ghezel, Moghani, and Baluchi breeds. To this purpose, different numbers of animals from each breed (58, 23, 18 and 17, respectively) were tested by specific primers and PCR products were then digested overnight using *AvaII* restriction enzyme. Resulting products were separated by electrophoresis on 7% polyacrylamide gel and was stained with silver nitrate. A 214bp fragment was amplified in all breeds. There was no difference between digestion patterns and all genotyped animals showed NN genotype. As such, two 130 bp and 84 bp fragments from enzyme digestion were observed for all animals indicating that the restriction site surrounding an A→G transition as functional mutation of *Callipyge* phenotype was absent in the experimental population of studied breeds. This preliminary result would assist in making decision on exploiting of *CLPG* gene in Iranian native sheep breeds.

Sexing of preimplantation mammalian embryos using PCR amplification

D.E. Ilie, C. Stanca and I. Vintila, USAMVB, Faculty of Animal Science and Biotechnology, Timisoara, 300645, Romania

Sexing of embryos is an objective that has been studied from several perspectives and more then that producing embryos of predetermined sex have a high importance in modern breeding. The sex determination assay should be capable of functioning with a minimal amount of DNA, such as that obtained from 1-2 cells (blastomers). Thus, successful sex determination requires that the assay has extreme sensitivity. In our study we detect the sex of the bovine and mouse embryos trough PCR based method from a small amount of DNA sample.

Sex-specific sequences of related species published were detected in the hope that in the future the preselection of offspring's of agriculturally important species can be apply at a high level. Sequences of several primers applied in bovine were analyzed by the Matthews M.E. and Reed K.C. (Genomics, 13 (4), 1267-73, 1992). For the mouse sexing the MZFX and MZFY primers were used to increase the sensitivity of the assay. Using the both type of primers under modified reaction conditions, bovine and mouse embryos could be sexed successfully. The technology used in the present study is very effective and can be accomplished in about 3 hours for a relative large number of samples. Hopefully, this method could be planned for use in one-cell PCR for embryo sexing and genotyping in modern breeding.

A mixture model analysis for ascites in broilers
S. Zerehdaran[1], E.M. van Grevehof[2], E.H. van der Waaij[2] and H. Bovenhuis[2], [1]Animal science Group, Gorgan Agricultural University, Gorgan, Iran, [2]Animal Breeding and Genetics Group, Wageningen University, PO Box 338, 6700 AH Wageningen, The Netherlands

The objective of the present study was to use bivariate mixture models to study the relationships between body weight (BW) and ascites traits. Birds were housed in 2 groups under different conditions. In the first group, BW, the ratio of right ventricular weight to total ventricular weight (RV:TV), and hematocrit value (HCT) were measured in 4,202 broilers under cold conditions; in the second group, the same traits were measured in 795 birds under normal temperature conditions. The distribution of RV:TV and HCT were skewed under cold conditions, suggesting different underlying distributions. Fitting a bivariate mixture model showed that there was only one homogeneous population for ascites traits under normal conditions, whereas there was a mixture of (2) distributions under cold conditions, non ascitic and ascitic birds. In the distribution of nonascitic birds, the phenotypic correlations of BW with RV:TV and HCT were close to zero (0.10 and −0.07, respectively), whereas for ascitic birds, the phenotypic correlations of BW with RV:TV and HCT were negative (−0.39 and −0.4, respectively). The negative correlations of BW with RV:TV and HCT in ascitic birds resulted in negative overall correlations of BW with RV:TV (−0.30) and HCT (−0.37) under cold conditions. The present results indicate that the overall correlations between BW and ascites traits are dependent

---

Non genetic and genetic analysis of some traits in rabbits. I - Non genetic factors
A.S. Khattab[1], A. Ghany[2], A. El Said, M. Abou-Zeid[1] and Rabab, M. Kassab[1], [1]Animal production Department, Faculty of Agriculture, Tnata University, Egypt and [2]Animal Production Research Institute, Ministry of Agric., Egypt*

This work was carried out at Sakha Experimental Rabbits, Kafr EL- Sheikh, Animal Production Research Institute, Ministry of Agriculture, Dokkki, Cairo, Egypt, during the period from October 2000 July 2003. Breeds used are four Foreign breeds (New Zealand White (NZW), Californian (CAL), Bauscat (Bas) and Flander (FLAN)). In addition, Baladi Black (BB) as local breed was also used. Traits studied are litter size at birth (LSB), litter weight at birth (LWB), bunny weight at weaning (WW) and marketing weight at 8 weeks(MW). Means of LSB were 8.37, 7.25, 6.93, 6.69 and 6.79 for FLAN, CAL, NZW, Bas and BB, respectively. Means of LWB were 395,349, 384, 366 and 335 gm, respectively. Means of WW were 292, 294, 357, 348 and 312 gm, respectively. Means of MW were 927, 875, 868, 867 and 750gm, respectively. Year and month of kindling had a significant effect on all traits studied for all breeds and rabbits born during December, January and February had the highest LSB, LWB, WW and Mw, while, those born during April, September and November had the lowest traits studied. Parity had a significant effect on all traits studied for all breeds and in most cases all traits studied increased with advance in parity and reached the maximum in 3rd parity. Sex had no significant effects on al traits studied.

Non genetic and genetic analysis of some traits in rabbits. II - Genetic analysis

A.S. Khattab[*1], A. Ghany[2], A. El Said, M. Abou-Zeid[1] and Rabab. M. Kassab[1], [1]Animal production
Department, Faculty of Agriculture, Tnata University, Egypt [2]Animal Production Research Institute,
Ministry of Agric. Egypt*

The present study was carried out at Sakha Experimental Rabbits, Kafr EL- Sheikh, Animal Production Research Institute, Ministry of Agriculture, Dokkki, Cairo, Egypt, during the period from October 2000 July 2003. Breeds used are four Foreign breeds (New Zealand White (NZW), Californian (CAL), Bauscat (Bas) and Flander (FLAN)). In addition, Baladi Black (BB) as local breed was also used. Traits studied are litter size at birth (LSB), litter weight at birth (LWB), bunny weight at weaning (WW) and marketing weight at 8 weeks(MW). Data were analysis by using Multi Trait Animal Model (MTAM). Each breed was analysis separately. The model included direct, maternal genetic, permanent environmental and residual as a random effects, month and year of kindling, parity and sex as a fixed effects. Direct heritability for LSB ranged from 0.01 to 0.13, for LWB ranged from 0.20 to 0.29, for WW ranged from 0.21 to 0.43 and for MW ranged from 0.31 to 0.44. Maternal heritability ranged from 0.20 to 0.30 for LSB, from 0.5 to 0.21 for LWB, from 0.01 to 0.17 for WW and from 0.02 to 0.10 for MW. Results show that maternal heritability for LSB are higher than those estimates at weaning and marketing. Phenotypic and genetic correlations among all traits and breeding values are also estimated.

Genetic study of milk production of Alpine goats in Morocco

*A. Lakhdar [1], N.Roudies [2], A. Tijani[1], [1]Ecole Nationale D'Agriculture, Meknès, Morocco, [2]Domiane
Agricole Douiet, Fès,Morocco*

The aim of this study was to investigate environmental effects on daily milk yields of French Alpine does raised in Morocco, and to estimate their breeding values for this trait. Data analysed were 13,206 records of test day milk yields from 1811 lactation of 624 does generated during the years (1996-2004) at Domaine Agricole Douiet farm in Fes region. Environmental effects were analysed by general linear model. Breeding values were estimated by BLUP Programs of Misztal, using a single-trait repeatability animal model. The genetic trend was determined by averaging the breeding values of does on birth year. Mean and standard deviation for test day milk yields were respectively 3.4 kg and 1.2 kg. A very highly significant (p<0.001) variation in test day milk yields was observed due to age of kidding-parity, type of kidding, year and season of kidding, test date, stage of lactation. Estimated heritability was 0.36.The genetic trend for test day milk yields was generally positive but a fluctuation was observed in the average breeding value for the whole period analysed. The results of our study suggest that the implementation of a genetic evaluation is likely to improve current selection decisions to optimise genetic gain.

A PCR-RFLP method for the analysis of Egyptian goat MHC class II DRB gene

S. Ahmed and O.E. Othman, Department of Cell Biology, National Research Center, Dokki, Egypt

This study aimed to analysis the genetic polymorphisms of MHC class II DRB gene in the Egyptian goat using PCR-RFLP method. The amplified fragment with size of 285-bp was digested by two restriction enzymes TaqI and PstI. Restriction digestion of PCR product by TaqI enzyme respresented two digested fragments at 122- and 163-bp (T restriction pattern) or undigested fragment at 285-bp (t restriction pattern). The freaquencies of TT, Tt and tt patterns of MHC class II DRB in the Egyptian goats were 29.5, 61.4 and 9.1%, respectively. After the PstI digestion of the 285-bp PCR products amplified by the Egyptian goat DNA, the results showed that the frequency of pp pattern (270-and 15-bp restricted fragments)was 29.5% and the frequency of heterozygous Pp pattern (270-, 226-, 44- and 15-bp restricted fragments)was 70.5% while the PP pattern (226-, 44- and 15-bp restricted fragments) was not displayed in tested goat animals. The results declared that the p pattern (64.8%) is dominant over the P pattern (35.2%) in Egyptian goat.

Genomic analysis of the CSN1S2 and B-Lg loci in two Czech goat breeds

Z. Sztankoova[1], V. Matlova[1], C. Senese[2], T. Kott[1], J. Soldat[1] and G. Mala[1], [1]Research Institute of Animal Production, Prague – Uhrineves, Czech Republic, [2] Dipartimento di Scienze delle Producioni Animal, Universita degli Studi della Basilicata, Potenza, Italy*

In goat species, three Ca-sensitive casein (αS1, β- and αS2) show alleles with strong differences in the level of expression resulting in differences not only in protein quality but also in the physico-chemical properties of milk. β-Lg is a globular protein belonging to the lipocalin family. It is one of the major whey proteins present in the milk of ruminants. The aim of present study was to describe allele at the goat CSN1S2 locus associated with a high, medium and null amount of this protein fraction, and allele at β-Lg locus, at promoter region in dairy goats kept in the Czech Republic. A total of 81 goats belonging to White Short-Haired (31, WSH) and Brown Short-Haired (50, BSH) breed were analyzed. At locus CSN1S2 F allele was predominant, in both breed (WSH, BSH), F=0.66%, N=0.34 % and F=0.88%, N=0.12%, respectively, where N=A, B, C, E and G. Variant -60C at β-Lg locus was predominant in both breed, (WSH, BSH), -60C= 0.81%, -60T=0.19%, and -60C= 0.72%, -60T=0.28%, respectively.

This work has been supported by the Ministry of Agriculture of the Czech Republic (Institutional Programme) MZE 0002172701401 and National Programme (IG57051).

Study of whole blood potassium polymorphism and its relationship with other blood electrolytes of Kermani sheep in Iran

H. Moradi Shahrbabak[1]*, M. Moradi Shahrbabak[1], G.Rahimi[2], H. Mehrabani Yeganeh[1], [1]Department of Animal Science, Faculty of agriculture, University of Tehran, P.O. Box 31587-11167,Karaj, Iran, [2]Department of Animal Science, Faculty of agriculture, University of mazandran, sari, Iran

The whole blood potassium concentration has shown the bimodal distribution in sheep, which has been classified into LK and HK types. Blood potassium concentration of 188 animals ranged from 8 to 44 meq/Iit, based on this factor animals to whole blood potassium concentration of 18 meq/lit and below were classified as low potassium (LK) type and those with 23 meq/lit of blood and above were classified as high potassium (HK) type. Mean and range of Blood potassium concentration for LK animals were 12.086 meq/lit and 8-18 meq/lit and for HK animals were 32.614 meq/lit and 23-44 meq/lit respectively. The frequency of HK gene was found to be 0.902. Concentration of sodium, calcium in whole blood were also determined, the mean and range of blood sodium concentration were 1737.36 PPM and 343-5000.04 ppm respectively. The relationship between potassium and sodium concentrations in whole blood of sheep was significant. And negative estimated correlation around –0.19 which was significant.

The mean of whole blood sodium concentration was 3020.9 ppm and 2672.5 ppm for LK and HK respectively. Remarkable differences in calcium and magnesium concentrations were not recognized between LK and HK types.

BW, WW, 6MW and YW did not differ significantly between the LK and HK in this study.

Study of whole blood potassium biochemical polymorphism and its relationship with other blood electrolytes of varamini sheep in Iran

H. Moradi Shahrbabak[1]*, M. Moradi Shahrbabak[1], H. Mehrabani Yeganeh[1], M,Soflaei[2], [1]Department of Animal Science, Faculty of agriculture, University of Tehran, P.O. Box 31587-11167,Karaj, Iran, [2]Kerman's Jihad-e-Agriculture Educational center, Iran

The whole blood potassium concentration has shown the bimodal distribution in sheep, which has been classified into LK and HK types; HK allele is recessive to LK with a single gene inheritance. Blood potassium concentration of varamini Sheep ranged from 6.5-28.06 meq/Iit, based on this factor animals to whole blood potassium concentration of 13.59 meq/lit and below were classified as low potassium (LK) type and those with 18.32 meq/lit of blood and above were classified as high potassium (HK) type. Mean and range of Blood potassium concentration for LK animals were 8.2 meq/lit and 6.5-13.59 meq/lit and for HK animals were 23.54 meq/lit and 18.32-28.06 meq/lit respectively. The frequency of HK gene was found to be 0.908. Concentration of sodium, calcium and magnesium also determined. The relationship between potassium with sodium and calcium concentrations in whole blood of sheep was significant. And negative estimated correlations were around –0.39 and -0.33 respectively which were significant.

The mean of whole blood sodium and calcium concentration were 5147.4 and 86.54 ppm for LK and 4490 and 78.33 ppm for HK respectively and all of differences were significant($p < 0.05$). The mean of whole blood magnesium concentration was 31.73 ppm for LK and 35.12 ppm for HK respectively and this differences wasnt significant($p < 0.05$).

Genetic porlymorphism of hemoglobin, transferrin and ablumin and investigation of their relationship with some productive traits in Kermani sheep of Iran

Z. Ghadiri[1], M. Moradi Shahrbabak[1], G. Rahimi[2], H. Moradi Shahrbabak[1]*, [1]Department of Animal Science, Faculty of agriculture, University of Tehran, P.O. Box 31587-11167, Karaj, Iran. [2]Faculty of agriculture, University of mazandran, sari, Iran

Data on 300 Kermani sheep from Kermani Breeding Center of Shahrbabak were used in this research. Horizontal polyacryle amide electrophoresis was used to detect polymorphism of Transferrin and Albumin, and Acetate cellulose was used for Hemoglobin. Three genotypes of Hemoglobin were observed(AA, AB and BB with frequency of 4.6, 43.8 and 51.6 respectively). Frequency of A allele of Hemoglobin in Kermani sheep was further than other investigated Iranian sheep. For Transferrin 8 Allels were determined, (B, C, A, D, E, G, Q and L in frequency 46.7, 27.11, 12.82, 9.16, 2.01, 1.28, 0.55 and 0.37 respectively). This alleles make 19 genotypes that BC genotype had the most frequency and AL & LD genotypes had the less frequency. Albumin Electrophoresis results showed two genotypes SS and SW with frequency of 98% and 2% respectively. Afther genotyping of all sheep for the proteins, effect of proteins on body weight, daily weight gain and wool production was investigate. Trensferrin and Hemoglobin effect were singnificant on 12 month weight (P<0.05), that sheep with BB Hemoglobin type and CE Transferrin type had a highest weight. Effect of proteins on other trait were not significant although some genotypic groups of hemoglobin and transferrin had a significant differences.

Dominant white in sheep (*Ovis aries* L.) by segregation analysis

C. Renieri[1]*, A. Valbonesi[1], V. La Manna[1], M. Antonini[2] and J.J. Lauvergne[3], [1]Dipartimento di Scienze Ambientali, Università di Camerino, 62032 Camerino, Italy, [2]ENEA C.R. Casaccia Biotec-Agro, 00060 S.M. di Galeria, Roma, Italy, [3]Departement de Génétique Animale, CRJ, 78350 Jouy-en-Josas, France

Coat color heredity patterns, from crosses and paternal backcrosses involving black and brown spotted Merinos rams and full white Merinos derived ewes, were investigated through segregation analyses. Statistical tests validate the hypothesis of a dominance with complete penetrance of: *i*) full white over pigmented, *ii*) black over brown, *iii*) uniform over spotted pattern, as well as, of a dominant epistasy of white/pigmented over either black/brown or uniform/spotted. Moreover, the pigmented pattern segregations are independent of the sex of the individuals. These results largely agree with those already reported in literature, yet they are in contrast with the Lauvergne's hypothesis of a codominance between full white and pigmented patterns. It was, thus, concluded that at least two different phenotypic models appear to exist in sheep.

Genetic potentials of Naeini sheep for pre-weaning growth rate
V. Edriss and M.A. Edriss, Isfahan University of Technology, Department of Animal Science, 84156 Isfahan, Iran*

Among 27 breeds of sheep, Naeini sheep is the widest spread breed of sheep in Iran. This breed has an important economic impact for central rural area of the country. In order to estimates genetic parameters 27 rams and 421 ewes were mated in Naeini breeding station in Natanz town which is located in central part of Iran. The lambs were weaned at 90 ± 10 days of age. Birth, weaning and daily gain weights were analyzed using different animal models. Results showed that average of birth and weaning weights for them were $3.1\pm.66$ and 25.45 ± 3.83 Kg which confirmed they are among small size category but average daily gain was 245 ± 41.5 g compare to Bakhtiary sheep which is one of the heaviest with good feedlot performance (248 ± 40 g) shown that Naeini sheep is a suitable breed for meat production. Tow regression models showed that by increasing birth weight (BW), weaning weight (WW) would increase (WW $=.876$BW$+23.37$), but increasing birth weight would decrease daily gain (DG $= -1.36$BW$+249.2$). Heritability estimates for birth and weaning weights were $.44\pm.01$ and $.19+.02$ respectively. When maternal environment effects were added to the model, they were $.34\pm.02$ and $.12\pm.09$ respectively. Substitution of additive genetic maternal effects for maternal environmental effects did not change WW heritability ($.12\pm.07$) while BW heritability was decrease to $.17\pm.01$. It shows maternal environment (may be milk production) is more important than genetic maternal effects for weaning weight.

The evaluation of Kleiber ratio as a selection criterion for ram selection in Torky-Ghashghaii sheep breed
B. Eilami and M. Safdarian, Fars Research Center for Agriculture and Natural Resources, P.O.Box 71555-617, Shiraz, Iran*

Mainly, the factors of birth weight, weaning weight and weight of six and twelve months are used as selection criterion for improvement of mutton and lamb production in sheep. In this study kleiber ratio was checked as an indirect selection criterion in Torky-Ghashghaii sheep breed. Kleiber ratio is defined as the average daily gain per one unit of metabolic weight in final test period (ADG/ $W^{0.75}$). The data were obtained from a feedlot experiment on 50, 50 and 60 head male lambs of Torky-Ghashghaii breed in three years respectively. Lambs were individually fed after post-weaning and the body weight, ADG, feed consumption and feed conversion ratio were measured every 15 days. The genetic and phenotypic parameters of birth weight, weaning weight, initial weight, final weight and average ADG during fattening period, feed efficiency and kleiber ratio were estimated using a sire model. The heritability estimates for kleiber ratio was 0.23 (± 0.02). The genetic and phenotypic correlation of kleiber ratio with daily gain was 0.96. The heritability of feed efficiency was 0.28 and had a high genetic correlation with ADG. The genetic and phenotypic correlation of kleiber ratio with birth weight, weaning weight, initial weight and final weight was lower than genetic correlation of ADG with these characters. Kleiber ratio showed high correlation with feed efficiency in this breed, thus kleiber ratio can be used as indirect selection criterion to improve feed efficiency.

First results on relationship between BLG polymorphism and milk fatty acid composition in milk of Massese Dairy ewes
G. Conte, M. Mele, A. Serra and P. Secchiari. University of Pisa DAGA-Scienze Zootecniche, via del Borghetto 80, 56100 Pisa, Italy

Beta-lactoglobulin (BLG) is the major whey protein in ruminant milk. BLG polymorphisms were detected in several species. In ewe, three co-dominant alleles are known (A, B, C), which differ by one amino-acid replacements: BLG-A$\rightarrow$BLG-B= TyrA$_{20}\rightarrow$HisB$_{20}$, while BLG-A$\rightarrow$BLG-C= GlnA$_{148}\rightarrow$ArgC$_{148}$. The A and B forms of BLG are ubiquitous, while the BLG-C isoform was reported in few ewe breeds. The influence of milk beta-lactoglobulin phenotype on the fat, protein and fatty acids of 163 individual milk samples from Massese ewes were considered. Sampling was conducted randomly on five dairy ewe farms located in the area of origin of the Massese breed. Three beta-lactoglobulin genotypes were detected: their frequencies were AA 20.25%, AB 63.80% and BB 15.95% respectively. In agreement with previous studies, parameters of milk chemical composition did not result significantly affected by milk beta-lactoglobulin polymorphism. For the first time, a significant effect of beta-lactoglobulin phenotypes on milk fatty acid composition was revealed. The study highlighted a co-dominant effect of beta-lactoglobulin alleles on the concentration of some milk fatty acids, particularly for *trans* fatty acid content, which was higher in heterozygous sheep. Milk from AB sheep also contained lower amount of saturated fatty acids and medium-chain fatty acids and higher levels of long-chain fatty acids.

Age correction factors for milk yield and fat yield of Friesian cows in Egypt
S.M. Yener[1], N. Akman[1] A.S. Khattab[2] and A.M. Hussein[3], [1]Animal Science, department, Ankara University, Turkey, [2]Animal production Department, Faculty of Agriculture, Tanta, Tnata University, Egypt, [3]Animal Production Research Institute, Ministry of Agric. Egypt

The effect of month and year of calving, parity and age at calving on 305 day milk yield (305 d MY) and 305 day fat yield (305 d FY) was investigated on 2181 lactation records of Friesian cows kept at Sakha farm belonging to the Ministry of Agriculture in Egypt. A least squares analysis of variance of the data showed significant effect of month and year of calving and age as a linear and quadratic terms on 305 d MY and 305 d FY. Estimates of partial linear and quadratic regression coefficients of 305 d MY and 305 d FY on age at calving were significant and being 7.65 (0.028) kg/mo and -0.089 (0.028) kg/mo^2, respectively for 305 d MY and 0.302 (0.095) kg/mo and -0.002 (0.0008) kg/mo^2, respectively for 305 d FY. A set of different age correction factors was derived by fitting a polynomial of second degree of production (305 d MY and 305 d FY) on age at calving. Factors used for adjusting 305 d MY could be applied to 305 d FY without substantial loss in accuracy.

Genetic aspects for productive and reproductive traits in Friesian cows in Egypt

S.M. Yener[1], N. Akman [1], A.S. Khattab*[2] and A.M. Hussein[3], [1]Animal Science, department, Ankara University, Turkey, [2]Animal production Department, Faculty of Agriculture, Tanta, Tnata University, Egypt, [3]Animal Production Research Institute, Ministry of Agric. Egypt

A total of 2818 normal lactation records of Friesian cows kept at Sakha Experimental farm, belonging to Ministry of Agriculture, during the period from 1996 to 2002 were used to estimate phenotypic and genetic parameters for productive traits (milk yield (MY), fat yield (FY), protein yield (PY), lactation period (LP), and days dry (DP) and reproductive traits (days open (DO) and Calving interval (CI)).Data were analyzed using multi trait animal model according to Boldman et al. (1995). Least squares analysis of variance showed significant effects of month, and year of calving and parity on all traits studied. Heritability estimates were 0.14, 0.11, 0.14, 0.15, 0.05, 0.06 and 0.03 for MY, FY, PY, LP, DP, DO and CI, respectively. Genetic correlations among MY, FY, PY and LP were positive and high and ranged from 0.60 to 1.00. Also, genetic correlations between DP, DO and CI and each of MY, FY and PY were positive and ranged from 0.21 to 0.49. While, genetic correlations between DP and each of DO and CI were negative. Phenotypic correlations among all traits are similar to genetic correlations.

Genotyping progeny tested bulls in the Irish dairy herd for polymorphisms in possible candidate genes for body condition

B.M. Kearney*[1], M. Daly[1], F. Buckley[2], T.V. McCarthy[3], P.R. Ross[1] and L. Giblin[1], [1]Moorepark Food Research Centre, Teagasc, Fermoy, Co. Cork, Ireland, [2]Dairy Production Research Centre, Teagasc, Fermoy, Co.Cork, Ireland, [3]Biochemistry Dept. UCC, Cork, Ireland

Dairy cows with poor body condition directly after parturition, experience reduced fertility and increased susceptibility to disease. Body condition in dairy cattle has a high heritability and much research is focused on understanding the genetic component of this complex phenotype. The aims of this study were to (1) screen for well documented polymorphisms in potential candidate genes (leptin, DGAT, thyroglobulin, mitochondrial transcription factor and PPARGC1) and (2) describe the leptin promoter, in 15 bulls ranked highly or lowly for Body Condition Score (BCS) based on evaluations of their daughter progeny. Six Holstein Friesian bulls ranked in the top 2.5% for BCS and nine Holstein Friesian bulls ranked in the bottom 3% for BCS, from a total number of 1462, were selected. Genomic DNA was isolated from semen straws. PCR-RFLP was used to screen for polymorphisms in the candidate genes. In this study, no association was found between these polymorphisms and BCS. Both novel and previously described polymorphisms were identified in 2.4kb of the leptin regulatory region. Although a number of these polymorphisms were located on, or close to, putative regulatory motifs (TESS predictive software); they did not associate with BCS in our experimental set.

Genetic relationships between shapes of lactation curve and changes of body weight during lactation in Holsteins

T. Yamazaki[1], H. Takeda[1], A. Nishiura[1], T. Beppu[2], S. Kawata[2] and K. Hagiya[2], [1]National Agricultural Research Center for Hokkaido Region, 062-8555, Sapporo, Japan, [2]National Livestock Breeding Center, 961-8511, Fukushima, Japan

We estimated genetic correlations between shapes of lactation curve and changes of body weight during lactation using daily lactation data. The data analyzed was consisted 687 first-lactation Holstein cows, which had 182,620 records of milk yield and 39,424 records of body weight during lactation. Four traits were defined for shapes of lactation curve, 305-d milk yield (M1), linear regression coefficient of milk yield from 5 to 35 DIM (M2), difference in average milk yield 36-95 DIM and 30-40 DIM (M3), and linear regression coefficient of milk yield from 100 to 240 DIM (M4). Two traits were defined for changes of body weight, changes of average body weight from 0-7 to 10-50 DIM (W1) and from 10-50 to 200-240 DIM (W2). Positive genetic correlations were estimated between M1, M2 and W1, and M1, M3 and W2, and negative genetic correlations were estimated between M4 and W1, and M2, M4 and W2. It was indicated that genetic relationships between shapes of lactation curve and changes of body weight are negative in early-middle lactation stage, on the other hand, are positive in later lactation stage.

Random regression analysis of cattle growth path - a cubic model

H.R. Mirzaei[1], A.P. Verbyla[2], M.P.B. Deland[3] and W.S.Pitchford[2], [1]University of Zabol, Iran, [2]University of Adelaide, Roseworthy Campus SA 5371 Australia, [3]SARDI, Struan, Naracoorte SA 5271 Australia*

A cubic random regression analysis was conducted to model growth of crossbred cattle from birth to about two years of age. Hereford (H) cows (581) were mated to 97 sires from Angus (AxH), Belgian Blue (BxH), Hereford (HxH), Jersey (JxH), Limousin (LxH), South Devon (SxH) and Wagyu (WxH), resulting in 1144 steers and heifers born over 4 years. The model for ln(wt) included fixed effects of sex, sire breed, age (linear, quadratic and cubic), as well as two-way interactions between the age parameters and sex or breed. Random effects were animal (genetic relationship), dam, residual between animal, linear age by animal, age by dam, management group, management by age as linear, quadratic and cubic. At birth, AxH, JxH and WxH were 7, 18 and 11% lighter and than purebred Hereford calves. Growth rate increased during the pre-weaning period with a change in the rank of breeds during the first 6 months. Relative growth rate was higher in the smaller breeds (JxH and WxH) than the heavy breeds. Genetic (r_G) and maternal permanent environmental (r_C) correlations between birth weights (constant) and linear growth rate was strong and negative (-0.87 ±.04 and -0.96 ± 0.04, respectively). The model is being used to predict the effect of growth path on carcass quality.

Estimation of genetic parameters for first lactation milk yield of holstein cows using random regression model

Neivein Gamal[1]*, Manal Elsayed[2] and E.S.E. Galal[2], [1]Cattle Department, Animal Production Research Institute, Nady Elseid st., Dokki, Giza, Egypt [2]Animal Production Department, Faculty of Agriculture, Ain Shams University, Shoubra El-Khaima, Cairo, Egypt

Data used in this study were collected from the Assiout private Farm in Assiout Governorate in the South of Egypt. In total, a data set of 3875 test-day milk yield (TDMY) records for the first lactation of 414 cows daughters of at least 66 sires and 197 dams were available from 1998 till 2004. Data were classified according to the month of calving into four seasons, winter, spring, summer and autumn. The statistical model included year-season, the linear and quadratic order of age, fixed regression, a random additive genetic effect for each animal, a random permanent environmental effect for each cow, and a random residual effect. The incomplete Gamma function (IGF) was chosen to describe the shape of the lactation curve. This function was fitted for each lactation for each cow. DFREML software was used to estimate the components of (co)variance of TDMY in a Random Regression model (RRM). Estimates of the phenotypic, additive genetic and permanent environmental correlations between daily milk yields ranged form 0.07 to 0.68, 0.18 to 0.98 and -0.6 to 0.99, respectively. Estimates of heritability varied from 0.03 for DIM 65 and 275 to 0.14 for DIM 185.

Investigations on two candidate genes on chromosomes 14 and 20 and their influence on milk production and functional traits in Bavarian cattle breeds

T. Seidenspinner[1]*, R. Fries[2] and G. Thaller[1], [1]Institute of Animal Breeding and Husbandry, Christian-Albrechts-University, 24098 Kiel, Germany, [2]Chair of Animal Breeding, Technical University of Munich, Weihenstephan, 85354 Freising, Germany

The goal of the study was to investigate the candidate genes *DGAT1* and *GHR* in the Bavarian dual purpose cattle breeds Fleckvieh and Gelbvieh. Granddaughter designs consisting of 60 (26) sires and 748 (56) sons were available for Fleckvieh (Gelbvieh). Gene substitution effects of the base substitutions K232A and F279Y on milk production and functional traits were simultaneously estimated by standard regression methodology based on estimated breeding values. The alleles K in *DGAT1* and Y in *GHR* were rare in the Fleckvieh and Gelbvieh population, respectively. *DGAT1* showed marked effects on milk production traits in both breeds, with alleles K increasing content and A increasing yield traits. In addition, favourable effects of the A allele were found for direct milking ease, direct and maternal calving ease and longevity on the nominal significance level in Fleckvieh. Milk production and functional traits in Fleckvieh were also influenced by the base substitution F279Y in the *GHR*. The F allele was, analogous to the K allele in *DGAT1*, superior for fat and protein content, but also inferior for the functional traits direct calving ease, direct stillbirth rate and longevity. In Gelbvieh, no significant effects of the F279Y polymorphism were detected.

Effect of the DGAT1 gene polymorphism on milk production traits in Hungarian Holstein cows

I. Anton, K. Kovács, L. Fésüs* and A. Zsolnai, Research Institute for Animal Breeding and Nutrition, Herceghalom, Hungary

The effect of the lysine/alanine (K232A) polymorphism in acylCoA-diacylglycerol-acyltransferase 1 (DGAT1) on dairy traits was confirmed by numerous authors, and significant differences were observed for milk fat, milk protein and milk yield in Holstein-Friesian cows. The objective of this study was to evaluate the effect of the DGAT1 locus on the milk production traits in the Hungarian Holstein Friesian population and to determine the distribution of the genotypes and allele frequencies. 250 blood samples have been collected from different Holstein Friesian herds and DGAT1 genotypes were determined by PCR-RFLP assay. Milk production data have been collected throughout three consecutive lactations and statistical analyses have been carried out to find association between genotypes and milk production traits.

The effect of milk protein genes on breeding values of milk production parameters in Czech Fleckvieh

J. Kucerova[1]*, A. Matejicek[2], E. Nemcova[1], O. M. Jandurova[3], M. Stipkova[1] and J. Bouska[1] [1]Research Institute of Animal Production, Pratelstvi 815, 104 01 Prague – Uhrineves, Czech Republic, [2]University of South Bohemia, Studentska 13, 370 05 Ceske Budejovice, Czech Republic, [3]Research Institute of Crop Production, RSV Karlstejn, 267 18 Karlstejn 98, Czech Republic.

Genotypes of milk protein genes (alpha$_{S1}$-casein, beta-casein, kappa-casein and beta-lactoglobulin) were detected in 440 individuals of Czech Fleckvieh breed using a method PCR-RFLP. Breeding values of genotyped animals were obtained from the Official Database of Progeny Testing. Statistical analysis was carried out in program SAS using the MIXED procedure and restricted maximum likelihood method (REML). The aim of this study was to find the relation between genotypes of genes and milk production parameters expressed by the breeding values of animals. The positive findings in frequencies of *CSN3* allele *B* (0.38) and genotype *BB* (13 %) were detected in observed population. Significant differences were found between genotypes of alpha$_{S1}$-casein, beta-casein and kappa-casein loci and breeding values for milk production parameters. Genotypes of beta-lactoglobulin locus were not significantly related to any milk production parameter. These findings contribute to better understanding the genetic background of Czech Fleckvieh breed and help with implementation the information to breeding process.
This study was supported by project NAZV 1G46086.

Potential effects of marker assisted selection on genetic gain in dairy cattle – results from model calculations
C. Edel and G. Thaller, Christian-Albrechts-University, Hermann-Rodewald-Strasse 6, 24118 Kiel, Germany

The potential of MAS to increase genetic gain in traditional progeny-testing schemes was analysed using deterministic simulations. Four paths of selection with overlapping generations were modelled. Selection was assumed to take place in one or several stages but not recurrently. Selection was on selection indices combining information on the EBVs of father and mother, phenotypic information on paternal half sibs, own performance and progeny (if available). Additional QTL-Information was modelled as a trait correlated to the breeding goal trait with a heritability of one. It was only used for the genetically typed candidate itself. The QTL as an information was introduced in a population both in information- and in Bulmer-equilibrium of a stable selection scheme. It was then used both in pre-selecting young bulls for progeny testing and in pre-selecting females to produce males. Comparison was to traditional schemes depending on pedigree information in these selection steps. Heritability of the breeding goal trait and the fraction of genetic variance explained by the QTL (P) were varied. Additional genetic gain reached from a 1% increase (h^2:.5, P=.05) up to more than 30% (h^2:.05, P=.5). Additional accuracy of selection can also be used to fix genetic gain but reduce testing capacities substantially between 25% and 100% depending on h^2 and P, indicating that in some cases half sib schemes might be preferable.

Clinical cytogenetics and its importance for genetic health in cattle
I. Nicolae, Research and Development Institute for Bovine, Balotesti, Romania

Clinical cytogenetics is one of the most important discipline of domestic animal cytogenetics. Applied to cattle, which represent the main important domestic species, clinical cytogenetics should receive much more attention because the preventive chromosome analysis are economically very motivated. The role of chromosome abnormalities as causes of reproductive failure is very important and involve both the chromosome number and the chromosome structure which are very ofen associated with developmetal anomalies, embryonic death and various levels of infertility. For this reason, it is necessary to keep breeding animals under cytogenetic control, particularly in cattle populations applying artificial inseminations, because the inherited aberrations can quickly become distributed in the next generations. Since reproductive performance is a very important characteristics of domestic animals, cytogenetics has been used as a diagnostic tool in conjunction with the development and refinement technologies such as *in vitro* fertilization. Another aspect is the ethical one. Artificial insemination should particularly stand for high quality breeding and it is ethically not right to use breeding animals with chromosome aberrations, which produce reduced number of offsprings, increased number of offsprings with fertility problems, high rates of stillbirth and malformation. The role of clinical cytogenetics in the genetic improvement of cattle, became very important in the new context of European integration. Much more, if we take into consideration the rules concerning the genetic health of domestic animals, this desideratum became a must.

Identification of a missense mutation in the bovine prolactin receptor (PRLR) gene and analysis of allele frequencies in four dairy cattle breeds

L. Fontanesi[1], E. Scotti[1], M. Tazzoli[1], S. Dall'Olio[1], R. Davoli[1], E. Lipkin[2], M. Soller[2] and V. Russo[1]*, [1]DIPROVAL, Sezione di Allevamenti Zootecnici, University of Bologna, Reggio Emilia, Italy, [2]Department of Genetics, The Hebrew University of Jerusalem, Jerusalem, Israel

Prolactin receptor (PRLR), a member of the superfamily of cytokine receptors, exerts its functions through binding prolactin, placental lactogen and growth hormone, that play major roles in a large number of biological processes like development, fetal growth, pregnancy, lactation and response to stress. The *PRLR* gene has been assigned to bovine chromosome 20 in which several QTL for milk production traits have been mapped in dairy cattle. Considering *PRLR* as a candidate gene for milk production as well as reproductive traits, we searched for polymophisms in this gene. Parts of the coding sequence of the bovine *PRLR* gene were investigated by PCR-SSCP and sequencing. A missense single nucleotide polymorphism (T→C) was identified by comparing animals from different breeds. Subsequently, allele frequencies were estimated in 217 animals of four dairy cattle breeds, two cosmopolitan breeds (Holstein and Brown Swiss) and two Italian local breeds (Reggiana and Modenese), using PCR-RFLP. Allele T was the most frequent in Holstein (0.97), Reggiana (0.75) and Modenese (0.69) and was about 0.50 in Brown Swiss. Further studies will be carried out to evaluate the effects of this mutation on production and reproduction traits in dairy cattle.

Detection of major genes for cortisol and creatinine levels in pigs divergently selected for stress responses

H.N. Kadarmideen*, S. Gebert, L.L.G. Janss and C. Wenk, Institute of Animal Sciences, ETH Zurich, CH-8092 Zurich, Switzerland

Pig populations (n=417) that were under ten years of divergent selection for their response to stress stimuli were investigated for segregating major gene in the inheritance of cortisol and creatinine levels. Cortisol is considered to be indicator for stress and involved in fat, protein and carbohydrate metabolism, affecting growth and meat quality. Exploratory analyses using Bartlett's test revealed marked differences between sire groups in cortisol and creatinine variance (heterogeneity of variance highly significant at $p<0.0001$) and showed that cortisol and creatinine are not controlled by the same major gene but rather two independent genes. A Bayesian segregation analysis (BSA) was performed using Monte carlo Markov Chain techniques. BSA with informative priors on and/or fixed variance for polygenic component was used. BSA model forcing complete dominance for cortisol estimated an additive effect of 86.9 and with a frequency of (cortisol increasing) recessive allele of 26%. Estimation of transmission probabilities confirmed that segregating gene is Mendelian. Similar results were found for creatinine levels, where the recessive genotype conferred high creatinine, but with a high population frequency of 93% and additive effect of 4.14. These results show strong evidence of major gene for cortisol and creatinine levels. These results would be useful in translational genomics of cortisol and creatinine in other species including humans.

Identification of polymorphisms and expression of selected porcine fetal skeletal muscle genes

K. Bilek, K. Svobodova, A. Knoll and P. Humpolicek*, Mendel University of Agriculture and Forestry Brno, Zemedelska 1, 613 00 Brno, Czech Republic

Investigation of the gene expression helps to understand mechanisms of organism or tissues development. This project is aimed at investigation of fetal skeletal muscles development in pig by means of quantitative real-time PCR. The results of subtraction hybridization analysis were used for this study. In that previous study we identified group of genes *(ACTB, ESD, H3F)* which were over-expressed. By means of real-time PCR method and SYBR Green chemistry, we exactly determined relative gene expression of individual genes. All products were verified using agarose gel electrophoresis and direct sequencing of PCR product. DNA sequences of Large White, Landrace, Piétrain and Meishan breeds were comparative analyzed and nucleotide differences were founded. This work is helpful for better understanding of skeletal muscle development of pig and the conclusions can be used for marker assisted selection in meat production.

Supported by the Czech Science Foundation no. 523/06/1302 and 523/03/H076.

Herbs and other functional foods in equine nutrition

Carey A. Williams and Emily D. Lamprecht, Rutgers University, Animal Science, New Brunswick 08901, USA

A majority of the many herbs and other functional foods on the market today have not been scientifically tested; this is especially true in equine research. The following paper will review literature pertinent to herbal supplementation in horses as well as other species. Common equine supplements like devil's claw, echinacea, garlic, ginseng, flaxseed, horse-chestnut, yucca, valerian root, etc. are not regulated and few studies have investigated the safe yet effective dose of these compounds. Ginseng is commonly studied and has been found to exert an inhibitory effect on IL-1β and IL-6 gene expression, decrease TNF-α production by macrophages, and decrease cyclooxygenase-2 expression, and suppress histamine and leukotriene release. On-going studies in horses are testing one dose of ginseng on anti-inflammatory effects post-exercise. Active ingredients in horse chestnut include a group of saponins and bioflavonoids eliciting antioxidant activity. Equine studies found that garlic fed at > 0.2 g/kg/d developed Heinz body anaemia. Glycyrrhetinic acid, found in liquorice, increased apoptosis in cells *in vitro* and elevated production of reactive oxygen species. Herbs can have a drug-like action that can interact with other components in the horses' diet. Some herbs contain prohibited substances like salicylates, digitalis, heronin, cocaine and marijuana. Drug-herb interactions are also common side effects and caution needs to be taken when determining which 'natural product' to use. Few herbs have had sufficient research to warrant concrete recommendations for efficacy and dosage in equines.

Session H30

Theatre 2

'Unusual' feeds for performance horses

P.A. Harris and H. Gee, Equine Studies Group, WALTHAM Centre for Pet Nutrition, Melton Mowbray, Leics LE14 4RT, UK

Equine feeding and stable management practices for horses kept around the world vary greatly, and, very few people feed just a simple single grain or compound feed plus roughage diet. In developed countries especially, many add additional separate feeding stuffs and supplements in order to personalise the diet, add that little bit extra to boost performance, improve health and / or correct an imbalance etc. Some of the supplementary feedstuffs that we add today have been included in the diets of horses for centuries: Garlic being one such example *'Before 1733 the diet [of racehorses] consisted of barley, beans, wheat-sheaves, butter, white wine and up to 25 cloves of garlic per feed'*. However, the feeding practices such as adding ale and dairy products is perhaps less prevalent today than in the 1700s when *new strong ale and the whites of 20 eggs or more, with no water'* was recommended. This paper will consider some of the less well discussed supplementary feedstuffs that are, or perhaps should be, added to performance horse's diets. This will include the role of certain spices such as fenugreek as palatants, and the potential health benefits of cinnamon and curcumin. It will also discuss the potential role of the 'ultra' trace and rare earth elements.

Session H30

Theatre 3

Regulation for feed additives(FA): Application to horses

A. Aumaitre* and W. Martin-Rosset, INRA Theix, 63122 Saint Genes Champanelle, France

Accurate recommendations on exact nutritional requirements of different categories of horses have not been established. It is particularly true for micronutrients, probiotics, natural products, mostly recommended on empirical basis. Thus the recommendations derive from the experience on "other species". Nutritionists and toxicologists have assessed the safety and the efficacy of these products before authorisation. An initial legislation raised by the EU derived from a compromise between the national legislation of the different countries. Presently and since 1990 a completely new legislation has been published as Directives and Regulations completing and repelling the initial legislation on the basis of the Food law 28 January 2002.
– A comprehensive list of FA has been established avoiding strictly drugs and antibiotics;
– The use of FA (added to the diet) is regulated on the risk assessment expressed by an independent source of advise (EFSA). It concerns the safety for the animals and the consumers, and welfare of animals;
– Guidelines for the assessment of the safety of FA have been established by a scientific panel (FEEDAP);
– Authorisation by the official authorities (EP-EC-CE) needs an application showing experimentally the safety and favourable effects. Maximum level of nutrients and toxicants are officially listed (Council Directive 1334/2003).
Despite the horse species is up to now poorly documented (except for enzymes and micro-organisms), a comprehensive summary of the recommendations for assessing the safety of FA for horses is proposed for discussion.

Herbal extract in the horse: A review of the Italian perspective

E. Valle and D. Bergero, University of Turin, Faculty of Veterinary Medicine, via Leonardo Da Vinci 44, 10095 Grugliasco, Italy*

The use of supplements containing herbal extract is widespread and increasing in Italy, both in human and veterinary medicine, in particular for the horse market. This fact can be due to different factors: the increased sensibility of owners for "natural" medicine, health and welfare of the horse; the increased number of anti – doping controls; the increased negative perception in the use of traditional drugs, to enhance horses' welfare and performances.

Therefore, more and more evidences have been produced about the importance of studying supplements based on herbal products for the horse's market. We analysed the main horse's herbal products that are present on the Italian market to consider their claimed properties. This paper in particular list and discuss the use of the most common herbals' supplements, containing garlic (*Allium sativum*), *Schizandra chinensis*, devils'claw (*Harpagophytum procumbens*), cat's claws (*Uncaria tomentosa*), *Boswellia serrata*, echinacea (*Echinacea angustifolia*) and Bromelain. Furthermore we discuss the importance to get more scientific information to evaluate the benefits and the risks about the use of these new products. The real effect of these substances in the horse must be clearly understood, because the common opinion of horses' owners that "natural" is a synonymous of "safe" is not always true. Legislation on this topic is also reviewed.

Estimation of nutrient digestion and faecal excretion of mixed diets in poines

A.S. Chaudhry, School of Agriculture, Food & Rural Development, University of Newcastle upon Tyne, NE1 7RU, UK.

Three diets were compared to improve digestion and reduce excretion in ponies in a 3x3 latin square experiment. These isonitrogenous diets contained 50% Lucerne (LN=diet A), or 23% LN plus 26% straw+10% hay (diet B), or 25% LN+20% brewers' grains (diet C), as the main fibre sources plus oil, barley, maize, wheat, peas and beans. About 5kg per diet was fed daily in two portions to the relevant pony from which total faeces were collected every 8 hours over 7 days. The faeces were dried and weighed to estimate faecal dry matter output at each collection time. The faeces were analysed together with diets to estimate nutrient digestibility per diet per pony. Statistical analysis showed that the diets A and C were significantly more digestible than diet B ($P<0.05$). While the faecal excretion patterns of diets were comparable, the rates and extents of faecal excretion varied with time. Diet B yielded the greatest faeces and so was least digestible and most wasteful. Clearly diet B contained highest fibre which was inefficiently utilised by the pony and so should be used with caution in horse diets. It was possible to reduce the faecal output and thus improve nutrient digestion by modifying the ingredient composition of a horse diet. However, it would be essential to optimise the level of fibrous ingredients which may otherwise benefit the horse gut and behaviour, help reduce faecal excretion and the cost of feeding a horse.

Effect of dental overgrowth on dry matter content and particle size in faeces from horses fed forages with high or low NDF content

U.P. Jørgensen[1], L. Raff[1], P. Nørgaard[1] and T.M. Søland[2], [1]The Royal Veterinary and Agricultural University, [1]Department of Basic Animal and Veterinary Sciences, [2]Department of Large Animal Sciences, 1870 Frederiksberg, Denmark

The aim of this experiment was to study the effects of forage type and dental overgrowth in horses on faeces characteristics. Four adult horses were fed rations with either low (LF) or high (HF) NDF content in a randomized cross-over design before and after dental correction. NDF made up 0.6 kg NDF from each of four forages in both rations. The rations were supplemented with chopped alfalfa hay and concentrates to maintenance level. The forage NDF content ranged from 34 to 61% and 64 to 81% of DM in the LF and HF rations, respectively. Faeces were collected during four days, washed in nylon bags with a pore size of 11 μm, freeze dried and sieved before scanning and image analysis. The overall arithmetic mean particle length (APL) and the 95 percentile of particle length (95PL) and width (95PW) were 4.7, 13.9 and 1.7 mm, respectively. Dental correction decreased faeces DM content, DM loss at washing, median particle length, 95PL and 95PW ($P < 0.03$), and tended to decrease APL ($P < 0.07$). The LF ration caused longer APL ($P < 0.03$), whereas faeces DM content was lower ($P < 0.001$).

Effect of removing dental overgrowth on chewing activity of trotter horses fed eight different forages

L. Raff[1], U.P. Jørgensen[1], P. Nørgaard[1]*, T.M. Søland[2], [1]Department of Basic Animal and Veterinary Sciences, [2]Department of Large Animal Sciences, The Royal Veterinary and Agricultural University 1870 Frederiksberg, Denmark

The aim of this study was to compare chewing activity in horses fed different forage meals before and after dental correction. Four trotters were assigned in a randomized split plot design with two forage rations of each four types of forage, before and after dental correction. Forage providing 0.6 kg forage NDF was allocated for 60 minutes at 9, 12, 15 and 18 h in a randomized design with four forage meals/day during four days in four periods and two dental conditions. Ration l: grass silage, haylage, dried grass 1, dried grass 2 (DM 47, 75, 90 and 90%; NDF 34, 52, 53 and 61%; lignin:NDF ratio 0.064 ± 0.016). Ration II: grass hay 1, grass hay 2, seed straw and barley straw (DM mean 90%; NDF 64, 69, 73 and 81%; lignin:NDF ratio 0.095 ± 0.014). Eating behaviour was observed, jaw movement (JM) oscillations and forage intake were recorded for each meal. Chewing time and number of JM per kg DM and NDF were affected by forage type and dental correction ($P < 0.01$). The chewing time after dental correction ranged between 56 to 98 min/kg NDF and dental correction decreased chewing time between 3 to 31% per kg NDF.

Feeding systems of fantasia horses in Morocco
A. Araba, O. Bendada, A. Ilham and H. AlYakine, Hassan II Institute of Agronomy and Veterinary Medicine, BP 6202, Rabat-Instituts, Rabat, Morocco*

Horse rearing in Morocco suffers from several handicaps representing a threat for its development. Feeding management is considered one of the main handicaps. The objective of this study is to characterise and analyse the feeding systems of fantasia horses in some areas of Morocco. To this end, 34 horse owners were investigated (average of 1 horse per owner). From February to May, the feeding system is based on the pasture of oat and/or barley or natural grass. The surface grazed per day can vary from 15 to 40 m^2 according to the duration of the daily pasture. Indeed, some owners let their horses graze during all the day, whereas others choose a grazing limited to the morning. During the remainder of the year, from June to January, the horses receive oats and/or barley hay, or straw if hay is not available. The concentrate feeds distributed are mainly made up from barley grains combined or not with other concentrates. Daily intake varies from 2 to 6 kg. To prepare the horses to fantasia festivities, the proportion of barley in the diet increases and those of bran, corn, dry sugar beet pulp, bean and sunflower meal decrease. The study includes a nutritive analysis of these rations and provides practical recommendations to improve them.

Effects of forage type and time of feeding on water intake in horses
U.P. Jørgensen[1], L. Raff[1], P. Nørgaard[1] T.M. Søland[2], [1]The Royal Veterinary and Agricultural University, Department of Basic Animal and Veterinary Sciences, [2]Department of Large Animal Sciences, 1870 Frederiksberg, Denmark*

This study compared water intake during forage meals. Four horses were assigned to a randomized cross-over design with two forage rations of each four forages, fed before and after dental correction. At 9, 12, 15 and 18 h, 0.6 kg forage NDF was allocated for 60 minutes in random order of four forages during four experimental days. Ration I: grass silage, haylage, dried grass 1 and dried grass 2 (DM 47, 75, 90 and 90%; NDF 34, 52, 53 and 61%). Ration II: grass hay 1, grass hay 2, seed straw and barley straw (DM mean 90%; NDF 64, 69, 73 and 81%). The horses had free access to water from a pail, and water intake was recorded after each meal. The log transformed intake of water per kg DM and per kg NDF, with and without including forage moisture, were affected by meal time ($P < 0.02$) and forage type ($P < 0.001$), whereas dental status did not affect water intake. Water intake (l/kg DM) was lower at the 15 h meal compared with 12 and 18 h meals ($P < 0.05$). The geometric mean water intake ranged from 5.3 for grass silage to 9.5 l/kg NDF for dried grass 2.

Fresh forage in dairy ass' diet: Effect on plasma fatty acid profile

B. Chiofalo[1]*, D. Piccolo[1], C. Maglieri[2], E.B. Riolo[1], E. Salimei[2], [1]Dept. MO.BI.FI.P.A., University of Messina, 98168 Messina, Italy, [2]Dept. S.T.A.A.M., University of Molise, 86100 Campobasso, Italy

The effect on the plasma acidic composition in lactating asses fed with fresh forage was studied. Eight Martina Franca asses were divided into two groups of 4 each one, fed with 8 kg/head·day of meadow hay (CTR group) and 20 kg/head·day of fresh herbage and 3 kg/head·day of meadow hay (TRT group); all the animals received 2.5 kg/head·day of mixed feeds. The trial lasted for 63 days. On the feeds and plasma samples, fatty acids were analysed by GC-FID and the quality indices calculated. Data were subjected to ANCOVA. In fresh forage, considerable higher percentages for n-3 PUFAs (39.85 vs. 14.33g/100 g total fatty acids), especially for linolenic acid (14.13 vs. 39.53) were observed, whereas the n-6 PUFAs were similar in feedstuffs (meadow hay: 28.79%; fresh forage: 25.81% g total fatty acids). Consequently, in the TRT group, the plasma levels of both n-3 PUFAs (7.09% vs. 4.97%) and linolenic acid (5.76 vs. 3.47) were significantly ($P<0.001$) higher, and plasma n-6 PUFAs showed no significant ($P=0.191$) differences between treatment(TRT: 38.90% vs. CTR: 37.37%). As a non ruminant herbivore, donkey's plasma fatty acids profile reflects the unsaturated dietary pattern; moreover, the investigated quality indices were significantly ($P<0.05$) lower in the TRT group, being IA= 0.32 vs. 0.37 and IT=0.75 vs. 0.87.

Qualification and assessment of work organization in livestock farms: The French experience

S. Madelrieux[1]* and B. Dedieu[2], [1]Cemagref, UR Développement des Territoires Montagnards, BP76, F-38402 St Martin d'Hères, [2]INRA, UMR Metafort, Theix F-63122 St-Genès-Champanelle

Farmers have to cope both with the society and market pressures on their manners of doing and with the enlargement of farms, off-farm opportunities and deep changes in workforce, working duration and rhythms expectations. Working conditions and the efficiency of work organization (WO) are critical issues nowadays. The bibliography shows that WO is mainly discussed by social sciences (notably ergonomics and sociology), but that livestock sciences have a significant contribution to the debate. Indeed, technical changes modify working calendars, priorities between tasks and the interchangeability between workers; technical adaptations are levers to solve problems of organization with equipments, buildings and the workforce. We present here French approaches of WO that take into account livestock management and its implications in WO. The Work Assessment method represents the WO and evaluates work durations and time flexibility for farmers. The Atelage model describes and qualifies the WO with its various regulations and time scales, integrating the other activities - economic or private - that farmers can carry on. Three principles underpin them: all workers are not interchangeable; tasks have different temporal characteristics (rhythms, postponement...); the year is a chaining of work periods that differ in their daily form of organization. We illustrate with concrete examples how these approaches contribute to accompany farmers in their thoughts to change.

Small-scale lamb fattening systems in Syria; future development and opportunities

B.W. Hartwell*[1,3], L. Iñiguez[2], W.F. Knaus[3], M. Wurzinger[3], [1]The Royal Veterinary and Agricultural University, Grønnegårdsvej 2, 1870 Frederiksberg C.,Denmark, [2]International Center for Agricultural Research in Dry Areas, POB 5466, Aleppo, Syria, [3]University of Natural Resources and Applied Life Sciences, Gregor-Mendel-Strasse 33, A – 1180 Vienna, Austria

Small-scale intensive Awassi lamb fattening production systems emerged the last decade in Syrian rural and peri-urban areas. The Gulf's affluent markets and enhanced local demand of Awassi sheep meat prompted this development. A study characterizing 241 lamb fattening systems found that lamb fattening is an efficient income-generating option for small-scale farmers and a family employment source, though largely constrained by feeding costs and animal health issues. Prompted by this finding a research strategy was formulated to produce least-cost rations as options for farmers to improve their economic returns. Cost-effective rations for fattening using locally available feedstuffs were tested in an on-station environment. The most promising least-cost ration was then implemented on-farm in collaboration with farmers. Cost-benefit analysis and carcass evaluations demonstrated the feasibility to produce at lower costs, though seasonal and yearly feed price fluctuations cannot be ignored in this regard. To ensure sustainable development of small-scale lamb fattening systems, further research is needed to learn more on least-cost rations, assess simple measures to curb disease-related problems, identify and articulate production and marketing policy issues required, and assess preliminary conditions to establish a profitable fattening enterprise as a technological option for small producers.

Animal and economic performances for large French suckler cattle herds: The case of Charolais

P. Veysset*, M. Lherm and D. Bébin, INRA Clermont-Theix, Economie de l'Elevage, 63122 Saint Genès Champanelle, France

To preserve their income, since more than twenty years, suckler cattle farmers have to extend their holding (area and herd) and to improve their labour productivity. From 1989 to 2003, a group of 69 charolais livestock farms from the centre of France increased its area and its livestock of more than 30% with constant workforce. We compared the evolution of the technical and economic results over 15 years of the 25% farms, from this 69 farms constant sample, which owned the largest herds per worker ("LH": 17 farms) with the others ("Others": 52 farms). The birth rate, the calves mortality rate and the losses of cattle were systematically in discredit of the "LH". In spite of that the meat production per LU was comparable in the two groups, the proportion of animal sold fattened being higher for the "LH". The herd costs were the same for both "LH" and "Other", but the "LH" owned more non premium eligible cows, so the bovine gross margin was slightly lower for the "LH". The income per worker was always positively correlated with the size; but before the first CAP reform the herd size was important because of the output that it generated, after 1992 the size has an effect on the farm income by the subsidies that it makes possible to perceive.

Management intensity and biodiversity conservation: Is farm size the key?

M. Tichit[1]*, P. Grené[2] and F. Léger[1], [1]INRA SAD, 16 rue Claude Bernard, 75231 Paris, France, [2]INRA Domaine Saint Laurent de la Prée, 17450 Fouras, France

In agricultural landscapes, habitat quality is a crucial issue for the conservation of biodiversity. Several kinds of data suggest that farm characteristics are important driver of habitat quality, but to date a few studies have investigated whether management intensity is related to farm size. The question whether biodiversity may benefit from a few large versus several small farms is still unclear. In a grassland landscape, with high priority for bird species conservation, we tackle this question by 1) analysing a large data set on management practices in 574 grassland fields within a large range of farm sizes and 2) assessing the impact of management on habitat quality. Three types of grassland managements were identified through multivariate analysis showing an increasing level of intensity. Higher habitat quality was related to medium and low management intensity. However, no relationship was established between farm size and management intensity. We conjecture that a threshold on farm size may however exist with potential detrimental effects on biodiversity due to homogenisation of habitats. Our results are discussed in the light of other farm functions in the local economies and we conjecture that the maintenance of a variety of farm sizes may be needed to reconcile productive, environmental and social functions assigned to agriculture.

Trajectories of evolution of cattle farming systems in Spanish mountain areas

A. García[1], A. Olaizola[2], J.L. Riedel[1], A. Bernués[1]*, [1]CITA Gobierno de Aragón, Apdo. 727,50080 Zaragoza, [2]University of Zaragoza, Miguel Servet 177, 50013 Zaragoza, Spain

The evolution of cattle farms in 3 valleys of the Spanish Pyrenees is analysed in the period 1990-2004. Data was collected from the same sample of 101cattle farms in both dates trough direct questionnaires to farmers. Using multivariate techniques (Principal Components PCA and Cluster Analysis) different trajectories of evolution were detected in terms of structural change, production orientation, use of natural resources and labour productivity. Also, some driving factors (physical and socio-economic environment, family characteristics, etc) associated to the trajectories were analysed using Logistic Regression.

In general, farms experienced drastic changes in the period of study. Both land and herd sizes increased notably (20 and 40%, respectively). The length of the grazing period and the utilization of mountains pastures also increased around 20%. Dairy completely disappeared and instead many farms have now fattening activities. Labour productivity (before premiums) remained constant. Four factors, identifying types of dynamics, were clearly revealed in the PCA: 1. increment of size and labour productivity; 2. extensification of management; 3. stocking rate on owned land (excluding communal pastures) and 4. orientation towards fattening calves. According to these factors 6 groups of farms, which showed distinct trajectories of evolution, were identified in the Cluster Analysis. Family and socio-economic variables associated to the different trajectories are also discussed in the paper.

A survey of smallholder cattle farming systems in the Limpopo Province of South Africa
A. Stroebel[1]*, F.JC. Swanepoel[1], A.N. Pell[2] and I.B. Groenewald[1], [1]Centre for Sustainable Agrculture, University of the Free State, P O Box 339, Bloiemfontein, 9300, South Africa, [2]CIIFAD, Cornell University, Ithaca, 14850, NY

Eighty-six smallholder cattle farmers in the Nzhelele District of the Limpopo Province of South Africa were surveyed. Data was collected by means of a structured questionnaire and standard Participatory Rural Appraisal (PRA) methods. The farmers owned between one and 67 cattle, with an average of 10.3 head of cattle per household. The average age at first calving was 34.3 months. The rates of calving, weaning, calf mortality, herd mortality and offtake were 49.4%, 34.2%, 26.1%, 15.6% and 7.8% respectively. Contrary to the situation in many other regions of Southern Africa, commercial enterprise, not social prestige, constituted the main reason for farming with cattle. A marked complimentarity in resource-use, i.e. crop residues as animal forage, has been demonstrated. The farmers use a variety of management aspects unique to their environment and natural resource-base. It is concluded that cattle production in smallholder farming systems are based on economic as well as sociological decisions by farmers to maintain and support family livelihoods.

Forage systems and work organisation in small-scale cattle farms in Brazilian Amazonia
N. Hostiou[1]*, J.F. Tourrand[2] and J.B. Veiga[2], [1]INRA Equipe TSE, 63122 Saint Genès Champanelle, France, [2] Embrapa, CEP 66090-100, Belém-PA, Brazil

In family farms holdings in Brazilian Amazonia, grassland is the basic all-year-round feed for dairy-beef cattle herds. Pasture perenniabiliy is essencial in the technical message with uniform recommendations. However, forage systems are basically managed using manual labour, with no equipment or mechanisation, and with work carried out by family members that is considered insufficient. The aim of this study is to explore how livestock farmers structure the technical management of their forage system and their work organisation over a year. The data, for seven dairy-beef farms, are taken from a monitoring of forage system management practices and working times quantified by the "Bilan Travail" method. Four logics for managing forage systems are identified, differentiated by working times and the workforce devoted to the grassland area and the land. Some livestock farmers aim to maintain pastures, cutting down weeds by hand and with wage earning workers. But others adopt simplified management characterised by very low working times. This study concludes on the necessity of taking account of the links between technical management and work organisation to produce recommendations appropriate to smallholder family systems in Amazonia.

Session L31   Theatre 8

Development trends in small-scale farms in transition sectors of CIS countries
Kurt J Peters, Institute for Animal Sciences, Humboldt Universität zu Berlin, Philippstr. 13 D 10115 Berlin, Germany

Changes in political system in CIS countries have led to deep-rooted reforms of the agricultural sector with the objective to move to a market oriented system. The privatization of agricultural land, the dismantling of collective farms, the liquidation of centralized planning and control structures related to input purchase and product marketing were most often among the set of new agricultural policies. This paper reports on statistical evidence of the transition process (Livestock populations, products, product prices, farm size) and the outcome of the transition policies.

The established Small-Scale farms and livestock holdings in general are not in a position to substitute for market supplies previously obtained from state farms. Small scale farms often remain in a state of subsistence and fail to be economically competitive due to inadequate access to appropriate technologies, low technical and managerial know-how, and the inappropriateness or absence of supporting sector institutions (marketing, extension, input supply, breeding and veterinary services). Domestic economic transition processes paired with the impact of global markets for agricultural and livestock products are of particular importance for a successful structural change. Policies for managing the import of cheap product from the global market as well as setting policies for the development and implementation of effective sector institutions, for improving the technical and managerial competence of small holder farmers, and for creating an enabling environment to let small scale farms develop into competitive production units continue to be important.

Session L31   Theatre 9

The economic and structural characteristics of sheep breeding in Turkey
İ. Dellal[1] and G. Dellal[2], [1]Agricultural Economics Research Institute, Ankara, Turkey, [2]Ankara University Faculty of Agriculture, 06110 Ankara, Turkey*

Sheep breeding is an important activity of Turkey's agricultural sector and economy. Sheep constitute 60% of the existing livestock number. Their products, including meat, milk and hides, contributed 24, 10 and 80 percent, respectively, of the value of these agricultural outputs. Sheep are kept mainly on the grazing lands of Turkey by small-scale family farms. The figures above clearly elucidate that sheep constitute the most significant livestock husbandry in Turkey.

In this paper, economic importance and structural characteristics of sheep breeding in Turkey will be examined at regional and national level. Structural and breeding characteristics like breed, production system, nomadic characteristics, sheltering, feed and water resources, lambing, milking and shearing managements of sheep breeding will be given.

Effects of Iodine supplementation on Iodine concentration in blood and milk of dairy cows

G. Flachowsky[1], F. Schöne[2], M. Leiterer[2], D. Bemmann[2] and P. Lebzien[1], [1]Federal Agricultural Research Centre (FAL), Institute of Animal Nutrition, Bundesallee 50, D-38116 Braunschweig, [2]Thuringia Research Centre for Agriculture, Jena, Germany

The objective of the study was to quantify the effects of increasing Iodine supply on the Iodine concentration in serum and milk of lactation cows as a contribution to assess the human Iodine supply via milk. Five lactating cows were fed during 4 periods of 14 days each for Iodine doses as calcium iodate-hexahydrate with the mineral feed (150g/d). The Iodine concentration of the basal diet amounted to 0.2 mg/kg DM. The supplements correspond to 0, 1.2, 5.5 and 10 mg/kg DM. Milk samples were individually taken on day 10, 12 and 14 of each period as aliquots from the morning and evening milking. Blood samples were taken 3 hrs after morning feeding to the same day. The Iodine analysis was carried out applying the ICP-MS-technique. Rising Iodine supplementation increased significantly the Iodine concentration in the serum (from 48, 66, 131 to 290 µg/kg) and milk (from 101, 343, 1215 to 2762 µg/kg for 0, 1.2, 5.5 and 10 mg I/kg DM resp.) of cows.
Such high I-concentration of milk may contribute to exceed the upper limits for I supply of people (600 µg/d). Therefore the EU decreased upper limit of I for dairy cows to 5 mg/kg feed.

NaCl and KCl intake affects milk composition and freezing point in dairy cows

J. Sehested[*] and P. Lund, Danish Institute of Agricultural Sciences, P.O Box 50, DK-8830 Tjele, Denmark

We hypothesised that freezing point of milk is affected by potassium (K), sodium (Na) and chloride (Cl) intake through electrolyte homeostatic mechanisms (e.g. aldosterone) in the udder.
A 4*4 Latin square experiment with 20 lactating dairy cows, 4 periods of 7 days and 4 dietary treatments was conducted. A total mixed ration based on maize silage was used on all treatments. Dietary treatments were: 1) Low-Na/Low-K; 2) Low-Na/High-K; 3) High-Na/High-K; 4) High-Na/High-K/High-fiber. Treatment levels were 12 or 35 g K and 1 or 10 g Na per kg dry matter by addition of chloride salts.
High versus low dietary KCl (both with low NaCl) significantly increased milk K (40.7±0.5 vs. 36.1±0.5 mM), Cl (26.0±0.7 vs. 20.1±0.7 mM), yield (30.5±0.7 vs. 27.5±0.7 kg) and protein (3.38±0.03% vs. 3.28±0.03%), but significantly decreased freezing point (-0.525±0.001 vs. -0.519±0.001 °C), fat (3.82±0.09% vs. 4.13±0.09%) and urea (2.9±0.1 vs. 4.4±0.1 mM).
High versus low dietary NaCl (both with high KCl) significantly decreased milk freezing point (-0.529±0.001 vs. -0.525±0.001 °C) and tended to decrease fat (3.62±0.07% vs. 3.82±0.09%, P=0.07), but did not affect milk yield, protein, lactose or urea. No significant effects was found on milk Na (mean; 23.5±0.4 mM).
In conclusion, milk composition and freezing point can be manipulated through dietary KCl and NaCl, and the udder seems to take part in the electrolyte homeostasis.

Milk fatty acid profile in goats receiving high forage or high concentrate diets supplemented, or not, with either whole rapeseeds or sunflower oil

Y. Chilliard*, S. Ollier, J. Rouel, L. Bernard and C. Leroux, Adipose Tissue and Milk Lipids Group, Herbivore Research Unit, INRA-Theix, F-63122 Ceyrat, France

Sixteen multiparous mid-lactation goats received 4 diets differing in either forage:concentrate ratio (HF-64:36 or LF-43:57) and/or lipid intake (0 *vs.* 130 g lipid/d, *i.e.* 4% of diet DM, from either whole rapeseeds, RS, or sunflower oil, SO) in a 4x4 Latin Square design. After 3 weeks of treatment, milk yield was lower (P<0.05) with HF-RS than other diets (3.85 *vs.* 4.24 kg/d) and milk fat content was higher with HF-RS or LF-SO compared with HF or LF diets (38.5 *vs.* 32.4 g/kg). Oleic acid in milk fatty acids (FA) was much higher with HF-RS than other diets (22.7 *vs.* 14.6%) whereas *trans*-18:1+*trans*-18:2 isomers were much higher with LF-SO (14.5%, *incl. 7.7% vaccenic and 3.5% rumenic acids*) than other diets (3.7%). Lipid supplementation decreased largely 12:0, 14:0 and 16:0 (-27%). HF-RS diet maximized the 18:2n-3/18:3n-6 ratio (0.46) whereas LF-SO minimized it (0.13). In conclusion, the studied dietary treatments strongly changed goat milk FA profile. High forage plus whole rapeseeds diet is very efficient to increase oleic acid and n-3/n-6 ratio, and to decrease saturated FA content of goat milk fat. Goat's responses to starch- and PUFA-rich diets differ markedly from cow's ones, particularly by the high milk fat yield response and the low level of *trans*10-18:1 and other "non-*trans*11" biohydrogenation intermediates. (*Work funded by the LIPGENE EU-FP6 Project, www.lipgene.tcd.ie*).

Effects of forage type on growth performance, carcass characteristics and meat composition of growing Friesian bulls

M. Ben Salem[1]* and M. Fraj[2], [1]INRA Tunisia, Laboratory of Animal and Forage Production, Rue Hédi Karray, 2049 Ariana, Tunisia, [2]ESA Mateur, 7030 Mateur,Tunisia

Twenty one growing Friesian bulls were used in a randomized design to study the effects of forage type on growth performance, carcass characteristics and meat composition over a 140-days fattening period. Animals were allocated, based on their body weight, to 3 homogenous groups. Each group was assigned randomly to one of the following treatments: 1) oat hay + concentrate (OH); 2) oat silage + concentrate (OS); 3) Corn silage + concentrate (CS). Results showed that total DMI did not differ (P> 0.05) between treatment groups. However, animals fed the hay diet consumed more concentrate (3.46 vs 3.07 and 3.16 kg DM for OH, CS and OS, respectively). Animals receiving corn silage had significantly higher (P<0.05) average daily gains (ADG) than those receiving OH and OS treatments (1359 vs 1218 and 1172 g/d for treatments CS, OS and OH, respectively). Feed conversion ratio was significantly higher (P<0.05) for animals fed the hay diet (7.58; 6.65 and 6.43 kg DM/kg of live weight gain for the OH, OS and CS treatments, respectively). No significant treatment effect (P>0.05) was observed for the dressing percentage. However, carcass composition varied with treatment. Corn silage resulted in carcasses with higher total fat and lower lean. Animals fed the corn silage diet had their meat significantly (P<0.05) higher in fat and lower in lean (25.21 and 17.9 Vs 18.42 and 20.4 and 14.74 and 20.8% for fat and protein contents respectively for CS, OS and OH).

Production of goat meat for the halal market

L.J. Asheim[1], L.O. Eik[2] and M.E. Sabrie[2], [1]Norwegian Agricultural Economics Research Institute, P. O. Box 8024, Dep., Oslo, Norway, [2]Institute of Animal Sciences, University of Life Sciences, 1432 Aas, Norway

This research project aims to increase production of Halal meat and in particular develop production systems to meet the off-seasonal demand for fresh meat of goat kids for religious celebrations. Using a linear programming model we investigate the economy of introducing cashmere goats on sheep farms. We hypothesize that by slightly altering kidding time, slaughter weights and use of outfield pastures or intensive feeding systems, a mixed production system can be sufficient flexible to meet fluctuating demand of fresh kid meat. Hence, the mixed system is expected to be less risky than a specialized sheep farming system with seasonal slaughtering of lambs in the fall. Production of Halal meat on surplus dairy goat kids is also discussed. Dairy goat farming is common in Norway and currently surplus kids are culled shortly after birth.

The Norwegian Muslim Community of about 100 thousand people as well as growing Muslim communities in neighbouring countries represent an interesting market for high quality goats' meat from Norwegian mountainous pastures. A food processing company (Alfathi) established by the Islamic Council, Norwegian Meat Ltd and University of Life Sciences has developed a range of Halal products such as pizzas, sausages and hamburgers. Rules of Halal slaughtering have been agreed upon with the veterinarian authorities.

Effect of dietary fatty acid profile on growth characteristics and lipid deposition in crossbred gilts

P. Duran[1], C. Realini[2], M. Gispert[2], M.A. Oliver[2] and E. Esteve-Garcia[1], [1]Departament de Nutrició Animal (I.R.T.A.), Mas Bové, P.O. Box 415, C.P. 43280, Reus (Tarragona), [2]Centre de Tecnologia de la Carn (I.R.T.A.), Granja Camps i Armet, C.P. 17121 Monells (Girona), Spain*

Seventy crossbred gilts (62 ±5 kg bw average) were divided in 10 animals x treatment. Pigs were fed one of seven treatments, a semi-synthetic diet formulated to contain a very low level of fat (NF) and six fat supplemented diets (10%) based on barley and soybean meal. The supplemental fats were tallow (T), high-oleic sunflower oil (SFHO), sunflower oil (SFO), linseed oil (LO), blend (B) (55% tallow, 35% sunflower oil, 10% linseed oil) and fish oil blend (FO) (40% fish oil, 60% linseed oil). Carcass fat content was clearly higher in NF diet (27.8%), with lowest values for diets with high content in polyunsaturated fatty acids (PUFAs), SFO 21.9% and LO 21.2%. Net fat synthesis was highest in NF treated animals. SFHO and T diets, with lower content in PUFAs, showed higher calculated Net fat synthesis compared to SFO and LO diets. Net fat synthesis showed high correlation with dietary PUFAs content. Results show that dietary PUFAs modify fat deposition in fattening pigs.

The effect of diet and aging on colour and colour stability of pork

K. Tikk[1], *J.F. Young*[1] *and H.J. Andersen*[2], [1]*Department of Food Science, Danish Institute of Agricultural Sciences, PO Box 50, DK-8830 Tjele, Denmark,* [2]*Arla Foods amba, Corporate R&D, Skanderborgvej 277, DK-8260 Viby J, Denmark*

The objective of the present study was to elucidate the effect of strategic finishing feeding containing a high ratio of rapeseed and grass meal compared with a conventional diet on pork colour and colour stability. Pork colour was measured after 1, 2, 4, 8 and 15 days of aging of M. *longissimus dorsi* (LD) and M. *semimembranosus* (SM), and the stability by subsequent measurements at 2, 4 and 6 days.

The strategic feeding resulted in a significant decrease in early *post mortem* temperature in LD, as well as a slight tendency to higher initial pH in SM. This resulted in higher colour scores in meat from the pigs given experimental diet compared with control diet, with the exception of the a*-value in SM. Moreover, these results were independent of days in retail display. Aging of the meat increased the L*-value and hue angle gradually whereas the a*- and b*-values initially decreased with subsequent increase after more than 4 days of aging. Finally, during retail display an increase in L*-value, b*-value, chroma and hue angle and a decrease in a*-value were measured in both muscles independent of day of aging. Change in colour during retail display was most pronounced from day 0 to 2 whereupon the rate of discoloration decreased.

Effects of dietary oils on lipid composition of raw and processed pork muscle

K. Nuernberg, U. Kuechenmeister, G. Nuernberg, M. Hartung, D. Dannenberger, K. Ender, Research Institute for the Biology of Farm Animals, Wilhelm-Stahl-Allee 2, D-18196 Dummerstorf, Germany

The aim of this study was to alter the fatty acid composition of porcine muscle and to assess changes of raw and grilled pig muscles from these animals after different storage periods. A total of 13 female and 12 castrated Pietrain x German Landrace pigs were fed a diet supplemented with 5 % olive or 5 % linseed oil during the growing-finishing period. An entire cut of the pork loin was stored at 5 °C for 48 h, 96 h or 144 h. Analyses of intramuscular fat were carried out on raw and grilled *longissimus* muscles following different storage intervals. Dietary inclusion of linolenic acid by linseed oil feeding increased the relative and absolute concentration of long-chain *n-3* fatty acids, whereas in the olive oil group the oleic acid in pork was higher. Storage time increased significantly the relative proportions of lauric acid, stearic acid, and oleic acid, while the percentage of linoleic, arachidonic, eicosapentaenoic acid, and the sum of polyunsaturated fatty acids, especially *n*-6 fatty acids, was decreased. Compared with raw muscle, grilling affected the relative fatty acid profile only slightly. Related to the original weight, storage and grilling increased the total fatty acid contents, the sum of saturated, *n*-6 and *n*-3 fatty acids of loin chops because of the water loss.

Effect of rapeseed and linseed in sheep nutrition on diurnal changes in the chemical composition of milk

B. Borys[1], A. Borys[2] and S. Grzeskiewicz[2], [1]National Research Institute of Animal Production, 32-083 Balice/Kraków, [2]Meat and Fat Research Institute, 4 Jubilerska St., 04-190 Warsaw, Poland*

A total of 12 nursing ewes in the 5th week of lactation were fed with forage, hay and concentrate mixtures. Standard mixture was used in the control group (C) and a mixture with whole rapeseed and linseed (100 and 50 g/animal/day, respectively) was used in the experimental group (RL). The animals were fed once daily. Milk was sampled for analysis 1 h after suckling. Samples were collected in 4 series during 24 h: before feeding and 6, 12 and 18 h after feeding.

Feeding oilseeds had a significant effect on the chemical composition of milk as it increased the content of DM (16.2 vs. 17.1 g/100 g in groups C and RL) and fat (5.6 vs. 6.9 g) and decreased the protein (4.5 vs. 4.2 g) and lactose (5.4 vs. 5.2 g). Of similar nature were the differences in the concentration of components in milk DM (17.2% higher for fat in the RL group, and 13.7 and 12.0% higher for protein and lactose in the C).

No significant differences were observed during 24 h in the chemical components of milk except the concentration of lactose in DM directly before feeding, which was significantly lower than in the other series of observations (30.8 vs. 32.5 g/100 g).

Performance of broilers fed dry peppermint (*Mentha piperita* L.) or thyme (*Thymus vulgaris* L.) leaf supplemented diet

N. Ocak, G. Erener, F. B. Ak. M. Sungu, A. Altop, A. Ozmen. Ondokuz Mayis University, Faculty of Agriculture, Department of Animal Science, Kurupelit, 55139, Samsun-Turkey

A study was conducted to determine the performance, carcass and gastrointestinal tract characteristics of broilers fed dry peppermint (*Mentha piperita* L.) or thyme (*Thymus vulgaris* L.) leaf as menthol and thymol sources supplemented diet. In the study, 312, one-wk-old broilers (Ross-308) were used in a randomised design. There were 3 dietary treatments, each consisting of 4 replicates (13 males and 13 females in each replicate). Control group was fed basal diet, while peppermint and thyme groups were fed a diet containing 0.2% peppermint or thyme (w/w), respectively. From 7 d to 35 d of age, the body weight gain was higher (P<0.05) in broilers fed peppermint-supplemented diet compared to control, but the effect of peppermint on body weight gain had disappeared at 42 d of age. Feed intake, feed efficiency and the characteristics of carcass and gastrointestinal tract were not significantly different among the 3 treatments. The peppermint or thyme increased (P<0.05) abdominal fat pad at 42 d of age. The present study shows that the feeding of either peppermint or thyme did not affect growth performance, feed efficiency, and carcass and gut characteristics of broilers, but an increase in abdominal fat should be taken into account for carcass quality and processing.

Substitution of yellow corn partially by sorghum in broiler ration and the effect of methionine and kemzyme supplementation on broiler performance,quality and nutrient digestibility

Samia M. Hashish and G.M. Elmallah, N.R.C.,Dokki Cairo Egypt

A study was conducted to evaluate the effect of substituting 50% yellow corn in broiler ration by sorghum grains and the effect of supplementing methionine and kemzyme(an enzyme preparation) on broiler performance,carcass quality and nutrient digestibility. 324 broiler chicks were devided into 6 groups,each contained 54 chicks in 3 replicates.

The first group was used as a control fed corn-soybean meal diet;the second contained 50% sorghum of yellow corn ;the third contained 50% sorghum of yellow corn + 0.05% kemzyme;the fourth included 50% sorghum + 0.1% kemzyme ;the fifth included 50% sorghum + 0.02% methionine and the six contained sorghum +0.02% methionine + 0.05% kemzyme. All growth parameters, were measured for 7 weeks period. Carcass quality measured and meat was chemicaly analysed. A digestibility trial was conducted to determine the influence of the treatments on broiler meat quality. The results indicated that the best growth parameters obtained(live body weight(gm.),body weight gain,feed conversion) of the group 6, significantly higher than the other groups. The lowest growth parameters found in group 2.

No significancy obtained of carcass edible parts chemichal analysis. No significancy in nutient digestibility of organic matter,protein,fat and N.F.E. was found.The results indicated that sorghum could successfuly substitute 50% of yellow corn in broiler ration specially when supplemented with 0.02% methionine +0.05% kemzyme.

Effect of processing technology on meat quality of pigs

Cs. Ábrahám[1], J. Seenger[1], M. Weber[2], K. Balogh[2], E. Szűcs and M. Mézes[2], Szent István University, [1]Department of Pig and Small Animal Husbandry,[2]Department of Animal Nutrition, H-2100 Gödöllő, Páter Károly u. 1.

The aim of our study was to investigate the effect of slaughtering technology as a complex factor on meat quality of pigs. In the experiment we compared a large-scale slaughtering technology with a small-scale one. 40 pigs were transported for 1.5 hours, and lairaged for 16 hours. The large-scale processing was a fully automated technology; the conventional one was carried out more or less manually. The cooling technology was the same in both cases; the carcasses were brought to cooler at 50-55. minutes after stunning. The following meat quality parameters were measured: pH_{45}, pH_{24}, L*, a*, b*, core temperature, drip loss.

The processing technology significantly influenced the meat quality parameters measured at 45th minute in the ham, such as the pH (A: 6.28±0.24; B: 6.47±0.14; P<0.05) and core temperature (A: 42.12±0.54; B: 40.93±0.40; P<0.001). However the pH_{24} and drip loss differed significantly at a low level, but no difference was found in case of parameters sensible for consumers, such as meat color.

According to our results, it can be stated that the large-scale slaughtering technology influences the intensity of the post mortem processes. The cooling technology has an outstandingly important role in the development of meat quality, and this operation can reduce the effect of all the previous factors.

Similar carcass-, meat- and eating quality of heavy young bulls produced in three different feeding systems

M. Vestergaard[1], *I. Clausen*[2] *and C.F. Børsting*[3]. [1]*Danish Institute of Agricultural Sciences, Tjele,* [2]*Danish Meat Research Institute, Roskilde,* [3]*Danish Cattle Research Centre, Tjele, Denmark*

The objective was to evaluate the potential of producing heavy young bulls in different feeding systems and to evaluate the consequences on quality characteristics. In total 45 Holstein bull calves purchased from 7 private farms at 3 months of age were randomly divided into three treatment groups of 15 calves. The control treatment (CON) comprised ad libitum access to a concentrate ration (9.07 MJ NE/kg DM) whereas the two other treatment groups (TMR) had ad libitum access to a maize silage-based ration (7.42 MJ NE/kg DM) from start of experiment until 10½ months of age, when one group changed to a barley-maize silage-based ration (TMR+) with high energy content (8.21 MJ NE/kg DM) and the other continued on the low TMR. From start (113± 4 kg) until 10½ mo ADG was 1.33 kg/d for CON and 1.07 kg/d for TMR (P<0.05). From 11½ mo until slaughter at 585 kg BW, ADG was 1.18, 1.32 and 1.48 kg/d and age at slaughter 476, 515, and 483 d for CON, TMR, and TMR+, respectively. Body condition at slaughter and carcass characteristics were not different although TMR+ tended to have highest EUROP conformation and dressing percentage (P<0.10). Extensive meat- and eating quality analyses showed only minor differences between treatments. The results show that it is possible to produce the same overall carcass and meat quality of heavy young bulls with all three feeding regimens.

Effects of humic acids on broiler performance and digestive tract traits

E. Ozturk and I. Coskun, Department of Animal Science, Faculty of Agriculture, University of Ondokuz Mayis, 55139-Kurupelit Samsun, Turkey

This study was carried out to investigate the effect of humic acids (HA) supplementation provided through drinking water on performance, carcass and some gut traits of broilers. In a randomized design, 480 mixed-sex Ross 308 broilers were allocated into four experimental groups consisted of four replicate (30 birds per replication). All birds were housed in pens over 42 day. All chickens were offered *ad libitum* a commercial diets in mash form. Treatments were: 1) 0 (H_0), 2) 1.7 (H_1), 3) 5.1 (H_2) and 4) 8.1 ppm (H_3) HA per liveweight supplemented in drinking water. Feed intake, body weight were recorded at 21st and 42nd days of trial. Data were analyzed by one way ANOVA. Live weight gain was higher in H_1 group than control group both on 21-42 and 0-42 day periods (P<0.05). Feed consumption in the H_1 group was higher than that in H_3 at 0–42 day period. Feed conversion ratio, carcass yield, and heart, gizzard, edible viscera, abdominal fat pat and gut weight were not affected by supplementation of HA (P>0.05). The H_2 treatment caused an increase in liver weight compared to control group. Gut length was higher in H_1 group than control group (P<0.05). These results show that 1.7 ppm HA supplementation increased liveweight gain and feed consumption without affecting feed efficiency.

Effect of feeding rapeseed and linseed to nursing ewes on diurnal changes in the lipid profile of milk

A. Borys*[1], B. Borys[2] and S. Grzeskiewicz[1], [1]Meat and Fat Research Institute, 4 Jubilerska St., 04-190 Warsaw, [2]National Research Institute of Animal Production, 32-083 Balice/Krakow, Poland

A total of 12 nursing ewes in the 5th week of lactation were fed with forage, hay and concentrate mixtures. Standard mixture was used in the control group (C) and a mixture with whole rapeseed and linseed (100 and 50 g/animal/day, respectively) was used in the experimental group (RL). The animals were fed once daily. Milk was sampled 1 h after suckling. Samples were collected in 4 series during 24 h: before feeding and 6, 12 and 18 h after feeding.

Feeding oilseeds did not cause differences in the cholesterol content of milk. The level of cholesterol was the highest 12 h after feeding.

Compared to C ewes, the milk fat of RL ewes contained significantly less SFA (60.8 vs. 63.8 g/100 g) with a higher level of C18:0 (15.1 vs. 9.8 g) and more MUFA (34.8 vs. 31.6 g). With a similar content of PUFA, the RL milk fat contained more C18:3 and Ω3 PUFA. The CLA content of RL and C milk fat was similar, while the CLA content of milk was higher for RL than C (33.5 vs. 27.5 mg/100 g).

There were significant differences in the SFA and MUFA content of milk fat during 24 h, with a stable level of PUFA.

Lipid profile of intramuscular fat in fattened kids depending on breed and age

B. Borys[1], *J.Sikora[1], A. Borys[2] and S. Grześkiewicz[2], [1]National Research Institute of Animal Production, 32-083 Balice/Krakow, [2]Meat and Fat Research Institute, 4 Jubilerska St., 04-190 Warsaw, Poland

Longissimus dorsi muscles of 88 Saanen [S] and Alpine [A] kids and Anglo-Nubian [N] × S crossbreds fattened to 90 and 180 days of age were investigated. The levels of cholesterol in muscles and fatty acids in intramuscular fat were analysed.

With no significant differences in the level of cholesterol, the fat of A goats had the highest level of SFA, lowest of MUFA, and similar PUFA (but the highest level of Ω3 PUFA). The fat of N×S kids contained more CLA then. S and A (61 vs. 45 mg/100g).

Compared to 90-day-old kids, the muscles of 180-day-old ones contained less cholesterol (66 vs. 76 mg) and CLA (44 vs. 56 mg/100 g fat). The fat of older kids contained less SFA (42.6 vs. 45.4 g), PUFA (8.6 vs. 13.7 g) and Ω3 PUFA (1.0 vs. 1.6 g), and more MUFA (48.4 vs. 40.5 g). Overall, the muscles of kids aged 180 days had a significantly more beneficial UFA:SFA ratio (1.351 vs. 1.233) and higher atherosclerosis risk (0.409 vs. 0.426), with a poorer PUFA:SFA ratio (0.202 vs. 0.308).

From the point of view of health-promoting quality, both factors analysed were generally found to have a clear but inconsistent effect on the lipid profile of muscle tissue in young goats.

Study of different levels of Digestible Undegradable Protein on the carcass characteristic of Kermani male lambs in Iran

M. Soflaei shahrbabak[1], Y. Rozbahan[2], M. Moradi shahrbabak[3], H. Moradi shahrbabak[3], [1]Dep. Animal Science,Kerman's Jihad-e-Agriculture Educational center, Kerman, Iran, [2]Faculty of Agriculture, Tarbiat Modaress, [3]Department of Animal Science, Faculty of agriculture, University of Tehran

Thirty six Kermani male lambs and their improved breed in Shahrbabak city were selected randomly with an initial live weight of 29 ± 2.5 Kg and the age of about 6-7 months. Three levels of digestible undegradable protein (DUP), 19.86 (1), 26.47 (2) and 33.08 (3) (g/Kg DM), and one level of metabolisable energy (10.5 Mj/Kg DM) based on the standard tables of male lambs feed (AFRC, 1995) were used. Animal were fed for a period of 95 days (14 days for adaptation and 81 days for period of trial) for 4 times in a day.. The data were statistically analyzed using completely randomized design with 3 diets (n=12).

The dietary had significant ($P<0.05$) effect on the weight and carcass efficiency and carcass length. Carcass efficiency is one of the criterions having differences and variety between breeds and different growth stages. Its improvement is of high value. The dietary had no significant ($P<0.05$) effect on the weight of parts of carcass and chess file interrupted surface. However, increase DUP in 2 and 3 rations causes slight increase in valuable carcass parts (fillet and ham). The results showed that the ration which contained 26.47g digestible undegradable protein (g/Kg DM) was economically viable

---

Effects of Syzygium aromaticum and Zingiber officinale essential oils on performance and some carcass, blood and intestinal parameters of broilers

Ahmet Tekeli, Hasan Rüştü Kutlu, Ladine Çelik, Murat Görgülü, Çukurova University Agricultural Faculty, Dept. of Animal Science, Adana-Turkey

The present study was conducted to determine whether *S. aromaticum* and *Z. officinale* essential oils, would affect performance, some carcass, blood and intestinal parameters of broilers. Ninety, 1-day-old broilers were allocated into 6 dietary treatments groups. Each essential oil supplementation to the basal diets was made at 120 ppm or 240 ppm alone or 60+60 ppm or 120+120 ppm in combinations. Each group was fed *ad libitum* for 42 days.

Essential oils did not affect body, carcass and abdominal fat weights, but significantly improved feed intake and feed efficiency ($P<0.05$). The group receiving 240 ppm *Z. officinale* converted feed into gain more efficiently than the others. Essential oils also affected some blood parameters and intestinal measurements and microflora ($P<0.05$). Glucose level was increased, while triglyceride was decreased by *Z. officinale* 240 ppm without any significant effect on cholesterol concentration. *Z. officinale* at 240 ppm decreased ceaceum weight and length, while increasing colon weight and length and also villi heights of the jejunum. It is concluded that essential oils, especially *Z.oficinale* at 240 ppm, exhibited positive effects on feed efficiency and also development and functions of digestive tract.

Financially supported by TUBİTAK (104O439) and Çukurova University Research Fund (D200431).

Effect of dietary roughage source on the fatty acid composition of cows' milk

S. Segato[1], C. Elia[1], C. Ossensi[1], S. Balzan[2], S.Tenti[1], S. Franco[2], M. Dorigo[1], L. Lignitto[1], E. Novelli[2] and I. Andrighetto[1], [1]Dept. of Animal Science and [2]Dept. of Public Health, Comparative Pathology and Veterinary Hygiene of Padova University, 35020 Legnaro (PD), Italy

According to a 2x2 cross over design, 14 Holstein dairy cows at 212±99 DIM were fed two isoenergetic (0.94 Unité Fouragère Lait/kg DM) and isonitrogenous (CP=14.8% DM) TMR based on grass hay (GH) or maize silage (MS). Milk samples were collected during the third week of each period and fatty acids (FA) profile (n=14) was detected by using gas-chromatography. Data were submitted to variance component ANOVA according to the PROC MIXED of SAS. Dietary treatment had no effect on DM intake (20.3 kg/d on average), milk yield (29.3 kg/d) and milk fat content (4.10%). Milk from GH-diet contained lower concentration of SFA (61.9 vs. 63.4% of total FA; $P<0.05$) and higher level of PUFA (6.1 vs. 5.8; $P<0.01$). Feeding more hay also increased CLA and n-3 FA and decreased C16:0, while C14:0 was similar. Increases in the PUFA and FA n-3 resulted in lower ($P<0.01$) atherogenic and thrombogenic indexes in GH-milk than in MS one. A complete substitution of maize silage with grass hay seems to improve milk FA profile, even if results could be also affected by different kind of concentrates used to balance the diets and/or different diet particle size distribution in the rumen.

Comparison of methods for the determination the fat content of meat

J.Seenger[1]*, K. Nuernberg[2], G. Nuernberg[2], M. Hartung[2], Cs. Ábrahám[1], E. Szűcs[1] and K. Ender[2], [1]Szent István University of Gödöllő, H-2103 Gödöllő, Páter K. u. 1. [2]Research Institute for the Biology of Farm Animals (FBN) D-18196 Dummerstorf, Wilhelm-Stahl-Allee 2, Germany

For the efficiency of the animal breeding, and also in the food science, the accurate determination of the components, including fat content of the meat is very important. The objective of this study was to compare different chemical and physical methods as well, which are widely used to determinate the fat content of the different tissues of the animals. In total 23 pigs and 19 cattle were included in the investigations. Different cuts of the carcasses like *longissimus* muscle, head, belly and breast, feet, ham, neck, loin, tenderloin and subcutaneous fats were used this comparison. During the investigation the following 3 chemical [Soxhlet method, automatic Fat extraction (ANKOM XT 15 Extractor), automatic fat extraction following hydrolyzing (ANKOM HCL Hydrolysis System, ANKOM XT 15 Extractor)] and 2 physical methods (Infratec1255 Food and Feed Analyzer, Tecator AB; FoodScan™ Lab, FOSS) were compared. For the accurate staistical analysis the different cuts were ordered into 3 groups. Data were analysed by the GLM procedure of SAS (1985), using the Least-Square-Method. The repeatability (θ) of the chemical methods for all samples was also calculated, it ranged between 0,9925-0,9977.

Influence of linseed oil and diet composition on conjugated linolic acids (CLA) in duodenal flow and milk fat of cows

G. Flachowsky[1], K. Erdmann[1], G. Jahreis[2] and P. Lebzien[1], [1]Federal Agricultural Research Centre (FAL), Institute of Animal Nutrition, Bundesallee 50, D-38116 Braunschweig, [2]Friedrich Schiller University Jena, Germany

The aim of the experiment was to investigate the influence of ration type and linseed oil supplementation on biohydrogenation in the rumen and its effect on CLA concentration in milk. Seven multiparous lactating cows fitted with cannulaes in the rumen and in the proximal duodenum were randomly assigned to four experimental periods applying a 2 x 2 Latin Square design. Ration types consist of 70% grass hay and 30% concentrate or 30% hay and 70% concentrate. Both rations were unsupplemented or supplemented with 200g linseed oil and fed for 4 weeks.

The CLA content in the duodenal digesta was very low (< 0.42% of fatty acid methyl ester; FAME) and not significantly influenced by feeding. The biosynthesis of trans fatty acids (tFA) in the rumen was significantly influenced by ration type/rumen pH and linseed oil supplementation. The t $FA_{18:1}$ increased from 5.3 (70% hay; - oil) to 20.0% (30% hay; + oil) of FAME in duodenal flow. The t $FA_{18:1}$ increased in milk fat from 8.7 (70% hay; - oil) to 47.3 g/kg (30% kg hay; + oil); the CLA concentration was only affected, when linseed oil was added (4.5 g/kg; - oil; 6.8 g/kg, + oil).

Effect of feeding regime in early life on fatty acid profile of meat from intensively finished beef bulls

I. Casasús[1], M. Tor[2], M. Joy[1], D. Villalba[2] and M. Blanco[1], [1]CITA-Aragón, [2]UdL, Spain

Two weaning strategies associated to different feeding regimes in mountain beef cattle production systems were compared to test the influence of diet composition in early life on meat nutritive value. Sixteen spring-born Parda de Montaña bull calves were assigned to two treatments: Early-weaned (EW) calves were weaned at 90 d of age and placed on a concentrate-based fattening diet, while Normally-weaned (NW) calves grazed with their dams on high mountain ranges with no supplement until 180 d of age. From d180 to slaughter at the age of one year calves from both treatments received the same grain-based finishing diet. Calf performance in the different periods was significantly affected by weaning strategy, EW having higher GMD from d90 to d180, and TW showing compensatory growth from d80 to slaughter. Despite the different growth paths, carcass and instrumental meat quality traits were not influenced. Fatty acid analysis of intramuscular fat samples was performed by gas chromatography. No differences between proportions of polyunsaturated (PUFA), monounsaturated (MUFA), saturated (SFA) fatty acids and the ratios PUFA:SFA, Hypercholesterolemic:Hypocholesterolemic and w-6:w-3 fatty acids were found associated to both treatments. These results suggest that although differences related to diet composition might have arisen by d180, the common six-month fattening period on a similar diet was sufficient to eliminate any possible previous difference in fatty acid profile.

Acorn meal use as a nutrient in diets of rainbow trout
G.H. Shadnoush, Agriculture and Natural Resources Research Center, Shahre-kord, Iran P.O. Box 415,8813657351

In order to use of different ration of acorn meal in diet of rainbow trout as a nutrient, the experiment was carried out in complete random in design. The treatment was 1,2,3 and 4, which contained of 0,2,4,6 percent of acorn meal in different diets respectively. The diets were formulated with local material which could providing nutritive requirement of rainbow trout in principle of NRC with UFFDA soft ware dietary, as diets contained of 0,2,4,6 percent acorn meal and other nutrient were equal. The feed were given 3 times a day on biomass basis of pounds. In during of experiment, once every two weeks, after anaesthetize 50 percent of fishes in each pound were measuring a bout weight and length and the data were taken. The results showed which final weigh increase more was in treat 4 and have been significant differences with others. The greatest length was observed in treat 4, which had significant differences with treat 1. Feed consumption were not significant differences between any treats, but was more in treat 4. Feed conversion ratio affected by treat, and the best was in treat 4 and had significant differ with treat 1. The result showed which acorn meal can use as a nutrient in rainbow trout diet, because improve of fattening performance without physiological effect like similar condition of this experiment.

Determination of conjugated linoleico acid (CLA) in milk of bovine, Ninth region Chile
J.P. Avilez[1], D. Sanchueza[1], F. Garcia[1], R. Matamoros[1], J. Perea[2]* and A. Garcia[2], [1]Catholic University of Temuco, Chile, [2]Department of Animal Production, University of Cordoba, Spain

With the aim to determine the level of conjugated linoleic acid (CLA) in bovine milk of the productive system in the IX region in Chile, milk samples were analyzed from 21 bovine cattle of Frison Negro breed, during 2004.
Within the analyses of the milk through gas chromatography, it was determined that during the summer two isomers of CLA with biologically active functions, cis-9, trans-11 and the cis-10, trans-12. In autumn, the isomers cis-10, trans-12; cis-10, cis-12 and a mixture isomers composed by cis-9, trans-11; cis-11, trans-9 were analyzed. Then, the most important quantity of cis-9, trans 11 and cis-12, trans-10 was registered in milk fat coming from the cattle. In both seasons of the study, the percentage of the isomers cis-9, trans-11 (1.44% in summer and 1.42% in autumn) was higher to the one found in isomer cis-10, trans-12 (1.07% in summer and 1.04% in fall) in the bovine milk fat. In summer, it was a total value of 2.50% in the milk fat for the isomers cis-9, trans-11 and cis-12, trans-10 existing a trend to diminish slightly in fall (2.46%), due to these differences it was not noted a season effect in the Vilcun are in the IX region of Chile.

Effects of supplementation of copper and garlic on egg yolk lipids in commercial laying hens

Farshad Zamani[1] and Sayed hamid Fariman[2], [1]Agricultural research center of Shaherkord, Iran, [2]Azad university of Shaherkord, Iran

160 white leghorn 38 weeks old hens were divided to four treatments, and were fed basal diet and different supplementation. The first treatment diet, were added 200 ppm cupric sulphate and 1% garlic powder. The second treatment, only used 200 ppm cupric sulphate and in the third treatment diet were added 1% garlic powder. The fourth treatment was considered as a control group. The length of feeding with these diets lasted six weeks and at the end of the 4th, 5th and 6th weeks, the treatments were sampled. Cholesterol and triacylglycerol of egg yolks were analyzed by commercial enzymatic kits by spectrophotometry and fatty acids were analyzed by HPLC. The content of total lipid and stearic acid show no significant difference in none of the treatments compared with the control group($P>0.05$). In the first treatment, there was a reduction in cholesterol and an increase in oleic acid, linoleic acid and α-linolenic acid of egg yolk($P<0.01$). But no significant differences were observed in the content of palmitic acid and yolk triacylglycerol. The most changes in the second treatment, are similar to the first treatment but the content of yolk triacylglycerol increased at the 5th week, and the content of palmitic acid shows significant increase at the 6th week ($P<0.01$). In the third treatment and at the 6th week, there was a reduction in cholesterol and increase in palmitic acid ($P<0.05$). The result indicated that dietary copper can decreased cholesterol and saturated fatty acids of egg yolk.

Objective measurement of coat colour varieties in the Hungarian Grey cattle

A. Radácsi[1], B. Béri[1], I. Bodó[1], [1]University of Debrecen, Department of Animal Breeding and Nutrition, 4032 Debrecen, Böszörményi Str. 138, Hungary*

When preserving genetic resources one of the most important tasks is to maintain the typical characteristics in order not to lose the available genetic variability. Therefore, traits without economic value at the moment should also be conserved. The colour of the coat is an important trait in forming the breed character. The Hungarian Grey cattle is characterized by a great variety of coat colours, which are genetically determined and vary within the given breed character not only by age but are affected by the season, as well.

Research has been done on the largest Hungarian Grey cattle stock, at the Hortobágy Society for Nature Conservation and Gene Preservation. For objective measurement of coat colour the Minolta Chromameter CR-410 was applied.

Hungarian Grey calves are reddish-coloured at the time of the birth. However, several variants can be distinguished: dark reddish, reddish and light reddish. Proportions of the different varieties and the mean L*a*b* values of the specific variants were determined. Colour of heifers' winter and summer coat was analysed. Generally, the front part of the body is darker which is proved by the lower mean L* values (L*$_{neck}$: 49.45, L*$_{belly}$: 52.37 and L*$_{thigh-croup}$ 53.61). Statistically significant differences ($P<0,05$) were found between the L*a*b* values of winter and summer coat.

Session C33

Similar growth rate and rumen development in weaned dairy calves fed an alternative compared with a traditional calf starter concentrate

M. Vestergaard*, N.B. Kristensen, S.K. Jensen and J. Sehested, Danish Institute of Agricultural Sciences, Tjele, Denmark

Even small amounts of high-starch concentrate produce low (<5.5) rumen pH in the juvenile reticulo-rumen of milk-fed calves. We hypothesized that a new feeding concept including a high fibre-low starch concentrate and introduced later could circumvent these drastic acidotic conditions and still develop the rumen. Eight Holstein calves (48 ± 1 kg BW at birth) implanted with ruminal cannulaes (20 mm ID) at d 7 ± 1 of age were allocated to either a control treatment (CON) with free access to a barley-based concentrate (319 g starch/kg) from wk 1 or an alternative treatment (ALT) without access to concentrate until wk 4 when they got free access to a low-starch concentrate (68 g starch/kg). All calves had free access to artificially-dried grass hay and water, and they were fed fixed levels of skim-milk-based milk-replacer (6.4 L/d). Calves were weaned at wk 8 and were sacrificed at wk 10. From wk 1 to 10, there were no treatment effects on ADG (802 ± 35 g/d) or total solid feed DMI (78 ± 4 kg). Intake of dried-hay was markedly higher both before and after wk 4 ($P<0.05$) and minimum ruminal pH was higher (6.14 vs. 5.57 ± 0.09, $P < 0.01$) in ALT compared with CON calves. There were no differences in reticulo-ruminal mass, or in length, shape and aggregation of ruminal papillae at wk 10. Thus it is possible to formulate a rumen-friendly concentrate with very low-starch content that is associated with high growth rates and normal ruminal development in milk-fed calves.

Session C33

Feed intake and energy balance – investigations with intent to integrate the energy status of bull dams into the performance test on station

H. Hüttmann*, E. Stamer, W. Junge, G. Thaller, E. Kalm, University of Kiel, Institute of Animal Breeding and Husbandry, D-24098 Kiel, Germany

Since 1st of September 2005 the individual roughage intake has been measured with a computerized scale system at the dairy research farm Karkendamm, where the individual performance test of the bull dams of a German breeding company is located. In conjunction with the routinely collected data of the established individual performance test the daily energy balance has been calculated as the difference between energy intake and calculated energy requirements for lactation and maintenance costs.

Several criteria of energy balance have been defined over different lactation stages: mean daily energy balance during the first 20, 50 and 100 days of lactation, days in negative energy balance and total energy deficits in the early stage of lactation. For these traits genetic parameters have been estimated. The resulting heritabilities and the genetic correlations between energy balance criteria and performance traits have been taken into consideration for the decision of which trait of energy status should be included in the individual performance test of bull dams.

Session C33   Theatre 4

Interest in and market opportunities for special regional products
M. Klopcic[1], E. Oosterkamp[2], G. Tacken[2], M. Kos-Skubic[3], A. Pflimlin[4] and A. Kuipers[5], [1]Biotechnical Faculty, Zoot.Depart., Domzale, Slovenia, [2]Agricultural Economic Research Institute, Wageningen, The Netherlands, [3]Ministry of Agriculture, Forestry and Food, Ljubljana, Slovenia, [4]Institut de l'Élevage,Paris,France, [5]Expertisecentre for Farm Management and Knowledge Transfer, Wageningen, The Netherlands

According to EU definitions special regional products are classified in 3 groups. Characteristics are: territoriality, typicality, tradition and collectivity. To be successful in marketing these products, the base quality must meet prescribed standards. However, the "overall quality" encompasses more elements as mentioned above. In this paper strong and weak points of marketing such products will be described based on a SWOT analysis performed in Slovenia: 14 interviews and 2 workshops were held. Also the interest of farmers to diversify their operation with this branch was assessed using a questionnaire as tool. Territoriality (region) and tradition are important characteristics of the product, of which traditional breed and cultural heritage are usually part of. The willingness of the individual producers to obey to certain rules is essential and not easy to achieve: a similar looking product with the same price, but without costs of logo and certification is an easy route to go. Surprising was that EU and quality logo was often not part of the packing of product and that producers did not see any competitors in the market. Production volume is often the limiting factor. Home sales, restaurants and speciality shops are main selling points. Promotion is a week point: co-operation of different producers groups would be an advantage.

Session C33   Theatre 5

Feeding systems for dairy cows and their constraints in the Central East region of Tunisia
K. Kraiem[*] and R. Aloulou, Ecole Supérieure d'Horticulture et d'Elevage de Chott Mariem, BP 47, 4042, Sousse, Tunisia

We conducted a survey in 194 dairy producers in the Central East region of Tunisia to identify the feeding systems and their technical limitations. The informations gathered concern the producers, their farms and techniques and ways of feeding cows. We calculated many feeding parameters and made comparaisons between 3 areas in the region.. We identified 2 systems : the no land system where only dry forages (DF) were fed with concentrates (CC) and the integrated system where green chopped forages (GCF) were fed along with DF and CC. The no land system was used by 22.2 % of producers with differences between the 3 areas. Total and feedstuff dry matter intakes were low for both systems (16.4 and 16.52 kg DM for total intake, 6.8 and 7.2 kg DM for feedstuff, respectively). In the integrated system GCF were fed by a large number of producers, but the quantities fed were low and the periods of distribution were limited. Concentrates, wheat bran and wheat straw were widely utilized by producers (100, 65.5 and 71.1 % of them, respectively). The quality of the main feedstuffs fed (wheat straw and hay) was low. Most of the rations were unbalanced. These limitations explain the poor milk production realised.

The effect of addition of methionine and lysine in the form of rumen protected tablets on the milk protein yield in high-producing lactating dairy cows

J. Třináctý[*1], M. Richter[1] and P. Homolka[2], [1]Agrovýzkum Rapotín, Ltd., Dept. Pohořelice, Vídeňská 699, 691 23 Pohořelice, Czech Republic, [2]Research Institute of Animal Production, Praha-Uhříněves, 104 01, Czech Republic

The aim of the study was to confirm the possibility of compensation of deficiency of Lys and Met in diets of high-producing dairy cows based on corn silage by addition of these amino acids in the form of rumen protected tablets. The diameter of the experimental tablets was 6 mm, protective layer was on the copolymer vinyl-pyridine-styrene basis. The tablets were applied orally after mixing into a part of the feeding mixture before each feeding. The experiment in the form of Latin square design (4x4) was carried out on 4 lactating dairy cows in four 14 d intervals with four levels of factor (control, Lys, Met, Met+Lys). The milk samples were collected for the last 4 days of each period. The group of cows supplemented with Lys+Met showed the significantly higher (P<0.05, n=16) milk yield (34.2 kg/d, SEM=1.41) in comparison with group Lys (32.5 kg/d, SEM= 1.54) or Met (32.2 kg/d, SEM=1.77). Daily milk protein yield was significantly higher (P<0.05, n=16) in the group Lys+Met (1054 g/d, SEM=48.2) in comparison with control (990 g/d, SEM=43.7), Lys (998 g/d, SEM= 44.6) and Met (968 g/d, SEM=54.0).
This study was supported by MSM2678846201.

Feeding of pellets rich in NDF to lactating cows in automatic milking system

I. Halachmi[1], E. Shoshani[2], R. Solomon[2], E. Maltz[1] and J. Miron[1], [1]ARO, [2]Extension Service, Israel

If the milking frequency in an automatic milking system (AMS) is increased, the intake of concentrated pellets in the robot may be raised accordingly. Consumption of a large quantity of starchy grains within a short time can impair the appetite, voluntary visits to the milking stall and intake of DM and NDF. The null hypothesis was: 'the conventional starchy pellets fed in the AMS can be replaced with pellets rich in digestible NDF without impairing the cows' motivation to visit a milking stall voluntarily'. Fifty-four cows were paired according to age, milk yield and days in milk, and were fed a basic mixture (BM) along the feeding lane (19.9 kg DM/cow/d) plus a pelleted additive (~5.4 kg DM/cow/d). The two feeding regimes differed only in the composition of the pelleted additive, 49% starchy grain vs. 25% starchy grain plus soy hulls and gluten feed, as replacement for part of the grain. Both diets resulted in similar rates of voluntary milkings (3.31-3.39 visits/cow/day), similar average yields of milk and percentages of milk protein. The results suggest that an alternative pellet composition can be allocated in the AMS in conjunction with BM in the feeding lane without negative effect on the appetite, milk yield and composition, or milking frequency. It opens the opportunity to increase yield and solids by increasing the amount of pellets allocated to selected high-yielding cows, via the AMS while maintaining a high-frequency voluntarily milkings.

A study on metabolic profile test in a Brown Swiss herd

A. Abavisani[1] *and M. Adibnishabouri*[2], [1]*Department of physiology, Veterinary faculty, Tehran University, Iran,* [2]*Department of veterinary, Khorasan Jihad-e-agriculture education complex*

Metabolic profile testing has generally been used as part of a multidisciplinary approach for dairy herds. This technique is widely considered a worthwhile method for ascertain nutrition and productive problems in production animals.

Our goal was to evaluate metabolic profile in a Brown Swiss herd and we wanted to determine relationship between herd problems with metabolic profile. We chose 21 cows in 3 groups based on milk production(7 dry cows, 7 intermediate milk producing cows and 7 high milk producing cows). Blood samples were collected and the following components were determined in blood serum: Glucose, Cholesterol, TG, TP, Albumin, Globulin, PCV, Hb, BUN, Ca, P, Mg, Na, K, SGOT and CBC. Health history, nutrition status and fertility situation were obtained at sampling time.

The results indicated that mean value of components in three group were at normal range. However, value of Glucose, Cholesterol, TG, Ca, P and Mg were marginal in high milk producing group and these values were less than two other groups.SGOT was more in high milk producing group in comparison of other groups. Also, many metabolites and electrolytes of one cow in high milk producing group were less than normal values. The cow had a history with milk fever and hypomagnesaemia.

These results suggested that this herd has not a serious nutritional or metabolic problem and nutrition status is good. Some present herd problems were individual problems.

Once daily milking does not change milk fatty acid profile in cows in equilibrated energy balance

Y. Chilliard[1], *D. Pomiès*[1], *P. Pradel*[2] *and B. Rémond*[3], [1]*Herbivore Research Unit, INRA-Theix, F-63122-Ceyrat, France,* [2]*INRA, 15190-Marcenat,* [3]*ENITAC, 63370-Lempdes, France*

Once daily milking (ODM) decreases milk yield and increases milk fat content (Anim. Res., 53:201-212), but its effect on milk fatty acid (FA) composition is not known. This trial compared cows while they were milked twice or once daily and were fed 2 levels of concentrate (normal or low [normal – 3 kg]) or three levels (normal, low or very low [normal – 6 kg]), respectively. Five groups of 11 cows in the declining phase of lactation were involved, and measurements done on pooled milk samples taken after 6 weeks on all the cows of each experimental group. ODM did not modify milk FA composition (major, minor, saturated, unsaturated, *trans* or conjugated). Decreasing concentrate level in cows milked twice daily increased milk oleic acid content (+2.2g/100g FA) and decreased 10:0 to 14:0 (-2.0g/100g FA), in agreement with simultaneous increase in blood plasma non-esterified FA concentration and decrease in calculated energy balance (J. Dairy Sci., 74:1844-54). However, decreasing concentrate level in ODM group had no or very few effects on these parameters, in agreement with the fact that these cows remained in positive or equilibrated energy balance. In conclusion, ODM does not change milk FA composition, except in restricted fed cows in which the lower milk energy yield allows animals to remain in positive energy balance, compared to changes observed in twice daily milked ones.

Reasons for culling in Iranian Holstein cows

A.A. Naserian[1], M. Sargolzaee[1], M. Sekhavati[1] and B. Saremi[2], [1]Agriculture College of Ferdowsi University of Mashhad, Iran, [2]Education center of Jihad-e Agriculture, Mashhad, Iran

Culling of dairy cows is probably one of the most complex decisions in dairy operations. The aim of this study was to determine the profiles of culled cows in order to access the possible contribution to economic losses due to health disorders in the dairy herds of Khorasan province. Data regarding all exits of cows from the herd were collected during a 5-year prospective survey in 15 large dairy commercial Holstein herds with over 4000 milking cows totally (From March 1999 to March 2004). All herds were recorded by an official milk-recording scheme. The management and feeding systems were almost similar in all herds. Rolling herd's averages were also similar in all herds; over 24000 lb. A polytomous stepwise logistic regression method was used because it allows the use of a non-ordinal categorical variable. The model was run (Procedure PR of BMDP). Table 1 shows the results of this study. The most frequent primary culling reasons were infertility and health disorders, 25.13, 28.57 of total cull respectively. Percentages of six groups of culling reasons for level of parity showed that the first parity level had more frequented for reproductive problems and health problems too. Therefore, in this study, more than one half of the cows were declared willed in view of health or reproductive related problems.

Environmental contamination by Hexachlorocyclohexane of bovine milk: A case study in Central Italy

B. Ronchi, P.P. Danieli, U. Bernabucci, Dipartimento di Produzioni Animali, Università della Tuscia, via S. C. de Lellis, 01100, Viterbo, Italy

Hexachlorocyclohexane contamination of agro-ecosystems is a World scale problem. At the beginning of 2005 a wide area of dairy cows farming in Central Italy was discovered to be polluted by Hexachlorocycloexane (HCH) (mean concentration in soils 0.078 mgHCH/kg). Lacking wide knowledge of HCH metabolism in dairy cows, this study was performed to assess the presence of HCH in milk of cattle after an event of chronic exposure to HCH. During the study, animals were fed on feedstuffs containing no detectable HCH amount. In five dairy farms, milk and blood samples were collected from July to September 2005 every three weeks from animals in different lactating stage. Pre-treated samples were analysed by Gas Chromatography using an Electron Capture Detector for quantification of HCH-isomers. Statistical analysis was performed by means of ANOVA and LR. Differences in milk contamination by β-HCH among dairy farms ($p<0.01$) and sampling time were found ($p<0.05$). Milk contamination levels exceeded at any time the EU limit (0.003 mgβ-HCH/kg), posing potential hazards for human consumption due to chronic exposure. A linear regression between blood serum and milk β-HCH content was found ($r^2=0.919$, $p<0.05$). Furthermore, β-HCH as a traces was detected in blood serum when milk levels were below the analytical limits indicating the usefulness of serum HCH content as an early indicator of animal exposure.

Some effects on weaning results of beef calves
F. Szabó, L. Nagyand Sz. Bene, University of Veszprém, Georgikon Faculty of Agriculture, Hungary, H-8360 Keszthely Deák F. str. 16.

Weaning weights of 469 beef calves, 232 male and 237 female born between 1988 and 2003 from cows of 9 breeds (Hungarian Fleckvieh, Hereford, Aberdeen Angus, Red Angus, Lincoln Red, Charolais, Limousin, Blonde d'Aquitaine and Shaver) kept in the same condition on peat-bog soil pasture at Keszthely were evaluated. The effect of breed and age of the dams, year and season of birth, and sex of the calves on weaning weight adjusted to 205 days of age was computed by multivariate analysis of variance (GLM), software SPSS 9.0.
Overall mean of the 205 days weaning weight of all calves was 198,6 kg. The evaluated factors had significant ($P<0.01$) effect on 205 day weight. Mean values of the adjusted weaning weights by breeds were 207, 164, 212, 211, 193, 178, 204 and 216 kg, respectively. According to the age of the dams the adjusted weaning weight increased up to 5 years, and after the maximum (214 kg) decreased. The minimum values were found in the group of 2 year old (188 kg) and 12 year old (187 kg) cows. In accordance with the birth year the highest weaning weight (241 kg) was observed in 2002, while the lowest (139 kg) in 1999. Due to the birth season, winter, spring, summer and autumn, the 205 day weaning weight were 185, 200, 222 and 188 kg, respectively. Male calves reached 206, while female calves 191 kg mean value of the adjusted weaning weight.

Body weight prediction using digital image analysis for different breeds of slaughtering beef cattle
Y. Bozkurt[*], S. Aktan and S. Ozkaya, Suleyman Demirel University, Faculty of Agriculture, Department of Animal Science, Isparta, Turkey, 32260

In this study, it was aimed to predict body weight of slaughtering beef cattle by using both traditional methods and digital image analysis system. Some digital images and body measurements such as body weight, body area, wither height, chest girth, body length, chest depth, hip width and hip height of different breeds of beef cattle; Holstein, Brown Swiss and their crosses were used and prediction models were developed. There were significant differences ($P<0.05$) between the body measurements obtained by traditional methods and digital image analysis system. Considering breed differences, the R^2 values of prediction equations were 76, 87.9 and 85.6% for Holstein, Brown Swiss and crossbred animals respectively. The regression equation which included only body area showed that the prediction ability of digital image analysis system was higher for Brown Swiss and cross breeds ($R^2=75$ and 78% respectively) while it was lower for Holsteins ($R^2=66.5\%$).
The results showed that the prediction ability of digital image analysis system was very promising to predict body weight. However, there is a need for further studies in order to develop better techniques to use for prediction.

Effects of an additive enriched with the first limiting amino acids on growing performances of double-muscled Belgian Blue bulls fed corn silage based diet

P. Rondia[1], E. Froidmont[1]*, Y. Beckers[2], V. Decruyenaere[3] and N. Bartiaux-Thill[1], [1]CRA-W, Animal Production and Nutrition Department, rue de Liroux 8, 5030 Gembloux, Belgium [2]FUSAGx, Unité de Zootechnie, Passage des Déportés 2, 5030 Gembloux, Belgium [3]CRA-W, Section Systèmes agricoles, rue du Serpont, 100, 6800, Libramont, Belgium

Two trials were conducted on double-muscled Belgian Blue bulls (BBb). The aims of the first one were (1) to determine the ruminal bypass of free amino acids (Met and Lys) administred intraruminally (once per day during the morning meal) and (2) to determine the lack of the first limiting amino acids (AA) of BBb for a corn silage based diet in order to formulate a specific additive that was investigated in the second trial. The first experiment was led on 6 fistulated animals according a 6*3 cross-over design. The results suggested that His, Met, Arg, Lys and Val were respectively the first limiting AA in our experimental conditions. The mean ruminal escape of Met and Lys was similar and particularly high (bypass = 37%) due to the low volume of rumen liquid and the high particle outflow rate. The second experiment was led on 24 animals in a randomised blocks design. The additive had only a significant effect on growth performances during the growing period of the BBb (370–430 kg) with an increase of ADG by 255 g/d.

Fatty acid profile of different muscles from Charolais and Simmental bulls supplemented with whole sunflower seed

L. Bartoň*, V. Kudrna, D.Řehák, V. Teslík, R. Zahrádková, D. Bureš, Research Institute of Animal Production, Přátelství 815, 104 00 Prague – Uhříněves, Czech Republic

We have examined the effects of cattle breed, dietary fat, and muscle location on the fatty acid (FA) profile of intramuscular fat. Forty-six Charolais (CH) and Simmental (SI) bulls (357±60 kg) were given two isonitrogenous and isocaloric diets supplemented with either whole sunflower seed high in C18:2n6 (SUN) or Megalac high in C16:0 (CON) as different sources of dietary fat. After slaughter, samples of m. longissimus thoracis and m. infraspinatus were collected and the concentrations of different fatty acids in total lipids were determined. The concentrations of C14:0 and C16:0 were higher in CH bulls (P<0.001) while C18:1n9 was higher in SI (P<0.001). As a result, muscles of SI contained a higher ratio monounsaturated/saturated FA (P<0.001). SUN diet increased C18:0, C18:2n6 (P<0.001), c9,t11CLA (P<0.01), and the ratio PUFA n6/n3 (P<0.001). M. infraspinatus generally contained more saturated and less polyunsaturated FA than m. longissimus thoracis, probably reflecting differences in the content of phospholipids in different metabolic fibre type muscles. The study was supported by the project MZE0002701403.

Session C33   Theatre 16

Isolation and identification of salmonellae serotypes from bovine and ovine meat which were slaughtered in kermanshah slaughterhouse

Amin Bahiraie*, A. Chale Chale and A. Nooriyan, Razi University, Faculty of Veterinary medicine, Dept of Meat Hygiene, Kermanshah, 67156-8-5414, 1451,Iran

Meat is one of the most important animal protein resources and has got major role in human heath and market economy.the unhygienic meat can make a lot of problems in regard to zoonotic disease as like as Salmonellosis. The present study was conducted to survey ovine and bovine meat contamination rate to salmonellae and to find out which Sero or bioserotype is dominant. Bacteriological examination was carried out on 500 bovine and 500 ovine meat samples in four season of year (2004).

Samples were examined in regarding routine bacteriological examinations and were stereotyped based on Kauffman-white Schema. Salmonellae were isolated from 52 of bovine samples (10.4 %) and 31out of 500 ovine carcasses (%6.2). The highest isolation rate in bovine 10.4 % in fall and of ovine was from summer with6.4 %. The least isolation rate in bovine samples with6.4 % in winter and from ovine also was from winter with 1.6 %. Stereotyping of salmonellae isolate from bovine carcasses was S.typhimorium31.6% and then S. entritidis17. %, S.doblin15.5 %, S. abortus ovis8.15 %, S cholerae_suis 3.9 %, S. paratyphi B3.9%, S. pullorum 3.3 %, And from ovine carcasses was S. typhimorium 28.7 %and then S. entritidis18.2 %, S. abortus ovis14.7%,S. doblin11.2%,S. cholerae. suis2.1 %, S. pullorum1.4 %, respectively.

Session C33   Theatre 17

Nutritional effects of choline on milk production, milk composition and blood metabolites of lactating dairy cows

A.H. Toghdory*[1], T.Ghoorchi[2] and A.A. Naserian[3], [1]Islamic Azad university of Gorgan, Iran, [2]Gorgan Agricultural Sciences and Natural Resources University, Iran, [3]Ferdowsi University, Mashhad, Iran

To investigate the nutritional effects of choline in dairy cows ration, eight multiparous Holstein dairy cows in early lactation with a mean milk production of 34.6±2.82 Kg/d used in a change over statistical design with four periods. Experimental treatments were 1) control, 2) rumen-unprotected choline 50 gr/d, 3) rumen- protected choline 25 gr/d and 4) rumen-protected choline50 gr/d. Diet were fed two times a day and choline top dressed in first meal. During last 7 and 2 days of each period milk yield recorded and sampled for determine milk composition. Blood sampling were in the last day of each period for determine plasma glucose, triglyceride, cholesterol and blood urea nitrogen. The results showed that rumen-protected choline (50 gr/d) increased Dry matter intake, milk yield, 4% fat corrected milk yield and fat production(P<0.05). DMI, milk composition and blood metabolites were not affected by rumen-protected choline 25 gr/d(P>0.05). Milk fat percentage in rumen-protected choline 25 gr/d significantly higher than Rumen-unprotected choline (P<0.05). Milk composition (percentage and production), DMI, efficiency of feed conversion and blood metabolites were not affected by rumen-unprotected choline(P>0.05). In conclusion, our results indicated that rumen-protected choline 50 gr/d had a significant effect on performance of dairy cows in early lactation.

Fertility trends in Holstein Friesian cows

E. Báder[1]*, Z. Gergácz[1], A. Muzsek[1], A. Kovács[1], I. Györkös[2] and P. Báder[1], [1]University of West Hungary, Faculty of Agriculture, 9200 Mosonmagyaróvár, Hungary, [2]Research Institute for Animal Breeding and Nutrition, 2053 Herceghalom, Hungary

We examined the trends of the time of the first two consecutive inseminations in stocks with high Holstein-Friesian (n=12820) blood-proportion in 6 randomly chosen large farms between 1990-2004.

The time of first insemination was between 54 and 91 days among the animals were inseminated at least two times. The time of the second insemination was between 109 and 147 days from the calving. Between the two inseminations passed 45-56 days (one heating cycle). 45-57% of the cows needed more inseminations, the days open was between 154-188 days, and the insemination index was 2,74-3,21.

The rate of inseminations was 29,7-38,8% in the first heating cycle, and 13,4-17,6% in the second heating cycle from the total second inseminations. The rate of insemination was about 10% in the third heating cycle. 4,3-6,4% of cows were inseminated again in the fourth cycle, therefore they were not inseminated at least for 76 days. It is a warning data that the 9,1-17 % of the second inseminations was after 88 days. The rate of the inseminations out of heating cycle was between 1,8-10,7% per cycle, but to sum up the four cycles, the rate of the inseminations out of cycle was between 16,3-33,7%, therefore one third of the inseminations was out of cycle.

Superovulatory response of Holstein dairy cows in Turkey

Z. Gocmez[1], U. Serbester[1], E. Sirin[2], E. Yazgan[1], O.G. Dundar[1], N. Cetinkaya[1], A.G. Onal[3], M. Kuran[2], [1]Institute of Cukurova Agrucultural Research, Adana, [2]Universities of Gaziosmanpasa, Tokat and [3]Mustafa Kemal, Hatay, Faculty of Agriculture, Department of Animal Science, Turkey

The total number of transferable embryos is a major limiting factor for the success of embryo transfer programmes. Recently, a MOET programme has been implemented to create a nucleus herd for the improvement of dairy cattle in Turkey. The present study report preliminary results of superovulation treatments within this programme. A total of 43 Holstein heifers were superstimulated with FSH (Foltropin V, AGTECH) which were injected at 12 h interval in a declining dose over a 4 days period (20 or 30 mg FSH) starting on day 9 following PRID (DIF, Turkey) insertion. $PGF_{2\alpha}$ (Dinolitik, Pfizer) was injected with the 5th FSH injection and PRID was removed 36 h after the $PGF_{2\alpha}$ injection. Following AI on 48 and 72 h after the $PGF_{2\alpha}$ injection, uterus of donors were flushed nonsurgically on day 7 after the AI. In result, the number of corpora lutea was 14.3±8.3 per donor and the total number of embryos/ova recovered and the number of trasferable embryos were 7.36± 1.1 and 6.1± 1.08 respectively. Recovery rate was 53%. In conclusion, considering the fact that it can be regarded as training period of the technicians, a reasonable number of embryos were obtained from the donors for the use in the MOET programme to improve dairy cattle in Turkey.

Pregnancy rates in multiparous dairy cows following Ovsynch and CIDR ovulation synchronization and timed artificial insemination protocols

M. Aali*[1] and R. Rajamahendran[2], [1]Kuwait Institute for Scientific Research, P.O Box 24885, 13109 Safat, Kuwait, [2]Faculty of Land and Food Systems, The University of British Columbia, 248-2357 Main Mall, Vancouver, BC V6T1Z4, Canada

The objective of this study was to compare pregnancy rates following Ovsynch and the CIDR ovulation synchronization protocols and timed artificial insemination (AI) in postpartum multiparous Holstein dairy cows. The cows were divided into the Ovsynch (n = 67) and the CIDR (n = 71) treatment groups. The Ovsynch treatment protocol consisted of an initial injection of GnRH (100μg), a $PGF_{2\alpha}$ injection (25mg) 7 d later and a second injection of GnRH (100μg) 48 h after $PGF_{2\alpha}$ injection. The CIDR treatment protocol consisted of insertion of CIDR vaginal device and injections of P_4 (100mg) and estradiol cypionate (0.5mg ECP) on the first day of treatment, a $PGF_{2\alpha}$ injection (25mg) 7 d later, removal of CIDR 24 h after $PGF_{2\alpha}$, and a second ECP injection 48 h after $PGF_{2\alpha}$ injection. Artificial insemination for Ovsynch and CIDR, respectively, was performed at 64 and 76 h after $PGF_{2\alpha}$ injection. Pregnancy rates (based on ultrasound at d 35 after AI) for the Ovsynch and CIDR, respectively, were 26.9.1±1.6% and 45.0± 2.2% (P<0.05).
The CIDR-based protocol numerically outperformed Ovsynch in this experiment and may be a breakthrough to enhance pregnancy in multiparous dairy cows.

The effects of post – insemination supplemental progestrone on pregnancy rate in dairy cattle

M. Rahimi*[1], H. Karamishabankareh[2] and H. Mirzaei[2], [1]Razi University, College of Veterinary Medicine, Kermanshah, Iran, [2]Faculty of Agriculture, Razi University, Kermanshah, Iran

The objectives of this study were to investigate the effects of post-insemination(PI) supplemental progesterone (P_4) on serum P_4 concentration and pregnancy rate in dairy cattle. Total number of 40 dairy cattle were randomly divided into two groups; one group as the treatment (T group, n=20), and the other one as the control group (C group, n=20). Cows in T group and C group were divided into two subgroups (n=10), as T_1, T_2, and C_1, C_2, respectively. One dose of 5 ml of an oily solution containing P_4 (25mgr/ml), were injected to the cows in T_1 subgroup on day 5 PI, and two doses of P_4 to the cows in T_2 subgroup on day 5 and day 10 PI. Cows in C_1 and C_2 subgroups were injected sterile normal saline on day 5 and days 5 and 10 PI, respectively.
Blood samples were collected on day 0, 5, 10, 16, and 19 PI, and Serum progesterone concentration was determined.The pregnancy was diagnosed by ractal exam on day 40 PI.
Treatment group had a significant increase in P_4 concentration compared to the control group.The pregnancy rate significantly increased in P_4 treated cows. In conclusion, the administration of P_4 on day 5 PI significantly increased the serum P_4 concentration and improved the pregnancy rate in dairy cattle.

Oestradiol benzoate – oxytocin don't alter peogesterone concentration in sheep: Application to transcervical embryo transfer
H. Karami Shabankareh, A. Sarvari and H. Hadjariyan, Razi University, Iran

Experiment was conducted to determine whether oestradiol benzoate (E2) and oxytocin (OT) can be used to improve transcervical (TC) embryo transfer (ET) procedures for sheep by dilating the cervix. It was important that the E2-OT treatment may alter luteal function in ewes. This study was performed out of breeding season. Sanjabi ewes (n=18) were assigned to treatments in a 2×3 factorial array (6 groups). Estrus synchronization was performed using fluorogestone acetate impregnated intravaginal sponges (Chrono-gest®, Intervet). Ewes were superovulating with 400 IU PMSG at the time of sponge removal. On d 10 after vaginal sponge removal, ewes received i.v. 0 (saline), 200 μg of E2; 9h later, ewes received i.v. 0 (saline), 200, 400 IU units of OT. To monitor luteal function, progesterone (p4) was measured in jugular blood collected from days 0, 2,8,9,11,13,15,17,19, after vaginal sponge removal. The treatments did not affect serum progesterone concentration ($p > 0.05$). Luteal function was prolonged and progesterone concentration levels started to decrease on d 13 of estrus cycle in control and treated ewes. On days 11–13 of the cycle the minimal luteotrophic support allows the increase in prostaglandin to induce luteolysis and consequently, progesterone concentrations fall. Luteolysis marks the beginning of the follicular phase. Results indicate that E2-OT given 10 d after pessary removal had not significant effect on luteal function ($p > 0.05$). These treatments may be useful for improving TC-ET in sheep.

The effects of gonadorelin (a GnRH analogue) Administration 5 and 10 days after insemination on serum progesterone, and pregnancy rate of dairy cattle exposed to heat stress
H. Karami Shabankareh and H. Mirzaie, Department of Animal Science, University of Razi, Iran

This study was carried out to evaluate the effects of GnRH post-insemination in dairy cattle. Cows in the treated group (n = 20) which had been inseminated on Day 0 were treated with 5ml Gonadorelin i.m. 5day after AI (each ml contain 5 μg luliberin-A) (GnRH-1 subgroup; n = 10) or days 5&10 after AI (GnRH-2 subgroup; n = 10). Cows in Control group (n = 20) received saline only day 5 after AI (Control-1; n = 10) or days 5&10 after AI (Control-2; n=10). Blood samples were collected on days 0, 5, 10, 16 and 19 (AI =Day 0) for analysis of serum P4 concentration. There was no significant difference between GnRH-1 subgroup and GnRH-2 subgroups in serum P4 concentration although both of them had significantly increase in P4 concentration versus Control group (*P<0.05*). Progesterone concentration in days 10, 16 and 19 was (4.75 ±1.15 versus 2.86 ±0.58 ng/ml), (5.89 ±1.08 versus 3.43±0.96 ng/ml) and (6.51 ±1.61 versus 2.62±1.85 ng/ml) in treated group and Control respectively. Pregnancy rate significantly (*P<0.05*) increased in treated cows (55%>35%). In conclusion, GnRH administration 5day after AI significantly (*P<0.05*) increased progesterone concentration and improved pregnancy rate.

The influence of embryo stage and season on the success of ET in beef cattle

Z. Hegedűšová[1,2], K. Fréharová[1,2], M. Slezáková[1], J. Kubica[1,2] and F. Louda[1,2], [1]Research Institute for Cattle Breeding, Ltd., Rapotin, Vikyrovice, Czech Republic, [2]Agrovyzkum Rapotín,Ltd., Vikyrovice, Czech Republic

We monitored influence of embryo development stages on conception rate (CR) during season after transfer of fresh and frozen embryos in beef cattle. Embryos were obtained during 1991 – 2004. Basic statistical analyse icluded 1339 embryotransfers - SAS 8.0. Obtained embryos were divided according to development stage after flushing. We monitored CR after transfer of particular stages at fresh and frozen embryos. CR 47,41 % was detected in early blastocysts, 40 % in blastocysts, and 48,87 % in morulas at fresh embryos. CR 43,48 % was detected in early blastocysts, 38,35 % in blastocysts, and 41,82 % in morulas by frozen embryos. CR was 44,64 % in recipients after transfer of total embryos (fresh and frozen) in early blastocyst, 39,05 % in blastocyst, and 44,78 % morula. Highest CR was detected in morula after transfer. Influence of fresh and frozen embryos development stage, on conception of recipients was statistical nonsignificant. Additionally, we evaluated level of embryo viability after transfer during particular annual season. Highest CR was detected after transfer of blastocysts and early blastocysts in spring (46,99 % and 46,22 %), and after transfer of morulas in summer (50,00 %). Influence of embryo development stage on conception of recipients (heifers) was statistical significant only in summer (P<0,01).

Utilization of biotechnologies in reproduction of high milk efficiency cows

A. Jezkova, L. Stadnik and F. Louda, Czech University of Agriculture in Prague, 165 21, Czech Republic*

The results of reproduction in herd of Holstein cows were evaluated. Methods of heat detection, reproduction characteristics, and method of heat synchronization in relation of health status and milk efficiency of cows were detected. The cows with reproduction disorders were treated by OVSYNCH method to heat synchronization and ovular dysfunction therapy in regular fortnightly intervals. Samples of cervical mucus for laboratory testing (arborisation and durability test) were sampled in this group of cows.

The average conception rate after all insemination were 40,92% at day milk yield 33,67 kg and 3,21% protein content. Motility of sperms was at average 21,74%, resp.12,92% and 7,79% in process of 30, resp.60 and 90 minutes durability test.

Cumulative reproduction failure frequency was 44,76 - 47,36% of cows in monitored herd. Conception rate of OVSYNCH treated cows was 46,30% after first insemination. Improvement of conception rate is 2,97-5,80% in comparison with untreated cows. The conception rate in relation of sperms motility in cervical mucus durability test was evaluated. Higher percentage of pregnancy is in case of higher motility and longer survival of sperms in cervical mucus durability test. Conception rate was only 1,95% in case of zero motility of sperms after 30 minutes durability test. The most stable results of sperms motility in all time of durability test in cases of ferny crystallization (the best stage for insemination) were detected.

Microflora of reproductive organs in dairy cows around calving

A. Jemeljanovs, I.H. Konosonoka, I. Sematovica and B. Puce, Latvia University of Agriculture, Research Institute of Biotechnology and Veterinary Medicine "Sigra", 1 Instituta Street, Sigulda LV-2150, Latvia*

The objective of the current study was to investigate the aerobic and facultative anaerobic microflora of the uterine exudate and udder secretion samples of cows around calving. For bacteriological examination, uterine exudate and udder secretion samples were aseptically taken from cows two to five days post partum. Samples were inoculated on different complex and selective microbiological media. Microorganisms were further identified using gram-positive and gram-negative kits of BBL Crystal Identification System. In total, 60 uterine exudate and udder secretion samples were investigated. The acquired data were analysed using analysis of variance. Microorganisms from the genera *Micrococcus, Staphylococcus, Enterococcus* and *Escherichia* were isolated both from uterine exudate and udder secretion samples. Other microorganisms as *Bacillus licheniformis, Streptococcus agalactiae, Lactobacillus spp.* and *Hafnia alvei* were isolated only from uterine exudate samples. The most frequently isolated microorganisms were from the genus *Staphylococcus*. 65.2 % of uterine exudate and 82.6 % of udder secretion samples contained representatives from this genus. 26.6 % and 21.7 % from all isolated *Staphylococcus* in uterine exudate and udder secretion samples, respectively, was the well-known pathogen *Staphylococcus aureus*.
Isolation of coagulase negative staphylococci in monocultures from uterine exudate samples, indicate, that these mastitis microorganisms are implicated also in endometritis of dairy cows around calving.

Effect of milking frequency on udder capacity and milk yield of dairy cattle under desert climate

H.S. Al-Jbeile, Qassim University Animal production dep., Buraidah, Saudi Arabia, P.O. Box 1482,51431

The effect of breed, season of the year and stage of lactation on the response of dairy cows (average producers) to thrice-daily milking was studied in 25 Friesian and 20 Jersey cows using the technique of half-udder study. The results indicated that there was a positive response to increased milking frequency in both breeds, with higher response by the Friesian than the Jersey and during the winter than during the early hot summer season at the same stage of lactation. Season of the year had no effect when the stage of lactation did not consider in comparisons for both breeds. Hourly secretion rate of milk was also increased due to a third milking, did not change by stage of lactation in both breeds. These increases in yield and rate of milk secretion was observed at the morning milking that next to the third milking (at mid-night) of the udder's half milked only three times, suggested a local regulatory mechanism for controlling milk secretion. This mechanism did not influence the composition of milk at any level, although a reduction in fat content was observed at the morning milking in both the twice and thrice daily milked glands. Maximum capacity of the udder (milk yield after 36-h accumulation) was not altered significantly due to increased milking frequency. Statistical analysis – GLM – LSMEANS procedures using SAS package

Effect of intensity of rearing to growth, body proportions and udder development of Holstein heifers

L. Stádník, F. Louda*, J. Dvořáková and O. Amerlingová, Czech University of Agriculture, Department of Animal Husbandry, Kamýcká 129, 165 21, Prague 6 – Suchdol, Czech Republic*

Trial and control groups were founded about 20 Holstein heifers. New system of adjustment of stable background were aplicated in trial group from time of preparing cows for calving. Differencies were in nutrition, system of hygiene and of prevention of health. Nutrition was intensive – native milk (first 3 weeks), milk mixture with high content of milk components (from 3rd to 8th week) and Eurolac based on soya, slanetz and maize (2 months) in trial group. Interlac-pectin was used for prevention of diarrhoea and Seche-etable was used for better microclimate in trial group. Live weight and body proportion were measured from 3rd month to 14th month of age. The SAS software was used for statistical evaluation. Trend of better growth abilities was detected in trial group from the 5th month of age. Optimum growth is not achieving in evaluated herd in comparison with growth standard for Holstein heifers. Relation between disease in first 2 month of age and next growth (live weight, body proportion) of heifers were measured. Heifers with disease had lower level of all evaluated traits. Udder development was evaluated with Aloka SSD 500 ultrasound using. This evaluation is possible from 9 months of age in first time. Differencies were detect in period between 10-14 moths especially.

Dynamism of outside and inside teat proportion changes in time of lactation

F. Louda, L. Stádník*, A. Ježková and M. Rákos, Czech University of Agriculture, Department of Animal Husbandry, Kamýcká 129, 165 21, Prague 6 – Suchdol, Czech Republic*

Changes of teat inside and outside proportion in time of lactation was measured in 70 dairy cows of Holstein and Czech Pied Cattle. Milk production, time of milking and somatic cells count was detected too. Outside measurement of teat was made with slide caliper. Inside measurement of teat – length of teat canal, area of teat end and wall tickness were measured on ultrasonographic frame – linear sound with 7,5 MHz frequency. Evaluation of values of features measured before, immediately after and 3 hours after milking. First measuring was made between 40th and 70th day of lactation and second measurement in period 200-230 days. Non-significantly higher values was detected in outside parameters of teat in second part of lactation. Inside proportions of teat, measured by sonography, were higher in second part of lactation. Length of teat canal was significantly higher about 0,15 cm before, 0,16 cm immediately and 0,13 cm 3 hours after milking (P≤0,01). Area of teat end measured before, after and 3 hours after milking was larger about 0,23 cm2, 0,28 cm2 and 0,37 cm2 in second part of lactation (P≤0,01). Differencies were non-significant in daily milk production and somatic cells count. We can expect confirmation of relations between this results and health of udder and milk production economy in next studies.

Composition characteristics and physico-chemical properties of Charolaise cows colostrum
A. Zachwieja[1], P. Nowakowski[1]*, J. Twardoń[2], T. Szulc[1], A. Tomaszewski[1] and A. Dobicki[1], Agricultural University of Wroclaw, [1]Institute of Animal Breeding, [2]Department and Clinic of Obsterics, Ruminanat Diseases and Animal Health, Chełmońskiego 38C, 51-630 Wroclaw, Poland

The aim of research was to characterise composition of Charolaise cows colostrum including immunoglobulins, somatic cells and number of microorganisms, as well as its technological properties (acidity, thermostability, coagulation). Colostrum samples were collected from 52 cows during the first full milking after parturition. Basic composition (Milco Scan 133B), somatic cell counts (Somacount, Bentley), acidity (pH, °SH), thermostability (alcohol test), coagulation, immunoglobulin content in colostrum and total number of microbes were evaluated.
Obtained results conform to data of other authors, but they are different from colostrum characteristics of dairy cows. Higher level of total protein (17,18%) as well as immunoglobulins (144,27g/l) was stated in Charolaise cows colostrum when compared to dairy cows. Higher value for SCC (2881 th) and total microorganisms (2200 th) and ph value (6,21) was also found in Charolaise cows while thermostability (1,32 ml) and acidity (16,15°SH) values were lower.

Function verification of ruminally protective layer of protein tablets
J. Třinácty[*1], M. Richter[1], P. Doležal[2], G. Chládek[2] and J. Szakács[3], [1]Research Institute for Cattle Breeding, Ltd., Dep. Pohořelice, Czech Republic, [2]Mendel University of Agriculture and Forestry Brno,613 00, Czech Republic, [3]Liaharenský podnik Nitra, company, Slovak Republic

The aim of the experiment was to confirm the functionality of layer protecting the protein tablets against the rumen activity using in vivo method. For this purpose as a marker amino acids (Met, Lys, His) supplementing the amino acid composition of soya protein added to tablets were used. In the experiment the effect of amino acids mixture either in the form of rumen protected tablets or as a powder on the flow of mentioned free amino acids through the duodenum was compared. 3 lactating dairy cows (15 – 20 kg milk/day) fitted with ruminal and duodenal cannulas divided into 2 groups (2+1) with rotation of experimental factor in four 14 d periods (cross-over design) were used. Experimental feeding supplements together with the marker of digesta flow (chromium oxide) were inserted through a ruminal cannula into the rumen. Determined flow of free amino acids was corrected on dry matter intake and expressed in mg AA/day/kg DMI. In the case of all three mentioned amino acids, significantly higher flow ($P < 0.01$, n = 5) in experimental group (tablets) was observed in comparison with control group (powder) of lactating dairy cows: (89.0 vs. 22.4), Lys (142.2 vs. 50.0), His (143.7 vs. 33.5).
This study was supported by NAZV 1B44037

The effect of soya protein enriched with essential amino acids added to rumen of dairy cows in a form of rumen-protected tablets on the amino acid profiles of casein

S. Hadrová[1], L. Křížová[1], J. Šterc[2], O. Hanuš[1], J. Třinácty[1], [1]Research Institute for Cattle Breeding, Ltd., Rapotín, Dept. Pohořelice, Czech Republic, [2]University of Veterinary and Pharmaceutical Sciences, Ruminant Clinic, Brno, Czech Republic*

The aim of this study was to determine the effect of purified soya-protein HP300 enriched with the essential amino acids lysine, methionine and histidine added to rumen of lactating dairy cows either in a form of rumen-protected tablets or non-tableted mixture on the casein content and yield and changes in the amino acid profiles of casein. The experiment was carried out on three lactating dairy cows H100 fitted with ruminal and duodenal cannulas. Cows were fed on a diet based on a corn silage, alfalfa hay and a supplemental mixture. The experiment was divided into 4 periods of 3 d (10 d preliminary and 3 d experimental period). In the first period one cow received the rumen-protected tablets (T group) and the others two received the non-tableted mixture (C group). In the subsequent period the rate of cows was antipodal. The casein content and yield was significantly higher in the T group in comparison to group C (2.70% and 477.96 g/d vs. 2.46% and 407.82 g/d, $P<0.05$). The increases in the casein yield resulted in significantly higher yields of individual amino acids in casein ($P<0.05$) in the T group. The relative content of threonine and proline in casein differ significantly between groups T and C ($P<0.05$).

This study was supported by NAZV 1B44037

Effect of growth hormone (GH) polymorphism on composition and properties of cow colostrum and level of immunoglobulins in calf's blood serum

A. Zachwieja[1], P. Nowakowski[1], T. Szulc[1], A. Tomaszewski[1], A. Dobicki[1] and J. Twardoń[2], Agricultural University of Wroclaw, [1]Institut of Animal Breeding, [2]Department and Clinic of Obsterics, Ruminanat Diseases and Animal Health, Chelmonskiego 38C, 51-630 Wroclaw, Poland*

The aim of research was to evaluate the effect of growth hormone (GH) polymorphism on composition and properties of cow colostrum. Colostrum samples were taken from the first full milking of 67 cows of Black-White breed. Contents of: dry matter, protein, fat, lactose (Milco-Scan 133B, Foss Electric), immunoglobulin concentrations of class G, M, A (RID Kit Binding Site), levels of Ca, P, Mg and Na as well as acidity (pH, °SH), thermostability (alcohol test) and coagulation (Schern test) were measured in colostrum. Calves were fed their dams colostrum twice daily.

Total protein and class G, M, A immunoglobulin concentrations were determined in calf's serum collected in the 3-rd day of life. Due to the result of GH genotyping (PCR method) colostrum was distinguished as LL (39 animals) and LV genotype (24 animals). Dry matter, protein, fat and lactose contents as well as thermostability, acidity and coagulation properties of colostrum were similar in both genotype groups. The level of immunoglobulins in colostrum and in calves' serum did not differ between different GH genotypes. Results of research showed no relations between polymorphism of GH and colostrum properties and the immunoglobulin concentrations in calves' blood serum.

Researches on milk yield and components of Holstein-Friesian cattle
N. Topaloglu and H. Gunes*, Istanbul University, Faculty of Veterinary Medicine, Department of Animal Breeding, 34320 Avcilar, Istanbul, Turkey

This study was conducted to determine the duration of lactation, lactation and 305 days milk yield, milk fat and protein percentage and somatic cell counts of Holstein-Friesian cattle selected randomly in five private farms in England and to investigate the effects of some environmental factors on these traits. In the statistical analysis of data, the GLM procedure in SAS programme package was used.

Average of the duration of lactation, lactation and 305 days milk yield, fat and protein percentage and somatic cell counts were 324.32 days, 7715.23 kg, 7218.62 kg, 4.028%, 3.333% and 137.948 ('000 cell/ml) respectively. During the study, the farm where the animals were kept and the year in which lactation started made significant effects at $P<0.001$ level on all traits, also the lactations turn has made a $P<0.05$ level effect on these traits except the $P<0.05$ level effect on the duration of lactation. However season, like other factors made significant effects on the lactations duration, milk fat and protein rates at a level of $P<0.001$ and on 305 days milk yield at a level $P<0.01$.

Results of study showed that the management conditions and the use of high yielding breeders accomplished the improvement in yields in different years. Two principle factors came forward in the study and highest determining factors were observed to be the farm and turn of lactation.

Impact of environmental and genetic effects on behaviour of cows in automatic milking systems
S. König, F. Köhn and M. Gauly*, Institute of Animal Breeding and Genetics, University of Göttingen, Germany

Milking frequencies measured at official test days were used with repeated measurement analysis to reveal environmental and genetic impact on the milking frequency of cows in automatic milking systems. Repeated measurements were 3 test day observations per cow within days in milk (DIM) classes from 1216 cows in DIM class 1 (day 0 to day 99), from 1112 cows in DIM class 2 (day 100 to day 199), and from 1004 cows in DIM class 3 (day 200 to day 299) kept in 15 farms. Milking frequencies decreased with increasing parities and were highest for first parity cows. High daily milk yield was associated with higher milking frequencies on the phenotypic scale. Heritabilities for milking frequency were 0.16, 0.19 and 0.22 in DIM classes 1, 2, and 3. Higher heritabilities in the later stage of lactation were due to a substantial reduction of the residual variance. Genetic correlations between test day milk yield and daily milking frequency were in a range of 0.46 to 0.57 and between milking frequency and SCS near zero. The inevitable improvement of labor efficiency as well as the effect of increasing robotic milking demand such cows going voluntarily in automatic milking systems. The definition of the breeding goal, i.e. to include different aspects for temperament can make a contribution towards farmers and animals welfare.

Association between milk production and live weight during the lactation of Holstein and Czech Fleckvieh cows

D. Řehák*, J. Volek, F. Jílek, M. Fiedlerová and L. Bartoň, Research Institute of Animal Production, Přátelství 815, 104 00 Prague – Uhříněves, Czech Republic

The objective of this study was to examine the effects of milk production, season, breed, parity and days in milk on the live weight in the first 300 days of lactation. A total of 171 Holstein (n=118) and Czech Fleckvieh (n=53) cows having calved from April 2004 to October 2005 at the experimental station were examined. The live weight of cows was significantly affected by season (P<0.001), breed (P<0.001), lactation number (P<0.001), and stage of lactation (P<0.001). A linear relationship (P<0.001) was determined between milk production and live weight with significant difference between cows at first and second lactations and at third lactation. This relationship was not significantly affected by breed.
The study was supported by the project MZE0002701402.

The dairy cow performances can be affected by inflammations occurring around calving

G. Bertoni*, E. Trevisi, A.R. Ferrari, A. Gubbiotti- Istituto di Zootecnica, UCSC, 29100 Piacenza, Italy

Calving is often preceded or followed by inflammatory phenomena, consequent to clinical problems (infectious or metabolic diseases) or not. To evaluate these inflammation effects, cows of 7 herds were routinely checked for: blood profile, BCS, milk yield, health problems and fertility. Cows without drug treatments in the peripartum period (138) were retrospectively divided in quartiles according to a new Liver Functionality Index (LFI) which include plasma albumin, lipoproteins (cholesterol) and bilirubin at calving as well as their pattern of changes during 1st month of lactation. These parameters are impaired in case of inflammation, when liver is strongly occupied to synthesize positive acute phase proteins.

Cows with low LFI (vs. high), besides a significant but temporary impairment of liver functionality (higher bilirubin and lower both paraoxonase and retinol binding protein), showed higher inflammation indices (globulins, haptoglobin and ceruloplasmin), as expected. Furthermore, they had worse metabolic indices, i.e. higher NEFA and BOHB, lower urea (suggesting a reduced feed intake), but analogous liver enzyme levels. At last, they showed lower milk yield, slightly higher BCS and impaired reproduction (20 vs. 82 points of a Fertility Status Index).

In conclusion, cows living in the same farm and management can show different performances (milk yield and fertility), that seem well related to the occurrence of inflammatory conditions around calving and that would be avoided as much as possible.

Evaluation of milk performance parameters of Red Polish cattle using Angler breed

J. Buleca Jr.[1], A. Felenczak[2], J. Szarek[2], E. Dudríková[1], J. Buleca[1] and Ľ. Bajan[1], [1]University of Veterinary Medicine, Komenského 73, 041 81 Košice, Slovakia, [2]Agricultural University, al. Mickiewicza 24/28, 30 059 Krakow, Poland

In the studied period the advantage of crossbreeds from Red Polish x Angler over Red Polish cows in the first 305-day lactation was as follows: 197–640 kg of milk, 14–30,5 kg of milk fat and 6–19 kg of proteins. Crossbreed cows displayed higher levels of milk fat (by aver. 0.11–0.26%) whereas Red Polish cows were characterised by high levels of milk protein.

The examinations were complemented by analyses of polymorphic system of kappa-casein. The lowest clotting time and higher thermal stability as well as higher levels of calcium and polymorphism of kappa-casein was characterised by higher levels of crude protein, casein and fat in the milk of cows with homozygotes BB k-Cn. The polymorphism of kappa-casein was characterised by higher levels of crude protein, casein and fat in the milk of cows with genotypes BB k-Cn and AB k-Cn. The beta-lactoglobulin polymorphism displayed a significant variation in clotting time and calcium level of milk. In a group of cows with homozygotes BB-Lg the shortest time of milk clotting and a significantly lower calcium level were found. The results obtained so far indicate the need for determining kappa-casein genotypes in Red Polish bulls and crossbreeds and utilizing kappa-caseins while selecting bulls for mating.

First results of a crossing experiment of American Jersey bulls with Holstein cows in Saxony

W. Brade[1] and E. Brade[2], [1]Landwirtschaftskammer Niedersachsen, Johannssenstr. 10, D- 30159 Hannover, Germany, [2]Sächsischer Rinderzuchtverband (SRV), Schlettaer Strasse 8, D- 01662 Meissen, Germany*

The search for a suitable crossing partner for the Holstein cattle is not easy.

In a comprehensive experiment in Saxony Jersey bulls (J) from North America were selected as a crossing partner for Holstein Cows (H).

More than 250 F1-cows (JxH) of more than 10 different fathers have calved till now.

The comparison of simultaneous companions in the same herds points: The milk protein yield of F1-animals is only a little less than of purebred Holsteins. However, the F1-animals have better food utilization due to the lower weight.

Mean frequencies of dystocia and stillbirth are lower in the F1 than in the purebred Holsteins (stillbirth rate of calves: F1-mother (JxH), Holstein father: 4.9%; purebred Holsteins (mother and father): 10.9%).

The F1-cows (JxH) are also more fertile than purebred Holstein cows. Pregnancy rates after 1st insemination (for a 2nd gestation) are 68.2% in F1 and 58.5% in Holsteins.

Cross-sucking behaviour and blood glucose levels in group housed Holstein Friesian calves

M. Flömer[1], G. Ude[2], E. Moors[1], H. Georg[2], F.-J. Bockisch[2] and M. Gauly[1], [1]Institute of Animal Breeding and Genetics, Albrecht Thaer Weg 3, University of Goettingen, 37075 Goettingen, Germany, [2]Institute of Production Engineering and Building Research, Federal Agricultural Research Centre, 38116 Braunschweig, Germany

Major health problems in milk calves can be caused by cross-sucking behaviour. In the study 30 Holstein Friesian calves (17 males, 13 females) were separated at the first day of life from their mothers and randomly allocated into three groups. Calves of the first group were brought twice a day for 30 minutes to their mothers to suckle, whereas animals in groups 2 (full cream) and 3 (conventional milk replacer) received milk by an automatic feeding system at the same time (4.40 am and 2.30 pm).

Blood glucose levels were measured 30 min before and 1, 7 and 15 minutes after suckling. Cross-sucking behaviour was observed and recorded for 20 minutes after suckling was finished.

No cross-sucking behaviour was observed in animals of group 1. Blood glucose levels tended to be higher in this animals 30 minutes before, 1 and 7 min after the determination of milk intake. Animals of group 2 (full cream) showed in tendency less cross-sucking activities when compared with calves of group 3 (33.2 ± 32.9 sec. vs. 57.0 ± 114.0 sec.).

It is suggested that blood glucose level, associated with the type of suckling, is correlated with cross-sucking behaviour.

Effect of milking procedures on somatic cell count in milk

Heli Kiiman, Elli Pärna, Tanel Kaart and Olev Saveli, Estonian University of Life Sciences Kreutzwaldi 64, 51014 Tartu, Estonia

Infection status is the most important factor affecting somatic cell count (SCC) in milk. A high milk somatic cell count is almost always the result of infection. Variation in SCC from day to day for healthy cows does result in a slightly elevated count for some cows, but seldom to extremely high levels. Mastitis is the most costly disease in animal husbandry. It has been found (Heald et al., 1998; Rogers et al., 1995) that 200 000 cells/ml is the most practical threshold to determine the profitability of dairy farms. Data were collected from five dairy cattle farms, where cows were milked with pipeline milking system. The duration of each element of the working process was recorded. The data about milk yield, fat and protein content and SCC of the milk were collected. Monitoring of the working operations of 24 milking operators was carried out. The maximum duration of over-milking was 103 seconds. All the basic milking procedures were influenced the milk somatic cell count. A significant connection was between over-milking time and somatic cell count as well as between udder preparation for milking and milk SCC (P<0.001).

Relationship of somatic cell count with yield and quality milk in Mediterranean Buffalo
F. Sarubbi, R. Baculo and L. Ferrara, CNR ISPAAM Via Argine 1085, 80147 Napoli, Italy

The mozzarella cheese from buffalo milk gained DOC denomination because of its organolectic and sensorial characteristics. Intramammary inflammation, frequently occurring as a cryptic form in buffalo, causes great economic losses: price of Buffalo milk is 100% more than cow milk. Somatic cell count (SCC) increases with infection and it is commonly used as an indirect measure of bacterial infection and milk quality in dairy cattle. Aim of this study was to verify the relationship among logarithm of somatic cell count (SCCt), milk yield and milk constituents (protein, lactose, fat, total solids) in lactating buffaloes. A total of 1320 records: daily milk yield, percentage of fat, protein, lactose, total solids and SCC were collected. The mixed model methodology was used according to a repeated-measures scheme, as the restricted maximum likelihood method. The SCCt had a significant correlation with the calving year ($P<0.001$), but not with calving order and months of lactation. Milk yield and protein content did not have significant correlation with any effect. The calving year was significant for the content of fat and lactose ($P<0.01$ and $P<0.001$, respectively). The effects of calving year ($P<0.01$) and months of lactation ($P<0.001$) was significant on the total solids. The correlation among SCCt, milk yield and milk constituents was negative for milk yield and lactose content ($P<0.01$), but the protein content was positively correlated with SCCt ($P<0.05$).

Dry-off treatment in dairy cows with CNH as monitored by behaviour sensor
E. Maltz[1], N. Silanikove[1], A. Antler[1] and G.Leitner[2], [1]A.R.O. The Volcani Center, P.O. Box 6, Bet dagan, 50250, Israel, [2]National Mastitis Reference Center, Kimron Veterinary Institute,P.O.B. 12, Bet Dagan 50250, Israel

Dry-off is a stressful period in the life of the dairy cow. CNH is a revolutionary treatment for drying off cows that eases down stress. Recently a new behviour sensor was developed by S.A.E. Afikim® (Kibbotz Afikim, Israel) that measures number of steps, accumulating lying time and lying bouts. In this study the state of the art technology (behaviour sensor) was employed to quantify the relief of stress made possible by the state of the art drying off substance (CNH). In twenty lactating cows the behaviour variables of the last tree days of lactation were the reference to the first seven days of dry off. Ten cows were dried off conventionally and ten with CNH. On average, lying time, and average lying bouts time, as well as the ratio of number of steps to lying time were all significantly ($P<0.05$) bigger in the cows dried off with CNH. The number of steps was smaller in the CNH dried off cows and the difference became significant after the third day of dry off. These results suggest that the CNH dried of cows were calmer than those dried off.

Live weight and body measurements of Hungarian Grey bulls and cows

B. Nagy[1], Sz. Bene[1], I. Bodó[2], I. Gera[3], F. Szabó[1], [1]University of Veszprém, Georgikon Faculty of Agriculture, Hungary, H-8360 Keszthely Deák F. Str.16., [2]Debrecen University, Faculty of Agriculture, Hungary, H-4032 Debrecen, Böszörményi út 138, [3]Association of the Hungarian Grey Cattle Breeders, Hungary, H-1134 Budapest, Lőportár u. 16

Live weight and body measurements of 23 Hungarian Grey bulls and 42 cows were taken, morever some body measurement indices and correlation coefficients were calculated. Calves were born in January 2005. Dams of the calves were kept in the some conditions in the herds of Hortobágy Gene Reserve Company before calving. Weight and body measurements of the calves were taken just offter the birth when they became dry. SPSS 9.0 statistical programme were used for data processing.

According to the results the average birth weight of bulls was 852±82,3 kg, while that of cows 562±54,1 kg. Body measurements as an average for bull were as follows: height at withers 149 cm, width of shoulder 57 cm, width of hip 46 cm, leg circumference 24 cm, body length 171 cm. Height index was 87,1 – 89,8 %. Phenotypic correlation between birth wight and body measurements was medium or strong positive (r=0.41 – 0,80 for bulls and r=0,45 – 0.62 for cows).

Maternal productivity of beef suckler cows with low and high residual feed intake

M. McGee and M.J. Drennan, Teagasc, Grange Beef Research Centre, Dunsany, Co. Meath, Ireland*

An alternative measure of feed efficiency is residual feed intake (RFI). A total of 56 spring-calving beef suckler cows comprising 2 genotypes (G) were individually offered grass silage *ad libitum* pre-partum. Cow intake, liveweight, body condition score (BCS), calving difficulty and calf birth weight was recorded. Using early-calving cows, intake post-partum, colostrum yield and immunoglobulin (Ig) concentration, calf Ig status, milk yield and calf daily gain were determined. Expected feed intake (FI) was calculated for each G separately by regressing average daily FI (kg DM) on average daily liveweight gain and mid-test liveweight$^{0.75}$ over a 64-d period in late pregnancy. The RFI for each animal was then calculated as actual FI minus the FI predicted from the regression analysis. Within each G the cows were ranked on RFI and 0.5 were classified as having low or high RFI. Compared to cows with low RFI those with high RFI did not differ (P>0.05) in liveweight, liveweight daily gain and BCS but had a higher daily FI pre- (1.3 kg, P<0.001) and post- (0.5 kg, P=0.10) partum and greater (P<0.05) calf birthweight. There was no effect (P>0.05) of RFI on calving difficulty, colostrum yield and Ig concentration, calf Ig status or on cow milk yield and calf daily gain. These results indicate that beef suckler cows with lower residual feed intake are more energy efficient.

Behaviour of Parda de Montaña cow-calf pairs in restricted suckling systems
J. Álvarez, J. Palacio, I. Casasús, R. Revilla, A. Sanz*, CITA-Aragón, Apdo. 727, 50080 Zaragoza, Spain

With the aim of knowing the maternal behaviour of Parda de Montaña breed, sixteen autumn-calving cows with their calves (9 males and 7 females) were allowed to suckle one (RS1; 0800) or two (RS2; 0800 and 1500) sessions per day (30 min/session). Throughout suckling periods, behaviour trends were registered (days 30, 66 and 100 postpartum), each observer controlling one cow-calf pair. Cows had similar BCS at calving (2.6) and live-weight throughout lactation. RS1 showed higher suckling duration per session (21.3 vs 17.0 min; $P<0.01$; correlated with butts number towards cow udder, $r= 0.53$, $P<0.001$), but lower duration per day (21.3 vs 33.9 min; $P<0.001$). Likewise, RS1 pairs were close (<0.5m) more time per session (26.5 vs. 22.2; $P<0.001$), but less time per day (26.5 vs 44.3 min; $P<0.001$). Calf ADG was not affected by suckling frequency (700 g). Number of suckling bouts per session was similar in RS1 and RS2 (2.7 and 2.3), and higher in RS2 per day (2.7 vs 4.7; $P<0.001$). Number of licks to calves was higher in RS1 (55.5 vs 22.5; $P<0.05$) but these cows licked less their-selves (0.1 vs 2.7; $P<0.05$) whereas butts number (44.5), cross-suckling period (1.6 min), and cow-calf sniffing acts (0.2 min) per session were similar in RS1 and RS2. Both suckling systems allowed the establishment of maternal bonding, without effect on cow-calf performances.

Effect of a Lactic probiotic on cattle growth
M.A. Galina, M. Delgado and M. Ortíz, FES-Cuautitlan UNAM, Ciencias Pecuarias, Km 2.5 Carretera Cuautitlan Teoloyucan, San Sebastian Xhala Cuautitlan, 54714, Mexico

110 Zebu steers were fattened 120 d 323 (±.23) kg. plus four cannulated steers were placed in two diets. All animals were fed 55% corn stubble and 45% slow intake urea supplement. T1 (n=56 plus two cannulated) 320 (±.35) kg/BW fed basal diet. T2 n=27 plus two cannulated diet was spread with a lactic probiotic (*Lactococcus lactis; Lactobacilus brevis; helveticus and delbrueckii; Leuconostoc lactis* and *Bifidus essences, molasses and poultry litter.* Growth was 1.877 g/d (±318) T1 and 2.145 g/d (±267) T2 ($P<0.05$). Total dry matter intake (DMI) were 8.313 ± 183 g/d for T1 and 9.939 ± 223 for T2 ($P<0.05$). Ammonia concentration were augmented in T2 ($P<0.05$). *In vivo* nitrogen digestibility was higher ($P<0.05$) in T2 diet (78.13%), T1 (61.12%). Fiber digestibility was higher ($P<0.05$) for T2. Digestion rate of NDF constant (k_d/h) favored T2 diet ($P<0.05$). Passage rate (k_p/h) for NDF was 0.063/hr for T1 to 0.078/hr for T2 ($P<0.05$). True digestibility was higher in T2, 52.32% from T1 35.12% ($P<0.05$). Time of disappearance of cellulose in T1 (16.16 hr) was less ($P<0.05$) than in T2 (32.31 hr). Digestion rate was higher ($P<0.05$) in T1. Passage rate higher in T2 (0.078/hr) from T1 (0.062/hr). True digestibility in T2, (52.37%) was higher than that of T1 (44.10%) ($P<0.05$). Half-time (t ½) disappearance for hemi cellulose was higher for T2 34.12 hr ($P<0.05$). It was concluded that probiotic supplementation increase growth and ruminal physiology.

No reduction in growth performance and carcass quality of rose' veal calves with TMR feeding compared with concentrate feeding

M. Vestergaard[1], *I. Fisker*[2], *C.F. Børsting*[3] *and N. Oksbjerg*[1], [1]*Danish Institute of Agricultural Sciences, Tjele,* [2]*Danish Cattle Federation, Skejby,* [3]*Danish Cattle Research Centre, Tjele, Denmark*

To obtain a premium payment for rose' veal meat in Denmark, the slaughter companies request a carcass weight of 185 to 200 kg, with EUROP conformation above 3 coming from a calf below 10 months of age. To fulfil these requirements a high growth rate is needed. The current feeding system is almost entirely based on ad libitum access to high-starch concentrate and barley straw. However, this feeding regime has some negative consequences for rumen function (acidosis) and for the development of liver abscesses. Consequently, we investigated two alternative feeding strategies; one using a TMR based on 80% concentrate and 20% maize silage (NE-basis) from weaning to 200 kg and 65:35 from 200 kg to slaughter. The maize silage was of high quality (7.17 MJ NE/kg DM). The other strategy replaced barley straw with artificially-dried grass-clover hay. In total, 71 Holstein bull calves completed the experiment and were slaughtered at 286±3 d of age. There was no differences (P>0.20) between treatments in initial LW (55±2 kg), LW at slaughter (376±3 kg), daily gain (1,306±23 g/d), carcass weight (194±2 kg), EUROP conformation (3.5±0.1) and fatness (2.4±0.1), meat/tallow colour (2.8±0.1), and in lightness (L* 40±0.5), redness (a* 19.0±0.4) and pigment content (3.7±0.1 mg/g) of longissimus muscle. The results show that alternatives to the traditional concentrate feeding exist and will not compromise high daily gain and premium payment.

Utilization of pistachio by product in the diet of finishing calves

H. Fazaeli[1] *and N. Frough-Ameri*[2], [1]*Animal Science Research Institute, P.O. Box 1483, 31585 Karaj, I.R.Iran,* [2]*Agricultural and Natural Resources Research Center of Kerman Provinces, I.R.Iran*

Pistachio processing residues (PPR) was ensiled with 5 percent sugar beet molasses. In a completely randomized design, 24 Holstein male calves, with initial weigh of 218 ±13 Kg, were used, in which 4 diets respectively contained: 1) 0.00; 2) 13.2; 3) 22.4 and 4) 40 percent of PPR silage were tested when 40 percent of all diets was made of silage (corn or PPR).

During the 100 days of feeding trial, average daily gain were 1195±151, 1299±137, 1205±89 and 946±227 g for the diets 1, 2, 3 and 4 respectively which were significantly (P<0.05) lower for the diet 4. The average dry mater intake were 8.25±0.53, 9.7±.71, 10.1±.55 and 9.25±1.03 kg for the diets 1 to 4 respectively which were significantly (P<0.05) different among the diets. The average feed conversion ratio were 6.82±1.2, 7.1±0.9, 7.8±0.5 and 9.8±2.8 that was significantly (P<0.05) different between diet 1 and 4. The final live weight of the animals were 322±21, 333±15, 325±14 and 303±28 kg that was significantly (p<0.05) lower for the treatment 4. In can be concluded that feeding performance of PPR silage may be similar to the corn silage in the diet of finishing calves, when it is used in an amount of 66 percent of corn silage or 22.4 percent of total diet.

Beef production from Holstein-Friesian, Montbeliarde and Norwegian Red young bulls
M.G. Keane, Teagasc, Grange Beef Research Centre, Dunsany, Co. Meath, Ireland

Montbeliarde and Norwegian Red dairy cattle have been imported into Ireland and their male progeny are reared for beef. This study compared Holstein-Friesian (HF), Montbeliarde (MB) and Norwegian Red (NR) young bulls for growth, feed intake and carcass traits. Young bulls (12 per breed) with a mean birth date of February 22 were reared to slaughter at 17 months. After indoor calf rearing they were put to pasture for 176 days from May 6. This was followed by a 119-day store period, a 39-day growing period and a 130-day finishing period on *ad libitum* concentrates. After slaughter, carcass weight, perinephric plus retroperitoneal fat weight, carcass grades and carcass measurements were recorded. Silage intake, measured during the store period when 1 kg/day concentrates was also fed, was 20.8, 18.1 and 19.5 (s.e. 0.52) g/kg live weight for HF, MB and NR, respectively (P< 0.05). Corresponding slaughter and carcass weights per day of age were 996, 969 and 1013 (s.e. 27.4) g, and 487, 517 and 521 (s.e. 16.1) g (P<0.10). Kill-out proportion for HF, MB and NR was 488, 514 and 533 (s.e. 3.3) g/kg (P<0.001), with corresponding carcass weights of 258, 280 and 284 (s.e. 8.1) kg (P< 0.05). HF carcasses were less compact than MB and NR carcasses. It is concluded that MB and NR have superior beef traits to HF, with MB superior to NR.

Carcass and meat quality of Slovak Pinzgauer and Slovak Spotted bulls
P. Polák*, J. Mojto, J. Huba, K. Zaujec, E. Krupa, M. Oravcová and J. Tomka, Slovak Agricultural Research Authority, Research Institute for Animal Production, Hlohovská 2, 949 92, Nitra, Slovakia

Carcass quality and meat quality analyses were done on 32 Slovak Pinzgauer and 81 Slovak Spotted bulls. Both breeds are considered as genetic resources in Slovakia. Average weight before slaughter was 481 kg for Pinzgauer and 498 for Slovak Spotted bulls. Significant differences between breeds were found out for weight before slaughter, water and intramuscular fat content, colour of meat and level of marbling. The highest correlation coefficients were found between intramuscular fat content and marbling level in breeds, 0.83 for Slovak Pinzgauer and 0.76 for Slovak Spotted. Similarly, the significant correlation coefficients between marbling level and weight before slaughter in both breeds were detected. Non significant but interesting tendency was in dependence between intramuscular fat and carcass conformation, positive for Pinzgauer and negative for Slovak Spotted bulls. It means that marbling level in Pinzgau cattle increases with increase of carcass conformation. Different results in these two breeds lead from different age at slaughter maturity. Slovak Pinzgauer bulls are middle framed and their average daily gain is lower than Slovak Spotted which is large framed breed. Slovak Spotted bulls are in this weight still growing but Slovak Pinzgauer growth intensity is finishing. Pinzgauer bulls started to imbed intramuscular fat earlier than Slovak Spotted.

Audit of carcase quality (SEUROP) of young bulls in Slovak republic
J. Mojto[1], J. Kožuch[2], K. Zaujec[1] and P. Polák[1], [1] Slovak Agricultural Research Centre, 949 92 Nitra, Slovak republic, [2]Ministry of Agriculture of the Slovak republic, 812 66 Bratislava, Slovak republic

These results on meatiness and fattiness of carcasses in the category of young bulls in the Slovak Republic were obtained from the audit that takes place every year since 2003. Results of the audit shall provide complex and true information about the carcass quality in fattening animals for managing, breeding and producer organizations with the aim to control production of beef to obtain permanently better quality. In individual years were taken into the audit following numbers of slaughter carcasses: in 2003 – 13 805 animals, in 2004 – 15249 animals and in 2005 – 14 111 animals. Slaughter carcasses were evaluated by trained classifiers after the Slovak standard elaborated on the basis of the order of the Commission (ECC) No. 1208/91. Conformation class S with regard to negligible frequency is not incorporated into the results. During the assayed years we found significant decrease of carcasses in conformation class E (2.36 – 1.03 %) and U (46.14 – 9.97 %), and increase first of all in class R (44.52 – 52.50 %) and O (6.06 – 31.07 %). We consider the results for the year 2005 for real and they show the real situation in practice. These changes are probably connected with introduction of control of the work of classifiers, who mostly overestimated the carcass quality before. In classes of fat cover we found slight shift of carcasses into class 2 (44.71 – 49.85 %) and class 3 (9.33 – 22.43 %), which is desirable state for meat processing industry that necessitates the class of fat cover 3 and 4 with some brand meats.

Beef quality in the organic farms of Latvia
A. Jemeļjanovs, J. Miculis, J. Nudiens, V. Sterna, I.H. Konosonoka and B.Lujane, Latvia University of Agriculture, Research Institute of Biotechnology and Veterinary Medicine „Sigra", 1 Instituta Street, Sigulda LV-2150, Latvia*

Beef production has always been part of the Latvian agricultural sector because there are much pastures and lands for production grass and feed grain. The basis for beef production in Latvia is the universal Latvian Brown (LB) breed cows crossed with beef breed bulls – Hereford (HE), Charolais (CA), Aberdeen Angus (AN) and Limousin (LI). The welfare requirements and organic farming preconditions were ensured in cooperative society „Zaubes kooperativs". Investigations were carried out of beef animals (pure and crossing forms) carcass (EUROP classification, fat settling) and meat quality: dry matter, crude protein, intramuscular fat, fatty acids and cholesterol content.
We estimated carcasses on musculature development and ascertained that in prevalence they corresponded to R class. LBxLI were better animals they corresponded to U class. Equally good degree of fats settling had LBxCA, LBxLI cross animals. Higher crude protein level had LBxLI cross animals – 21.3% (154 samples) (p>0.05), the highest level of intramuscular fats had LBxLI and LBxAN cross animals -1.30% (p<0.10). Biological value of meat was determined according to ratio of amino acids tryptophane and oxyproline. The best results had meat samples of LBxLI cross animals (p<0.01). Evaluated that sum of unsaturated fatty acids was highest in the meat samples of HE (45.65±1.82% of fatty acids content) and LI (44.82±2.02%) breeds but it did not differed significantly (p>0.05).

The content of sodium and potassium in cattle liver
M.V. Strizhkova and O.S. Korotkevich, Novosibirsk State Agrarian University, The Research Institute of Veterinary Genetics and Selection, 160 Dobrolubov Str., 630039 Novosibirsk, Russia

Liver gives rise to the transformation of excessive glucose into glycogen. Sodium plays the main part in the transmission of glucose through cell membrane.
The liver content of potassium and sodium was studied in Black-and- White cattle aged 18 months. The concentration of macroelements in organs was determined by the method of plasma atom emission spectrometry. The average population value of potassium and sodium content was established in the liver of healthy White-and-Black cattle. The content of potassium in the animals' liver was 3.3 times higher than that of sodium (P<0.001). Great variation of the sodium level is 2 times higher than that of potassium. Great phenotypic differences are observed in the animals for the content of sodium and potassium. In individual animals the level of potassium was 99 times higher than that of the other ones. At the same time little individual differences were identified in the animals for the content of potassium in liver. The phenotypic variation of sodium is 2.8 times higher than that of potassium (P<0.001). This can be elicited from the intensity of metabolic processes which occur in liver.

Meat quality characteristics of beef from Charolais and Simmental bulls fed different diets
D. Bureš*, L. Bartoň, V. Teslík and R. Zahrádková, Research Institute of Animal Production, Přátelství 815, 104 00 Prague - Uhříněves, Czech Republic

The objective of the study was to compare physical, chemical, and sensory characteristics of *m. longissimus thoracis* from 46 Charolais (CH) and Simmental (SI) bulls. Within the breed, the animals were allocated to two dietary treatments and given two isonitrogenous and isocaloric diets based on maize silage, alfalfa hay, straw, and concentrates supplemented with either whole sunflower seed (EXP) or Megalac (CON) as a source of dietary fat. The bulls were slaughtered at the average live weight 640±38 kg and age 546±28 days. The statistic analysis was performed using the general linear model with breed and diet as fixed effects. The colour of meat was significantly lighter (P<0.001) and less reddish (P<0.05) in CH compared to SI bulls. No breed or diet effects were found for the chemical composition of muscle except for a higher hydroxyproline content in CH (P<0.001) than in SI. Sensory evaluation performed by a trained panel using a 7-point scale revealed a higher score (P<0.001) for texture in CH than SI while no differences (P>0.05) were shown between the dietary treatments. As indicated by triangle tests, the panellist were, however, mostly able to detect differences between breeds and dietary treatments. The study was supported by the project MZE0002701403.

Production of ripened Buffalo Bresaola: Evolution of chemical, physico-chemical and microbiological parameters
C. Diaferia*, G. Pirone, L. La Pietra, M. de Rosa and V. Magliano, Stazione Sperimentale per l'Industria delle Conserve Alimentari 43100 Parma, Italy

Bresaola was produced from the mass muscle of the buffalo thigh, correspond to "fesa", the posteromedial portion of the thigh muscle including rectum muscle, the adductor muscle and the *Semimenbranosus* muscle, and to "girello" which includes the *Semitendinosus* muscle. Twelve "girello" and "fesa" pieces were hand-salted, after massaging the pieces were placed in a closed container so as to obtain high relative humidity values. During the different processing steps, temperature and relative humidity parameters were recorded; additionally, on both external and internal portions of individual cuts, chemical physico-chemical and microbiological analyses were performed. The dehydration technique adopted, first with a cold process then at ambient temperature, allowed an homogeneous in the external and in the internal fractions reduction of water activity. Mass transfer occurs at constant rate during the first days of water evaporation from product surface, then it decreases during ripening when water diffusion from the inside to the surface regulates dehydration. The micro-organisms prevailing in bresaola produced according to the technological parameters and the initial hygienic conditions, belong to the group of lactic acid bacteria and micrococci non pathogenic staphylococci. No growth was observed of spoilage bacteria such as enterobacteria, oxidase-positive Gram negatives and Vibrio spp. or pathogenic bacteria.

The economics of competitive breeding programs
P.R. Amer, AbacusBio Limited, PO Box 5585, Dunedin, New Zealand

The principles of economics are relevant to the weighting of selection emphasis applied to breeding values for genetic traits and specific genetic loci. They are also relevant to the optimal design of breeding programs. In the 1970's and 1980's, theoretical development in this area focused on national improvement programs with a view of maximising the efficiency of production within the domestic industry. At that time, the globalisation of genestock supply was rapidly increasing, led by the most intensive livestock farming industries, and this has spread to intensive dairy cattle industries, and to a lesser extent, the extensive sheep and beef industries. With this globalisation, even national genetic improvement programs and agencies have come under increasing pressure to be competitive, in terms of both price and quality of genestocks supplied. As governments become increasingly reluctant to support national schemes competing against private entities (both domestically and internationally), there has been a shift in government and industry investment towards facilitating technologies, such as supporting national database and genetic evaluation developments, and blue skies research into new genomics technologies. This paper provides examples from the author's experiences of how economic principles, and their extension to market research techniques, can be applied to help breeding programs be more competitive. The additional roles of industries and governments in overcoming conflicts among industry stakeholders that arise due to market failure and short planning horizons are also discussed.

Economics of genetic improvement programmes
L. Dempfle, Department of Animal Science, TU-München, Alte Akademie 12, 85350 Freising, Germany

Genetic improvement of large animals is an extremely slow process, often being in the range of about 1% of the mean per year. In addition starting a breeding scheme from scratch there is a long time lag until first returns are obtained. It should also be considered that genetic improvement of the main traits have sometimes undesirable side effects.

On the other hand, however, genetic improvement has some unique features. At first approximation it is permanent, cumulative and multiplicative. If a fitness neutral trait is changed we get new allele frequencies and the new mean is permanent even without further selection. If we repeat selection the result is just added (cumulates up). Modern breeding schemes are structured as Nucleus -> Multiplier(s) -> Commercial stage. Breeding efforts are concentrated to the small nucleus where e.g. in pigs there might be three lines of hundreds sows each and it is affecting million of slaughter pigs. Multiplication can be very efficient (dairy cattle with AI) but also very inefficient (cattle without AI e.g. in developing countries). Looking at genetic improvement in an economic way the worth of an investment can be calculated by e.g. the Net Present Value. However, there are several problems: i) what time horizon to consider? With cattle in developing countries hardly any return is obtained in a decade. ii) what about a free rider competitor? (little protection by patents or intellectual property rights). iii) what costs are really attributable to genetic improvement?

A multivariate approach to derive economic weights for production and functional traits in dairy cattle
B. Lind, S. König and H. Simianer, Institute of Animal Breeding and Genetics, Albrecht-Thaer-Weg 3, 37075 Göttingen, Germany*

In Europe the political and economic conditions of dairy production are changing rapidly. For this reason the profitability of animal production has to increase and breeding goals need to be derived under expected future conditions. Many relevant parameters either have a biological variation (like most production and functional traits) and/or can be assigned a variation reflecting the uncertainty about future conditions (e.g. future prices or quota systems). We suggest to derive economic weights as the first derivative of the benefit function w.r.t. the trait over a multidimensional grid of value combinations for the variable parameters. In the two-dimensional case, w_{ij} is the economic weight obtained for the combination of, say, milk production level i and milk price scenario j, where the joint probability of this constellation is p_{ij}. Then, the expected overall weight $E(w)$ can be calculated as $E(w) = \sum_{i,j} p_{ij} w_{ij} / \sum_{i,j} p_{ij}$. With the same approach, it is also possible to give the standard error of the estimates as well as functional relationships of the economic weight with technical or biological variables. The method is illustrated in a case study on the derivation of economic weights for the German Simmental population. When calculated without quota limitations the economic weights per standard deviation were highest for protein yield, fat yield, longevity and somatic cell score, and lowest for calving ease and stillbirth.

Breeding program for German Fawn dairy goats in Germany

B. Zumbach[1] and K.J. Peters[2], [1]Department of Animal and Dairy Science, University of Georgia, Athens, GA, 30602; USA, [2]Humboldt University Berlin, Dep. Animal Breeding in the Tropics and Subtropics, Philippstr.13, Haus 9, 10115 Berlin, Germany

Using a common data base for small ruminants allows breeding value estimations based on an animal model. This study presents a suitable breeding programme for the German Fawn Goat with an estimated population size of 60,000 with 5,000 under performance testing The two tier programme is based on AI and NM with the following assumptions: 40 young AI sires in the elite tier, 1000 semen portion per young sire, success rate of 1st insemination 0.6; natural mating relation 1:40 with a success rate of 0.9; weaning rate 1.6 animals, i.e. 0.8 males per litter; pre-selection factor of 3. 250 elite dams are required mated with the 10 best young sires.

Young sires are selected based on dam, dam's and sire's half-sibs performances, for goats additionally own performance is included, resulting in accuracies of 0.39, 0.62, 0.32 and 0.56 and for sires to breed young sires, dams to breed young sires, sires to breed dams, and dams to breed dams, respectively. Generation intervals are 2, 3.5, 2.3, and 3 years, respectively.

Assuming heritability for milk production of 0.25 and estimated phenotypic standard deviation between 173 and 279 kg of milk, an annual genetic response of 20 to 30 kg milk is expected.

Elimination of the allele rendement Napole in Danish Hampshire pigs has reduced genetic gain in production traits and increased inbreeding

A.M. Closter, The Royal Veterinary and Agricultural University, Denmark

In this study, I tested that elimination of the allele Rendement Napole (RN$^-$) in Danish Hampshire pigs has reduced genetic gain in the production traits and increased the rate of inbreeding. I tested this premise by measuring the difference between the realised genetic gain and inbreeding after elimination and the predicted genetic gain and inbreeding that would have been realised if RN$^-$ had not been eliminated. The data used was obtained from the Danish National Database and included pigs born in the period 1992-2004. The elimination of the RN$^-$ allele decreased the genetic gain of production traits. The rate of inbreeding has increased more dramatically doing the elimination period than before the elimination. There is a cost associated with eliminating individual alleles. The cost of the elimination of a single allele can be reduced by lengthening the period of elimination.

Derivation of economic values of important traits for minimizing production system costs in silkworm (*Bombyx mori* L.)

M. Ghanipoor[1], S.Z. Mirhosseini[1]*, A.R. Seidavi[2] and A.R. Bizhannia[1], [1]Iran Silkworm Research Center, [2]Islamic Azad University, Rasht Branch, Iran

The present investigation has been carried out on three commercial varieties which were produced Iran including 110, 107 and 101433. Potential magnitude of variation in economic values was estimated when calculated for some alternative perspectives including animal unit, product unit and return unit.

The absolute economic values of production and reproduction traits were different in three scenarios while relative E.V.s of traits was stable against change in perspective. Economic values of cocoon weight and reproduction characters are related to fixed costs per moth. Also they have inverse relation with the trait mean. On the other hand, profitability due to increased trait mean will be resulted by decreased fixed costs. E.V.s of these traits have inverse relation with total cocoon produced per moth (X) in product unit perspective and R in return unit one. The absolute and relative E.V.s of cocoon shell weight and cocoon shell percentage are related to best cocoon percentage, shell price per gram and costs per moth. Relative E.V.s of reproduction traits have a direct relation with cocoon weight mean and inverse relation with trait mean. In the lines with higher cocoon weight mean, the breeding objective is focused on improving cocoon shell percentage and reproductive traits and selection emphasis on cocoon weight decreases.

Study on genetic-economic responses to index selection in commercial silkworm lines

S.Z. Mirhosseini[1]*, M. Ghanipoor[2], A.A. Shadparvar[1] and A.R. Seidavi[3], [1]Guilan University, Animal Science Department, Iran, [2]Iran Silkworm Research Center, [3]Islamic Azad University, Rasht Branch, Iran

This study was conduct to estimate genetic parameters, additive genetic values and economic values of cocoon traits to design selection index program and evaluate genetic-economic responses to index selection. The data used were approximately 8800 records including cocoon weight (CW), cocoon shell weight (CSW) and cocoon shell percentage (CSP) obtained from six generations. After estimating (co)variance components by REML method, additive genetic values of animals were predicted using Best Linear Unbiased Prediction (BLUP). The lines Xinhang1, 101433 and Y revealed the highest genetic improvement for CW, CSW and CSP respectively. The line 107 showed the lowest genetic gain for all the traits indicating low heritability for cocoon characters. The maximum and minimum genetic gain percentage over mean belonged to CSW and CSP respectively. CSW showed high level of genetic advance and heritability values implying that this may be predominantly under the control of additive gene action. Hence individual selection for this trait appears to result in better response. The results obtained demonstrate that genetic parameters and economic values are different in commercial lines. Therefore, genetic and economic responses to selection vary among them. Knowledge of actual economic response to selection can help to silkworm breeder to use high economically efficient lines in hybridization program to maximize cocoon producers' profitability in the future economic conditions.

Estimation of efficiency alterations and technological improvement in broiler husbandry industry

J. Azizi[1]*, A. Mohammadinejad[2], [1]Islamic Azad University, Rasht Branch, Iran, [2]Islamic Azad University, Science and Research Branch, Iran

Meat is main source for human protein supply. Protein importance is being to it is one of four necessary nutrients at human regimes. It has much amino acid at its structure. Meat consumption at Iran must be 60 kg/year/person. But meat consumption at Iran is around 28kg/year/person. It is very less in compare with world meat consumption which is around 37 kg/year/person. At past decade, broiler husbandry industry progress very much. At this study, it is investigated efficiency of broiler husbandry industry during 1990-2005. All data determined using border Production functions and mathematics relations of Malmquist index and analyzed to its components. Results was showed that during 1990-2005 and following on state. Interference eliminating at production and distribution of broiler meat, efficiency of production resources improved and thus technical efficiency improved at broiler husbandries. During 1993-1996 it is showed an 8.9% technological jump at broiler husbandry industry. But during 1996- 2005, it is showed a constant period for efficiency and technical parameters and indices.

Heterosis, general combining ability (GCA) and special combining ability (SCA) for resistance traits against disease in some lines of silkworm *Bombyx mori* L.

A.R. Seidavi[1]*, A.R. Bizhannia[2], S.Z. Mirhoseini[2], M. Mavvajpour[2] and M. Ghanipoor[2], [1]Islamic Azad University, Rasht Branch, Iran, [2]Iran Silkworm Research Center

Combining ability analysis is the most widely used biometrical tool in determining promising silkworm parents and hybrids and detecting relative magnitude of genetic variability. It observed that 3-7 additive genes play more important role in the inheritance of some economical characters. Significance of GCA at the larval and pupal resistance characters in Japanese lines (even though the mean of these characters are low) indicated the additive effects of genetical control on these characters. Therefore it could be expected that with the selection of Japanese lines with the better resistance characters as a maternal breeds for combination, resistance of the hybrids will considerably increase. In contrast, in Chinese lines (although they have higher resistance characters) the resistance characters have lower additive genetical variance and it is expected that in the resulting hybrids considerable improvement would not occur. Parents possessing high GCA are generally considered for population development and for initiation of pedigree breeding, as it is heritable and can be fixed. Parents with high GCA produce high heterosis as GCA consists of non-additive effects, dominant effects and other interactions. SCA is not heritable and therefore it cannot be utilized in pure line breeding.

Determination of selection indices for six commercial silkworm lines
M. Ghanipoor[1], S.Z. Mirhosseini[1]*, A.R. Seidavi[2] and A.R. Bizhannia[1], [1]Iran Silkworm Research Center, [2]Islamic Azad University, Rasht Branch, Iran

Genetic parameters and economic values of some important traits in six commercial silkworm varieties (110, 107, 101433, Xinhong1, Koming1 and Y) were analyzed and selection indices for these traits were constructed. Data obtained from 6 generations were used in order to estimate genetic and environmental parameters of cocoon weight (CW), cocoon shell weight (CSW) and cocoon shell percentage (CSP). In each line and generation, 24 families were reared. Heritability of CW and CSW were higher than heritability of CSP. A positive and higher relationship between CW and CSW was observed. The highest correlation between these traits belonged to 101433 (0.95). Genetic correlation between CW and CSP was negative in the varieties 107, Koming1 and Y and positive in 110, 101433 and Xinhong1. The highest correlation between these traits belonged to the variety Xinhong1 (0.22). High positive genetic correlation was observed between CSW and CSP. The highest correlation between these traits belonged to Koming1 (0.68). The expected economic efficiency as the result of family index selection was the highest. Thus, family index is suggested for selection. The maximum and minimum selection accuracy, expected economic gain and relative efficiency of individual to family index belonged to Xinhong1 and 107, respectively. It is suggested that more studies should be carried out to know effect of index selection on commercial hybrid performance.

Determination of diseases natural resistance in Romanian honeybees (A.m. carpatica) by means of specific assays in order to implement an optimal breeding methodology
Eliza Cauia[1]*, Cecilia Radoi[1], A. Siceanu[1], I. Guresoaie[1], B. Poncea Andronescu[1], Dana Tapaloaga[2], P. Tapaloaga[2], Florentina Mihalache[2] and G. Prefac[2], [1]The Beekeeping Research and Development Institute, 42 Ficusului Blvd, Bucharest,Romania, [2]The University of Agronomical Sciences and Veterinary Medicine, 59 Marasti Blvd, Bucharest, Romania

In the last years, in internationally context, was proved that long-term chemotherapy could have negative impact, increasing the resistance of bee diseases to the medication with consequences from economic points of view especially on the quality of beekeeping products. In this context, the strategy of bee diseases control was reanalyzed, the researches being reoriented trough finding the alternative and non-polluted solutions, which could be integrated in an efficient system of bee diseases control, depending on breed, climate, and maintaining systems. Thus, one of the research directions is the ameliorative selection of bees for natural diseases resistance combined with the other desired morpho-productive and behavioral characteristics, in this way a series of specific test being already established. The preliminary researches carried out in the breeding apiaries of the Beekeeping Research and Development Institute (Bucharest) aimed to establish the natural diseases resistance degree of local bee populations (A. m. carpatica), by applying specific tests to identify the bee colonies with high hygienic behavior. Parallel, a series of traits were evaluated in a breeding protocol (brood viability, temperament, honey production, etc). The selected strains formed a multiplication apiary the base for implementing a closed population breeding program.

Economic weights for litter size and fur coat traits of arctic fox in Poland

H. Wierzbicki[1], A. Filistowicz[2] and P. Przysiecki[3], [1]Wroclaw University of Agriculture, Kozuchowska 7, 51-631 Wroclaw, Poland, [2]Wroclaw University of Agriculture, Institute of Animal Breeding, Chelmonskiego 58c, 51-630 Wroclaw, Poland, [3]State School of Higher Education, Mickiewicza 5, 64-100 Leszno, Poland

Economic and productive data from Polish arctic fox farming were used to develop a bio-economic deterministic model simulating an average fox farm. The model was used to estimate economic weights for direct (body size – BS, fur quality – FQ, and colour type – CP) and maternal (litter size – LS) traits. A 10-year investment period and two alternative discount rates: 0% and 6% were assumed when computing the number of discounted expressions (NDE). Marginal economic values (MEV) were calculated per one female purchased. Highest MEV were obtained for LS (88.304 PLN), followed by FQ (86.548 PLN) and BS (39.024 PLN). Much lower MEV (10 times lower than for BS and 20 times lower than for LS and FQ) was computed for CT (3.944 PLN). The relative economic weights derived for the alternative of 0% interest rate per year were 0.48 for LS, 0.15 for BS, 0.35 for FQ and 0.02 for CT, and were comparable to those estimated for the situation of 6% interest rate per year (0.46, 0.15, 0.36 and 0.03, respectively).

Influence of different feedlot types on economic weights of current and predicted systems for Charolais breed using bioeconomical approach

E. Krupa, D. Peškovičová[*], Z. Krupová, P. Polák and J. Tomka, Slovak Agriculture Research Authority, 94992 Nitra, Slovakia

The economic weights (EW) of 15 production and reproduction traits for Charolais cattle raised in Slovakia were calculated. Four different fattening systems were simulated (intensive feedlot in bind technology-Fed01, Intensive, free-Fed02, Extensive, free-Fed03, Extensive, pasture-Fed04). EW for current production system (2005) and for the future production system (2010) were calculated. The economic importance of evaluated traits was little bit higher in all feedlots types for production system in 2010 year. The different influence of feedlot types on EW were observed only for reproduction traits and average daily gain during fattening period. Other differences were very low. The EW for average daily gain during fattening were 61.59 Sk (37,5 Sk = 1 Euro), 63.53 Sk, 150.42 Sk 150.42 Sk per 10 g in Fed01, Fed02, Fed03 and Fed04 in 2005 year, respectively. Influence of Fed01 and Fed02 on EW of mature weight of cows (-8.76 Sk, -9,54 Sk) was stronger than influence of Fed03 or Fed04 (-3.56 Sk, -3.56 Sk per kg). The EW of mean class of calving performance and losses of calves (at calving and till 48 hour after calving) were much higher in Fed03 and Fed04 than in Fed01 and Fed02 feedlot systems. The highest relative importance was found for EW of average daily gain during fattening period for all feedlot types.

Breeding values for auction prices – a total merit index in dairy cattle?
S. König, S. Schierenbeck, B. Lind* and H. Simianer, Institute of Animal Breeding and Genetics, University of Göttingen, Germany

The multitude of information sources determining sales on auction raises the idea to consider the auction price as a unique trait in a total merit index for dairy cattle. A disadvantage in this case is due to the fact that observations for auction prices are limited to a small percentage of cows. However, the impact of other traits of interest on the auction price can be used to derive their economic value in a combined breeding goal. Auction prices, test day production records, scores for 17 linear scaled type traits and notes for 4 type composites (angularity, body, feet and udder) were available from 1565 Holstein cows marketed on auction after their first calving. Analysis of variance carried out significance (p<0.01) of test day production records, SCS, age of first calving and of all conformation traits on the market price. Applying the stepwise procedure, udder traits were most important followed by traits describing feet and legs. The heritability for auction price was 0.27 and genetic correlations between auction price and notes for angularity, body, feet and udder were 0.10, 0.21, 0.55, and 0.55, respectively. The moderate positive genetic correlation (0.19) between auction price and total merit index (RZG) indicates the possibility to utilize auction prices for the derivation of economic values for type traits in a combined breeding goal.

Genetic and economic response of breeding policy in dairy herds
P. Šafus*, J. Přibyl and Z. Veselá, Research Institute of Animal Production, P.O. Box 1, 104 01 Prague-Uhříněves, Czech Republic

The objective of the present study was to compare the genetic and economic response of some breeding measures in a commercial dairy herd within an agricultural enterprise. Simulations for use of breeding arrangements and their consequences over a fifty-year period were carried out for the above models using the gene flow method.

Genetic and economic responses to genetic gain were evaluated for combination the breeding arrangements: single use of bulls under testing and negative selection of cows in the herd (30% of animals are discarded from reproduction and the animals are left in the herd and used for breeding by beef bulls); single use of bulls under testing and negative selection of first-calvers in the herd (25% of animals are discarded for slaughter); single use of proved bulls (selection intensity 1% of the best bulls) and negative selection of cows in the herd; single use of proved bulls and negative selection of first-calvers in the herd.

Only minimum changes will occur since the 25th year of observation. A comparison of the particular models showed the highest gain on average per cow for the whole observed period for single use proved bulls and negative selection of cows in the herd. Single use of bulls under testing and selection of cows in the herd resulted in the second highest cumulative genetic gain.

Conservation and genetic improvement of Tunisian native dairy sheep productivity by Sarda genes using intra-uterine artificial insemination

M. Djemali[1]*, S. Romdhani[2], L. Iniguez[3], A. Bedhiaf[1,] L. Sâadoun[4] and I. Inounou[3], [1]Laboratoire des Ressources Animales et Alimentaires, 1082 Cité Mahrajène, Tunisia,[2]Laboratoire des Productions Animales et Fourragères, Rue Hédi Karray, 2049 Ariana, Tunisia, [3]ICARDA. P.O. Box 5466, Aleppo, Syria, [4]OEP,30 rue Alain Savary, 1002 Tunis, Tunisie

It has been identified that the Sicilo-Sarde breed, the only indigenous milking sheep in the country had recognized a considerably population reduction due to indiscriminate crossing. The objective of this study was to use frozen semen from Sarda rams via intra uterine artificial insemination to improve its productivity under southern Mediterranean conditions. A total of 600 ewes from 10 flocks were synchronized in April 2005 by inserting hormonal sponges. Sponges were removed after 14 days and 500 IU of PMSG were injected. Intra uterine insemination using laparoscopy was used 48 hours later. Results showed that fertility, prolificacy and mortality rates were on average 53%, 166% and 5%, respectively. All F1 females are kept as replacement and F1 males are actually managed in a selected ram program. Important facts interacted in the current production conditions of this breed: a well defined market with an unsatisfied demand for increased production, the involvement of the private sector that has installed capability to absorb the production and processes the milk, and the recently formed association of producers of this breed in the need of technological improvement.

Adaptations to pregnancy are influenced by maternal nutrition around conception

M.H. Oliver*, A.L. Jaquiery, F.H. Bloomfield, C. Rumball and J.E. Harding, Liggins Institute, University of Auckland, New Zealand

Maternal nutrition during pregnancy has important effects on survival, size and health of the offspring. The potential to alter growth, health and welfare throughout the productive life of the postnatal offspring requires more study. In many species maternal undernutrition, or limitation of specific macro- and micro-nutrients, results in reduced size at birth, glucose intolerance, elevated blood pressure and altered hypothalamo-pituitary-adrenal axis function in later life. Nutritional influences on birth size and carcass development is an important focus of agricultural research. However, altered maternal nutrition may alter physiology of the offspring independent of any effect on size at birth. Coordinated research between biomedical, agricultural and veterinary sectors is required. Our biomedical research currently focuses on the role of maternal nutrition before and around the time of conception. Moderate undernutrition of ewes 60d before to 30d after mating alters fetal growth trajectory and its response to subsequent nutritional insults during later gestation. Fetal metabolic and endocrine responses were also perturbed, and 50% of lambs of undernourished mothers delivered early and died. At least some of these effects may be a consequence of impaired maternal metabolic responses to pregnancy, and preliminary data suggest that undernutrition in the period before mating may be most critical in this regard. The effects of periconceptional undernutrition on the survival, growth, health and welfare of the offspring remain to be determined.

Intrauterine metabolic programming and postnatal growth and energy metabolism in a life time perspective

M.O. Nielsen*, S. Husted, A. Kiani, A. Chwalibog and M.P. Tygesen, The Royal Veterinary & Agricultural University, Gronnegaardsvej 7, DK-1870 Frederiksberg C, Denmark

From rodent, human and sheep studies it is known that maternal feed restriction may compromise provision of nutrients for the growing foetus. This may induce so-called foetal metabolic programming, which is associated with altered foetal development and function of a number of organs, including endocrine organs responsible for regulation of growth and development. As a result post-natal growth capacity, muscle development and metabolic adaptability in sheep are altered in a way which is not consistent with optimized animal production performance. Lowered metabolic rate and more efficient glucose sparing mechanisms may however improve the immediate chances of survival of the newborn, when feed resources are scarce. Consequences of foetal metabolic programming in relation to animal production are not well characterized. It is likely to play a role in extensive animal production systems, in parts of the world where the dry season and scarcity of food coincides with the time of the year, where nutritional requirements of the pregnant female are at a maximum, e.g. in mid-late gestation. The newborn mammal relies on delivery of milk by the dam for its nutrition. Maternal malnutrition during late gestation interferes with maternal endocrine systems involved in regulation of mammary growth, and may negatively impact nutritional supply to the off-spring not only during gestation but additionally also post-natally.

Mitofusin 2 (Mfn2): A switch to stop mitogenesis in insulin-dependent myogenesis *in vitro*

P. Pawlikowska and A. Orzechowski*, Department of Physiological Sciences, Warsaw Agricultural University, Nowoursynowska 159, 02-776 Warsaw, Poland

The prerequisite for muscle differentiation is to withdraw cells from cell cycle. Being able to evoke two opposing effects insulin is both mitogen and differentiation factor because it stimulates cell proliferation and myotube formation in skeletal muscle myogenesis. Our previous results have shown that mitochondrial activity increased in response to insulin in differentiating muscle cells. Moreover, protein kinase kinase/extracellular-signal-regulated kinase (MAPKK/ERK - MEK) inhibitor PD98059 accelerated, whereas either the phosphatidyl-inositol 3-kinase (PI-3K) inhibitor LY294002 or blockade of mitochondrial respiration both abrogated insulin-mediated myogenesis. This study points to the mitochondrial transmembrane protein called hyperplasia suppressor gene/ mitofusin2 (HSG/Mfn2) which regulates both mitochondrial fusion (demonstrated by perinuclear mitochondria clustering) and insulin-dependent myogenesis *in vitro*. The molecular mechanism of this phenomenon is unknown, although immunoprecipitation studies indicate that during insulin-mediated myogenesis Ras protein (upstream activator of MAPK/ERK1/2 cascade) interacts with HSG/Mfn2 in muscle cells. Interaction of Ras with Mfn2 continued unless insulin was present and was blunted after PD98059 co-treatment. It indicates that insulin-mediated myogenesis is augmented by inhibition of MEK, most likely by lack of mitogenic signals opposing muscle differentiation. We suggest, that insulin stimulates Mfn2 protein expression which in turn binds to Ras and inhibits MEK-dependent signalling pathway. At the same time PI-3K-dependent signalling pathway is boosted, mitochondrial respiration increased and the rate of myogenesis accelerated.

In vitro variation in primary satellite cell proliferation and differentiation within litters of pigs

P.M. Nissen* and N. Oksbjerg, Danish Institute of Agricultural Sciences, Department of Food Science, P.O. Box 50, DK-8830 Tjele, Denmark

Within litters of pigs the fastest growing pig has a faster growth rate of the individual muscle fibres compared with the slowest growing pig. Differences in growth rate of individual fibres partly reflect the rate of proliferation and differentiation of the satellite cells (SC). Thus, the aim of this project is to identify differences in proliferation and differentiation of SC within litters of pigs. SC was isolated from semimembranosus from the lightest (L), middle (M) and heaviest (H) weight female pig within 8 litters. Cells were seeded in 96-wells plates and grown in proliferation medium until 80% confluence. The number of viable cells was estimated by addition of WST-1 and measuring the absorbance after 4 h. At 80% confluence medium was change to differentiation medium, and differentiation was estimated by measuring the CPK-activity at 10 time-points. SC from M and H pigs grew at a faster rate and therefore reached confluence one day before cells from L pigs. For differentiation, M pigs showed an initial faster differentiation than L and M pigs, whereas at later time-points differentiation rate was only slower for L than M and H pigs. Thus, SC from L pigs has both a slower proliferation and differentiation rate than SC from H and M pigs, when grown under the same in vitro conditions.

Plasma metabolites and BCS in Italian Friesian heifers on different rearing schemes

F. Abeni[1]*, L. Calamari[2], L. Stefanini[3], F. Calza[1], M. Capelletti[1] and G. Pirlo[1], [1]CRA Istituto Sperimentale per la Zootecnia, 26100 Cremona, Italy, [2]Istituto di Zootecnica, Facoltà di Agraria UCSC, 29100 Piacenza, Italy, [3]Azienda Sperimentale "V. Tadini", 29027 Gariga di Podenzano (PC), Italy

The aim of this work was to evaluate two rearing schemes on 60 Italian Friesian heifers, which were allotted on two experimental feeding groups to obtain a moderate (0.70 kg/d; M) or a high (0.90 kg/d; H) average daily gain from the 5th to the 15th month of age. Every 28 d, heifers were weighed, scored for BCS, and measured for: wither height, hip height, body length, and heart girth. Blood samples were collected at 9 and 15 month of age and analyzed for plasma metabolites and enzymes. Actual average daily gain was 0.77 kg/d in M and 0.90 kg/d in H reaching a body weight of 359 and 406 kg, respectively. The BCS increased in both groups throughout the trial and significant differences between groups were observed starting at 9 month of age. Higher values ($P<0.05$) of plasma urea, Ca, albumin and γ-GT were observed in H group at 9 month of age. These parameters showed similar differences between groups at 15 month of age but only that of γ-GT was still significant. These results indicate that rearing schemes based on high daily gains can affect some plasma parameters influencing body development of the heifers.

Energy availability (glucose) affects proliferation and differentiation in porcine muscle satellite cell cultures
N. Oksbjerg and P.M. Nissen, Danish Institute of Agricultural Sciences, Department of Food Science, P.O. Box 50, DK-8830 Tjele, Denmark

Glucose availability is of great importance for muscle development and growth both during foetal and postnatal growth and deficiency may have long lasting consequences for growth and health. Components of the IGF system and myostatin affect myogenesis and may interact with nutrients on myogenesis. The purposes of the present study were to examine the influence of various concentrations of glucose on proliferation and differentiation of porcine satellite cell cultures and the gene expression of IGF-I, IGF-II, the type I IGF-receptor and myostatin. Satellite cell cultures were established from the semimenbranosus muscle (fast-twitch) of 6-week old pigs. For measuring proliferation and differentiation, cells were grown in 96-well plates and for proliferation cells were grown to approximately 80% confluence and number of cells were measured by a WST-1 proliferation kit. For measuring differentiation, cells were grown to 80% confluence and then swifted to a serum-free medium and differen-tiation was indicated by creatine phoshpokinase (CPK) following 24, 48 and 72 hours of incubation. Proliferation increased by increasing concentrations of glucose and this occurred either when cells were grown in foetal serum or in serum-free medium. Differentiation was dependent on the time. Thus, in early differentiation (24 hours) CPK decreased, whilst in late differentiation (48 and 72 hours), CPK increased by increasing addition of glucose. These results are discussed in relation to gene expression analyses.

The functional state of the abomasum in calves in the first month of postnatal life
Edīte Birģele, Aija Ilgaža, Dace Keidāne and Arnis Mugurēvičs, Preclinical Institute, Faculty of Veterinary Medicine, Latvian University of Agriculture, K.Helmaņa street 8, Jelgava, LV-3004, Latvia

The dynamics of pH in the abomasum of calves from birth until 4 weeks of age was investigated. The pH of the abomasum was measured by means of a two–electrode pH–probe inserted through the fistulae. The uninterrupted pH measurements lasted on average 8 hours – 4 hours before and 4 hours after feeding. The results of these studies proved that hydrochloric acid concentration in the abomasum of a newborn calf was already high – pH 1.6 ± 0.21. After the first feeding the acid level in the abomasum decreased rapidly, reaching pH 6.2–6.3. The acid concentration in the abomasum increased on average to pH 2.83 ± 0.05 after 7 hours from the first feeding, but in four weeks old calves – after 4 hours. In a newborn calf, starting from the second day of its life, the conditioned reflex to "eating time" appeared. It is well-known that secretion of gastric acid is regulated neurally by stimulation of *n.vagus* and humorally by activation of histamine H_2 receptors in parietal cells. For stopping *n.vagus*, atropine sulphate (0.06 mg/kg, intravenously) was used, and for blocking humoral regulation – cimetidine (0.16 mg/kg, intravenously). The results clearly showed that atropine sulphate and especially cimetidine had an inhibitory influence on HCl production in the abomasum of calves in the first month of postnatal life.

Session Ph35

The effect of feed on the intraruminal and intra-abomasal pH dynamics in goats

Edīte Birģele, Dace Keidāne, Aija Ilgaža and Arnis Mugurevičs, Preclinical Institute, Faculty of Veterinary Medicine, Latvian University of Agriculture, K.Helmaṇa street 8, Jelgava, LV-3004, Latvia

The effect of feed on the intraruminal and intra-abomasal pH dynamics in goats was investigated. Chronic fistulas were operated in the rumen and abomasum. The pH dynamics in adult animals was estimated in each part of the stomach separately after feeding the concentrated mixed feed, fodder beet and hay, or after feeding the concentrated mixed feed and hay simultaneously. The intraruminal and intra-abomasal pH dynamics in three-month-old kids were estimated after feeding them the concentrated mixed feed, hay and mother's milk. All the experimental animals were kept under similar conditions and fed with equally balanced feed. Physiological investigations were started at 6 a.m. in the morning prior to animal feeding and continued 4-7 hours after feeding. Multielectrode pH probes, Oakton glass electrodes and pH meter were used. We conclude that the intraruminal and intra-abomasal pH dynamics in goats depend on the content of feed, the time passed after the animal is fed, and the age of the animal. The intraruminal and intraabomasal pH is different in three-month-old kids and in adult goats; in young animals, in the morning prior to feeding the pH of the content of the rumen is lower (the content is more acidic) than that of the adult animals, and at the same time the intra-abomasal medium is slightly less acidic in comparison with that of the adult goats.

Session H36

Horse breeding in Turkey

*C. Özbeyaz[1], Ö. Gücüyener[1] and *S.M. Yener[2], [1]Ankara University Faculty of Veterinary. 06110 Ankara, Turkey, [2]Ankara University Faculty of Agriculture, 06110 Ankara, Turkey*

In this paper the following aspects of the horse breeding in Turkey were treated: historical development, horse breeds imported to and bred in Turkey, their numbers, purpose of breeding and the herdbook system. The numbers of horses kept at the State Farms and the numbers of Arabian and Thoroughbred horses were tabulated by years. Also included in this presentation were information on the laws concerning horse breeding, equine federations and clubs, the numbers of equestrian and traditional sports clubs and their distribution by provinces.

Session H36

Theatre 2

Turkish equestrian sports and clubs

S.M. Yener[1], D. Alic[1] and K. Ural[2], [1]Ankara University Faculty of Agriculture Department of Animal Science, 06110 Ankara, Turkey, [2]Ankara University Faculty of Veterinary Medicine Department of Internal Medicine, 06110 Ankara, Turkey

The first equestrian school was established in 1911 in Bakırköy, Turkey. Since then horse-riding has been one of the most popular sport branches and various activities have been carried out for improvement. The Turkish Equestrian Federation, as an officially active foundation, under which the horse-riding clubs in our country has been organized, has been conducting activities and arranging organizations.

As in all sports branches, horse-riding requires technical knowledge and education. At the first place, this education has officially been given by Turkish Equestrian Federation and in addition by private establishments. With the support of Turkish Equestrian Federation, 21 horse-riding clubs are active in Ankara, Istanbul, Izmir, Antalya, Bursa, Kocaeli, Eskisehir, Adana, Konya provinces and one riding club is soon going to be active in Sivas province.

In the present study, it was aimed to put forward the historical evolution of Turkish Equestrian Federation and the member riding clubs, the education levels of grooms, riders and trainers and competition activities.

Session H36

Theatre 3

Studies on horses in Turkey

S. Koçak[1], M.Tekerli[1] and C. Özbeyaz[2], [1]Afyon Kocatepe University Faculty of Veterinary medicine, 03100 Afyon, Turkey, [2]Ankara University Faculty of Veterinary medicine 061100 Ankara, Turkey

Horse was used as an essential transportation tool in both Turkey and world for long years after domestication. It was also significant object in the life of Turks. Majority of ancient Turkish poems and tales includes horse and horsemanship. Cirit, some kind of game being played with horse, was most favorite sport of Turks ancestors.

A number of studies were undertaken in the branches of horse breeding, diseases and cures after declaration of Republic in Turkey. In this paper, the above mentioned studies belonging to zootechny, genetics, microbiology, virology, parasitology, surgery, internal diseases and reproduction subdivisions of Veterinary science are summarized.

Possibility of improving racing traits of Arabian and Thoroughbred horses in Turkey

B. Ekiz, O. Kocak and A. Yilmaz*, Istanbul University, Faculty of Veterinary Medicine, Department of Animal Husbandry, Avcilar, Istanbul, Turkey

The aim was to evaluate the possibility of improving racing traits of Arabian and Thoroughbred horses by selection in Turkey. Racing records were obtained from Turkish Jockey Club.

Dataset for racing time (RT), best racing time (BRT) and rank were edited and analysed according to dirt, turf races and entire dataset. Genetic parameters were estimated by REML procedure using DFREML programme.

Heritability estimates of entire dataset for Arabian and Thoroughbred horses were 0.280 and 0.317 for RT, 0.281 and 0.467 for BRT, 0.069 and 0.132 for rank, 0.139 and 0.194 for annual earnings (AE), 0.174 and 0.291 for earnings per start (EPS), 0.152 and 0.188 for logAE, 0.171 and 0.341 for logEPS. Repeatability estimates were 0.417 and 0.359 for RT, 0.430 and 0.500 for BRT, 0.133 and 0.215 for rank, 0.301 and 0.318 for AE, 0.384 and 0.447 for EPS, 0.342 and 0.308 for log EA, 0.324 and 0.409 for logEPS.

Heritability estimates of racing traits in Thoroughbred horses were higher than those of Arabian horses. Hence, selection to improve racing performance would result in higher genetic response in Thoroughbred than Arabian horses.

Among the heritability and repeatability estimates of racing traits for Arabian and Thoroughbred horses, the highest estimates were obtained for BRT. Therefore, selection for this trait might result in more genetic improvement than other investigated traits.

Connectedness between seven European countries for horse jumping competition, the "Interstallion Pilot Project II"

C. Ruhlmann[1]*, E. Bruns[2], E. Fraehr[3], E. Koenen[4], J. Philipsson[5], M. Pierson[6], K. Quinn[7] and A. Ricard[1], [1]SGQA, INRA 78352 Jouy en josas, France, [2]University of Goettingen 37075 Goettingen, Germany, [3]CentreNational Department of Horse Breeding Udkearsvej 15, SkejbyDK-8200 Aarhus Denmark, [4]IR&D NRS,P.O. Box 454, 6800 AL Arnhem, The Netherland, [5]Swedish Univ. of Agricultural Sciences, Box 7023, S-750 07 Uppsala, Sweden, [6]Stud book SBS, Avenue Prince de Liege 103 bte 4, 5100 Namur, Belgique, [7]Irish Horse Board Agriculture House, Kildare street, Dublin2, Ireland

Data on jumping breeding values of stallions and pedigree up to 3 generations was provided by 7 countries, including 6317 stallions and 22324 different horses. Identification of horses between countries was done as there was no unique identification number. About 1000 stallions and 5000 ancestors were provided by each country. Common stallions between countries varied from 0 to 308 and common horses from 130 to 1166. "Genetic Similarities" were not the correct measurement of connectedness as size of progeny by country by stallion was not in equilibrium. Correlation between standard errors of country effects was computed in a model including also the genetic value of horses (h^2=0.20, all relationship included). For Germany, Belgium, France and The Netherlands, these correlations were 0.32 to 0.51 corresponding to a balanced schema with 11 to 19 progeny per stallion per country. Other correlations reached 0.08 to 0.27. These results allow continuing the project with calculation of genetic correlations.

Session H36

Theatre 6

Survival analysis of the length of competition life of Standardbred trotters in Sweden

T. Árnason, International Horse Breeding Consultant AB, Knubbo, S-74494 Morgongava, Sweden

Current genetic evaluations of Swedish Standardbred trotters involve racing performance traits based on the results accomplished as 3- to 5-year-old. In Sweden the trotters are allowed to race up to the age of twelve years. Within this restriction stallions and geldings generally compete as long as they are competitive and sound while good mares may go to breeding earlier. The material in this study included 32504 male progenies of 398 sires. The mean number of competition years for 27412 males with uncensored records were 4.2. The corresponding number of right censored records were 5092 (15.7%) with the average censoring time of 3.8 years. The data on the length of competition life were analysed by the survival model of Prentice and Gloeckler which is included in the Survival Kit programs. The estimated sire variance component was 0.0235 which corresponds to the effective heritability, h^2_{eff}=0.09. The estimated sire effects were transformed into standardized indices. The correlation between the sires' BLUP indices of racing performance as 3- to 5-year-old and the survival analysis estimate of competition life length was r=0.51. Several models were tested in attempt to eliminate voletary reasons related to the length of the competition life. Further investigations are needed before this procedure can be applied for genetic evaluations of Swedish trotter stallions for soundness and stayability.

Session H36

Theatre 7

Selection of racehorses on jumping ability based on their steeplechase race results

A. Bokor, C. Blouin and B. Langlois, INRA-CRJ-SGQA 78352 Jouy-en-Josas France

The aim of this study was to detect Thoroughbred mare families and sire lines in France, in the United Kingdom and Ireland whose offspring may be successful in steeplechase races and can be recommended therefore for sport horse breeding. Race results were collected from all steeplechase races in these countries between 1998 and 2003 and contained the results of 17 355 horses, 12 861 dams and 2 452 sires. In France non-thoroughbred horses were also included in the analysis because they race and mate together with Thoroughbreds. Performance was measured using two criteria: earnings and ranks after some mathematical transformations. The effects of year, sex and age were considered as fixed, animal, permanent environment and maternal as random. For ranks, pre-correction for the race effect was introduced to overcome computational difficulties. The maternal environmental components for ranks were 0.021 in France and zero 0.000 in the United Kingdom and Ireland. Estimated heritabilities for the ranking criteria were 0.18 (repeatability 0.33) in France and 0.06 (repeatability 0.19) in the United Kingdom and Ireland. The high genetic correlation between the two traits (0.94 and 0.97 resp.) gives the opportunity to choose the most suitable criteria for breeding value estimation. The ranking value which is normally distributed by construction was preferred because it has a great advantage for comparison between countries.

Genetic parameters for endurance ride in the Spanish Arab Horse

I. Cervantes[1]*, M. Valera[2,] M.D. Gómez[1], C. Medina[1] and A. Molina[1], [1]Department of Genetics.,University of Cordoba, [2]Department of Agro-forestal Sciences, University of Seville, Spain

The aim of this study was to estimate the genetic parameters for endurance ride aptitude in Spanish Arab horses and to evaluate the influence of the main non-genetic factors.

The data set comprised 1,721 records of 787 horses from 109 endurance rides held between 2000 and 2005. The total number of records for the Spanish Arab horse was 547 entries of 249 horses. The breed, ride-year, geographic zone and season effect were the significant factors.

Ride time (as deviation from the best time obtained within each ride) and final rank were the best traits. The average time obtained within each ride and the number of participants were included as covariate (L,Q) for the ride time evaluation and for the final rank evaluation respectively.

Estimates for heritability were 0.130 ± 0.02 for ride time and 0.176 ± 0.02 for final rank. A BLUP animal model was used to estimate breeding values using VCE software package v.5.0.

The evolution of breeding values showed a slight genetic progress with the masal selection carried out until now. Therefore we consider that to make a Genetic Selection Scheme by means of BLUP evaluation would be useful to improve endurance aptitude.

Genetic evaluation of show jumping performance in young Spanish sport horses

M.D. Gómez[1], I. Cervantes[1], E. Bartolomé[1], A. Molina[1] and M. Valera[2], [1]Department of Genetics, University of Cordoba, [2]Department of Agro-forestal Science, University of Seville, Spain

Young horse competitions (for animals between 4 and 6 years old) have been developed in Spain since 2004. Their main aim is to contribute to equine selection programs in dressage, eventing and show-jumping performance.

A multivariate BLUP Animal Model, using Groeneveld's VCE (version 5) software package was developed to estimate the breeding values in jumping performance of Spanish young horses. Negative points obtained during the competition (caused by mistakes and times exceeding the allowed) and final ranking of 486 young animals were considered.

The data file had 2016 racing performances collected from a total of 34 different competitions. The pedigree file had a total figure of 3,280 animals obtained from the Stud-book of the different participant breeds (four generations of ancestors). Age (4, 5 and 6 years old), level of stress (measured as transport, in 2 levels), training level (measured as training intensity grouped in 4 levels related to number of hours in training/week), event location (15 levels), season (3 levels), and rider (158) were included as fixed effects in the model. The random variables were individual additive and permanent environmental effects.

The heritability of both traits and the breeding value of each one for the participant animals were estimated. These estimated values were used to rank potential breeding animals for jumping performance.

Genetic evaluation of dressage performance in Spanish Purebred horses

M. Valera[1], M.D. Gómez[2], I. Cervantes[2] and A. Molina[2], [1]Department of Agro-forestal Science, University Seville, [2]Department of Genetics, University Cordoba, Spain

Young horses have participated in specific competitions of dressage, eventing and show-jumping since 2004 in Spain. The scores obtained in these competitions were used to estimate the breeding value of the animals for each discipline, since it is greatly simplified and is easy to manage for disciplines having many participants. The evaluation is based on partial (each reprise) and final (average) scores.

The breeding values of Spanish Purebred (SPB) horses for dressage were calculated by a repeatability multivariate BLUP Animal Model, using Groeneveld's VCE software package v.5.0. Final and partial scores of 387 SPB young animals were included.

Type of racetrack (grouped in 3 levels), ambient (measures like temperature*humidity, grouped in 7 levels) and level of stress before the competition (using transport*duration of the journey*rest time before the competition, 33 levels) were included as fixed effects in the model. The random variables were judge, rider, individual additive and permanent environmental effect.

The data file included dressage performances collected in 30 different competitions between 2004-2005. The pedigree file was created including four generations of the participant horses, obtained from the Stud-book of this breed, and had a total of 2,753 animals.

The values of heritability and the breeding values were estimated. The results were published using a total selection index, in which the different breeding values were weighed up.

Development of a BAC-based physical map of the horse genome

H. Blöcker[1], M. Scharfe[1], M. Jarek[1], G. Nordsiek[1], F. Schrader[1], C. Vogl[2], B. Zhu[3], P.J. De Jong[3], B.P. Chowdhary[4], T. Leeb[5,6] and O. Distl[6], [1]German Research Centre for Biotechnology, 38124 Braunschweig, Germany, [2]Institute of Animal Breeding and Genetics, The Vienna University of Veterinary Medicine, 1210 Vienna, Austria, [3]Children's Hospital Oakland, Oakland, CA 94609, USA, [4]Department of Veterinary Integrative Biosciences, Texas A&M University, TX 77843, USA, [5]Institute of Genetics, University of Berne, 3012 Berne, Switzerland, [6]Institute of Animal Breeding and Genetics, University of Veterinary Medicine Hannover, 30559 Hannover, Germany*

For the horse a whole genome shotgun (WGS) sequence will become available. For the accurate long-range assembly of WGS sequences high-resolution BAC-based physical maps have been proven to be an essential component. In a Lower Saxonian effort a physical map of the horse genome will be created. The horse physical map will be based upon a combination of BAC fluorescent fingerprinting and BAC end sequencing of the CHORI-241 library. Fluorescent fingerprints of 150,000 BAC clones (~10x genome coverage) will be obtained by using the 4-restriction enzyme 4-color technique and separating the resulting fragments on capillary sequencers. The fingerprinted clones will also be end sequenced. The BAC end sequences (BESs) will enable the anchoring of the emerging BAC contigs to the equine RH map as well as the comparative analysis with respect to the human genome. At the time of writing the first ~16,000 BESs have been produced and submitted to the public databases.

Genetic analysis of allergic eczema in Icelandic horses

K. Grandinson[1]*, S. Eriksson[1], L. Lindberg[1], S. Mikko[1], H. Broström[2], R. Frey[3], M. Sundquist[4] and G. Lindgren[5], [1]Dept of Animal Breeding and Genetics, SLU, Box 7023, 75007 Uppsala, Sweden, [2]Dept of Clinical Sciences, SLU, 750 07 Uppsala, Sweden, [3]Norrsholms Animal Hospital, 602 37 Norrköping, Sweden, [4]Östra Greda Research Group, 387 91 Borgholm, Sweden, [5]Dept of Medical Biochemistry and Microbiology, Uppsala University, 751 24 Uppsala, Sweden

Allergic eczema in Icelandic horses mainly reflects a hypersensitivity reaction against bites from gnats of the genus *Culicoides*. The aim of this study was to estimate genetic parameters for allergic eczema. The analysis included records from 1210 horses born in Sweden during 1991-2001. These horses were sired by 33 stallions selected for having more than 50 offspring each. Owners classified clinical signs of allergic eczema as unaffected, mild, moderate, or severe eczema. Close to 8% of the horses showed clinical signs of allergic eczema and offspring of dams suffering from eczema had higher risk of developing eczema. Frequency of eczema in the different paternal half-sib groups varied between 0-29%. Variance components were estimated using REML animal and sire models, including fixed effects of age and geographic location of the horse. The heritability for allergic eczema was estimated at 0.14 (SE 0.06). In contrast to age of the horse, different geographic areas were significantly associated with severity of the eczema. We conclude that genetic selection could decrease the frequency of allergic eczema among Swedish born Icelandic horses.

Biometrical and breeding analysis of Hafling horses in the Czech Republic

J. Navratil, V. Kutilova, P. Kutilova and F. Louda*, Czech univerzity of agriculture 16521, Czech Republic

The breed of Hafling horses is in our country dynamical grow up for their modesty, resistence and good character traits. The aim of this work was to map part of the czech population of hafling horse, we have obtained data of 166 mares and 18 stallions (summary population 355 mares and 25 stallions). We made an analysis of representation of all seven stallions lines, distribution of mares into the sections of stud book and an age structure of population. They were engaged in measuring 19 body proportions. Out of gained values we calculated nine indexes of physique. Out of all values of body proportions and indexes of physique were calculated basic statistical characteristics.By using correlation analysis we valued the dependence between the lenght of the head to the end of the turbinate bone and the lenght of the head to upper lip. For determining of degree of body proporcionality (by Dusek).We compiled a net of physique. The calculated degrees of proporcionality were compared by using correlation analysis with the points which horses got for their exterior. Considering that no simmilar researches have been realized yet. Research data of basic boddy proportions (mares 141,34/179,54/19,08, stallions 143,19/183,84/20,34) will be used for improvement of breeding standarts. Mapping of population of hafling horse to get a completed set of data will be continued.

Tori Stud Farm 150, founder of Tori Horse Hetman 120

Heldur Peterson, Estonian Unuversity of Life Sciences, Institute of Veterinary medicine and Animal Sciences

In 2006, Estonian horse breeders celebrate two anniversaries. There was founded tory Stud Farm in Estonia in 1856. A turning point in the improvement of the native Estonian horse breed occurred in 1892, the Tory Stud hired a stallion Hetman from Fr.G. M v. Berg, the owner of the Sangaste manor. Since then, the breed group of horses bred at Tori has been named the Tori Roadster type, some later as tori Horse. the foundation sire Hetman (155-183-22,0) was born in Poland in 1886. He descends from stallion Stuart of the Norfolk - Roadster breed and an unknown English Hunter mare. More than 1,000 mares were crossed with Hetman. his descendants are horses comnining both utility and sporting qualities. the versatile horse were suitable for riding, field work and transportation. Hetman's sons have been crossed with 11,781 mares, so he is recognized as the progenitor of the breed.

Reproduction parameters on the Caspian miniature colts

Shahram Dordari, Morteza Rezaee and Amini Fereydoun, Tehran Agricultural Research Center, Iran

The Caspian miniature horse is an ancient breed of horse, which was thought to be extinct for many years. The objectives of this study were to: 1 - Determine age at puberty; and 2 - Characterize seminal characteristics and sexual behaviour, of the Caspian miniature horse. Data for sexual behaviour represented all colts at puberty. The sexual behaviour was recognized as 4 degrees: 1 - without sexual behaviour; 2 - Erection and Flehman reflex; 3 - Mounting and Intromission; and 4 – Ejaculation. Seminal collections were attempted every 2 weeks from 50 to 140 weeks of age. For all collections, times to erection, mount and ejaculation and seminal characteristics were recorded. Age at puberty was defined as the first ejaculate containing 50 million spermatozoa with >10% motile.

Age of puberty: 24.6 ± 8.2 months. Weight at the puberty: 189.8 ± 24.07 kg. Semen concentration at puberty: $57 \pm 0.32 \times 10^6$ cc. Semen volume: 18.2 ± 6.2 cc. Gel volume: 1.5 ± 1.2 cc. Motility: $24.2 \pm 14\%$. pH: 7.88 ± 0.7. Testosterone concentration: 0.54 ± 0.2 ng/mL.

In this study the differences and similarities between this breed and other ponies were determined. In our study we found that puberty in the Caspian miniature colt was at 24.6 months.

Physical and farming system characteristic of Karabakh horses
P. Mosapoor, A.M. Aghazadeh and Mahpeykar, Urmia University, Department of Animal Science, 81110 urmia Iran

Azerbaijan is well known for indigenous breeds of horses in Iran. Livestock census from 1992 to 2004 revealed a marked decrease (with annual reduction of 3.9%) in horse population, due to the changes in agriculture, transport and communication.Due to this drastic reduction in recent years and the fact that no information was available in literature on farming system and breed characteristic of these horses under natural habitat. A questionnaires was developed and used to collect information from door- to-door survey at Varzagan, Ahar, Ardabil, Kali bar, districts. The different physical characteristics and 21 body measurements were recorded on 52 (49 stallions and 3 mares) adult above 3 years Karabakh horses, from Varzagan (63.46%), Ahar (19.23%) Ardabil (9.7%) and Kali bar (5.8%) districts. The Karabakh was developed in Nagorny Karabakh in Azerbaijan between the Arras and Kura rivers. The average measurements (in cm) of stallions were : face length 64.93 ±0.19, face width 17.71± 0.27, ear length 15.29± 0.25, ear width 8.49±0.12,height at withers 142.93± 1.26, height at knee 45.79± 1, height at arm 98.40± 0.36, cannon bone (circumference) 18.37± 0.14, chest width 36.85 ± 0.4, heart girth 160.94± 0.79, back girth (paunch girth) 167.04± 1, height at croup 139.44± 0.62, height at hock 56.64± 0.47.

Effect of the sex on physical characteristics of horse meat produced and marketed in Sicily, Italy. Preliminary results
L. Liotta[1], S. D'Aquino[1], L. Sanzarello[1] and V. Chiofalo[1,2], [1]Dept. MOBIFIPA, Uiversity of Messina, 98168 Messina, Italy, [2]Consortium of Meat Research, Sicilian Region, Italy

On the carcasses of 40 half-breed horses, 22 males and 18 females, from 2 to 9 years old, were taken out samples of *Longissimus dorsi* muscle between the 16th and 17th rib. Four days after slaughter, on each sample, pH, colour, cooking loss and tenderness (WBS) were determined. Results were subjected to ANCOVA statistical analysis using the following model: $J_{ij} = \mu + sex + b * age_{ij} + \varepsilon_{ij}$. The physical characteristics were not significantly influenced by sex (male vs. female). The pH mean values showed a good acidification of the meat (male 5.69 vs. female 5.64); Chroma (male 24.28 vs. female 25.49) as well as Hue (male 0.61 vs. female 0.65) testified an intense colour of the meat; Luminosity (male 56.92 vs. female 57.56) showed high values which could be due to the young age of most of the horses, not older than 24 months. The cooking loss values, within the normal range, were very similar between the sexes (male 26.81% vs. female 26.88%). The low values observed for the WBS (male 3.16 kgf/cm^2 vs. female 3.54 kgf/cm^2) classify tender these meats. This preliminary study that need to be further investigated, has pointed out that the horse meat produced and marketed in Sicily show good rheological characteristics and a high homogeneity in relation to the sex of the animals.

Authors index

Memisi, N.	217	Molenda, J.	41
Mendizábal, J.A.	223	Molina, A.	110, 111, 114, 245, 246, 351
Merzaei, H.R.	64	Molnár, A.	58, 92, 224
Mestre, R.B.	188	Moloney, A.	25
Metges, C.C.	133	Momani Shaker, M.	90, 214
Meunier, B.	48	Mondal, M.	184
Meydan, H.	27	Montazer Torbati, F.	270
Meyer, U.	125	Montironi, A.	26, 241
Mézes, M.	300	Moors, E.	68, 69, 223, 327
Mezoszentgyorgyi, D.	42	Morad, A.A.	142, 143
M'hamdi, N.	3	Moradi Shahrbabak, H.	130, 176, 275,
Miari, S.	233		276, 303
Miceikiene, I.	30	Moradi Shahrbabak, M.	12, 24, 275, 276, 303
Micol, D.	55	Mordak, R.	78
Miculis, J.	333	Moreira, O.C.	101
Migdal, W.	239	Morel, P.C.H.	198
Mihok, S.	244	Morfin-Loyden, L.	132
Mijic, P.	31, 109	Morgan, C.A.	172
Mikko, S.	353	Moris, A.	70
Milan, D.	268	Moro, S.	241
Milán, M.J.	80	Morris, S.T.	198
Milerski, M.	199, 200, 215	Morrow, J.L.	75
Mioc, B.	109	Mosapoor, P.	355
Miraei-Ashtian, S.R.	107	Mostafaloo, Y.	163
Miraei-Ashtiani, S.R.	106	Mostafalou, Y.	139
Miraglia, N.	69	Motamedi, G.	43
Miranda, H.	264	Mottet, A.	98
Mirhadi, S.A.	166	Moulin, C.H.	96
Mirhoseini, S.Z.	102, 108, 175, 338, 339, 340	Mountzouris, K.C.	197
Mirhosseini, S.Z.	338	Mourad, M.	222
Miró, F.	245	Mousavi, M.A.	149
Miron, J.	310	Mousavi, S.	148
Mirzaei, F.	194, 195, 280	Mudrik, Z.	121, 167
Mirzaei, H.	317, 318	Mudrunková, R.	216
Mirzaei Aghsaghali, A.	152	Mugurevics, A.	346, 347
Misztal, I.	61, 265	Munoz-Gutiérrez, M.	10
Mitloehner, F.M.	75	Murai, M.	120
Moafi, M.	160	Muratovic, S.	135
Modesto, M.	72	Murray, M.	173, 179
Moeini, M.	124, 201, 225	Musapuor, A	130
Moghadam, A.	201	Muscio, A.	77
Moghaddam, A.	202	Mutlu, F.	33
Mohammadi, A.R.	43	Muwalla, M.M.	85
Mohammadinejad, A.	339	Muzsek, A.	316
Mohammadzade, M.	270	Myers Hill, G.	50
Mohammadzadeh, M.	265	Mylyszová, J.	125
Moini, M.	270		
Mojto, J.	332, 333		
Mokhtar, M.M.M.	199		

Safus, P.	37, 342	Seba, K	37
Saghi, D.A.	201, 226, 265, 270	Sebsibe, A.	206
Saghirashvili, G.	262	Secchiari, P.	278
Sahana, G.	267	Sedda, P.	233
Saive, P.	129	Seegers, H.	40, 260
Sala, G.	60	Seenger, J.	300, 304
Salajpal, K.	57	Segato, S.	304
Salazar-Ortiz, J.	13	Sehested, J.	117, 295, 308
Salem, M.A.I.	199	Seidavi, A.R.	102, 165, 175, 338, 339, 340
Salimei, E.	290	Seidenspinner, T	281
Salomon, R.	187	Seitpan, K.	89
Saltalamacchia, F.	42	Seker, M.	87
Samie, A.	130, 138	Sekhavati, M.	312
Sánchez, M.	224	Sematovica, I.	320
Sanchueza, D.	306	Sencic, D.	132, 227
Sanjabi, M.R.	122	Senese, C.	274
Sanli, E.	209	Sepehri, A.	252
Santa-María, M.	195	Serbester, U.	316
Santos, V.	59, 214, 215	Seric, V.	227
Santos-Silva, M.F.	112	Seroussi, E.	48
Sanz, A.	58, 76, 253, 330	Serra, A.	278
Sanzarello, L.	355	Serrano, M.	111
Sara, P.	169, 187	Sevi, A.	77
Saremi, B.	312	Seyedabadi, H.R.	103
Sargolzaee, M.	312	Shaat, I.	78
Sariozkan, S.	170, 261	Shadnoush, G.H.	306
Sarubbi, F.	151, 328	Shadparvar, A.A.	338
Sarvari, A.	318	Shafiei sabet, S.	123
Sasaki, O.	235	Shaker, M.M.	216
Satoh, M.	234	Shalloo, L.	256
Sauerwein, H.	53	Sharifi, S.D.	186
Saveli, O.	327	Shawrang, P.	116, 138, 139, 140
Savic, M.	99	Sherzhantova, A.I.	241
Sayadnejad, M.	24	Shiri, S.A.	201, 226, 265
Scaramuzzi, R.J.	10	Shiri, S.A.K.	270
Scardigli, S.	19	Shivazad, M.	129, 151, 164
Scharfe, M.	352	Shojaei, B.	176
Schatzmann, U.	254	Shoshani, E.	75, 310
Schiavina, S.	21	Siceanu, A.	340
Schichowski, C.	223	Siculella, L.	182
Schierenbeck, S.	342	Sikora, J.	302
Schneider, J.	9	Silanikove, N.	248, 328
Schöne, F.	295	Silva, F.C.	152
Schrader, F.	352	Silva, S.	59, 214, 215
Schuh, M.	171	Simai, Sz.	42
Schwörer, D.	230	Simianer, H.	17, 19, 336, 342
Scollan, N.D.	25	Simoni, A.	69
Scott, K.	40	Sirin, E.	189, 316
Scotti, E.	284	Sitzia, M.	188

Printed in the United States
by Baker & Taylor Publisher Services